Chemistry
Foundation

for Class **IX**

Atul Singhal

Professor, Institute of Science and Technology
Kansas, Missourie, USA

Universities Press

CHEMISTRY FOUNDATION FOR CLASS IX

UNIVERSITIES PRESS (INDIA) PRIVATE LIMITED

Registered office
3-6-747/1/A & 3-6-754/1, Himayatnagar, Hyderabad 500 029, Telangana, India
info@universitiespress.com; www.universitiespress.com

Distributed by
Orient Blackswan Private Limited

Registered office
3-6-752 Himayatnagar, Hyderabad 500 029, Telangana, India

Other offices
Bengaluru, Chennai, Guwahati, Hyderabad, Kolkata,
Mumbai, New Delhi, Noida, Patna, Visakhapatnam

© Universities Press (India) Private Ltd 2023
First published 2023

ISBN: 978-93-93330-30-7

Cover and book design
© Universities Press (India) Private Ltd 2023

Typeset in Times New Roman PS Std 10 *by*
SRS Publishing Services, Puducherry

Printed in India by
B.B. Press, Noida 201 301

Published by
Universities Press (India) Private Limited
3-6-747/1/A & 3-6-754/1, Himayatnagar, Hyderabad 500 029, Telangana, India

Disclaimer
Care has been taken to confirm the accuracy of the information presented in this book. The publisher and author, however, cannot accept any responsibility for errors or omissions or for consequences from the application of the information in this book, and make no warranty, express or implied, with respect to its contents.

Dedicated to
My Grandparents, Parents and Teachers

Preface

The Chemistry Foundation series comprises the recommended preparatory books for students aspiring to crack the prestigious examinations such as NEET, JEE (Main and Advanced), NTSE, KVPY and the Olympiads. They provide class-tested material and practice problems that help aspirants in understanding important theories, concepts and as a consequence develop problem-solving skills to attempt these examinations with confidence. The series is written in lucid language and aims to assist students in understanding the concepts even without the help of an instructor as most of the questions have been solved.

The series strictly adheres to the latest pattern of both CBSE and ICSE syllabi and the content is developed in such a way that a student will not only be able to understand all concepts but also apply them in the entrance examinations. All the chapters in this series also conform to the pattern laid out in the NCERT textbooks. The objective of this series is to provide the course material in a structured sequence that will help students prepare in an organised manner. The unique feature of this series is its huge repository of multiple choice questions that cover all important topics in every chapter. Each practice exercise is further divided into two sections—***Practice Questions:*** Analyse Your Concepts (School Exam Based) and ***Competition Window:*** Objective Type [for NEET, JEE (Main and Advanced), NTSE, KVPY and the Olympiads].

This book contains a supporting mobile App that provides chapter-wise practice tests for every chapter. It also includes Objective Olympiad Questions (IOQJS) from 2014 to 2020.

Salient Features of the Series

- Structured as per the syllabi of CBSE, ICSE and the topics required for the entrance examinations.
- Based on the latest exam patterns.
- Includes True–False, Fill in the blanks, Match the Following, Very short, Short, Long answer questions including higher order thinking questions, case study based and practical-based questions for school exams.
- Includes chapter-wise NCERT exercises and exemplar questions with solutions.
- Includes objective type questions; topic-wise, miscellaneous and advanced MCQs (single and multiple choice); comprehension passage or study case based, integer type and matrix match.

The purpose of this series is two-fold—to strengthen the foundation in chemistry and thus build the students' confidence to ace the examinations. I welcome suggestions and constructive criticism from the readers. Students can share their feedback at singhal.atul1974@gmail.com.

Atul Singhal

Acknowledgements

The contentment that accompanies the successful completion of my work would remain essentially incomplete if I fail to mention the people whose constant support has encouraged me. I am grateful to all my revered teachers, especially, Late J K Mishra, Dr D K Rastogi, Late A K Rastogi and my honourable guide, Dr S K Agarwala. Their knowledge and wisdom assisted me in no small measure in presenting this work. I express my immense gratitude to my colleagues for collaborating so nicely and constructively during the various stages of this project. I am extremely thankful to Ronit Singh Kushwaha for typing out the manuscript.

I am indebted to my father, B K Singhal, mother Usha Singhal, brothers Amit Singhal and Katar Singh, who have been my motivators at every step. Their never-ending affection has provided me with moral support and encouragement while writing this book. Last but not the least, I express my deepest gratitude to my wife Urmila, my daughters Khushi and Shanvi and my little, but witty-beyond-years son Shashwat, who always supported me during my work.

I would like to express my heartfelt gratitude to Dr Gita S Dattatri for her brilliant editing and upgradations. I would also like to show my appreciation to Kallol Das, Thomas Mathew Rajesh and Madhu Reddy for providing all the necessary help and assistance.

Dr Atul Singhal
singhal.atul1974@gmail.com

Contents

Chapter 6: The Periodic Table — 6.1

Chapter 7: Chemical Bonding — 7.1

Chapter 8: Gaseous State — 8.1

Chapter 9: Water — 9.1

Chapter 10: Hydrogen — 10.1

Chapter 11: Atmospheric Pollution — 11.1

Appendices — A.1

CHAPTER

1 Matter in Our Surroundings

LEARNING OBJECTIVES

After studying this unit, you will be able to understand:

- Matter and its type
- Properties of solids, liquids and gases
- Changes in states of matter
- Effect of temperature and pressure on state of matter
- Melting, freezing, vaporisation, condensation, and so on
- Types of latent heat
- Sublimation and liquefaction

Introduction

When we look at our surroundings, we see a vast variety of things possessing different shapes, sizes, textures, and so on. Everything in our universe is made up of material which is called *matter*. The air we breathe, the food we eat, clouds, stars, plants and animals, even a small drop of water or a particle of sand—everything is matter. When we look around, we observe that all these occupy space, that is, they have volume and mass. Early Indian philosophers classified matter in the form of five basic elements – the *panch tatva* – air, earth, fire, sky and water. According to them, everything, living or non-living, was made up of these five basic elements. Ancient Greek philosophers had arrived at a similar classification of matter. Modern day scientists have evolved two types of classification of matter, based on their physical properties and chemical nature. In this chapter we shall learn all about matter based on its physical properties. The chemical aspects of matter will also be taken up in subsequent chapters.

Matter

Anything that occupies space and has mass is known as *matter*. Everything around us is made up of tiny pieces or particles. The particles that it form it are atoms or molecules. For example, food, water, air, clothes, table, chair, plants and trees (Fig. 1.1). On the basis of their chemical properties, matter is classified as elements, compounds and mixtures.

Characteristic Features of Matter

In nature, many forms of matter can be seen. Many characteristics are common among these.

1. **Mass:** Any form of matter has mass. Mass is the amount of matter present in any species. For example, even though air is invisible, it has some mass.
2. **Space:** Any form of matter also occupies space. For example, a chair in a room, a book on a bookshelf, a ring in a box.

Fig. 1.1 Examples of states of matter

3. **Property of inertia:** Matter has the property of inertia. For example, a football can move only when it is pushed by an external force by a player; a book lying in any place will remain as it is unless some external force disturbs its position, by picking it up or pushing it.
4. **Influence of gravity:** When anything is thrown upwards with a force, it comes to the ground automatically due to the force of attraction exerted by the earth—*gravity*. Anything composed of matter is under the influence of gravity. For example, the fruits always fall down from the tree; water always flows from a higher level to the lower level.
5. **Matter cannot be destroyed or created:** We can say that matter cannot be destroyed or created. During all physical and chemical changes, the total mass of the matter before and after the change remains constant (Lavoisier's *law of conservation of mass*).

Physical Nature of Matter

Every substance in nature consists of a special set of properties that allows us to recognise and distinguish it from other substances. Such properties of matter can be categorised as *physical* or *chemical* (Fig. 1.2). The measurement of physical properties is possible without changing the identity and composition of the substance. Such properties are colour, hardness, odour, boiling point, melting point, density, and so on.

Chemical properties represent the way by which a substance can change or react to give other substances. Flammability is such a property and is the ability of a substance to burn in air or in the presence of oxygen (O_2).

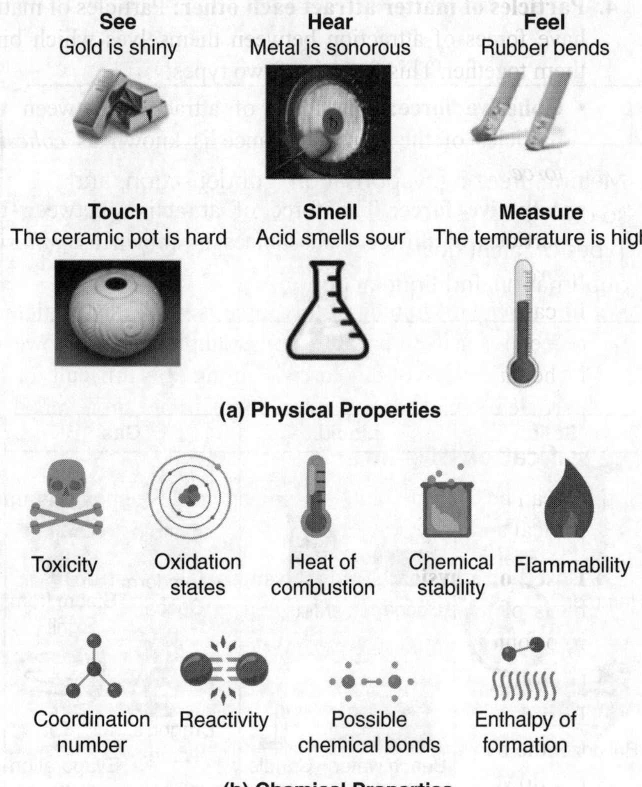

Fig. 1.2 Physical and chemical properties

Particle Nature of Matter

As per as ancient Indian and Greek philosophers, 'matter is composed of very tiny particles that cannot be further sub-divided'. Scientifically, John Dalton was first to explain the nature and composition of matter.

Based on the results of diffusion and Brownian motion experiments (Fig. 1.3), the existence of particles in matter and their motion can be proved. Here are a few properties which prove the particle nature of matter.

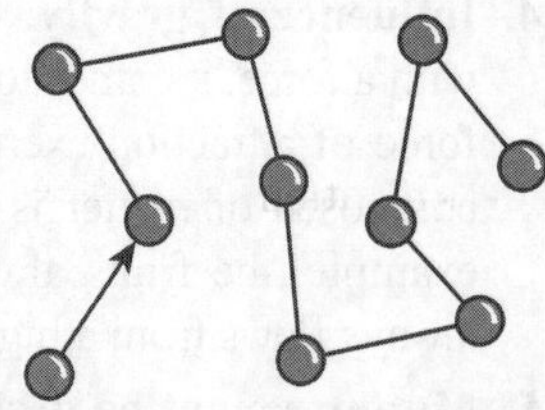

Fig. 1.3 Brownian motion

1. The movement of pollen grains in water is very rapid and occurs in a zig zag manner since pollen grains are hit by fast moving water molecules. As a result, pollen grain molecules can move in the net force direction on the surface of water (similar to Brownian motion).

2. When a solid, for example $KMnO_4$, is dissolved in a liquid like water, the aqueous solution becomes coloured (purple) even if a very small amount or just a crystal of $KMnO_4$ is added in water. It means even one crystal of $KMnO_4$ (matter) is composed of tiny particles. These tiny particles

- **Diffusion:** The spreading out and mixing of a substance with another substance due to the motion of its particles is known as diffusion. It is a property of matter which is based on the motion of its particles.

  Diffusion is the flow of molecules from the side with higher concentration to the side with lower concentration. For example, perfume spray diffuses in all possible direction in room.

- **Osmosis:** Osmosis is the flow or movement of solvent particles from a dilute solution (less solute side) to a concentrated solution (more solute side) through a semi-permeable membrane. For example, raisins swell in water but shrink in sugar water.

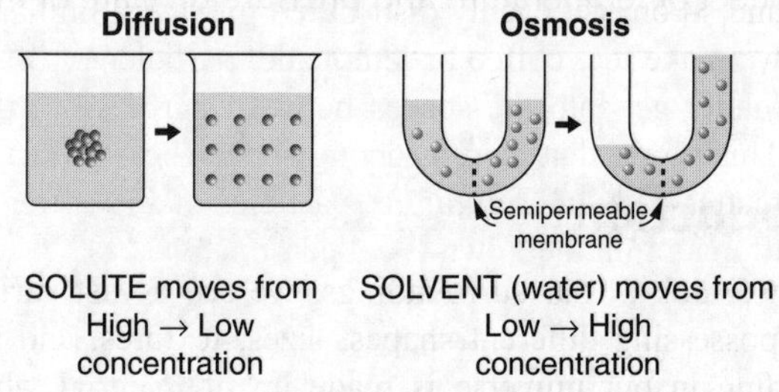

Fig. 1.4 Diffusion vs osmosis

- **Intermolecular force of attraction:** Intermolecular forces are the forces of attraction or repulsion which act between neighbouring particles (atoms, molecules or ions). These forces are weak compared to the intramolecular forces, such as the covalent or ionic bonds between atoms in a molecule.

  Forces between molecules are responsible for the state of matter—solid, liquid or gas, affect the melting and boiling points of compounds as well as the solubility of one substance in another.

Table 1.1 Types of intermolecular forces

Type	Present in	Molecular perspective
Dispersion	All molecules and atoms	$\delta-$ $\delta+\cdots\delta-$ $\delta+$
Dipole–dipole	Polar molecules	$\delta+$ $\delta-\cdots\delta+$ $\delta-$
Hydrogen bonding	Molecules containing H bonded F, O or N	$\delta+$ $\delta+\cdots\delta-$ $\delta+$ $\delta-$ $\delta-$
Ion–dipole	Mixtures of ionic compounds and polar compounds	$\delta-$ $\delta-$ $\delta-$ + $\delta-$ $\delta-$ $\delta-$

undergo diffusion as they have kinetic energies due to which they mix very well in aqueous solution to make it purple coloured.

Characteristics of Particles of Matter

1. **Particles of matter are very, very small:** Matter is composed of very tiny particles. This can be easily understood when we add a crystal of $KMnO_4$ or copper sulphate (blue vitriol) in water. The aqueous solution becomes purple or blue and the size of crystal becomes smaller and smaller. Finally, the crystal decomposes into a number of tiny particles as the colour becomes uniform all around.

2. **Particles of matter have spaces between them:** As we know, particles of sugar, salt, potassium permanganate, and so on, get evenly distributed in water. Similarly, when we make tea, coffee or lemonade, particles of one type of matter get into the spaces between particles of the other. This shows that there is enough space between particles of matter. This intermixing of particles of two different types of matter on their own is called *diffusion*.

Fig. 1.5 Dissolution of salt in water

When we dissolve salt in water, the particles of salt get into the spaces between particles of water.

3. **Particles of matter are constantly moving:** Particles of matter are continuously moving, that is, they possess what we call *kinetic energy*. As the temperature rises, the particles move faster. So, we can say that with an increase in temperature, the kinetic energy of the particles also increases.

4. **Particles of matter attract each other:** Particles of matter have forces of attraction between themselves which bind them together. This force is of two types:
   - **Cohesive force:** The force of attraction between the particles of the same substance is known as *cohesive force*.
   - **Adhesive force:** The force of attraction between the particles of different substances is known as *adhesive force*.

   In case the magnitude of this force is weak, the particle or object is easily breakable, for example, chalk. However, if the magnitude of this force is strong, it is difficult for the particle or object to be broken, for example, an iron nail.

Classification of Matter

Matter can be classified into several categories depending upon its physical and chemical nature.

1. **Based on physical state (physical classification):** On the basis of their physical states, all matter can be classified into four groups (Fig. 1.6).
   (i) Solids   (ii) Liquids   (iii) Gases   (iv) Plasma

2. **Based on purity/composition (chemical classification):** On the basis of purity, matter can be classified as shown in the flowchart (Fig. 1.7).

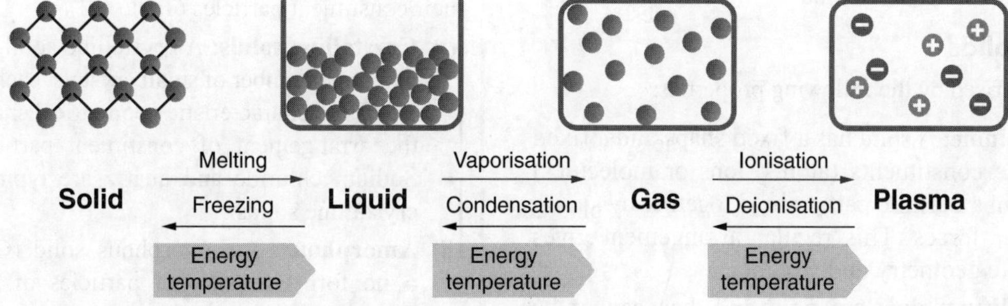

Fig. 1.6 Physical classification

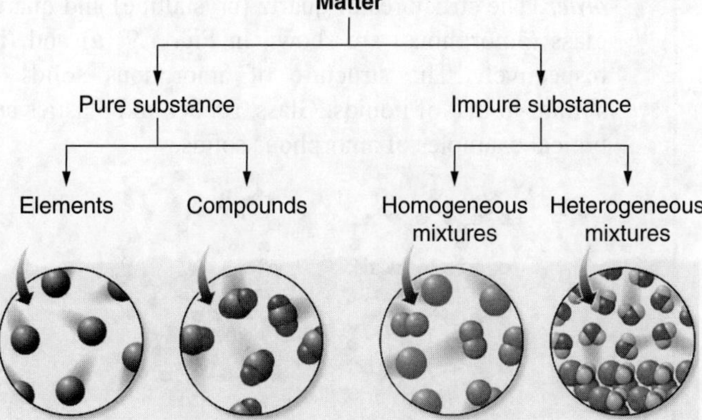

Fig. 1.7 Chemical classification

KNOWLEDGE BOOSTER

Now five states of matter have been identified—solid, liquid, gas, plasma and Bose–Einstein condensate.

Illustrations

1. Which of the following can be labelled as matter?

   Table, air, love, chocolate, smell, hate, almonds, thought, cold, soft-drink, smell of perfume.

   Solution: Table, air, chocolate, smell, almonds, soft-drink and smell of perfume are classified as matter.

2. A diver is able to cut through water in a swimming pool. Which property of matter does this observation show?

   Solution: The phenomenon of cutting through water by the diver shows that matter has space between its particles.

3. Classify the following into osmosis/diffusion:
   (a) Swelling up of a raisin on keeping in water.
   (b) Spreading of virus on sneezing.
   (c) Earthworm dying on coming in contact with common salt.
   (d) Shrinking of grapes kept in thick sugar syrup.

   Solution:
   (a) Osmosis   (b) Diffusion
   (c) Osmosis   (d) Osmosis

Solids

The *solid state* is one of the four fundamental states of matter. The molecules in a solid are closely packed together and contain the least amount of kinetic energy. A solid is characterised by incompressibility, structural rigidity, high mechanical strength and resistance to a force applied to the surface. In a solid, the constituent species are closely packed or held together by strong forces; so, in between the constituents, there are very small gaps and they can neither cannot move nor flow (Fig. 1.8). For example, ice, wood, coal, stone, iron, steel, brick, and so on.

Fig. 1.8 Solid

Properties of Solids

Solids are characterised by the following properties.

1. **Shape and volume:** A solid has a fixed shape and a fixed volume as the constituents (atoms, ions or molecules) are arranged in a definite pattern held together by strong intermolecular forces. This regular arrangement gives them a definite geometry and volume.

2. **Compressibility:** Solids are rigid and they cannot be compressed easily as the constituents are closely packed; intermolecular distance or space in them is very low and cannot be further reduced.

3. **Hardness and density:** Solids are quite hard and they have high densities since they have close packed arrangement.

4. **Diffusion:** A solid cannot diffuse into another solid easily as very strong intermolecular forces of attraction are present in solids.

   Examples: (a) If we write something on a blackboard and leave it uncleaned for a considerable period of time, we will find that it becomes quite difficult to clean the blackboard afterwards. This is due to the fact that some of the particles of chalk have diffused into the surface of blackboard. (b) If two metal blocks are bound together tightly and kept undisturbed for a few years, then the particles of one metal are found to have diffused into the other metal.

5. **Melting points:** Solids have a high melting point as a huge amount of energy is required to overcome the strong intermolecular forces present in their constituents.

6. **Motion:** Solids do not flow, so they cannot acquire the shape of the vessel in which they are kept. They do not show any motion as the constituents are strongly held at their respective positions by strong intermolecular forces. They can only vibrate.

KNOWLEDGE BOOSTER

Types of Solids

Solids can be classified as *crystalline* or *amorphous* on the basis of the nature of order present in the arrangement of their constituent particles (Table 1.2).

(i) **Crystalline solids:** A crystalline solid usually consists of a large number of small crystals, each of them having a definite characteristic geometrical shape. In a crystal, the arrangement of constituent particles is ordered. Sodium chloride and quartz are typical examples of crystalline solids.

(ii) **Amorphous:** An amorphous solid (Greek *amorphos* = no form) consists of particles of irregular shape. The arrangement of constituent particles (atoms, molecules or ions) in such a solid has only *short range order*. The structures of quartz (crystalline) and quartz glass (amorphous) are shown in Fig. 1.9 (a) and (b) respectively. The structure of amorphous solids is similar to that of liquids. Glass, rubber and plastics are typical examples of amorphous solids.

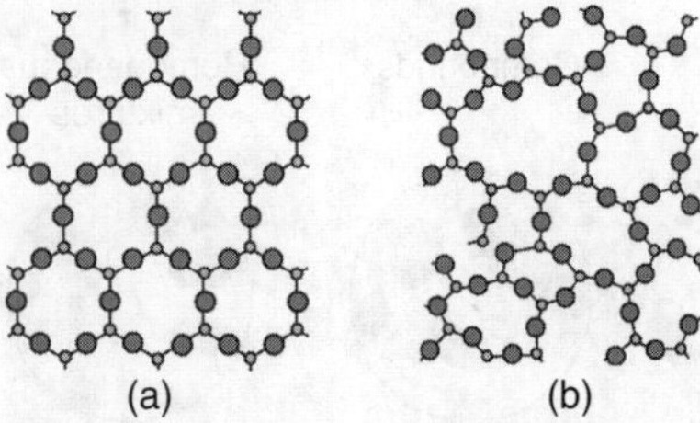

Fig. 1.9 Two dimensional structure of (a) quartz and (b) quartz glass

Table 1.2 Distinction between crystalline and amorphous solids

Property	Crystalline solids	Amorphous solids
Shape	Definite characteristic geometrical shape	Irregular shape
Melting point	Melt at a sharp and characteristic temperature	Gradually soften over a range of temperatures
Cleavage property	When cut with a sharp-edged tool, they split into two pieces and the newly generated surfaces are plain and smooth	When cut with a sharp-edged tool, they cut into two pieces with irregular surfaces
Heat of fusion	They have a definite and characteristic heat of fusion	They do not have a definite heat of fusion
Anisotropy	Anisotropic in nature	Isotropic in nature
Nature	True solids	Isotropic in nature
Order in arrangement of constituent particles	Long range order	Pseudo solids or super cooled liquids, only short range order

Illustrations

1. Give reasons:

   (a) A wooden chair should be called a solid.

   (b) We can easily move our hand in air but to do the same through a solid block of wood we need a karate expert.

   Solution:

   (a) There is a strong force of attraction between the molecules of wood and the intermolecular space is the least. So, a wooden chair has a definite shape and volume and it should be called a solid.

   (b) Air molecules are extremely far from each other due to negligible force of attraction between them. So, our hand gets sufficient space to move in air and we also displace some air molecules by applying force.

However, a solid block of wood has closely packed molecules so there is no question of the movement of hand through it in the absence of suitable force in the proper direction.

2. Give reasons for the following:

   (a) Quartz has a sharp melting point but glass does not.

   (b) Amorphous solids are called super-cooled liquids.

   Solution:

   (a) Quartz has a sharp melting point as it is a crystalline solid but glass being an amorphous solid does not have a sharp melting point.

   (b) Amorphous solids are called super-cooled liquids as they can flow slowly at room temperature, like liquids.

Liquids

A liquid is one of the four fundamental states of matter and is the only state with a definite volume but no fixed shape. A liquid is a nearly incompressible fluid that conforms to the shape of its container but retains a constant volume independent of pressure. In liquids, the molecules are more closely held than in gases due to strong intermolecular forces (but less than in solids). Liquids have more vacant space than solids (Fig. 1.10). In liquids, molecules can move or flow.

For example, water, milk, fruit juice, ink, groundnut oil, kerosene and petrol.

Fig. 1.10 Liquid

Properties of Liquids

Liquids are characterised by the following properties.

1. **Shape and volume:** A liquid will have a fixed volume but cannot have a fixed shape. A liquid takes the shape of the vessel in which it is placed as the liquid molecules are not held strongly enough to have a fixed position or definite geometry. However, the intermolecular force is strong enough for them to have a definite volume.

2. **Compressibility:** Like solids, liquids cannot be compressed much because the vacant space is not much; however, they can be compressed more than solids but less than gases.

3. **Hardness and density:** Liquids have moderate to high densities. They are usually less dense than solids but more dense than gases.

4. **Diffusion:** Liquids are more diffusible than solids but less than gases as the vacant spaces between their molecules is more than that of solids but less than that of gases. They do not fill their container completely. Diffusion in liquids is slower than in gases as the particles in liquids move slower as compared to the particles in gases. The rate of diffusion in liquids is much faster than that in solids because the particles in a liquid move much more freely, and have greater spaces between them as compared to particles in the solids.

For example, (a) the spreading of purple colour of potassium permanganate into water on its own, is due to the diffusion of potassium permanganate particles into water. (b) The spreading of blue colour of copper sulphate into water on its own, is due to the diffusion of copper sulphate particles into water.

5. **Melting points:** Liquids have lower melting points than solids but higher than that of gases as the intermolecular forces are weaker than in solids but stronger than in gases.

6. **Motion:** Liquids generally flow or show motion easily as there is enough space between their molecules. However, their motion is less than in gases.

7. **Evaporation:** Liquid molecules can evaporate and change into vapour at room temperature as the kinetic energy of liquid molecules can overcome the intermolecular force present between molecules. The rate of evaporation is not the same for all liquids as the kinetic energies of liquid molecules differ.

Illustrations

1. Give two reasons to justify the following.

   (a) Water at room temperature is a liquid.

   (b) An iron almirah is a solid at room temperature.

   Solution:

   (a) Water is a liquid at room temperature because it has a tendency to flow. It takes the shape of the container in which it is filled, but its volume remains the same.

   (b) An iron almirah is a solid at room temperature because its shape and volume are definite. It is hard and rigid. Its density is high.

2. Explain the following:

   (a) Liquids generally have lower density compared to solids. But you must have observed that ice floats on water.

   (b) When a liquid is transferred from a small vessel to a large vessel at the same temperature, what will be the effect on vapour pressure?

   Solution:

   (a) Liquids have lower density than solids. Water is also a liquid so it should also have less density than ice. However, this is not so because of the cage-like structure of ice. That is, the presence of vacant spaces between water (H_2O) molecules when they link with each other in ice. The number of these spaces is comparatively less in water. Being more porous than water, ice is lighter than water and floats over the surface of water.

   (b) There will be no effect on vapour pressure when a liquid is transferred from a small vessel to a large vessel at same temperature as vapour pressure does not depend on the size of the vessel.

KNOWLEDGE BOOSTER

- **Vapour pressure:** The pressure exerted by the vapour in equilibrium with the liquid state at a given temperature is known as the vapour pressure of the liquid. It depends upon the nature and temperature of the liquid but not upon the size of the vessel.

$$\text{Vapour pressure} \propto \text{Temperature}$$
$$\propto \frac{1}{\text{Intermolecular force of attraction}}$$

- **Surface tension:** The property of the surface of a liquid that allows it to resist an external force, due to the cohesive nature of its molecules. The cohesive forces between liquid molecules are responsible for the phenomenon known as surface tension (Fig. 1.11).

  For example, the property that is responsible for the spherical shape of liquid drops is called surface tension. The liquid drops assume a spherical shape as spheres occupy a minimum surface area because of surface tension. Surface tension is produced because of the cohesive forces that exist between various liquid molecules. These forces cause water droplets to take a spherical form.

- **Capillary action:** Capillary action is important for moving water (and all of the things that are dissolved in it) around. It is defined as the movement of water within the spaces of a porous material due to the forces of adhesion, cohesion and surface tension (Fig. 1.12).

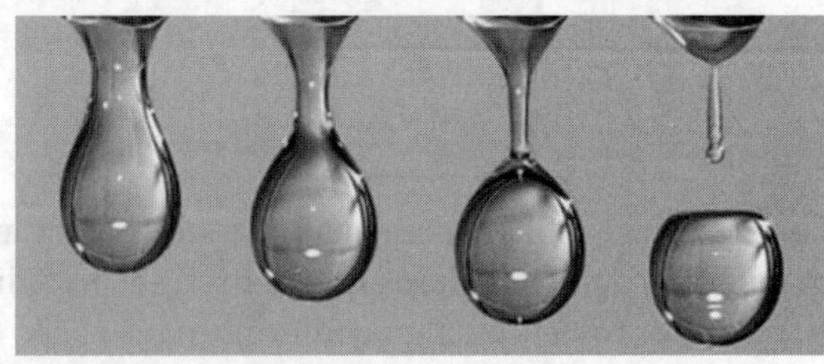

Fig. 1.11 Drops falling in spherical shape

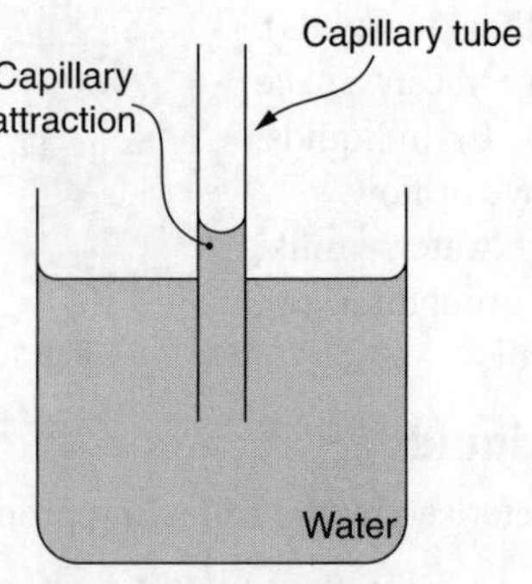

Fig. 1.12 Capillary action

Gases

The gaseous state is one of the four fundamental states of matter. A gas has no definite volume and no definite shape (Fig. 1.13). Gases are compressible and diffusible as in their molecules have large spaces in between them and are not closely packed due to weak interactions. A pure gas may be made up of individual atoms, elemental molecules made from one type of atom or compound molecules made from a variety of atoms. A gas mixture, such as air, contains a variety of pure gases. Examples of gases are air, oxygen, hydrogen, nitrogen, helium, and so on.

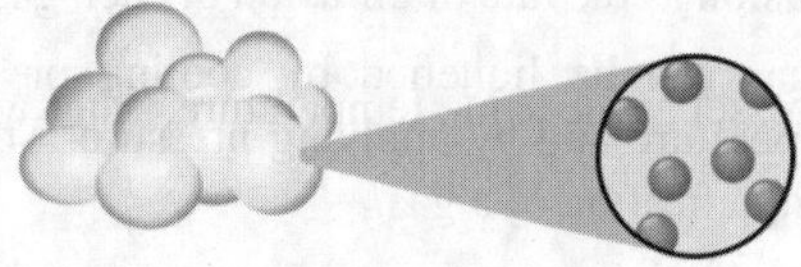

Fig. 1.13 Gas

Properties of Gases

Gases are characterised by the following properties.

1. **Shape and volume:** Gases have neither a fixed shape nor a fixed volume. They acquire the shape and volume of the vessel in which they are stored. This is because gaseous molecules have very weak interactions amongst each other and can thus move freely and occupy any amount of space.

2. **Compressibility:** Gases can be compressed easily by applying external pressure as the space between the gas molecules is large due to weak intermolecular forces between them. During compression, the gas molecules come closer to each other and occupy less space.

3. **Density:** Gases have very low densities and they are very light.

4. **Diffusion:** Gases can diffuse easily as the gaseous molecules are weakly held and have enough kinetic energy to diffuse. The gases can fill their container completely. For example, if LPG (Liquified Petroleum Gas) leaks, its molecules can spread all around the kitchen and it can be detected by the smell.

For example, (a) When someone opens a bottle of perfume in one corner of a room, its smell spreads through the whole room quickly. The smell of perfume spreads due to the diffusion of perfume vapours into air. (b) When we light an incense stick (agarbatti) in a corner of a room, its fragrance spreads through the whole room very quickly. The fragrance spreads all around due to the diffusion of its smoke into the air.

5. **Homogeneous nature:** Gases have similar composition in all parts, so they are homogeneous in nature. A gaseous mixture is always homogenous as it is has only the gas phase.

6. **Liquefaction:** A gas can be liquefied by cooling and by applying pressure.

   Temperature in °C, kelvin (K) and fahrenheit (F)

   Temperature in kelvin (K) = $t°C + 273$ K

   Temperature in fahrenheit (F) and °C relation: $\dfrac{F-32}{9} = \dfrac{C}{5}$

   For example, 14°F in °C is $\dfrac{14-32}{9} = \dfrac{C}{5}$

   On solving, C = –10°C

   Temperature in kelvin (K) = $-10° + 273 = 263$ K

Exertion of Pressure

Solids exert pressure only in the downward direction, liquids exert pressure downward as well as to the sides while gases exert pressure in all directions (Fig. 1.14). For example, a balloon. This pressure is due to the bombardment of the gas particles against the walls of the container. Hence, gases exert pressure equally in all directions.

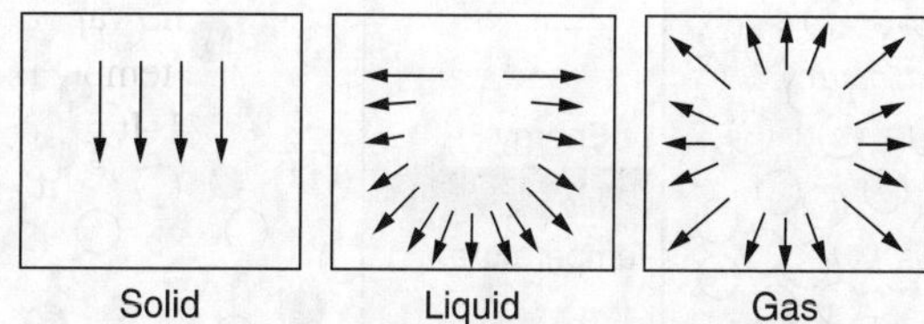

Fig. 1.14 Exertion of pressure

Table 1.3 Comparison of characteristic properties of solids, liquids and gases

No	Property	Solids	Liquids	Gases
1	Shape	Definite	Attain the shape of the container, but do not necessarily occupy all of it	Attain the shape of the container by occupying the entire space available to them
2	Volume	Definite	Definite	Attain the volume of the container
3	Compressibility	Almost nil	Very low	Very large
4	Fluidity or rigidity	Rigid	Fluid	Fluid
5	Density	High	Low	Very low
6	Diffusion	Generally do not diffuse	Diffuse slowly	Diffuse rapidly
7	Free surfaces	Any number of free surfaces	Only one free surface	No free surface

Illustrations

1. Give reasons:

(a) A gas completely fills the vessel in which it is kept.

(b) A gas exerts pressure on the walls of the container.

Solution:

(a) The forces of attraction between the molecules of gases is negligible. So, molecules of gases occupy the maximum space available to them. High kinetic energy possessed by their molecules also helps.

(b) The motion of particles is random and occurs at very high speed in the gaseous state. Due to this random movement, the particles hit each other and also the walls of the container. The pressure exerted by the gas is due to this force exerted by the particles per unit area on the walls of the container.

2. (a) How can you separate gases from a gaseous mixture?

(b) How can a gas be liquefied?

Solution:

(a) Gases can be separated from the gaseous mixture by diffusion as the rate of diffusion of each gas is different.

(b) A gas can be liquefied by cooling or lowering of temperature and by applying pressure. This is known as liquefication of a gas.

KNOWLEDGE BOOSTER

Plasma: Plasma is one of the four fundamental states of matter and was first described by chemist Irving Langmuir in the 1920s. It consists of a gas of ions – atoms which have some of their outer orbital electrons removed – and free electrons. Plasma can be artificially generated by heating a neutral gas or subjecting it to a strong electromagnetic field to the point where an ionised gaseous substance becomes increasingly electrically conductive. The resulting charged ions and electrons become influenced by long range electromagnetic fields, making the plasma dynamics more sensitive to these fields than a neutral gas.

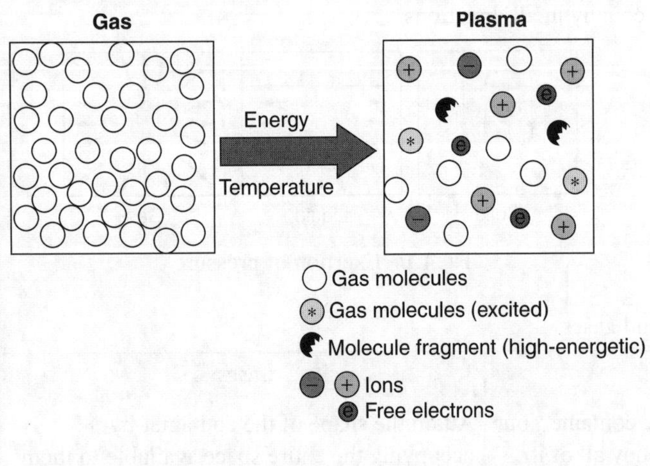

Examples of plasma state:

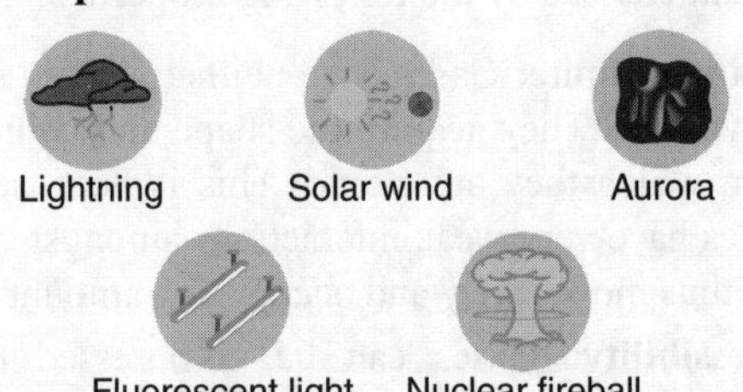

Bose–Einstein condensates: In 1995, two scientists, Cornell and Weiman, finally created this new state matter. Two other scientists, Satyendra Bose and Albert Einstein, had predicted it in the 1920s. They did not have the equipment and facilities to make it happen. If plasmas are super-hot and super-excited atoms, the atoms in the Bose–Einstein condensate (BEC) are total opposites. They are super-unexcited and super-cold atoms. BEC, a state of matter in which separate atoms or subatomic particles, cooled to near absolute zero (0 K, $-273.15°C$, or -459.67 F; K = kelvin), coalesce into a single quantum mechanical entity; that is, one that can be described by a wave function on a near-macroscopic scale.

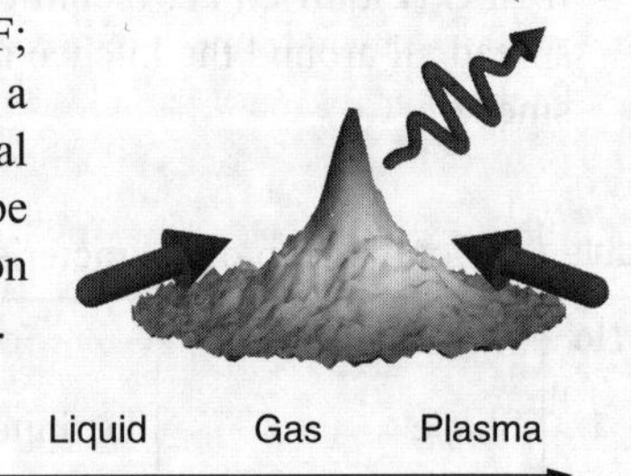

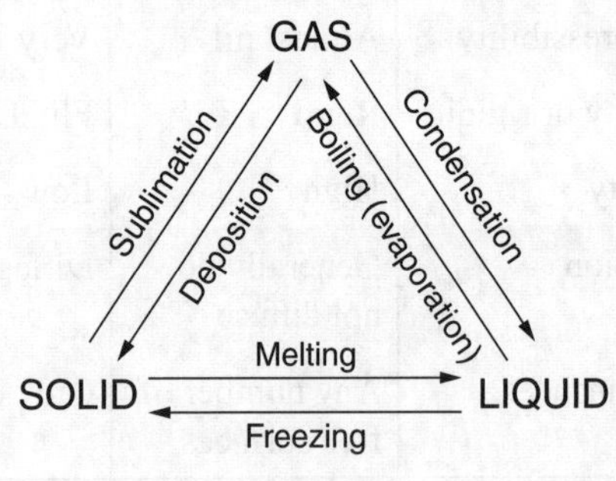

Change of State of Matter

A substance may exist in any of the three states of matter (solid, liquid or gas) depending upon the conditions of temperature and pressure. By changing the temperature and pressure, the state of a substance can be changed. A solid on heating usually changes into a liquid which on further heating changes into gas. Similarly, a gas on cooling condenses into a liquid which on further cooling changes into a solid.

The most familiar and common example is water. It exists in all the three states—solid: ice, liquid: water and gas: water vapour. Ice is a solid and may be melted to form water (liquid) which on further heating changes into steam (gas). These changes can also be reversed on cooling.

$$\text{Ice} \underset{\text{Cool}}{\overset{\text{Heat}}{\rightleftharpoons}} \text{Water} \underset{\text{Cool}}{\overset{\text{Heat}}{\rightleftharpoons}} \text{Steam}$$

(Solid)   (Liquid)   (Gas)

Effect of Temperature Change

By increasing the temperature (heating), a solid can be converted into the liquid state, and the liquid can be converted into the gaseous or vapour state. By decreasing the temperature (cooling), a gas can be converted into the liquid state, and a liquid can be converted into the solid state.

Solid State to Liquid State (Melting)

The process by which the solid state of a substance is changed into its liquid state on heating is known as melting or fusion. Melting is done by providing heat to the solid substance.

- **Melting point and melting:** The temperature at which a solid substance melts and changes into a liquid at atmospheric pressure is known as the *melting point* of the substance. The melting point of a solid is a measure of the force of attraction between its particles. The higher the melting point of a solid substance, the greater will be the force of attraction between its particles. For example, Melting point of ice = 0°C or 273 K, Melting point of wax = 63°C or 336 K and Melting point of iron = 1535°C or 1808 K.

  When a solid is heated, its particles absorb heat energy and they become more energetic; the kinetic energy of the particles increases. Due to the increase in kinetic energy, the particles start vibrating more strongly with greater speed. The energy supplied by heat overcomes the intermolecular forces of attraction between the particles. As a result, the particles leave their mean position and break away from each other. When this happens, the solid melts and a liquid is formed. This process is known as *melting*.

- **Factors affecting the melting point:** Although every substance has a specific melting point under atmospheric conditions, it can however be affected by the following factors.

  (a) **Effect of pressure:** The effect of pressure on the melting point varies according to the nature of solid.

  - The melting points of those solids which expand during melting, increase with increase in pressure as increase in pressure opposes expansion. For example, Cu, Au, Ag, paraffin wax.
  - The melting points of those solids which contract during melting, decrease with an increase in pressure as increase in pressure favours contraction. For example, ice, brass, cast iron.
  - Ice is a solid which contracts during melting. Therefore, when pressure is applied on two ice

cubes, the ice at the interface melts and when the pressure is released it solidifies again. The ice cubes rejoin at this spot. This phenomenon is known as *regelation*.

(b) **Effect of impurities:** If impurities are associated with a solid, its melting point decreases. This causes it to melt at a lower temperature. For example, in the case of rose metal [alloy of Sn (mp 239.1°C), pb (mp 327 °C), Bi (mp 271°C)], melting occurs at 94.5°C.

Liquid State to Gaseous State (Boiling or Vaporisation)

The process by which the liquid state of a substance is changed into its gaseous state on heating, is known as boiling or vaporisation.

- **Boiling point:** The temperature at which a liquid boils and changes rapidly into a gas at atmospheric pressure, is called the *boiling point* of the liquid. The boiling point of a liquid is a measure of the force of attraction between its particles. The higher the boiling point of a liquid, the greater will be the force of attraction between its particles. For example, Boiling point of water = 100°C or 373 K, Boiling point of alcohol = 78°C or 351 K and Boiling point of mercury = 357°C or 630 K.

  In a liquid most of the particles are close together. When we supply heat energy to the liquid, the particles of the liquid start vibrating fast. Some of the particles become so energetic that they can overcome the attractive forces of the particles around them. Hence, they become free to move and escape from the liquid. When this happens, the liquid evaporates, that is, starts changing into gas. At the boiling point, the particles of a liquid have sufficient kinetic energy to overcome the forces of attraction holding them together and separate into individual particles. The liquid then boils to form a gas.

- **Factors affecting boiling point:** Although every liquid substance has a specific boiling point under atmospheric conditions, it can be affected by the following factors.

  (a) **Effect of pressure:** On increasing the pressure, the boiling point of the liquid increases. For example, at higher altitudes water boils at a lower temperature than 373 K. In a pressure cooker, the water is heated in a closed vessel in a confined space. So when steam is formed, the pressure increases. As a result, the boiling point of water also increases and the food gets cooked faster and earlier.

  (b) **Effect of impurities:** If impurities are associated with a liquid its boiling point increases and it will boil at a higher temperature. For example, water containing common salt (NaCl) boils at higher temperature than 373 K.

Liquid State to Vapour State (Evaporation)

The process of converting a liquid into vapour at any temperature below its boiling point is known as *evaporation*. It is a surface phenomenon as it is confined to only surface molecules. It is a slower process than boiling.

- **Factors Affecting Evaporation:**

(a) **Temperature:** The rate of evaporation increases on increasing temperature. With an increase in temperature, more number of particles get enough kinetic energy to go into the vapour state.

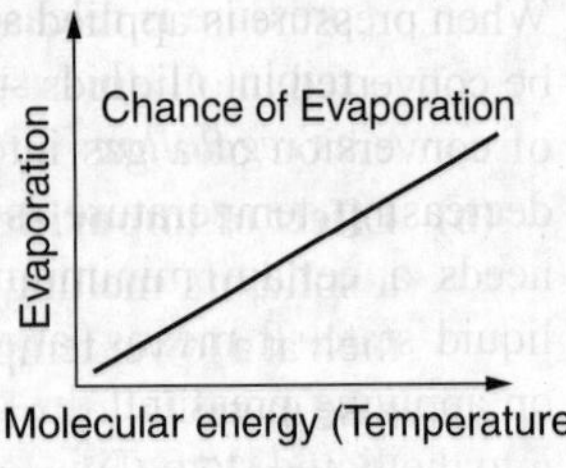

Rate of evaporation ∝ Temperature

For example, drying of clothes takes place more rapidly in summer than in winter. Water evaporates more quickly when it is heated; as the water boils it turns into steam.

(b) **Surface area:** The rate of evaporation increases on increasing the surface area of the liquid.

Rate of evaporation ∝ Surface area

For example, if the same liquid is kept in a test tube and in a china dish, then the liquid kept in the china dish will evaporate more rapidly as more of its surface area is exposed to air.

(c) **Humidity:** Humidity is the amount of water vapour present in air. Air around us cannot hold more than a definite quantity of water vapour at a given temperature. If the amount of water in air is already large or humidity is more, the rate of evaporation decreases. Hence, the rate of evaporation increases with a decrease in humidity in the atmosphere. For example, drying of clothes on a humid day takes a longer time.

Rate of evaporation ∝ 1/Humidity

(d) **Pressure:** The rate of evaporation also increases with a decrease in the pressure of the liquid as the molecules prefer to move from the side with the higher pressure to the side with the lower pressure.

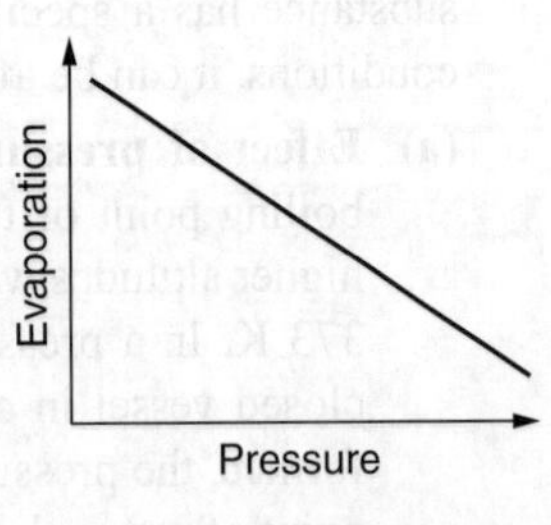

(e) **Wind speed:** The rate of evaporation also increases with an increase in the speed of the wind. With an increase in the speed of wind, the particles of water vapour move away with the wind, resulting in a decrease in the amount of vapour in the atmosphere. For example, clothes dry faster on a windy day.

- **Cooling produced due to evaporation:** Liquid molecules can absorb energy from the surroundings, overcome the force of attraction and change into the vapour form. In this case, the surroundings lose energy and become cold. This principle is quite useful in our daily activities. Some example are:

- cotton clothes produce a cooling effect during summer,
- a cool sensation is felt when alcohol is poured on the palm,
- cooling of water is possible in earthen pots (pitchers) during summer, and
- formation of water droplets can be seen on the outer surface of a glass containing ice cold water.

KNOWLEDGE BOOSTER

Difference between boiling and evaporation

(i) Boiling can take place only at a particular temperature for a liquid while evaporation can occur at all temperatures.

(ii) Boiling is a bulk phenomenon. That is, bubble formation occurs even below the surface. Evaporation is surface phenomenon in which bubble formation occurs only on the surface of liquid.

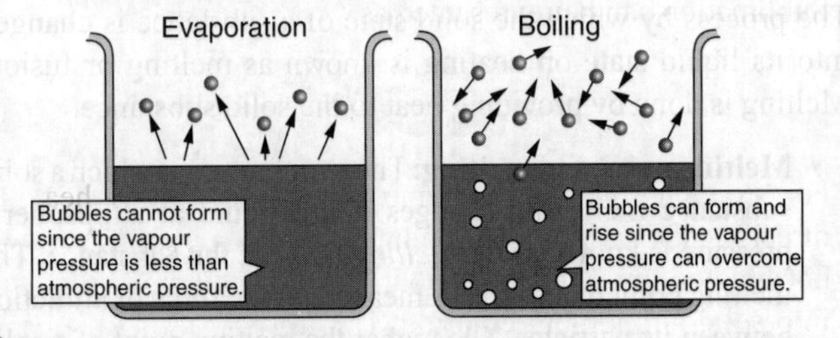

Liquid State to Solid State (Freezing)

The process of converting a liquid into a solid by cooling is known as freezing. Freezing means solidification. Freezing is the reverse of melting. So, the freezing point of a liquid is the same as the melting point of its solid form. For example, Melting point of ice = 0°C or 273 K and Freezing point of water = 0°C or 273 K.

- **Freezing point:** The temperature at which a liquid gets converted into a solid at atmospheric pressure is known as the *freezing point* of a liquid.

When a liquid substance is subjected to cooling, heat is extracted from the liquid. Therefore, the kinetic energy of the molecules decreases which results in a decrease in temperature. On reaching a certain temperature, further extraction of heat from the liquid results in a decrease in potential energy instead of decreasing the kinetic energy. As a decrease in potential energy leads to an increase in intermolecular forces of attraction due to decrease in intermolecular space, the molecular arrangement of the liquid changes to that of a solid.

- **Factors affecting freezing point:** Like the melting point, the freezing point is also affected by temperature and the presence of impurities. For example, the presence of impurities decreases the freezing of the liquid to some extent. This concept is used in making freezing mixtures. For example, by mixing 3 parts of ice with 1 part of common salt, we get a freezing mixture which can lower the temperature to −21°C. Such freezing mixtures are used for the preservation of food stuffs such as meat, fish, and so on, as well as vaccines!

Gaseous State to Liquid State (Condensation)

The process of conversion of a gaseous substance into its liquid state is known as *condensation*. It can be carried out by cooling the gas below a particular temperature. When heat is extracted from the gas, the kinetic energy of the molecules decreases so the temperature decreases. When a sufficiently low temperature is reached, further extraction of heat from the gas does not reduce the kinetic energy. As a result, an increase in intermolecular forces of attraction brings the molecules closer; thus, at this point, the gas passes into the liquid state.

Solid State to Gaseous State (Sublimation)

The process of changing of a solid substance directly into its gaseous or vapour form on heating is known as *sublimation*, and of the vapour form into solid on cooling is known as *deposition* or *desublimation*. The solid substance which undergoes sublimation is said to *sublime* while the solid deposit of a substance which has sublimed is known as a *sublimate*. Sublimation can be represented as:

$$\text{Solid} \underset{\text{Cooling}}{\overset{\text{Heating}}{\rightleftharpoons}} \text{Vapour (or gas)}$$

For example, when solid ammonium chloride is heated, it directly changes into ammonium chloride vapours. And when hot ammonium chloride vapour is cooled, it directly changes into solid ammonium chloride. The following substances also undergo sublimation—iodine, camphor, naphthalene and anthracene.

Effect of Pressure Change

The three states of matter differ in the intermolecular forces and intermolecular distances between the constituent particles (atoms, ions, molecules). Gases are said to be compressible since the space between the gaseous particles decreases on applying pressure. A substance may exist in any of the three different states of matter depending upon the conditions of temperature and pressure.

- If the melting point of a substance is above the room temperature under atmospheric pressure, it is said to be a solid.
- If the boiling point of a substance is above the room temperature under atmospheric pressure, it is classified as a liquid.
- If the boiling point of the substance is below the room temperature under atmospheric pressure, it is called a gas.

Liquefaction

When pressure is applied and temperature is reduced, gases can be converted into liquids—gases will be liquefied. The process of conversion of a gas into a liquid by increasing pressure or decreasing temperature is known as *liquefaction*. Every gas needs a certain minimum temperature for passing into the liquid state. It means, above this minimum temperature even on applying pressure, the gaseous state cannot be transformed into the liquid state. This temperature above which a gas cannot be liquified by applying pressure is known as the *critical temperature*. Every gas has a specific critical temperature. For example, the critical temperatures of CO_2 is $31.1°C$ and H_2 is $-165°C$.

$$\text{Liquification of a gas} \propto \text{Critical temperature}$$

Latent heat

The word 'latent' means 'hidden'. The heat energy that has to be supplied to change the state of a substance is known as its *latent heat*. Latent heat does not raise the temperature but latent heat always has to be supplied to change the state of a substance since every substance has some forces of attraction between its particles that hold them together. Thus, if the state of a substance has to be changed, it is necessary to break these forces of attraction between its constituents by supplying heat. The latent heat does not increase the kinetic energy of the particles of the substance.

Types of Latent Heat

There are two types of latent heat.

(i) **Latent heat of fusion:** The heat needed to convert a solid substance into its liquid state is known as *latent heat of fusion*. In other words, the latent heat of fusion of a solid is the quantity of heat in joules required to convert 1 kilogram of the solid to liquid, without any change in temperature. For example, the latent heat of fusion of ice is 3.34×10^5 J/kg or 80 cal/g.

(ii) **Latent heat of vaporisation:** The heat needed to convert a liquid substance into its vapour state is known as *latent heat of vaporisation*. In other words, the latent heat of vaporisation of a liquid is the quantity of heat in joules required to convert 1 kilogram of the liquid to its vapour or gaseous state, without any change in temperature. For example, the latent heat of vaporisation of water is 22.5×10^5 J/kg or 540 cal/g.

Illustrations

1. Fill in the blanks:

   (a) Evaporation of a liquid at room temperature leads to a ______ effect.

   (b) At room temperature, the forces of attraction between the particles of a solid substances are ______ than those in the gaseous state.

   (c) The arrangement of particles is less ordered in the ______ state. However, there is no order in the state.

   (d) ______ is the change of gaseous state directly to solid state without going through the ______ state.

   (e) The phenomenon of change of a liquid into the gaseous state at any temperature below its boiling point is called ______.

Solution:

(a) Cooling (b) Stronger

(c) Liquid, gaseous

(d) Desublimation or deposition, liquid

(e) Evaporation

2. Name A, B, C, D, E and F in the following diagram showing change in its state.

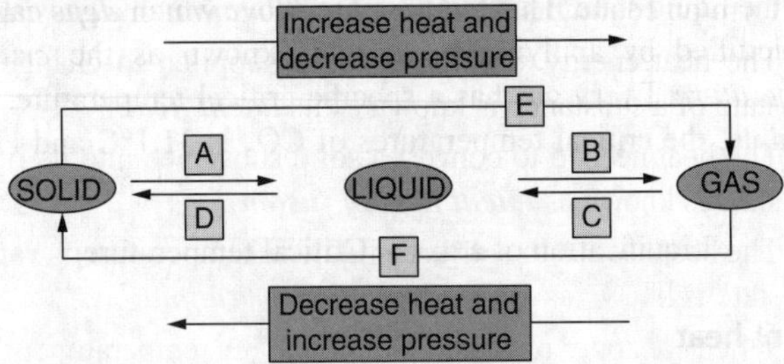

Solution:

A = Melting or fusion, where solid changes into liquid.

B = Evaporation or vaporisation, where liquid changes into gas.

C = Condensation or liquefaction, where gas changes into liquid.

D = Freezing or solidification, where liquid changes into solid.

E = Sublimation, where solid directly changes into gas without passing through the liquid state.

F = Desublimation or deposition, where gas changes into solid without passing through the liquid state.

3. Explain these observations.

(i) Khushi was making tea in a kettle. Suddenly she felt intense heat from the puff of steam gushing out of the spout of the kettle. She wondered whether the temperature of the steam was higher than that of the water boiling in the kettle.

(ii) Water as ice has a cooling effect, whereas water as steam may cause severe burns.

Solution:

(i) The temperature of both boiling water and steam is 100°C but steam has more energy because of its latent heat of vaporisation. Hence, steam is hotter than boiling water.

(ii) Water in the form of ice has low energy since water freezes at a lower temperature. When ice comes in contact with the body it draws heat from the body and produces a cooling effect. In the case of steam, the water molecules have high energy. The high energy of steam is transformed into heat and may cause severe burns.

4. Look at the figure below and suggest in which of the vessels – A, B, C or D – will the rate of evaporation be the highest? Explain.

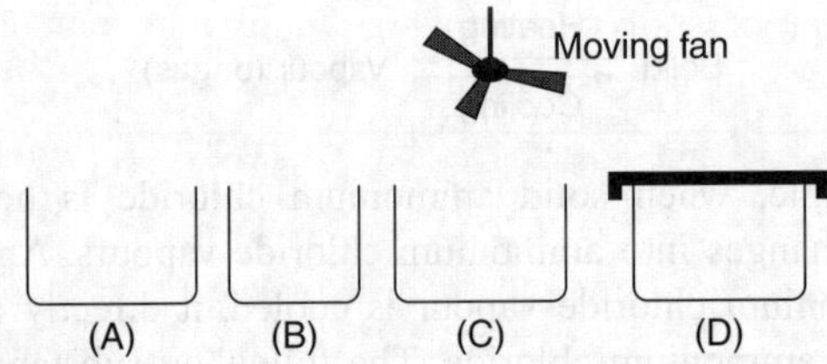

Solution:

The rate of evaporation depends on the surface area of the container. The more the surface area, the more is the evaporation. It also depends on the speed of the wind. If speed of the wind is more, more number of particles will evaporate from the surface. Hence, Fig. (C) in which both these factors, surface area and moving fan figure, the rate of evaporation will be maximum.

CHAPTER AT A GLANCE

- Everything in our universe is made up of material which has been named *matter*.

- Anything which occupies space and has mass is known as *matter*. The particles that make up matter are *atoms* or *molecules*.

- Matter has mass, occupies space, inertia, is influenced by gravity, and so on.

- During all physical and chemical changes, the total mass of the matter before and after the change remains constant (Lavoisier's law of conservation of mass).

- Diffusion and Brownian motion prove the *existence of particles* in matter and their motion.

- *Diffusion* is the flow of molecules from the side with higher concentration to the side with lower concentration.

- *Osmosis* is the flow or movement of solvent particles from the dilute solution (less solute) to the concentrated solution (more solute) through a semi-permeable membrane.

- Intermolecular forces are the forces of attraction or repulsion that act between neighbouring particles (atoms, molecules or ions).

- On the basis of their physical states, matter can be classified into four groups—solids, liquids, gases and plasma.

- Solids having the most closed packed arrangement of particles, hard with definite shape, volume and with least compressibility and diffusibility.

- In a crystalline solid, arrangement of constituents has long range ordered while in an amorphous solid, there is only short range order.

- A liquid has a fixed volume, no definite shape, less compressibility and diffusibility than a gas and less closely packed than solids but more than gases.

- A gas has no definite shape or volume but has more compressibility and diffusibility.

- Temperature in kelvin (K) = $t°C + 273$ K.

- *State conversion*: State conversion is possible as follows:

$$\text{Ice} \underset{\text{Cool}}{\overset{\text{Heat}}{\rightleftharpoons}} \text{Water} \underset{\text{Cool}}{\overset{\text{Heat}}{\rightleftharpoons}} \text{Steam}$$

$$\text{(Solid)} \qquad \text{(Liquid)} \qquad \text{(Gas)}$$

- The temperature at which a solid substance melts and changes into a liquid at atmospheric pressure is known as the *melting point* of the substance and this process is known as *melting*.

- The temperature at which a liquid boils and changes rapidly into a gas at atmospheric pressure is called the *boiling point* of the liquid and this process is called *vaporisation*.

- The process of converting a liquid into vapour at any temperature below its boiling point is known as *evaporation*.

- The temperature at which a liquid gets converted into a solid at atmospheric pressure is known as the *freezing point* of a liquid.

- The process of conversion of a gaseous substance into its liquid state is known as *condensation*.

- The process of changing of a solid substance directly into its gaseous or vapour form on heating is known as *sublimation*. The process of changing of a vapour directly into its solid form on cooling is known as *desublimation* or *deposition.*

- The heat energy which has to be supplied to change the state of a substance is known as its *latent heat*.

- The heat needed to convert a solid substance into its liquid state is known as *latent heat of fusion*.

- The heat needed to convert a liquid substance into its vapour state is known as *latent heat of vaporisation*.

- When we apply pressure and reduce the temperature, gases can be converted into liquids, that is, gases get liquefied. The process of conversion of a gas into a liquid by increasing pressure or decreasing temperature is known as *liquefaction*.

PRACTICE QUESTIONS

Analyse Your Concepts (School Exam Based)

Fill in the Blanks

Instructions: Complete the following statements with an appropriate word/term to be filled in the blank spaces.

1. Solid, liquid and gas are called the three _______ of matter.

2. Among solid, liquid and gaseous states, _______ state is least rigid.

3. The smell of perfume gradually spreads across a room due to _______

4. _______ have no definite volume and no definite shape.

5. Intermolecular space in gases is _______ than that in liquids.

6. Liquid and _______ states are known as fluid states.

7. Gases have a _______ rate of diffusion than liquids.

8. The temperature 177°C on the kelvin scale is equal to _______

9. The temperature at which a liquid changes into a gas/vapour is known as _______

10. Change of liquid state to solid state is known as _______

11. The boiling point of water is _______ K and the melting point of ice is _______ K.

12. Change of state directly from the solid to the gas without passing through the liquid state is known as _______

13. Change from the vapour state to the liquid state is called _______

14. The amount of heat needed to convert 1 kg of solid into liquid at its melting point is known as _______

True or False

Instructions: Read the following statements and write your answer as true or false.

1. On the basis of chemical properties, matter is classified as elements, compounds and mixtures.

2. Intermolecular space in solids is less than that in liquids.

3. The movement of pollen grains in water is very slow and in a zig zag manner.

4. The colour of an aqueous solution of $KMnO_4$ is purple due to diffusion.

5. Perfume spray spreads in all possible directions in a room due to osmosis.

6. The particles of matter attract each other.

7. Elements and compounds are the impure form of a substance.

8. Crystalline solids have the same repeated arrangement for a long range order.

9. Amorphous solids undergo clean cleavage.

10. Falling drops are spherical in shape due to surface tension.

11. A solid exerts pressure only in the downward direction.

12. In the presence of impurities, the melting point always increases.

13. The rate of evaporation increases with humidity.

14. The boiling point of water on the kelvin scale is 373 K.

Very Short Answer Type Questions

1. Name any three types of intermolecular forces.

2. An aqueous solution of copper sulphate becomes blue. Which phenomenon is it due to?

3. Two liquids X and Y have boiling points 380 K and 390 K respectively. Which of the two has greater intermolecular forces of attraction?

4. Which state of matter has definite shape and definite volume?

5. Name the physical state of matter
   (a) Which can be easily compressed
   (b) Which is most brittle
   (c) Which can flow but cannot fill the vessel completely

6. Name any three substances that can sublime.

7. Name the process for the following changes:
   (1) Liquid $\longrightarrow$ Gas
   (2) Solid $\longrightarrow$ Liquid
   (3) Gas $\longrightarrow$ Liquid

8. Which will have more volume—ice or steam?

9. What is the impact of increase of surface area on the rate of evaporation?

10. If fish is being fried in a neighbouring home, you can smell it sitting in your house. Name the process that brings this smell to us.

11. The kelvin temperature is 190 K. What is the corresponding celsius scale temperature?

12. What is the chemical name of dry ice?

13. Suggest a method to liquefy atmospheric gases.

Short Answer Type Questions

1. What are the characteristics of the particles of matter?

2. Compare the three states of matter in terms of following in decreasing order:
   (a) Compressibility
   (b) Hardness
   (c) Energy of molecules
   (d) Intermolecular space

3. What do you understand by the term *latent heat*? What are the two types of latent heat?

4. Define *melting point* of a substance. What is the melting point of ice?

5. How do solids, liquids and gases differ in shape and volume?

6. What type of clothes should we wear in summer?

7. Convert the following temperature to the celsius scale.
   (a) 300 K   (b) 573 K

8. It is a hot summer day, Rahul and Ravi are wearing cotton and nylon clothes respectively. Who do you think would be more comfortable and why?

Long Answer Type Questions

1. Comment on the following statements:
   (a) Evaporation produces cooling.
   (b) Rate of evaporation of an aqueous solution decreases with increase in humidity.
   (c) A sponge, though compressible, is a solid.

2. Explain the following:
   (a) Gases exert pressure
   (b) Evaporation causes cooling
   (c) Solids can be converted to liquids
   (d) Gases diffuse rapidly

3. (a) When a gas jar containing air is inverted over a gas jar containing bromine vapour, the red–brown bromine vapour diffuses into the air. Explain how bromine vapour diffuses into the air.
   (b) When a crystal of potassium permanganate is placed in a beaker of water, its purple colour spreads throughout the water. What does this observation tell us about the nature of potassium permanganate and water?

4. Which characteristic of the particles of matter is illustrated by these observations?
   (a) When sugar is dissolved in water, there is no increase in the volume. Which characteristic of matter is illustrated by this observation?
   (b) A piece of chalk can be broken into small particles by hammering, but a piece of iron cannot be broken into small particles by hammering.

NCERT Corner

In-text Questions

1. Give reasons for the following observation.
   The smell of hot sizzling food reaches you several metres away, but to get the smell from cold food, you have to go close.

2. The mass per unit volume of a substance is called density (Density = mass/volume). Arrange the following in order of increasing density—air, exhaust from chimneys, honey, water, chalk, cotton and iron.

3. What is the physical state of water at (a) 250°C (b) 100°C?

4. For any substance, why does the temperature remain constant during the change of state?

5. Why does a desert cooler cool better on a hot dry day?

6. How does the water kept in an earthen pot (matka) become cool during summer?

7. Why does our palm feel cold when we put some acetone, petrol or perfume on it?

8. Why are we able to sip hot tea or milk faster from a saucer rather than from a cup?

Exercises

1. Convert the following temperatures to the celsius scale.
   (a) 293 K   (b) 470 K

2. Convert the following temperatures to the kelvin scale.
   (a) 25°C   (b) 373°C

3. Give reasons for the following observations.
   (a) Naphthalene balls disappear with time without leaving any solid behind.
   (b) We can get the smell of perfume sitting several metres away.

4. Arrange the following substances in the increasing order of forces of attraction between the particles—water, sugar, oxygen.

5. What is the physical state of water at (a) 25°C (b) 0°C (c) 100°C?

6. Why is ice at 273 K more effective in cooling than water at the same temperature?

7. What produces more severe burns, boiling water or steam?

Exemplar Problems

Short Answer Type Questions

1. A sample of water under study was found to boil at 102°C at normal temperature and pressure. Is the water pure? Will this water freeze at 0°C? Comment.

2. A student heats a beaker containing ice and water. He measures the temperature of the content of the beaker as a function of time. Which of the following figures would correctly represent the result? Justify your choice.

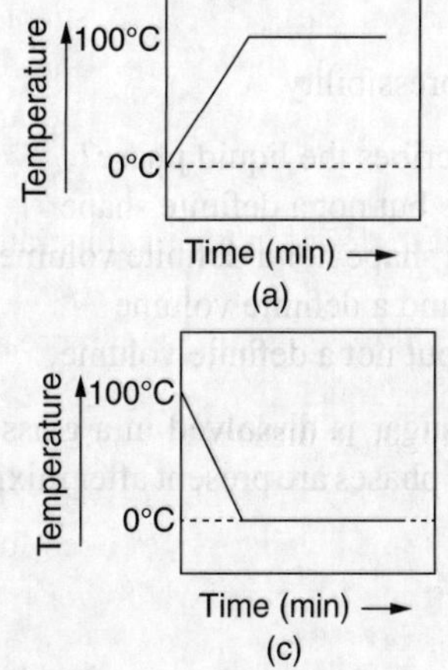

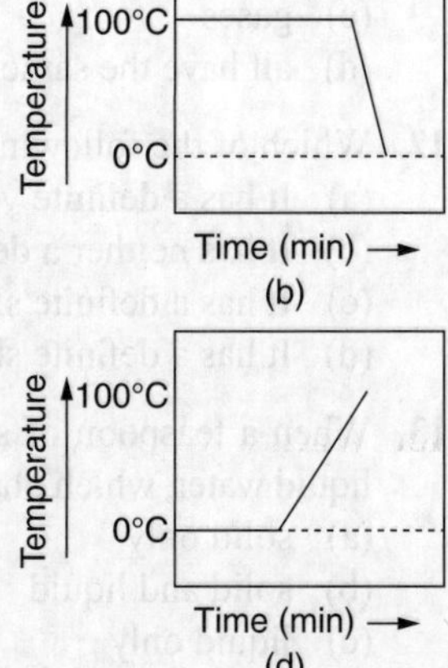

3. Match the physical quantities given in column A to their SI units given in column B.

Column (A)	Column (B)
(a) Pressure	(i) cubic metre
(b) Temperature	(ii) kilogram
(c) Density	(iii) pascal
(d) Mass	(iv) kelvin
(e) Volume	(v) kilogram per cubic metre

4. Osmosis is a special kind of diffusion. Comment.

5. Classify the following into osmosis/diffusion.
   (a) Preserving pickles in salt.
   (b) The smell of a cake being baked spreading throughout the house.
   (c) Aquatic animals using oxygen dissolved in water during respiration.

6. A glass tumbler containing hot water is kept in the freezer compartment of a refrigerator (temperature < 0°C). If you could measure the temperature of the content of the tumbler, which of the following graphs would correctly represent the change in its temperature as a function of time?

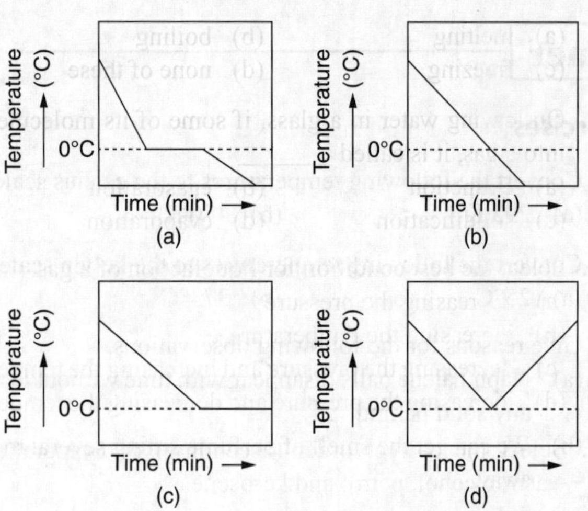

Long Answer Type Questions

1. You are provided with a mixture of naphthalene and ammonium chloride by your teacher. Suggest an activity to separate them with a well labelled diagram.

2. You want to wear your favourite shirt to a party, but the problem is that it is still wet after a wash. What steps would you take to dry it faster?

3. Why does the temperature of a substance remain constant when it is at its melting point or boiling point?

COMPETITION WINDOW (OBJECTIVE TYPE)

[For NEET, JEE (Main and Advanced), NTSE, KVPY and Olympiads]

Topic-wise MCQs

Matter and its type

1. Matter can be anything that
   (a) has mass
   (b) is occupying space
   (c) has both (a) and (b)
   (d) has none of these

2. The quantity of matter present in an object is known as its
   (a) mass   (b) density
   (c) weight   (d) gram

3. When we put some crystals of potassium permanganate in a beaker containing water, we observe that after sometime whole water has turned pink. This is due to
   (a) melting of potassium permanganate crystals
   (b) diffusion
   (c) boiling
   (d) sublimation of crystals

4. The state of matter having super energetic particles in the form of ionised gases is known as
   (a) liquid state
   (b) gaseous state
   (c) plasma state
   (d) Bose–Einstein condensate

5. The force that holds the particles of matter together is known as
   (a) intermolecular force   (b) nuclear force
   (c) intermolecular space   (d) bond

6. Which of the following is the densest state of matter?
   (a) gas   (b) solid
   (c) plasma   (d) liquid

7. The one in which interparticle forces are strongest, is
   (a) ethane   (b) carbon dioxide
   (c) potassium chloride (d) hydrogen

8. On the basis of chemical classification of matter, which of the following is considered matter?
   (a) element   (b) compound
   (c) mixture   (d) all of these

9. Which of the following statements is correct?
   (a) the particles of matter are very, very small
   (b) the particles of matter are constantly moving
   (c) the particles of matter attract each other
   (d) all of these

Solid, liquid and gas

10. All liquids have different
    (a) density   (b) viscosity
    (c) solubility   (d) all of these

11. Which of the following have the highest compressibility?
    (a) solids
    (b) liquids
    (c) gases
    (d) all have the same compressibility

12. Which of the following describes the liquid phase?
    (a) It has a definite volume but not a definite shape
    (b) It has neither a definite shape nor a definite volume
    (c) It has a definite shape and a definite volume
    (d) It has a definite shape but not a definite volume

13. When a teaspoon of solid sugar is dissolved in a glass of liquid water, which phase or phases are present after mixing
    (a) solid only
    (b) solid and liquid
    (c) liquid only
    (d) none of these

14. The boiling point of alcohol is 78°C. What is this temperature in the kelvin scale?
(a) 273 K (b) 351 K
(c) 373 K (d) 88 K

15. A gas can be compressed to a fraction of its volume. The same volume of a gas can be spread all over a room. The reason for this is that
(a) gases consist of molecules having very large intermolecular spaces which can be reduced or increased under ordinary conditions
(b) gases consist of molecules which are in a state of random motion
(c) the volume occupied by molecules of a gas is negligible compared to the total volume of the gas
(d) none of these

16. Non-reacting gases have a tendency to mix with one another. This phenomenon is called
(a) diffusion (b) effusion
(c) explosion (d) chemical reaction

17. Which has the least energetic molecules?
(a) gases (b) plasmas
(c) solids (d) liquids

18. In which phase of matter would you expect alcohol to exist at room temperature?
(a) gas (b) plasma
(c) solid (d) liquid

19. The SI unit of temperature is
(a) fahrenheit (b) kelvin
(c) celcius (d) both (b) and (c)

Change of state of matter

20. At higher altitudes, generally
(a) boiling point of a liquid decreases
(b) boiling point of a liquid increases
(c) melting point of a solid increases
(d) no change in boiling point

21. In which phenomenon does water change into water vapour below its BP?
(a) boiling (b) condensation
(c) evaporation (d) both (a) and (c)

22. The liquid that has the highest rate of evaporation is
(a) water (b) alcohol
(c) petrol (d) nail-polish remover

23. The change of a liquid into vapour is called
(a) sublimation (b) vapourisation
(c) solidification (d) none of these

24. What is the term used to describe the phase change of a liquid to a gas?
(a) condensation (b) melting
(c) boiling (d) both (a) and (c)

25. What term is used to describe the phase change of a solid to a liquid?
(a) melting (b) boiling
(c) freezing (d) none of these

26. On leaving water in a glass, if some of its molecules turn into a gas, it is called ________
(a) extinction (b) egasoration
(c) solidification (d) evaporation

27. Select the best condition for liquefaction of a gas.
(a) decreasing the pressure
(b) increasing the temperature
(c) decreasing the pressure and increasing the temperature
(d) increasing the pressure and decreasing the temperature

28. Select the correct order (ascending) of evaporation for water, alcohol, petrol and kerosene.
(a) petrol > alcohol > kerosene > water
(b) petrol > alcohol > water > kerosene
(c) alcohol > petrol > water > kerosene
(d) water > alcohol > kerosene > petrol

29. Which of the following processes is known as fusion?
(a) change of liquid to vapour
(b) change of gaseous state to solid state
(c) change of liquid to solid
(d) change of solid to liquid

30. When a gas is compressed keeping temperature constant, it results in a/an
(a) increase in collisions among gaseous molecules
(b) increase in the speed of gaseous molecules
(c) decrease in the speed of gaseous molecules
(d) decrease in collisions among gaseous molecules

Miscellaneous

1. Which of the following is a characteristic feature of matter?
(a) mass and space (b) property of inertia
(c) influence of gravity (d) all of these

2. Osmosis involves flow of
(a) solute molecules
(b) solvent molecules
(c) both solute and solvent molecules
(d) none of these

3. Which of the following represents intermolecular forces?
(a) dipole–dipole interaction
(b) ionic bond
(c) dispersion forces
(d) both (a) and (c)

4. Which of the following statement is correct regarding solids?
(a) solids have a fixed shape and a fixed volume
(b) solids have a high diffusion rate
(c) solids are quite hard
(d) both (a) and (c)

5. Which of the following are examples of amorphous solids?
(a) quartz (b) glass
(c) rubber (d) both (b) and (c)

6. Select the correct statement
   (a) the spreading of purple colour of potassium permanganate into water is due to diffusion
   (b) raisins swell in water due to osmosis
   (c) water can exist only in the liquid state
   (d) both (a) and (b)

7. Which of the following factors does not affect vapour pressure?
   (a) temperature
   (b) size of vessel
   (c) intermolecular forces of attraction
   (d) nature of solvent

8. 20 K is equal to
   (a) 283°C
   (b) −253°C
   (c) 293°C
   (d) −263°C

9. Which one does not sublime?
   (a) iodine
   (b) potassium Iodide
   (c) table salt
   (d) sugar

10. Dry ice is
   (a) ice that has been dried
   (b) solid carbon dioxide
   (c) ice having no water of crystallisation
   (d) none of these

11. Which of the following statement(s) about solids is correct?
   (a) expand when the temperature rises
   (b) have a low density
   (c) change shape easily
   (d) generally solids have a high density

12. Which of these choices is not defined as 'standard pressure'?
   (a) 760 torr
   (b) 1 atm
   (c) 14.7 psi
   (d) 1 pascal

13. The correct statement(s) amongst the following is/are
   (a) diffusion of a liquid and a gas is known as intimate mixing
   (b) diffusion also takes place in liquids
   (c) gases diffuse at different rates
   (d) some liquids diffuse at rates equal to that of gases

14. The property to flow is unique to fluids. Which one of the following statements is correct?
   (a) only gases behave like fluids
   (b) gases and solids behave like fluids
   (c) gases and liquids behave like fluids
   (d) only liquids are fluids

15. During summer, water kept in an earthen pot becomes cool because of the phenomenon of
   (a) diffusion
   (b) transpiration
   (c) osmosis
   (d) evaporation

16. Which of the following can exert pressure in all directions?
   (a) solids
   (b) liquids
   (c) gases
   (d) all of these

17. The state of matter can be changed by
   (a) change of temperature
   (b) change of pressure

   (c) both of these
   (d) none of these

18. Rate of evaporation depends upon
   (a) temperature
   (b) pressure
   (c) humidity
   (d) all of these

19. Which of the following is not correctly given here?
   (a) gas into liquid — liquefaction
   (b) liquid into solid — condensation
   (c) liquid to vapour — evaporation
   (d) solid to liquid — melting

20. Four gases P, Q, R and S, have critical temperatures 31.1°C, −165°C, −34°C and 79°C. Which is the most easily liquifiable gas here?
   (a) gas P
   (b) gas Q
   (c) gas R
   (d) gas S

Advanced and Olympiads

Single Choice

1. Which one of the following sets of phenomena would increase on raising the temperature?
   (a) diffusion, evaporation, compression of gases
   (b) evaporation, compression of gases, solubility
   (c) evaporation, diffusion, expansion of gases
   (d) evaporation, solubility, diffusion, compression of gases

2. Seema visited a Natural Gas Compressing Unit and found that the gas can be liquefied under specific conditions of temperature and pressure. While sharing her experience with friends she got confused. Help her to identify the correct set of conditions.
   (a) low temperature, low pressure
   (b) high temperature, low pressure
   (c) low temperature, high pressure
   (d) high temperature, high pressure

3. A few substances are arranged in the increasing order of 'forces of attraction' between their particles. Which one of the following represents a correct arrangement?
   (a) water, air, wind
   (b) air, sugar, oil
   (c) oxygen, water, sugar
   (d) salt, juice, air

4. On converting 25°C, 38°C and 66°C to the kelvin scale, the correct sequence of temperatures will be
   (a) 298 K, 311 K and 339 K
   (b) 298 K, 300 K and 338 K
   (c) 273 K, 278 K and 543 K
   (d) 298 K, 310 K and 338 K

5. Choose the correct statement of the following.
   (a) Conversion of solid into vapours without passing through the liquid state is called vaporisation.
   (b) Conversion of vapours into solid without passing through the liquid state is called sublimation.
   (c) Conversion of vapours into solid without passing through the liquid state is called freezing.
   (d) Conversion of solid into liquid is called sublimation.

6. The boiling points of diethyl ether, acetone and *n*-butyl alcohol are 35°C, 56°C and 118°C respectively. Which one of the following correctly represents their boiling points in the kelvin scale?
   (a) 306 K, 329 K, 391 K
   (b) 308 K, 329 K, 392 K
   (c) 308 K, 329 K, 391 K
   (d) 329 K, 392 K, 308 K

7. Which condition out of the following will increase the evaporation of water?
   (a) increase in the temperature of water
   (b) decrease in the temperature of water
   (c) less exposed surface area of water
   (d) adding common salt to water

8. In which of the following conditions, will the distance between the molecules of hydrogen gas increase?
   (i) Increasing pressure on hydrogen contained in a closed container.
   (ii) Some hydrogen gas leaking out of the container.
   (iii) Increasing the volume of the container of hydrogen gas.
   (iv) Adding more hydrogen gas to the container without increasing the volume of the container.
   (a) (i) and (iii)   (b) (i) and (iv)
   (c) (ii) and (iii)   (d) (ii) and (iv)

Multiple Choice

9. Which of these choices will change the state of matter?
   (a) pressure   (b) temperature
   (c) heat   (d) crushing a crystal

10. Select the one that is 'matter'
    (a) humidity   (b) feeling of heat
    (c) water   (d) smoke

11. Which is/are correct statement/s?
    (a) Of the three states of matter, the one that is most compact is the solid state.
    (b) Matter is continuous in nature.
    (c) The density of a solid is generally more than that of a liquid.
    (d) In the solid state, inter-particles space (that is, empty space) is minimum.

12. Which of the following properties are similar for solids, liquids and gases?
    (a) density
    (b) mass of the substance
    (c) particle size of the substance
    (d) movement of molecules

13. Which of these statement(s) is/are not true about gases?
    (a) Gases have high density.
    (b) Gases can be compressed more than solids.
    (c) Gases have very specific shapes.
    (d) Gases undergo diffusion fastest.

14. Which of these choices is/are example(s) of plasma?
    (a) neon sign
    (b) aurora borealis
    (c) fluorescent light bulb
    (d) incandescent light bulb

15. Which of the following statement(s) is/are applicable for amorphous solids?
    (a) Include glasses
    (b) Do not have specific melting points
    (c) Are more flexible at higher temperatures
    (d) Have sharp melting points

16. Which of the following statement(s) is/are true for gases?
    (a) Volume of the gas is almost equal to the volume of the container confining the gas.
    (b) Confined gas exerts uniform pressure on the walls of its container in all directions.
    (c) Gases do not have a definite shape and volume.
    (d) Mass of the gas cannot be determined by weighing a container in which it is enclosed.

Comprehension Type

Comprehension I: Matter is composed of tiny particles. it can be further divided into solid, liquid and gas on the basis of their physical state, and into elements, compounds and mixtures on the basis of their chemical composition. Solids, liquids, gases can be easily distinguished due to many differences in their structure and properties. Recently plasma and Bose–Einstein condensate have also been observed as states of matter. These forms can be interchanged into one another by changing the conditions of temperature, pressure, and so on.

17. Which of the following states of matter has the highest energy?
    (a) solid   (b) liquid
    (c) gas   (d) all have equal energies

18. Which of the following statements is incorrect?
    (a) the strongest intermolecular force is present in solids
    (b) gases diffuse fastest among solid, liquid and gas
    (c) intermolecular space is maximum in solids
    (d) gases are most compressible among solid, liquid and gas

Comprehension II: Evaporation is the process of conversion of liquid into vapour at any temperature, but below the boiling point. It is a surface phenomenon and slower than boiling. It is affected by temperature, pressure, surface area, humidity and wind speed. It has great significance in our daily life; for example, cotton clothes are good to wear in summer.

19. Select the correct statement here
    (a) boiling can take place only at a particular temperature for a liquid
    (b) evaporation can occur at any temperature
    (c) evaporation is a surface process
    (d) all of these

20. Which of the following plots is correct here?

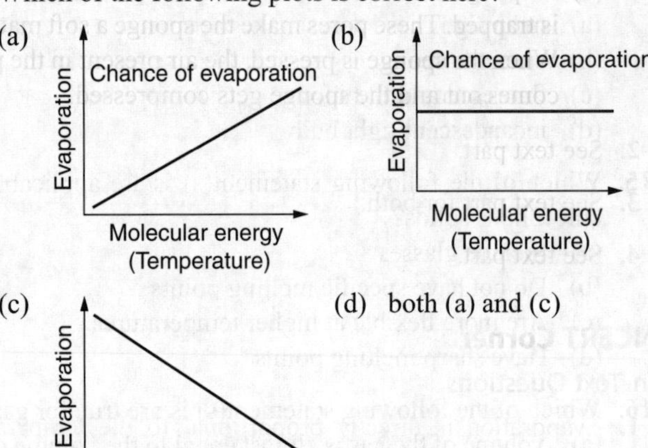

21. Which of the following involves evaporation?
- (a) cooling of water is possible in earthen pots (pitcher) during summer
- (b) cotton clothes produce a cooling effect during summer
- (c) formation of water droplets can be seen on the outer surface of a glass containing ice cold water
- (d) all of these

Matrix Matching

22. The non-SI and SI units of some physical quantities are given in column I and column II respectively. Match the units belonging to the same physical quantity.

Column I	Column II
(a) celsius	(p) kilogram
(b) centimetre	(q) pascal
(c) gram per centimetre cube	(r) metre
(d) bar	(s) kelvin
(e) milligram	(t) kilogram per metre cube

23. Match the following

Column I	Column II
(a) Solid into liquid	(p) Sublimation
(b) Liquid into gas	(q) Liquefaction
(c) Gas into liquid	(r) Vaporisation
(d) Solid into gas	(s) Melting

Integer Type

24. Which of the following properties represent physical properties of matter?
Seeing, toxicity, hearing, feeling, heat of combustion, touch, smell, reactivity, measurement

25. Which of the following properties represent chemical properties of matter?
Toxicity, oxidation states, hearing, heat of combustion, touch, flammability, coordination number, reactivity, measure, enthalpy of formation

26. How many of these are examples of the plasma state?
Lightning, soft drink, solar wind, aurora, plastic, fluorescent light and nuclear fireball

27. Which of the following processes can be used to obtain the gas or vapour state from the solid or liquid state?
Evaporation, boiling, sublimation, condensation, liquefaction

28. The boiling point of a gas is 282 K. If it is $X°C$, find the value of X.

29. The value of pressure in mm of mercury is 3040. If its value in atm is X atm, what is the value of X here?

30. For W kg of water, the latent heat of vaporisation was found to be 67500 kJ/kg. What is the value of W here?

Answer Keys

Fill in the Blanks

1. states	2. gas
3. diffusion	4. gas
5. more	6. gaseous
7. larger	8. 450 k
9. boiling point	10. solidification
11. 373 k, 273 k	12. sublimation
13. condensation	14. latent heat of fusion

True or False

1. T	2. T	3. F	4. T	5. F
6. T	7. F	8. T	9. F	10. T
11. T	12. F	13. F	14. T	

Very Short Answer Type Questions

1. Dipole–dipole force, ion–dipole force and hydrogen bonding.

2. Aqueous solution of copper sulphate becomes blue. It is due to diffusion.

3. Y. As Y has a higher boiling point than X, Y has stronger intermolecular forces of attraction.

4. Solid state has definite shape and definite volume.

5. (a) gas (b) solid and (c) liquid

6. Camphor, ammonium chloride and naphthalene.

7. (a) vaporisation (b) melting and (c) condensation

8. Steam has more volume but less density than ice.

9. On increasing the surface area, the rate of evaporation increases.

10. It is due to diffusion.

11. $190\ K = 273 - 190 = -83°C$

12. Solid carbon dioxide is known as dry ice.

13. By applying pressure and reducing the temperature, atmospheric gases can be liquefied.

Short Answer Type Questions

1. Characteristics of particles of matter are given below:
   (i) Particles of matter have space between them.
   (ii) Particles of matter are continuously moving.
   (iii) Particles of matter have forces of attraction between them.
   (iv) Particles of matter are very small in size.

2. (a) Compressibility: Gas > Liquid > Solid
   (b) Hardness: Solid > Liquid > Gas
   (c) Energy of molecules: Gas > Liquid > Solid
   (d) Intermolecular space: Gas > Liquid > Solid

3. See text part.

4. See text part.

5. See text part.

6. We should wear light coloured cotton clothes in summer because,
   (i) Cotton is a good absorber of water/sweat. It provides more surface area for the sweat to evaporate.
   (ii) Light colours absorb less heat.
   So, wearing light coloured cotton clothes helps us to feel cool and comfortable.

7. By using the given formula, we can convert the temperatures from kelvin to celsius.
   $T\ K - 273 = t°C$
   (a) $300\ K - 273 = 27°C$
   (b) $573\ K - 273 = 300°C$

8. Rahul is wearing cotton clothes which are more comfortable in the summer because cotton absorbs the sweat which causes cooling on evaporation. Ravi is wearing nylon clothes which do not absorb sweat. Hence, Ravi will be uncomfortable.

Long Answer Type Questions

1. (a) Evaporation is a surface phenomenon. The particles from the surface of the liquid take energy from the surroundings and change into vapour which results in the decrease in energy of the surroundings. Hence, a cooling effect is produced during evaporation.
   (b) The amount of water present in the air is known as humidity. If the water vapour present in air is already large, it is not able to take up more water through evaporation. Hence, the rate of evaporation of water will decrease. On a dry day, the air absorbs water more readily hence, the rate of evaporation is high on a dry day.

(c) A sponge is a solid but it has minute pores in which air is trapped. These pores make the sponge a soft material. When the sponge is pressed, the air present in the pores comes out and the sponge gets compressed.

2. See text part.

3. See text part for both.

4. See text part.

NCERT Corner

In-Text Questions

1. Evaporation is directly proportional to the temperature, which means, hot food evaporates easily. Diffusion of hot food vapour with air becomes very fast and can reach a distant place within a very short time.

2. We can solve this question by keeping this concept in mind. The correct order of density for gas, liquid and solid are:

   Gas < Liquid < Solid

   Thus, Air, exhaust from chimneys (gas)

   water, honey (liquid)   cotton, chalk, iron (solid)

   $\longrightarrow$

   Increasing order of density

3. As the boiling point of water is 100°C,
   (a) at 250°C, the state of water will be steam or water vapour, that is, gaseous state.
   (b) at 100°C, there will be a transition of the liquid state into the gaseous state. So, at this temperature, the state is/may be liquid as well as gaseous.

4. During the change of state of a substance, the temperature remains constant. This can be understood with the help of an example. When a solid is heated to its melting point, the temperature first rises and becomes constant when reaches its melting point. Now, on further heating, the heat energy provided to the substance helps to break the forces of attraction between the solid molecules. This heat is called latent heat. That is why, the temperature does not rise.

5. A desert cooler functions on the basis of evaporation. The rate of evaporation increases with an increase in temperature and a decrease in humidity. As evaporation increases when the day is hot and dry, the desert cooler functions better on such a day.

6. A large number of tiny pores is present on the surface of the earthen pot (matka). The water stored in the earthen pot (matka) evaporates faster through these pores due to the increased exposed surface area. As the process of evaporation causes cooling, the stored water inside the earthen pot (matka) becomes cool.

7. Acetone, petrol and perfume, being volatile, evaporate very fast when exposed to larger surfaces. During the process they absorb the required latent heat of vaporisation from the palm (if kept on palm). So, the process causes cooling and the palm feels cool.

8. A liquid has a larger surface area in a saucer than in a cup. Thus, it evaporates faster and cools faster in a saucer than in a cup. For this reason, we able to sip hot tea or milk faster from a saucer rather than a cup.

Textbook Exercises

1. In order to covert temperature from the kelvin to the celsius scale, we have to subtract 273 from the given value because $K - 273 = °C$.
  (a) $293\ K - 273 = 20°C$
  (b) $470\ K - 273 = 197°C$

2. To convert temperature from the celsius to the kelvin scale, add 273 to the given values because
$$°C + 273 = K$$
  (a) $25°C + 273 = 298\ K$
  (b) $373°C + 273 = 646\ K$

3. (a) Naphthalene is a substance which directly changes from solid to gas on heating by the process of sublimation. So, the naphthalene balls disappear with time as they sublime due to the heat of the surroundings.
  (b) The smell (aroma) of perfume reaches several metres away due to the fast diffusion of the gaseous perfume particles through the air.

4. The forces of attraction are the strongest in solids and the weakest (or negligible) in the case of gases. Sugar is a solid, water is in the liquid form and oxygen is a gas. Therefore, the order of forces of attraction is oxygen < water < sugar.

5. At 0°C water (liquid) starts getting converted into its solid form (ice) and at 100°C, water (liquid) starts to change into water vapour. Between 0° and 100°C, it remains in the liquid state. Thus,
  (a) liquid state
  (b) solid or/and liquid state (transition state)
  (c) liquid or/and gaseous state (transition state)

6. Ice at 273 K has less energy than water (although both are at the same temperature). Water has additional latent heat of fusion, hence, at 273 K ice is more effective at cooling than water.

7. Steam causes more severe burns than boiling water. The reason is that it releases the extra amount of heat (latent heat) which it has already absorbed during vaporisation.

Exemplar Problems

Short Answer Type Questions

1. The sample of water boils at a higher temperature which shows that the water is not pure. The impurities present in it cause the water to boil at a higher temperature. This water will freeze below 0°C.

2. (d) Since ice and water are in equilibrium, the temperature would be zero. When we heat the mixture, the energy supplied is utilised in melting the ice and the temperature does not change till the ice melts because of the latent heat

of fusion. On further heating, the temperature of the water would increase. Therefore (d) is the correct option.

3. (a) (iii)   (b) (iv)   (c) (v)
  (d) (ii)   (e) (i)

4. In diffusion, the particles move from a higher concentration to a lower concentration without separation by a semi-permeable membrane. In osmosis, the particles move from a lower concentration to a higher concentration (solvent to solution) when the two solutions are separated by a semi-permeable membrane. Hence, osmosis is a special kind of diffusion involving movement of particles.

5. (a) Osmosis   (b) Diffusion   (c) Diffusion
  (a) The hot water in the glass tumbler kept in the freezer will first become cold and the temperature will drop till 0°C. At 0°C, the water loses heat equal to the latent heat of fusion till the entire water freezes to form ice at 0°C. During this change of state from liquid to solid, the temperature remains constant.
  On further cooling, the temperature of ice slowly falls with time. Therefore, the correct option is (a).

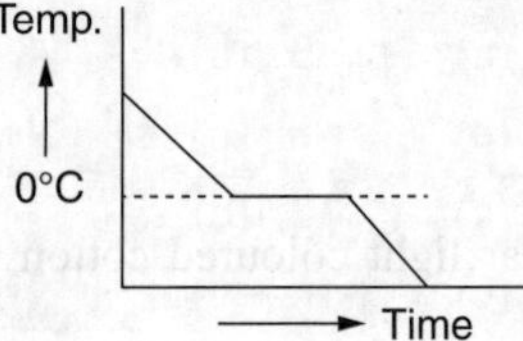

Long Answer Type Questions

1. A mixture of naphthalene and ammonium chloride can be separated as follows:
  Step 1: Put the mixture in a beaker and add water to it. Stir with a glass rod. Ammonium chloride being soluble in water gets dissolved, leaving behind the insoluble naphthalene.
  Step 2: Filter the solution. Naphthalene remains on the filter paper while ammonium chloride is obtained as filtrate.
  Step 3: Evaporate the filtrate to get back the ammonium chloride.

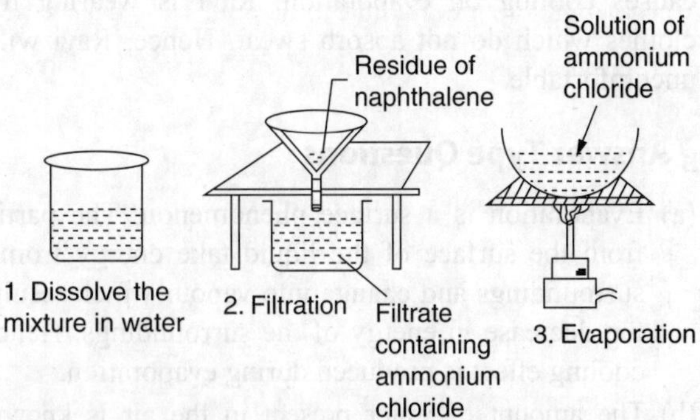

2. The process of drying the shirt can be made faster in the following ways:
  (i) Spread the shirt to increase the surface area which will increase the rate of evaporation.

(ii) Put it in the sun to increase the temperature to increase the rate of evaporation.

(iii) Keep it under the fan to increase the wind speed which increases the rate of evaporation.

3. When a substance melts, it absorbs heat for the conversion of the solid into the liquid state. As we continue heating, the heat supplied is used up in converting the solid into the liquid state by overcoming the forces of attraction between the particles and there is no change in temperature till the whole solid is converted into liquid. This heat that is absorbed by the solid which does not result in an increase in temperature, is called the latent heat of fusion. When a liquid is heated, it starts getting converted into the vapour. Further heat given to the liquid is used in changing the state and there is no increase in the temperature till the liquid starts boiling. This heat is known as the latent heat of vaporisation. Hence, the temperature of a substance remains constant at its melting point or boiling point until all the substance melts or boils.

Competition Window (Objective Type)

Topic-wise MCQs

1. (c)	**2.** (a)	**3.** (b)	**4.** (c)	**5.** (a)
6. (b)	**7.** (c)	**8.** (d)	**9.** (d)	**10.** (d)
11. (c)	**12.** (a)	**13.** (c)	**14.** (b)	**15.** (a)
16. (b)	**17.** (c)	**18.** (d)	**19.** (a)	**20.** (a)
21. (c)	**22.** (c)	**23.** (b)	**24.** (c)	**25.** (a)
26. (d)	**27.** (d)	**28.** (a)	**29.** (d)	**30.** (d)

Miscellaneous

1. (d)	**2.** (b)	**3.** (d)	**4.** (d)	**5.** (d)
6. (d)	**7.** (b)	**8.** (b)	**9.** (d)	**10.** (b)
11. (d)	**12.** (d)	**13.** (a)	**14.** (c)	**15.** (d)
16. (c)	**17.** (c)	**18.** (d)	**19.** (b)	**20.** (d)

Advanced and Olympiads

Single Choice

1. (c)	**2.** (c)	**3.** (c)	**4.** (a)	**5.** (b)
6. (c)	**7.** (a)	**8.** (c)		

Multiple Choice

9. (abc)	**10.** (acd)	**11.** (acd)	**12.** (bc)	**13.** (ac)
14. (abc)	**15.** (abc)	**16.** (abc)		

Comprehension Type

I. 17. (c) **18.** (c)

II. 19. (d) **20.** (d) **21.** (d)

Matrix Matching

22. (a)—(s); (b)—(r); (c)—(t); (d)—(q); (e)—(p)

23. (a)—(s); (b)—(r); (c)—(q); (d)—(p)

Integer Type

24. 6	**25.** 7	**26.** 5	**27.** 3	**28.** 9
29. 4	**30.** 3			

Hints and Solutions

Competition Window (Objective Type)

Topic-wise MCQs

2. The mass of an object is the fundamental property of the object, a numerical measure of inertia, a fundamental measure of the amount of matter in the object.

3. This is due to diffusion of particles of potassium permanganate into the intermolecular spaces in water.

10. Different liquids have different density, viscosity and solubility.

13. Sugar molecules occupy the intermolecular spaces present in water molecules. Thus, after dissolution only liquid is present.

14. $K = 273 + t°C$
$K = 237 + 78 = 351\ K$

16. Effusion is the process in which individual molecules flow through a hole without collisions between molecules.

17. The order of energy for different states of matter is the following:
Bose–Einstein condensate < solids < liquids < gases < plasmas

18. Alcohol, because of the presence of intermolecular forces of intermediate magnitude is a liquid at room temperature.

20. At higher altitude, the atmospheric pressure decreases. Therefore, at a lower temperature, the vapour pressure becomes equal to the atmospheric pressure. Thus, the boiling point of a liquid decreases.

21. The boiling point of water is 100°C whereas the evaporation of water into water vapour occurs at room temperature.

22. As intermolecular forces are the least in the case of petrol, it has highest rate of evaporation.

23. During vapourisation, energy is supplied to the liquid in the form of heat. This liquid reaches its boiling point and get converted into gas (vapour).

Miscellaneous

6. Water can exist in all three states.

13. The rate of diffusion of a gas depends upon its molar mass and each gas has a different molar mass. Mix a liquid of any colour with water, after some time, the entire mixture becomes coloured, showing that liquids show the phenomenon of diffusion. Moreover, no liquid diffuses as fast as a gas.

14. Gases and liquids flow due to the lower magnitude of intermolecular forces between the molecules. Gases and liquids take the shape of the container in which they are stored.

15. An earthen pot has small pores through which water keeps evaporating and evaporation causes cooling.

19. Conversion of a liquid into a solid is not condensation, it is freezing.

20. Liquefaction of a gas $\propto$ critical temperature.
    So, the order of liquefaction is S > P > R > Q.

Advanced and Olympiads
Single Choice

1. Rate of evaporation, diffusion and expansion of gases increases with an increase in temperature.

2. Gases can be compressed under low temperature and high pressure. Under these conditions, the particles of gases come closer and liquefy.

3. The forces of attraction between the particles increase in the order of gases < liquids < solids. Hence, the correct arrangement is oxygen, water, sugar.

4. $K = 25°C + 273$
   Hence $\quad 25°C = 273 + 25 = 298$ K
   $\qquad\quad 38°C = 273 + 38 = 311$ K
   $\qquad\quad 66°C = 273 + 66 = 339$ K

5. Conversion of vapour into the solid state without passing through the liquid state is called desublimation or deposition.

6. $35°C = 273 + 35 = 308$ K
   $56°C = 273 + 56 = 329$ K
   $118°C = 273 + 118 = 391$ K

7. Rate of evaporation increases with an increase in the temperature of water.

8. (ii) If some hydrogen gas is leaked from the container, the remaining gas will occupy the whole space and the distance between the molecules will increase.
   (iii) If the volume of the container is increased, the same number of molecules will occupy that space. Hence, the distance between the molecules will increase.

Multiple Choice

9. A crystal is a solid and even on crushing it you will get smaller crystals (solid).

13. Gases have very low intermolecular forces between their molecules. Thus, they can be compressed and undergo rapid diffusion.

14. Plasma is a super-energetic state of matter consisting of super-excited atoms.

15. Amorphous solids like glass and plastic do not have a sharp melting point.

16. The mass of a gas can be determined by weighing the container, filling it with gas and again weighing this container. The difference between the two weights gives the mass of the gas.

Integer Type

24. Seeing, hearing, feeling, touch, smell and measurement are physical properties of matter.

25. Toxicity, oxidation states, heat of combustion, flammability, coordination number, reactivity and enthalpy of formation are chemical properties of matter.

26. Lightning, solar wind, aurora, fluorescent light and nuclear fireball.

27. Evaporation, boiling and sublimation can be used for this purpose.

28. $X°C = 282 - 273 = 9°C$
    So, $X = 9$

29. As 760 mm of mercury = 1 atm
    3040 mm of mercury = 3040/760 = 4 atm
    Hence X = 4

30. 22500 kJ/kg Latent heat of vaporisation = 1 kg of water
    67500 kJ/kg Latent heat of vaporisation = 67500/22500 = 3 kg of water
    Hence, W = 3

Is Matter Around Us Pure?

After studying this unit, you will be able to understand:
- Matter and pure substances
- Elements and their types
- Compounds and mixtures and differentiate between them
- Solutions and their types
- Physical and chemical changes and differentiate between them
- Various methods of separation of mixtures

Introduction

How do we judge whether milk, ghee, butter, salt, spices, mineral water, juice (Fig. 2.1) and other things that we buy from the market are pure or not?

Fig. 2.1 Some cosumables

For us, 'pure' signifies no adulteration. However, from a scientist's perspective, all these things are actually mixtures of different substances and hence they are not pure! For example, milk is actually a mixture of water, fat and proteins. When a scientist claims that something is pure, it means that all the constituent particles of that substance are the same in their chemical nature. A pure substance consists of only a single type of species (particles). When we look around us, we see that most of the matter exists as mixtures of two or more pure components. For example, sea water, minerals and soil are all mixtures. All the matter around us is not pure. The matter around us is of two types—pure and mixture (Fig. 2.2).

In Chapter 1, we learnt about the three states of matter (solid, liquid and gas). Before understanding the chemical nature of matter, it is important for us to understand the scientific meaning of the term *chemical substance*. In scientific terms, *a substance is a type of matter that cannot be separated into other kinds of matter by any physical process*. Then, that substance is a *pure form of matter* and not a mixture of many different kinds of matter. Most of the things that we use in our day-to-day life are

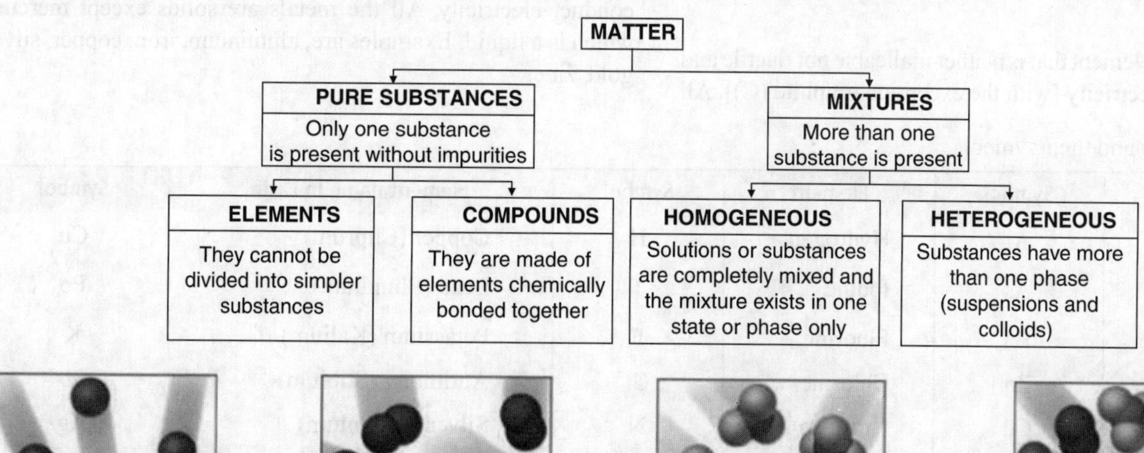

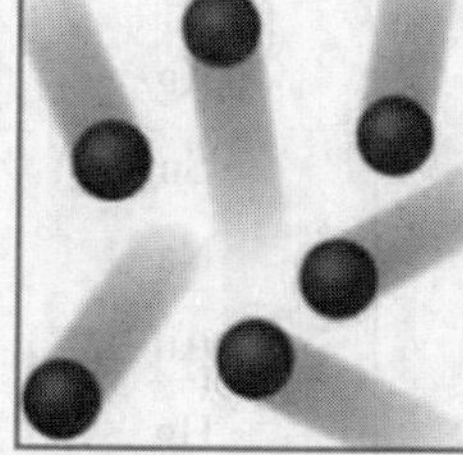
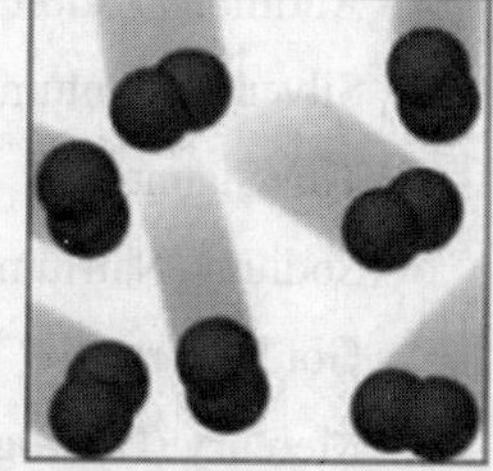
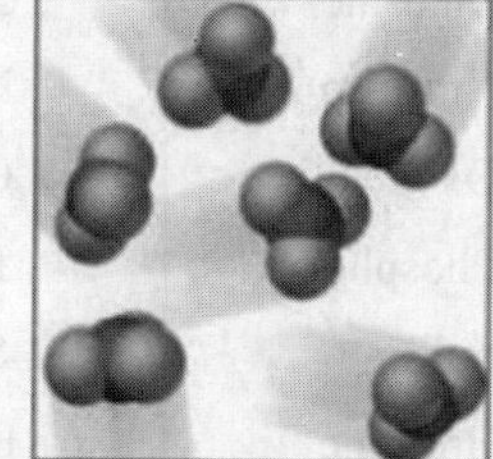
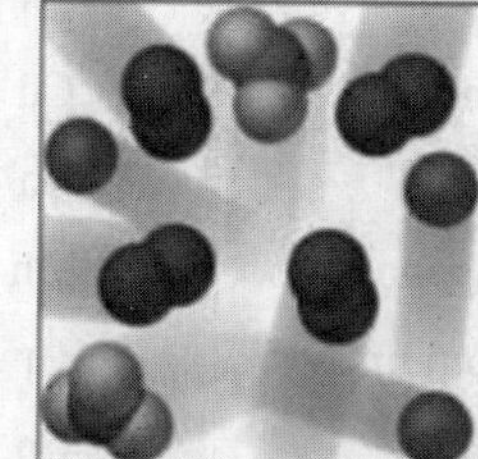
(i) Atoms of an element
(ii) Molecules of an element
(iii) Molecules of a compound
(iv) Mixture of elements and a compound

Fig. 2.2 Pure substances and mixtures

in the form of mixtures. Pure substances are rare. For example, dissolved sugar can be separated from its solution by using a physical process such as evaporation or distillation. Thus, sugar is a pure substance and cannot be separated by any physical process into its constituents. Similarly, common salt (sodium chloride), silver, gold, iron, mercury, magnesium oxide and hydrochloric acid are examples of pure substances.

Pure Substances

A pure substance has only a single type of particle and pure substances are always homogeneous. All the elements and compounds are pure substances as they have only one type of particle. A pure substance cannot be separated into other kinds of matter by any physical process. A pure substance is characterised by a fixed composition as well as a fixed boiling point and melting point.

Examples are hydrogen, nitrogen, oxygen, iron, copper, silver, gold.

Classification of Pure Substances

Pure substances can be divided into two types—elements and compounds.

Elements

An element is a substance that cannot be split up into two or more simpler substances by the usual chemical methods of applying heat, light and electric energy, because an element is made of only one type of atom. For example, oxygen is an element as it cannot be split up into two or more simpler substances by the usual methods of applying heat, light and electricity.

Types of Elements

Elements can be divided into three types—non-metals, metalloids and metals (Table 2.1).

Non-metals

A non-metal is an element that is neither malleable nor ductile and cannot conduct electricity [with the exception graphite (C)]. All the non-metals are gases or solids, except bromine which is a liquid non-metal at room temperature. Examples are, hydrogen (H), nitrogen (N), carbon (C), oxygen (O), phosphorus (P), sulfur (S), chlorine (Cl), bromine (Br), iodine (I), helium (He), neon (Ne), argon (Ar), krypton (Kr) and xenon (Xe).

Properties of Non-metals

The physical properties of non-metals are just the opposite of the physical properties of metals.

- They may be solids, liquids or gases at the room temperature.
- They are not lustrous or shiny but are dull in appearance.
- They are not malleable and are brittle.
- They are not sonorous.
- They are not ductile, that is, they cannot be drawn into wires.
- They are easily snapped on stretching.
- They are bad conductors of heat and electricity. However, graphite can show electrical conductance due to the presence of free electrons in its hexagonal structure.
- They are generally soft, not strong and have low tensile strength.
- They have comparatively low melting points, boiling points and densities.
- They may have many different colours.

Metalloids

The elements which show some properties of metals and some other properties of non-metals are known as metalloids. Their properties are intermediate between the properties of metals and non-metals. Metalloids are also known as semi-metals. Examples are, arsenic (As), selenium (Se), antimony (Sb), boron (B), silicon (Si) and germanium (Ge).

Metals

A metal is an element that is malleable and ductile and can conduct electricity. All the metals are solids except mercury, which is a liquid. Examples are, aluminium, iron, copper, silver, gold, zinc.

Table 2.1 Elements and their symbols

Element	Symbol	Element	Symbol	Element name in Latin	Symbol
Aluminium	Al	Hydrogen	H	Copper (Cuprum)	Cu
Arsenic	As	Iodine	I	Lead (Plumbum)	Pb
Barium	Ba	Fluorine	F	Potassium (Kalium)	K
Zinc	Zn	Chlorine	Cl	Antimony (Stibium)	Sb
Cadmium	Cd	Nitrogen	N	Silver (Argentum)	Ag
Calcium	Ca	Oxygen	O	Tin (Stannum)	Sn
Manganese	Mn	Phosphorus	P	Sodium (Natrium)	Na
Chromium	Cr	Sulfur	S	Gold (Aurum)	Au
Cobalt	Co	Bromine	Br	Mercury (Hydrogyrum)	Hg
Magnesium	Mg	Neon	Ne	Iron (Ferrum)	Fe

Properties of Metals

- Metals are solids at the room temperature. For example, iron, copper, aluminium, silver and gold are solids at the room temperature. Only mercury is in the liquid state at the room temperature.

- Metals are sonorous, so a metal makes a ringing sound when it is struck. For example, wires or strings for stringed musical instruments such as violins, guitars and sitars, and plate-type musical instruments such as cymbals.

- Metals are malleable so they can be beaten into thin sheets without breaking. For example, aluminium metal is quite malleable and can be converted into thin sheets (aluminium foil) which are used for packing food items like chocolates, medicines and cigarettes.

- Metals are ductile so they can be drawn or stretched into thin wires. However, all the metals are not equally ductile. For example, copper and aluminium metals are highly ductile and can be drawn into thin wires which are used in electrical wiring.

- Metals are good conductors of heat and electricity. Silver is the best conductor of heat and also has the highest thermal conductivity. Examples:
  - Electric wires are made of copper and aluminium as they are very good conductors of electricity.
  - The cooking utensils and water boilers are usually made of aluminium or copper as they are very good conductors of heat.

- Metals are shiny or lustrous and can be polished. This property of a metal containing a shining surface is known as metallic lustre. This property of metals makes them useful in making jewellery and decorative pieces. For example, gold, platinum and silver are used for making jewellery as they are bright and shiny. The shiny surface of metals also makes them good reflectors of light. For example, silver.

- Metals are generally hard; however, all the metals are not equally hard. The hardness varies from metal to metal and they cannot be cut with a knife. The exceptions are lithium, sodium and potassium which are soft metals.

- Metals are usually strong and have a high tensile strength so they can hold large weights without breaking. For example, iron (as steel) is very strong and has a high tensile strength so it can be used in the construction of railway lines, girders, machines, vehicles, bridges, buildings and chains

- Metals generally have high melting points and boiling points so most of the metals melt and vaporise at high temperatures. For example, iron has a melting point of 1808 K; solid iron melts and turns into liquid iron on heating to 1808 K. Tungsten has the highest melting point among all metals.

- Metals have high densities so they are heavy substances. For example, the density of iron metal is 7.8 g/cm^3. Iridium (Ir) and osmium (Os) have the highest densities while lithium (Li) has the lowest density.

- Metals are usually silver or grey in colour with the exception of copper (reddish-brown) and gold (yellow) (Table 2.2).

- A base metal is a metal that oxidises or corrodes easily and reacts with dilute HCl to give hydrogen. For example, Fe, Ni, Zn, Cu.

- A noble metal is a metal that is resistant to oxidation or corrosion. For example, Au, Ag, Pt, Rh, Ta.

Metallic Bonding

The bonding or interaction that holds the metal atoms firmly together by strong forces of attraction between the metal ions and the mobile electrons is known as *metallic bonding*.

By X-ray analysis of metal crystals, it has been revealed that each atom in a metal crystal is surrounded by 8 or 12 other metal atoms. In metal atoms, the valence electrons are very few, that is, mainly 1, 2 and 3. So, it is not possible for a metal atom to form 8 to 12 covalent bonds with neighbouring atoms. It was assumed that the atoms in metals are bonded with each other with a specific type of bonding called *metallic bonding*. Drude in 1900 proposed the theory of metallic bonding which was modified by Lorentz. According to them, metals having 1, 2 or 3 electrons in valence shells, being electropositive, can lose their electron(s) readily due to their low IE values to form free electrons and a stable kernel or core with positive charge(s). These free electrons are mobile in nature and can move from one kernel to another. The kernels are closely packed in a regular fashion throughout the crystal lattice. Thus, the metal crystal is represented by an arrangement of positively charged kernels in a sea of mobile electrons (Fig. 2.3) shared by all of them. As the shared electrons are delocalised, metallic bonds have neither direction nor saturation.

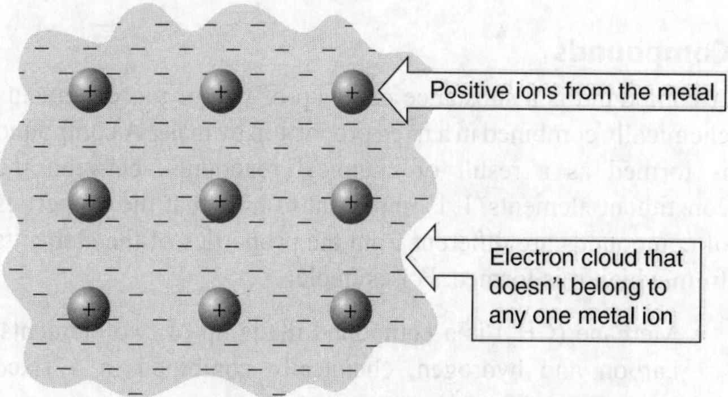

Fig. 2.3 Electron sea model

Conditions for the Formation of Metallic Bonds

There are two essential conditions for the formation of metallic bonds.

(i) The metal atoms should have low ionisation energy.

(ii) There should be a sufficient number of vacant orbitals.

The strength of the metallic bonds increases with an increase in the number of valence electrons and the charge on the nucleus.

$$\text{Strength of metallic bond} \propto \text{Number of valence electrons}$$
$$\propto \text{charge on metal ions}$$

Now it is clear why alkali metals are soft and have low melting and boiling points while transition metals are hard and have high melting and boiling points. The transition metals possess a higher number of valence electrons as well as a higher charge on the nucleus whereas alkali metals have just 1 valence electron and +1 charge.

Metallic bonding also explains the electrical and thermal conductance, metallic lustre, malleability, ductility, melting point, boiling points and hardness in metals.

Table 2.2 Difference between metals and non-metals

Metals	Non-metals
(i) They are generally solids at room temperature	(i) They may be solids, liquids or gases at room temperature
(ii) They are lustrous or shiny and can be polished	(ii) They are non-lustrous or dull and cannot be polished
(iii) They are sonorous and make a ringing sound when struck	(iii) They are not sonorous
(iv) They are malleable and ductile and they can be hammered into thin sheets and drawn into thin wires	(iv) They are brittle and are neither malleable nor ductile
(v) They are strong and tough and also have high tensile strength	(v) They are not strong and have low tensile strength
(vi) They are good conductors of heat and electricity.	(vi) They are bad conductors of heat and electricity

Illustrations

1. Classify the given elements as non-metal, metal or metalloid.

   Br, Ge, S, Sn, Pb, Al, B, Ga, Ar, Sb

   Solution:

   Non-metals: Br, S, Ar

   Metals: Sn, Pb, Al, Ga

   Metalloids: Ge, B, Sb

2. What is the effect of temperature on metallic conductance?

   Solution:

   On increasing the temperature, metallic conductance decreases as mobile electrons are pushed away due to the vibration of the slippery kernels.

Compounds

A compound is a substance made up of two or more elements chemically combined in a fixed proportion by mass. A compound is formed as a result of chemical reaction/s, between the constituent elements. It is important to note that the properties of compounds are different from the properties of the elements from which it is formed. For example:

- Methane (CH_4) is a compound made up of two elements, carbon and hydrogen, chemically combined in a fixed proportion of 3 : 1 by mass.
- Water (H_2O) is a compound made up of two elements, hydrogen and oxygen, chemically combined in a fixed proportion of 1 : 8 by mass.

Types of Compounds

on the basis of their properties, compounds can be further divided into three classes—acids, bases and salts. For example,

- Hydrochloric acid is an acid, potassium hydroxide is a base whereas potassium chloride is a salt.
- Sulfuric acid is an acid, sodium hydroxide is a base whereas sodium sulfate is a salt.

Features of a Compound

- A compound is always homogeneous in nature.
- In a compound, the constituents are present in a definite proportion by mass.
- The properties of a compound are different from the properties of its constituents.
- A compound has a fixed melting point and boiling point.
- The constituents of a compound cannot be separated by simple physical processes.
- The formation of a compound is generally accompanied by an evolution of energy in the form of heat or light. For example, burning of a candle gives CO_2 and water, and liberates heat and light energy.

KNOWLEDGE BOOSTER

Stoichiometric and non-stoichiometric compounds: Compounds having a definite atomic composition (whole number ratio) are known as stoichiometric compounds. For example, H_2O, CO_2, NH_3, CH_4. Compounds having a variable atomic composition (non-whole number ratio) are known as non-stoichiometric compounds. They are formed mainly by d and f-block elements. For example, $Fe_{0.93}O$, $Ni_{0.98}O$, $Cu_{1.7}S$.

Mixtures

Mixtures are constituted by more than one kind of pure form of matter, that is, a mixture is a substance which consists of two or more elements or compounds not chemically combined together. All the solutions are mixtures and the various substances present in a mixture are known as 'constituents or components of the mixture'. For example, lemonade (*nimbu pani*) is a mixture of water, lemon juice, sugar and salt.

A substance cannot be separated into other kinds of matter by a physical process. Sodium chloride is a substance and cannot be separated by any physical process into its chemical constituents. Similarly, sugar is a substance because it contains only one kind of pure matter and its composition is the same throughout.

A mixture has two or more different types of particles possessing different chemical nature. A mixture may be homogeneous or heterogeneous. All the mixtures are impure substances and a mixture does not have a fixed composition or a fixed melting point and boiling point.

Types of Mixtures

Depending upon the nature of the components that form a mixture, we can have different types of mixtures. Generally, mixtures are of two types—homogeneous mixtures and heterogeneous mixtures.

Homogeneous Mixtures

Those mixtures in which the substances are completely mixed and are indistinguishable from one another and form only one phase are known as *homogeneous mixtures*. All the homogeneous mixtures are known as solutions. They may be further divided into three types:

- **Solid–liquid mixture:** A mixture of sugar in water is a homogeneous mixture as all the parts of the sugar solution have the same sugar–water composition and appear to be equally sweet and there is no visible boundary of separation between sugar and water particles in a sugar solution (that is, have one 1-phase system).

- **Solid–solid mixture:** Alloys (mixture of two or more metals) such as steel (80% Fe + 12% Cr + 8% Ni), brass (59.2% Cu + 36.4% Zn + 3.26% Pb) and bronze.

- **Liquid–liquid mixture:** Rectified sprit (alcohol + gasoline), mixture of benzene and toluene.

Heterogeneous Mixtures

The mixtures in which all the substances remain separate and one substance is spread throughout the other substance as small particles, droplets or bubbles are known as heterogeneous mixtures (they have distinct phases). For example, a mixture of sugar and sand is a solid–solid heterogeneous mixture as different parts of this mixture will have different sugar–sand compositions and there is a visible boundary of separation between sugar and sand particles. The suspensions of solids in liquids are also examples of solid–liquid heterogeneous mixtures. A mixture having two or more immiscible liquids is also a case of liquid–liquid heterogeneous mixture.

Properties of Mixtures

- A mixture is usually heterogeneous.
- A mixture shows the properties of all the constituents present in it.
- A mixture can be separated into its constituents by physical processes.
- The composition of a mixture is variable; the constituents can be present in any proportion by mass.
- A mixture does not have a definite melting point and boiling point.
- As energy is usually neither given out nor absorbed in the preparation of a mixture, the formation of a mixture involves only a physical change.

Illustration

1. Classify the given substances into compounds and mixtures

   Cloud, iodised table salt, steam, sucrose, steel, aerated drink, dry ice, milk, 22 carat gold, marsh gas, petrol, smoke

   Solution:

   Compounds: Cloud, steam, sucrose, dry ice, marsh gas

   Mixture: Iodised table salt, steel, aerated drink, petrol, smoke, milk, 22 carat gold

---

Solutions and Their Types

Solutions

A solution is a homogeneous mixture of two or more pure substances. For example, lemonade and soda water are solutions. In a solution, there is homogeneity at the particle level. For example, lemonade tastes the same throughout. This shows that particles of sugar or salt are evenly distributed in the solution. A solution consists of two components—a solvent and a solute. The component of the solution that dissolves the other component in it (usually the component present in larger amount) is known as the *solvent* while the component of the solution that is dissolved in the solvent (usually present in lesser quantity) is known as the *solute*. For example, (a) a solution of sugar in water is a solid in liquid solution. Here, sugar is the solute and water is the solvent. (b) A solution of iodine in alcohol, known as *tincture of iodine*, has iodine (solid) as the solute and alcohol (liquid) as the solvent.

Different Types of Solutions

We can divide solutions into different categories.

Based on Amount of Solute in the Solution

Depending upon the dissolution of the solute in the solvent, solutions can be categorised into unsaturated, saturated and supersaturated solutions.

- **Unsaturated solution:** An unsaturated solution is a solution in which a solvent is capable of dissolving more solute at a given temperature.
- **Saturated solution:** A saturated solution can be defined as a solution in which a solvent is not capable of dissolving any more solute at a given temperature.
- **Supersaturated solution:** A supersaturated solution comprises a large amount of solute at a particular temperature. When the temperature is reduced, the extra solute will crystallise quickly.

Based on the Nature of Solvent

The solutions are of two types, depending on whether the solvent is water or not.

- **Aqueous solution:** When a solute is dissolved in water the solution is known as an *aqueous solution*. For example, salt in water, sugar in water and copper sulfate in water.
- **Non-aqueous solution:** When a solute is dissolved in a solvent other than water, it is known as a non-aqueous solution. For example, iodine in carbon tetrachloride, sulfur in carbon disulfide, phosphorus in ethyl alcohol.

Based on Amount of Solute

Depending on the amount of solute added to the solvent, solutions can be dilute or concentrated.

- **Dilute solution:** A dilute solution contains a small amount of solute in a large amount of solvent.
- **Concentrated solution:** A concentrated solution contains a large amount of solute dissolved in a small amount of solvent.

Based Upon Solute Particle Size

Solutions are of three types—true solutions, suspensions and colloids (Table 2.3).

1. **True Solution:** A true solution is a solution in which the particles of the solute can be broken down to such a fine state that they cannot be seen even under a powerful microscope.

   (a) Properties of true solutions:
   - A solution is always homogeneous in nature.
   - The properties of the solute are retained in a true solution. For example, a sugar solution is sweet in taste and a solution of salt in water is saline in taste.
   - The particles of a solution are smaller than 1 nm (10^{-9} m) in diameter. So, they cannot be seen by the naked eye.
   - A true solution does not show the Tyndall effect because of very small particle size. (Tyndall effect is the scattering of a beam of light passing through a solution so that the path of light is visible in the solution.) That is, true solutions are transparent to light.
   - The solute particles in a true solution can easily pass through a filter paper. Hence the solute particles cannot be separated from the mixture by the process of filtration.
   - The solute particles do not settle down when left undisturbed, that is, the solution is stable.

   (b) Types of true solutions: Usually we think of a solution as a liquid that contains either a solid, liquid or a gas dissolved in it. However, we can also have solid

Table 2.3 Comparison of true solutions, colloids and suspensions

S. No.	Property	True solution	Colloidal state	Suspension
1	Nature	Homogeneous	Heterogeneous	Heterogeneous
2	Particle size	Less than 1 nm (10^{-7} cm or 10 Å)	1 to 100 nm or 1×10^{-7} to 1×10^{-5} cm or 10–1000 Å	More than 100 nm or more than 1×10^{-5} cm or 1000 Å
3	Effect of gravity on particles	No effect, particles do not settle	No effect, particles do not settle	Particles settle on standing
4	Filterability	Pass unchanged through filter paper as well as animal or vegetable membranes	Pass unchanged through filter paper but not through animal or vegetable membranes	Do not pass through filter paper or animal and vegetable membranes
5	Diffusion	Diffuse rapidly	Diffuse rapidly	Do not diffuse
6	Visibility	Particles are completely invisible and thus do not scatter light	Particles are invisible to the naked eye, but they scatter light	Particles are visible to naked eye or under a microscope and they scatter light
7	Appearance of solution	Clear and transparent	Generally clear and transparent	Opaque
8	Colligative properties	Affect colligative properties	Do not affect colligative properties	Do not affect colligative properties
9	Tyndall effect	Do not exhibit	Exhibits	Do not exhibit
10	Coagulation	They can be coagulated by adding suitable electrolytes	They can be coagulated by adding suitable electrolytes	They are not coagulated

solutions (alloys), gaseous solutions (air), and so on (Table 2.4).

(c) Solution of a solid in a solid: Metal alloys are solutions of a solid in a solid type. For example, brass is a solution of zinc in copper. It is prepared by mixing molten zinc with molten copper and cooling the mixture.

(d) Solution of a liquid in a liquid: Vinegar is a solution of acetic acid in water.

(e) Solution of a gas in a liquid: Aerated drinks like soda water are gas in liquid solutions. These contain carbon dioxide (gas) as the solute and water (liquid) as the solvent.

(f) Solution of a gas in a gas: Air is a mixture of a gas in a gas and it is a homogeneous mixture. Its two main constituents are, oxygen (21%) and nitrogen (78%), and the other gases (CO_2, Ar) are present in very small quantities.

Table 2.4 Types of solutions

Type of solution	Solute	Solvent	Common examples
Gaseous solutions	Gas	Gas	Mixture of oxygen and nitrogen gases
	Liquid	Gas	Chloroform mixed with nitrogen gas
	Solid	Gas	Camphor in nitrogen gas
Liquid solutions	Gas	Liquid	Oxygen dissolved in water
	Liquid	Liquid	Ethanol dissolved in water
	Solid	Liquid	Glucose dissolved in water
Solid solutions	Gas	Solid	Solution of hydrogen in palladium
	Liquid	Solid	Amalgam of mercury with sodium
	Solid	Solid	Copper dissolved in gold

2. Suspensions: A suspension is a heterogeneous mixture in which the small particles of a solid are spread throughout a liquid without dissolving in it. In a suspension the solute particles do not dissolve but remain suspended throughout the bulk of the medium. For example, chalk–water mixture, muddy water, milk of magnesia, sand particles suspended in water, and flour in water.

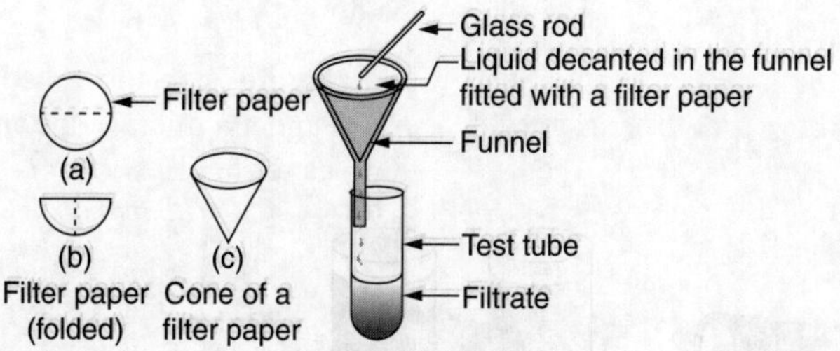

Fig. 2.4 Suspension

Properties of suspensions:

- A suspension is a heterogeneous mixture.
- The size of particles in a suspension is greater than 100 nm in diameter.

- The particles of a suspension can be seen with the naked eye or with the help of a simple microscope.
- As the particles of a suspension do not pass through a filter paper, it is possible to separate them by ordinary filtration.
- As the particles of a suspension settle down when a suspension is left undisturbed, a suspension is unstable.
- A suspension is not transparent to light as the particles of a suspension scatter a beam of light passing through it and make its path visible.

3. Colloids: A colloid is a type of solution in which the size of the solute particles is intermediate between those in true solutions and those in suspensions. The size of the solute particles in a colloid is larger than that of a true solution but smaller than that of a suspension. The particles of a colloid are uniformly spread throughout the solution. Due to the relatively smaller size of particles as compared to that of a suspension, the mixture appears to be homogeneous but actually, a colloidal solution is a heterogeneous mixture. For example, milk. A colloid consists of dispersed particles and the dispersion medium.

(a) Dispersed particles: The solute particles are known as *dispersed particles*.

(b) Dispersion medium: Solvents are known as the *dispersion medium*.

(c) Classification of colloids: Colloids are classified according to the physical state of the dispersed phase (solute) and the dispersion medium (solvent). Most of the colloids can be classified into the following eight groups (Table 2.5).

Table 2.5 Types of colloids

Dispersed phase	Dispersion medium	Colloidal system	Examples
Gas	Liquid	Foam or froth	Soap sols, lemonade froth, whipped cream
Gas	Solid	Solid foam	Pumice stone, styrene foam, foam rubber, dried sea foam
Liquid	Gas	Aerosols of liquids	Fog, clouds, mist, fine insecticide sprays
Liquid	Liquid	Emulsions	Milk, cold cream, tonics
Liquid	Solid	Gels	Cheese, butter, boot polish, table jellies
Solid	Gas	Aerosols of solids	Smoke, dust, haze
Solid	Liquid	Sols	Paints, starch dispersed in water, gold sol, muddy water, ink
Solid	Solid	Solid sols	Ruby glass, minerals, gem stones

(d) Types of colloids: A sol is a colloid made out of solid particles in a liquid medium. Based on their interaction or affinity of phase, colloids are of two types (Table 2.6).

- **Lyophillic colloids (suspensoid):** The colloidal systems in which the particles of the dispersed phase have a great affinity for the dispersion medium, are called *lyophilic* (solvent-loving) colloids. Some common examples of lyophilic colloids are gum, gelatin, starch, rubber, protein.
- **Lyophobic colloids (emulsoid):** The colloidal systems in which the particles of the dispersed phase have no affinity for the dispersion medium are called *lyophobic* (solvent-hating) colloids. Some examples of lyophobic colloids include sols of metals and their insoluble compounds such as sulfides and oxides.

Properties of Colloidal Solutions

Physical Properties

- **Heterogeneity:** Colloidal solutions are heterogenous in nature and consist of two phases—the *dispersed phase* and the *dispersion medium*. Experiments like dialysis and ultrafiltration indicate the heterogenous nature of the colloidal system.
- **Filterability:** Colloidal particles can pass through ordinary filter papers as the size of the pores of the filter paper is larger than that of the colloidal particles. however, they cannot pass through parchment paper, an animal membrane or an ultrafilter.
- **Non-setting nature:** Colloidal solutions are quite stable as the colloidal particles remain suspended in the dispersion medium indefinitely, that is, there is no effect of gravity on the colloidal particles.
- **Colour:** The colour of the colloidal solution is not always the same as the colour of the substances in the bulk. For example:
  ○ Finest gold is red in colour. When the size of the particles increases, it becomes purple.
  ○ Dilute milk gives a bluish tinge in reflected light while a reddish tinge is the transmitted light.
- **Stability:** Colloidal solutions are quite stable.
- **Visibility:** Due to the small size of colloidal particles, we cannot see them with the naked eye. As these particles can easily scatter a beam of visible light, they become visible when viewed through an ultramicroscope.

Mechanical Properties

- **Brownian movement:** Colloid particles show a ceaseless, random and swarming zig-zag motion in the dispersal medium known as *Brownian movement* (shown for the first time by Robert Brown, Fig. 2.5). The particles of the dispersal medium collide with the particles of the dispersed phase; as a result, the particles of the dispersed phase move in the direction of the net force. This motion increases the stability of colloids.

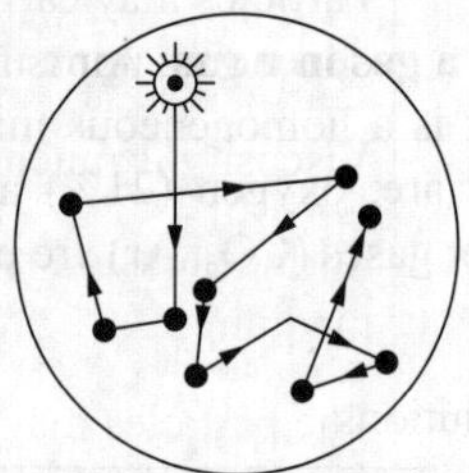

Fig. 2.5 Brownian movement

- **Sedimentation:** The heavier sol particles tend to settle down very slowly under the influence of gravity. This is called *sedimentation*. This rate of settling down or sedimentation can be accelerated by the use of a high-speed centrifuge called an ultracentrifuge.
- **Diffusion:** The colloidal particles have a tendency to diffuse from the high concentration side to the low concentration side. However, the rate of diffusion of colloidal particles is less than that of true solutions.

Optical Properties (Tyndall Effect)

When a strong beam of light is passed through a colloidal solution, its path becomes visible (bluish light) when viewed at right angles to the beam of light (Fig. 2.6). The Tyndall effect was used by Zsigmondy and Siedentopf in devising the ultramicroscope. It is due to the scattering of light by colloid particles that the sky looks blue, dust particles can be seen in the light coming from a ventilator in a dark room, the tail of a comet is visible, and so on.

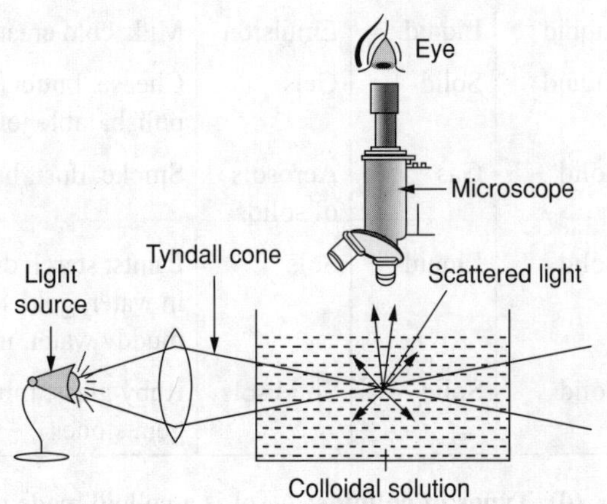

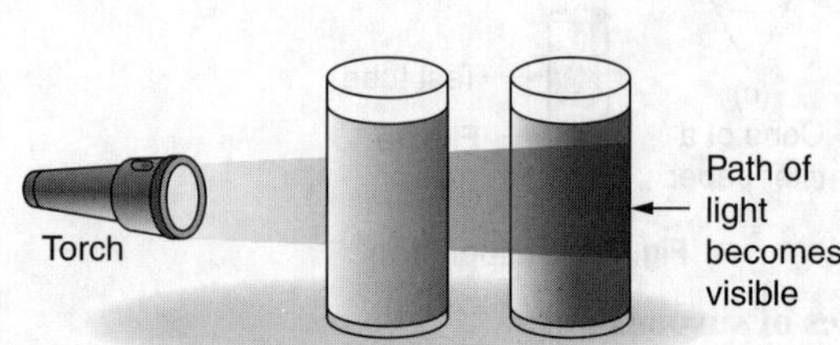

Fig. 2.6 Tyndall effect

Table 2.6 Differences between lyophilic and lyophobic sols

S. no.	Property	Lyophilic sols	Lyophobic sols
1	Nature	Reversible	Irreversible
2	Preparation	They can be prepared very easily by shaking or warming the substance with the dispersion medium. They do not require any electrolyte for stabilisation.	They can be prepared by special methods and addition of stabiliser is essential for their stability.
3	Stability	They are very stable and are not easily coagulated by electrolytes.	They are generally unstable and get easily coagulated on the addition of electrolytes.
4	Charge	Particles may carry no charge or very little charge depending upon the pH of the medium.	Colloidal particles have a characteristic charge (positive or negative).
5	Viscosity	Viscosity is much higher than that of the medium.	Viscosity is nearly the same as that of the medium.
6	Surface tension	Surface tension is usually less than that of the medium.	Surface tension is nearly the same as that of the medium.
7	Migration in electric field	The particles may or may not migrate in an electric field.	The colloidal particles migrate either towards the cathode or the anode in an electric field.
8	Solvation	Particles are highly solvated.	Particles are not solvated.
9	Visibility	The particles cannot be seen under an ultramicroscope.	The particles though invisible, can be seen under an ultramicroscope.
10	Tyndall effect	Less distinct.	More distinct.
11	Action of electrolyte	Large amount of electrolyte is required to cause coagulation.	Small amount of electrolyte is sufficient to cause coagulation.
12	Conductivity	They show a high conductivity which can be measured.	Due to their sensitivity in an electrolyte, conductivity can rarely be measured over a considerable range of concentration.
13	Examples	Most of the organic substances (starch, gums, proteins, gelatin).	Generally inorganic substances (metal sols, sulfide and oxide sols).

Illustrations

1. Smoke and fog are both aerosols. In what way are they different?

   Solution:

   In smoke, the dispersed phase is solid and the dispersion medium is a gas. In fog, the dispersed phase is liquid and the dispersion medium is a gas.

2. Give some examples of Tyndall effect observed in your surroundings.

   Solution:

   Examples of Tyndall effect:

   (i) When light rays enter into a dark room through a hole or a small window.

   (ii) Sunlight passing through a group of trees in a forest.

   (iii) Path of light rays seen in front of the projector in a cinema hall.

Concentration of Solutions

Concentration

The concentration of a solution is the amount of solute present in a given mass or volume of solution. It can also be defined as the amount of solute dissolved in a given mass or volume of solvent.

Concentration of solution = **Amount of solute/Amount of solution**, or, **Amount of solute/Amount of solvent**

$$C = \frac{w\ (g)}{V\ (L)} = \frac{w\ (g)}{V\ (mL)} \times 1000$$

There are various ways of expressing the concentration of a solution, but here we will learn only two methods.

The most common way of expressing the concentration of a solution is the 'percentage method'.

Mass Percentage (w/W%)

w% by mass% indicates that w grams of solute is taken in 100 grams of solution. For example, a 5% solution of glucose means that 5 grams of glucose is present in 100 grams of the solution.

The concentration of a solution in terms of mass percentage of solute is given by the following formula.

$$\text{Concentration of solution} = \frac{\text{Mass of solute}}{\text{Mass of solution}} \times 100$$

$$\text{Mass of solution} = \text{Mass of solute} + \text{Mass of solvent}$$

So, we can obtain the mass of solution by adding the mass of solute and the mass of solvent.

In the above example,

$$\text{Mass of solute (salt)} = 5 \text{ g}$$

$$\text{Mass of solvent (water)} = 95 \text{ g}$$

So, $\text{Mass of solution} = \text{Mass of solute} + \text{Mass of solvent}$

$$= 5 + 95 = 100 \text{ g}$$

$$\text{Concentration of solution} = \frac{5}{100} \times 100 = 5\% \text{ (by mass)}$$

Percentage Strength (w/V%)

It is known as percentage by strength w/V%. By strength, signifies that w grams of solute is taken in 100 mL of solution. For example, a 10% solution of glucose by strength signifies that 10 grams of glucose is present in 100 mL of the solution.

The concentration of a solution in terms of percentage by strength of solute is given by the following formula.

$$\text{Concentration of solution} = \frac{\text{Mass of solute}}{\text{Volume of solution}} \times 100$$

Volume Strength (v/V%)

This is used in the case of a liquid solute dissolved in a liquid solvent. The concentration of a solution is defined as volume strength. v/V% by strength means v mL of the solute is present in 100 mL of the solution. For example, a 42% v/V solution of ethyl alcohol means that 42 mL of ethyl alcohol are present in 100 mL of the solution.

$$\text{Concentration of solution} = \frac{\text{Volume of solute}}{\text{Volume of solution}} \times 100$$

KNOWLEDGE BOOSTER

Mole fraction (X): It is defined as the ratio of the number of moles of one component, solute or solvent, to the total number of moles of all the components (solute plus solvent) present in the solution. It is denoted by 'X'.

So, $$X_{\text{solute}} = \frac{\text{Moles of solute}}{\text{Moles of solute} + \text{Moles of solvent}}$$

For example, in a solution containing n_1 moles of solute and n_2 moles of solvent, the mole fraction of the solute (X_{solute}) and the mole fraction of the solvent (X_{solvent}) are

$$X_{\text{solute}} = \frac{n_1}{n_1 + n_2}, \quad X_{\text{solvent}} = \frac{n_2}{n_1 + n_2}$$

The sum of the mole fractions of all the components present in the solution is always one (unity). Mole fraction is always temperature independent as the mass of solvent is used here.

Solubility

The maximum amount of a solute that can be dissolved in 1 litre of a solution at a specified temperature is known as the solubility of that solute in that solvent. It can be given in g/L or mol/L.

Factors Affecting Solubility

(i) Effect of temperature

- The solubility of solids in liquids generally increases on increasing the temperature; and decreases on decreasing the temperature.
- The solubility of gases in liquids generally decreases on increasing the temperature; and increases on decreasing the temperature as the dissolution of a gas in a solvent is exothermic (heat is evolved).

(ii) Effect of pressure

- The solubility of solids in liquids remains unaffected by the changes in pressure.
- The solubility of gases in liquids increases on increasing the pressure; and decreases on decreasing the pressure as the solubility of a gas is directly proportional to the partial pressure of the gas.

(iii) Nature of solute

- Polar compounds (HF, NH_3) dissolve more in polar solvents like water whereas non-polar compounds (F_2, Cl_2) dissolve more in non-polar solvents like CCl_4, CS_2.
- Gases having more intermolecular forces of attraction between the molecules are more soluble. For example, CO_2 is more soluble than oxygen in water.

(iv) Nature of solvent: Solvents having a high value of the dielectric constant (such as water) can dissolve ionic and polar compounds more than solvents with low dielectric constants.

(v) Size of solute particles: Solute having a smaller size of particles are more soluble. For example, powdered sugar dissolves more than granules in water.

Illustrations

1. A syrup is prepared by dissolving 125 g of sucrose in 175 g of water. Find the mass% of sucrose in this syrup.

Solution:

Mass of solute (sucrose) = 125 g

Mass of solvent = 175 g

So, Mass of solution = Mass of solute + Mass of solvent

$$= 125 + 175 = 300 \text{ g}$$

$$\text{Concentration of solution} = \frac{\text{Mass of solute}}{\text{Mass of solution}} \times 100$$

$$\text{Concentration of solution} = \frac{125}{300} \times 100 = 41.66\% \text{ by mass}$$

2. A solution is prepared by dissolving 90 g of glucose in 200 mL of water. Find the percentage strength of the glucose solution.

Solution:

Mass of solute (glucose) = 90 g

Volume of solvent = 200 mL

$$\text{Concentration of solution} = \frac{\text{Mass of solute}}{\text{Volume of solution}} \times 100$$

$$\text{Concentration of solution} = \frac{90}{200} \times 100 = 45 \text{ by strength}$$

Physical and Chemical Changes

In order to understand the difference between a pure substance and a mixture, let us understand the difference between a physical and a chemical change (Table 2.7).

Table 2.7 Differences between physical and chemical changes

	Physical change	Chemical change
1	No new substance is formed	A new substance is formed
2	It is a temporary change	It is a permanent change
3	It is easily reversible	It is usually irreversible
4	The mass of the substance does not alter	The mass of the substance does alter
5	Very little heat or light energy is generally absorbed or evolved	A lot of heat or light energy is absorbed or evolved

Physical Change

In the previous chapter, we learnt about a few physical properties of matter. The properties that can be observed and specified, such as colour, hardness, rigidity, fluidity, density, melting point and boiling point are the physical properties. The interconversion of states is a physical change because these changes occur without a change in composition and there is no change in the chemical nature of the substance. A change in which no new substances are formed but the physical form of the substance changes, is known as a *physical change*. The product formed in such changes is chemically identical to the starting substance.

For example, when ice is heated, it changes into liquid water and on further heating, water changes into steam. However, water in the solid form (ice), liquid form or in the gaseous form (steam), is chemically the same substance (H_2O).

- This transformation represents a physical change and such a change can be reversed easily. For example, steam on cooling forms liquid water, which on further cooling changes into ice.
- The following processes involve physical change. Making of ice cream, sublimation of camphor, melting of wax, breaking of a glass pane, a rock rolling down a hill, bending of a glass tube by heating.

Chemical Change

A change in which one or more substances change into new substance(s) is known as a *chemical change*. It cannot be reversed easily. This process causes a change in the chemical composition which brings about a change in the chemical properties also. A chemical change is also called a *chemical reaction*.

For example,

- Both water and cooking oil are liquids but their chemical characteristics are different. They differ in odour and inflammability. Oil burns in air whereas water extinguishes fire. So, it is the chemical property of oil that makes it different from water (burning is a chemical change).
- When electricity is passed through water, it decomposes into two new substances, hydrogen and oxygen. Similarly rusting of iron and calcination of limestone are also examples of chemical changes.
- Blood clotting, burning of charcoal, digestion of food, rusting of iron, formation of biogas, burning of fuel like petrol, ripening of fruits, drying of paint and burning of magnesium ribbon in air.

KNOWLEDGE BOOSTER

The following processes involve both physical and chemical changes—sublimation of NH_4Cl, heating of zinc hydroxide, heating of sodium nitrate and burning of a candle.

$$\boxed{\textbf{Illustrations}}$$

1. Classify the following as physical or chemical changes.

(i) Dissolving salt in water (ii) burning of magnesium ribbon in air (iii) crystallisation of copper sulfate (iv) conversion of milk into curd (v) burning of coal (vi) electrolysis of sodium chloride solution by passing current (vii) rusting of iron nails

Solution:

(i) Physical change (ii) Chemical change (iii) Physical change (iv) Chemical change (v) Chemical change (vi) Chemical change (vii) Chemical change

2. Classify each of the following as a physical or a chemical change. Give reasons.

(a) Drying of a shirt in the sun.

(b) Rising of hot air over a radiator.

(c) Burning of kerosene in a lantern.

(d) Change in the colour of black tea on adding lemon juice to it.

(e) Churning of milk cream to get butter.

Solution:

(a), (b), (e): Physical changes because there is no change in chemical composition.

(c), (d): Chemical changes because new substances are formed.

Separation of Mixtures

Most of the materials around us are mixtures and not pure substances. Such mixtures have two or more substances mixed in them. So, it may not be possible to use a mixture as such in homes and in industries. We might require only one or two separate constituents of a mixture for our uses. So, such mixtures have to be separated into their individual constituents to make them useful. Various methods of separation are used to get individual components from a mixture. Separation makes it possible to study and use the individual components of a mixture. Heterogeneous mixtures can be separated into their respective constituents by simple physical methods like handpicking, sieving and filtration. Some special techniques can also be used for the separation of the components of a mixture (Fig. 2.7).

liquid solvent while the other constituent is insoluble. This difference in the solubilities of the constituents of a mixture can be used to separate them. For example, sugar is soluble in water while sand is insoluble, so a mixture of sugar and sand can be separated by using water as the solvent.

- **Separation by sublimation:** The process of sublimation can be used to separate those substances which sublime on heating, from a mixture. The solid substance obtained by cooling the vapour is called the sublimate. For example, iodine, naphthalene, anthracene, camphor and ammonium chloride can be separated from a mixture by sublimation.

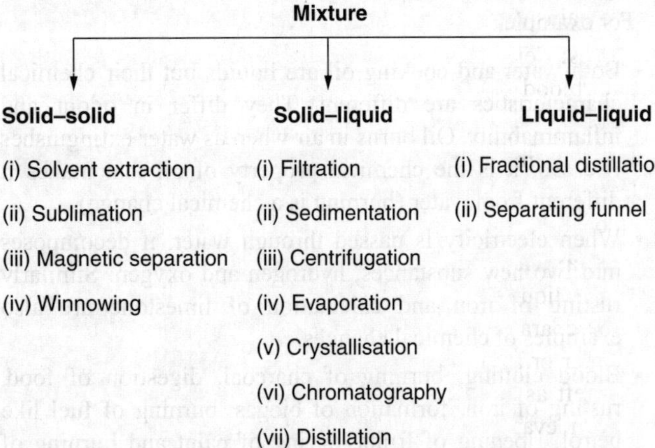

Fig. 2.7 Types of mixtures and methods to separate them

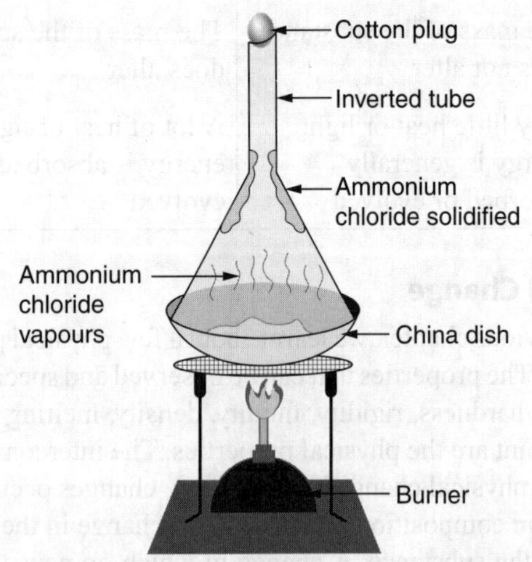

Fig. 2.8 Sublimation of ammonium chloride

Separation of a Mixture of Two Solids

All the mixtures having two solid substances can be separated by one of the following methods.

- **Separation by a suitable solvent:** In a few cases, one of the constituent of a mixture is soluble in a specific

- **Separation by a magnet:** Iron is attracted by a magnet; this tendency of iron is used to separate it from a mixture having iron as one of the constituents. For example, a mixture of iron filings and sulfur powder can be separated by using a magnet as iron filings are attracted by a magnet while sulfur is not (Fig. 2.9).

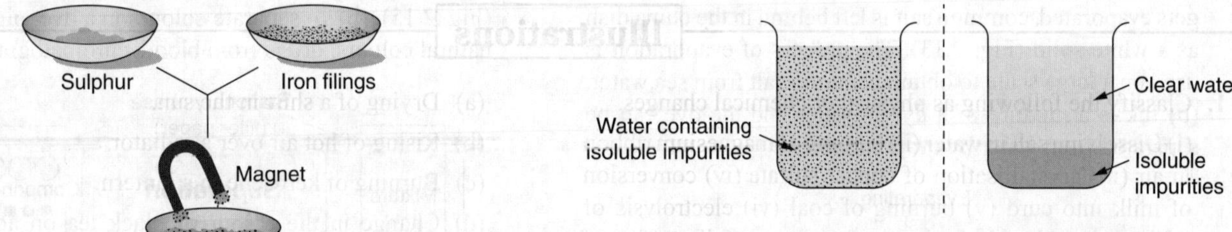

Fig. 2.9 Separation using a magnet

Fig. 2.11 Sedimentation followed by decantation

- **Winnowing:** This method is used to separate lighter particles from heavier particles by blowing air/by the force of wind. For example, separation of rice grain from husk. In its simplest form, it involves throwing the mixture into the air so that the wind blows away the lighter chaff while the heavier grains fall back.

Separation of a Mixture of a Solid and a Liquid

All the mixtures having a solid and a liquid can be separated by any of the following processes.

- **Separation by filtration:** The process of removing insoluble solids from a liquid by using a filter paper is known as filtration. Filtration can be used for separating insoluble substances from a liquid. Here, the liquid passes through the filter paper and gets collected in the beaker kept below the funnel. As solid particles do not pass through the filter paper, they remain behind on the filter paper as residue and the clear liquid obtained is the filtrate. For example, a mixture of chalk and water is separated by filtration (Fig. 2.10).

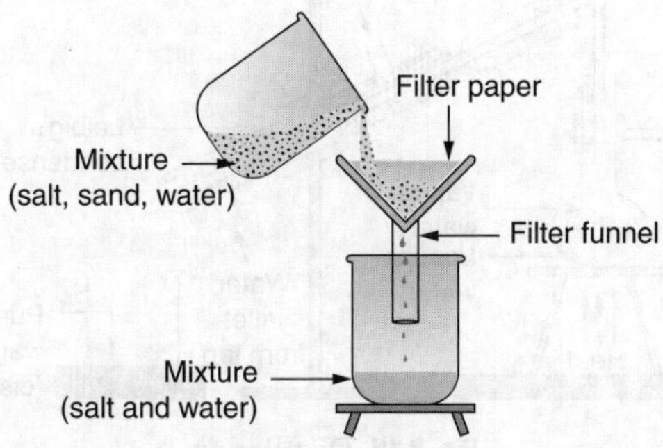

Fig. 2.10 Filtration

- **Sedimentation and decantation:** This method is applicable for a mixture having one solid and one liquid component and the solid must be insoluble in the liquid. In the sedimentation process, heavier components of the mixture settle at the bottom due to gravity. For example, settling of mud particles in water. Decantation follows sedimentation and this process involves pouring the clear, upper liquid out of the container, without disturbing the sediment (Fig. 2.11). For example, after the tea leaves have settled down, the clear liquid tea from the top can be poured into a cup. Hence the transfer of clear tea is a case of decantation.

- **Separation by centrifugation:** the suspended particles of a substance in a liquid can be very rapidly separated by the centrifugation process with the help of a machine called a centrifuge (Fig. 2.12). Here, the mixture of fine suspended particles in a liquid is taken in a test tube and the test tube is placed in a centrifuge machine and rotated rapidly for some time. As the mixture rotates around rapidly, a centrifugal force acts on the heavier suspended particles and brings them down to the bottom of the test tube. The clear liquid, being lighter, remains on top.

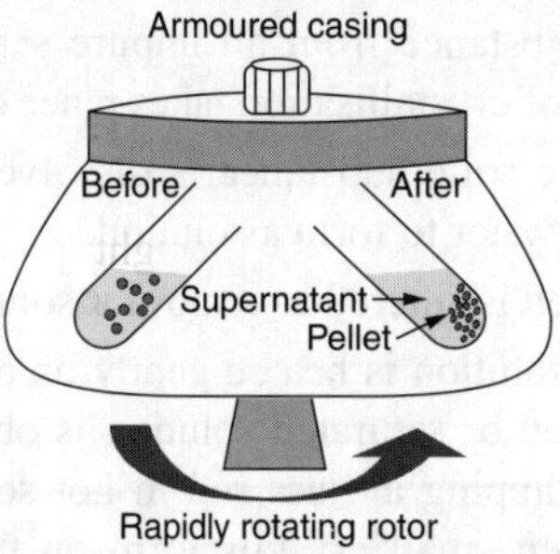

Fig. 2.12 Centrifuge

For example, the clay particles suspended in water can be very rapidly separated by centrifugation; the clay particles settle down at the bottom of the test tube and clear water remains at the top.

Uses: Centrifugation can be used in diagnostic laboratories for blood and urine tests, in dairies and at home to separate butter from cream.

A mixture of common salt and water cannot be separated by filtration or centrifugation as common salt is completely dissolved in water.

- **Separation by evaporation:** The process of changing of a liquid into vapour is called evaporation. It is used to separate a solid substance that has been dissolved in water or any other liquid solvent. The dissolved substance is left as a solid residue when all the water or liquid has been evaporated. This method is based on the fact that liquids vaporise easily whereas solids do not vaporise easily. Evaporation of a liquid can take place slowly even at room temperature, but the solution is heated in order to hasten the process. For example, (a) common salt (NaCl) dissolved in water can be separated by evaporation. The solution of common salt and water is taken in a china dish and heated gently by using a burner so the water present in the salt solution will form water vapour and escape into the atmosphere. When all the water present in the solution gets evaporated, common salt is left behind in the china dish as a white solid (Fig. 2.13). The process of evaporation is used on a large scale to obtain common salt from sea water. (b) Ink is a mixture of a dye in water and the dye can be separated from water by evaporation.

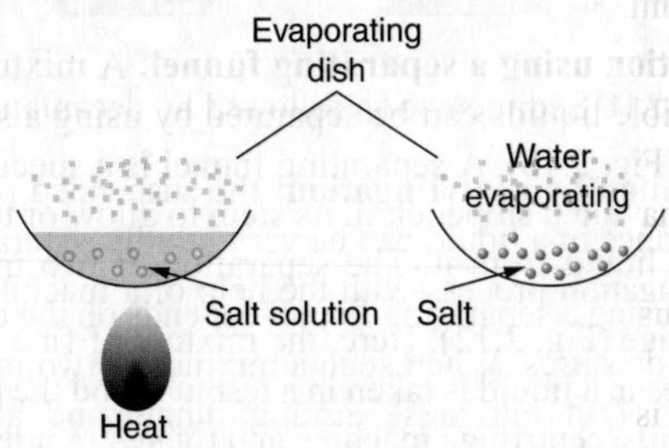

Fig. 2.13 Evaporation

- **Purification by crystallisation:** The process of cooling a hot, concentrated solution of a substance to obtain crystals is known as crystallisation. It can be used for obtaining a pure solid substance from an impure sample (Fig. 2.14). The process of crystallisation takes place as follows.
  - The impure solid substance is dissolved in a minimum amount of water to form a solution.
  - The solution is filtered to remove insoluble impurities.
  - The clear solution is heated gently on a water bath till a concentrated or saturated solution is obtained. It can be tested by dipping a glass rod in hot solution again and again. When small crystals form on the glass rod, the solution is saturated.
  - Now allow the hot, saturated solution to cool down slowly.
  - Crystals of pure solid are formed while impurities remain dissolved in solution.
  - Now separate the crystals of pure solid by filtration and dry them.

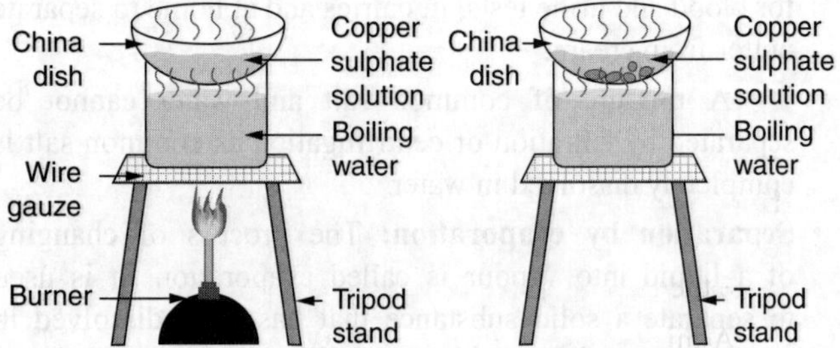

Fig. 2.14 Crystallisation

- **Separation by chromatography:** Chromatography is the technique of separating two or more dissolved solids which are present in a solution in extremely small quantities or those that cannot be separated by other methods. By using paper chromatography, we can separate two or more different substances present in the same solution. It is based on the fact that though two or more substances are soluble in the same solvent, their solubilities may be different. For example, (a) black ink is a mixture of several coloured substances which can be separated by paper chromatography

(Fig. 2.15). (b) To separate colours in a dye, pigments from natural colours, drugs from blood, chromatography is used.

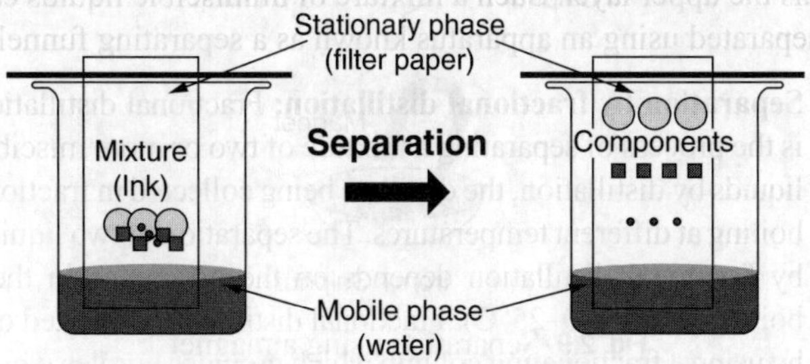

Fig. 2.15 Paper chromatography

- **Separation by distillation:** Distillation is the process of heating a liquid to form vapour and then cooling the vapour to get back liquid. Distillation can be shown as:

$$\text{Liquid} \xrightleftharpoons[\text{Cooling}]{\text{Heating}} \text{Vapour (or gas)}$$

The liquid obtained by condensing the vapour is known as the *distillate*. When the homogeneous mixture of a solid and a liquid is heated in a closed distillation flask, the liquid, being volatile, forms vapour. The vapours of the liquid are passed through a condenser where they get cooled and condense to form the pure liquid. This pure liquid is collected in a separate vessel (Fig. 2.16). The solid, being non-volatile, remains behind in the distillation flask. For example, salt solution can be separated into salt and water, benzene and toluene can be separated, chlorobenzene and bromo benzene can be separated.

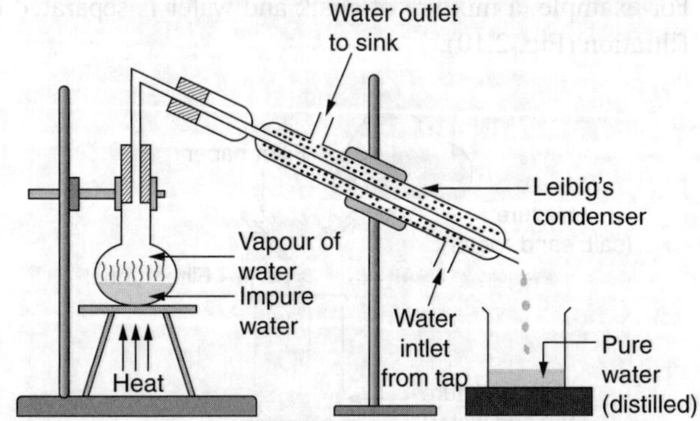

Fig. 2.16 Distillation

Separation of a Mixture of Two or More Liquids

Any mixture having two or more liquids can be separated by two methods—by fractional distillation or by using a separating funnel.

Miscible Liquids

The liquids which can be mixed in any proportion and form a single layer are known as *miscible liquids*. For example, alcohol and water are miscible liquids and form a single layer on mixing. Such a mixture of miscible liquids can be separated by the process of fractional distillation.

Immiscible Liquids

The liquids which do not mix with one another and form separate layers are known as *immiscible liquids*. For example, oil and water are immiscible liquids and form separate layers on mixing. Water being heavier, forms the lower layer and oil being lighter forms the upper layer. Such a mixture of immiscible liquids can be separated using an apparatus known as a separating funnel.

- **Separation by fractional distillation:** Fractional distillation is the process of separating a mixture of two or more miscible liquids by distillation, the distillate being collected in fractions boiling at different temperatures. The separation of two liquids by fractional distillation depends on the difference in their boiling points (10–25°C). Fractional distillation is carried out by using a fractionating column which provides hurdles which increases the cooling surface area. For example, ethyl alcohol and water are two miscible liquids. The boiling points of ethyl alcohol and water are 78°C and 100°C respectively. As their boiling points are quite different, they can be separated by fractional distillation (Fig. 2.17). This mixture is heated in a distillation flask fitted with a fractionating column. When the mixture is heated, both ethyl alcohol and water form vapours as their boiling points approach. The ethyl alcohol vapour and water vapour rise up in the fractionating column. The upper part of the fractionating column is cooler, so as the hot vapours rise up in the column, they get cooled, condense and trickle back into the distillation flask.

The more volatile liquid (less boiling point) distils over first and the less volatile liquid (more boiling point) distils over later. That is, ethanol distils out first.

Fractional distillation is also used in the separation of gasoline, kerosene, diesel, and so on, from crude petroleum.

- **Separation using a separating funnel:** A mixture of two immiscible liquids can be separated by using a separating funnel (Fig. 2.18). A separating funnel is a special type of funnel having a stopcock in its stem to allow or to stop the flow of liquid from it. The separation of two immiscible liquids using a separating funnel depends on the difference in their densities. When such a mixture of two immiscible liquids is put into a separating funnel and allowed to stand for some time, the mixture separates into two layers according to the densities of the liquids. The heavier (denser) liquid forms the lower layer while the lighter liquid forms the upper layer. On opening the stopcock of separating funnel, the lower layer (heavier liquid) comes out first and is collected in a beaker. When the lower layer of heavier liquid has completely run off, the stopcock is closed. The lighter liquid in the upper layer is collected in a separate beaker by opening the stopcock again. For example, water and kerosene are two immiscible liquids, so they can be separated by using a separating funnel.

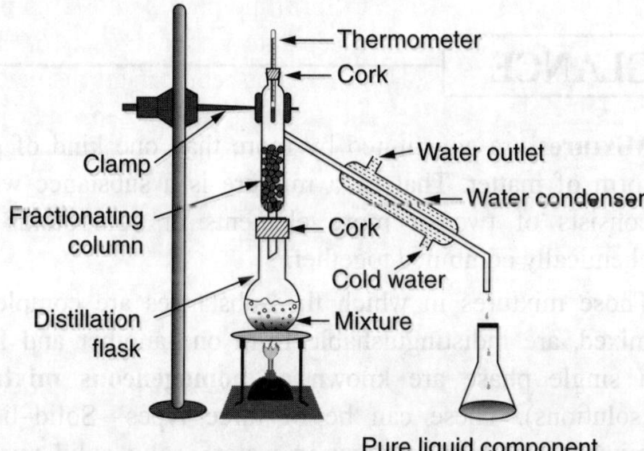

Fig. 2.17 Fractional distillation

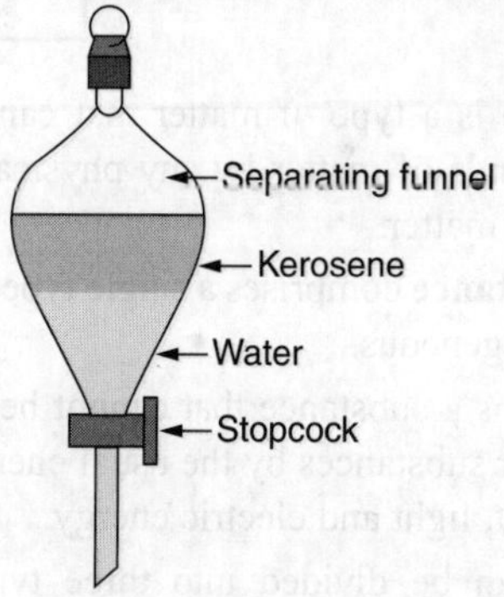

Fig. 2.18 Separating funnel

Illustrations

1. Name the process used to separate the given mixture.

    (i) A mixture of methanol (338 K) and acetone (329 K)

    (ii) Camphor having traces of common salt

    (iii) Kerosene having water

    Solution:

    (i) By fractional distillation as boiling points are close

    (ii) Sublimation

    (iii) Solvent extraction using a separating funnel

2. Name the process used to separate the given mixture.

    (i) A mixture of liquid X (366 K) and liquid Y (353 K)

    (ii) A mixture of liquid P (355 K) and liquid Q (415 K)

    (iii) 0.1 g of a liquid having 3 components in very minute amounts.

    Solution:

    (i) By fractional distillation as boiling points are close (differ by only 12 K)

    (ii) By simple distillation as boiling points differ by a large difference of 60 K

    (iii) Column chromatography

KNOWLEDGE BOOSTER

In the extraction of iron from its ore, the lighter slag is removed from the top by sublimation to leave the molten iron at the bottom in the furnace. The principle is that immiscible liquids separate out in layers depending on their densities.

Separation of gases from air

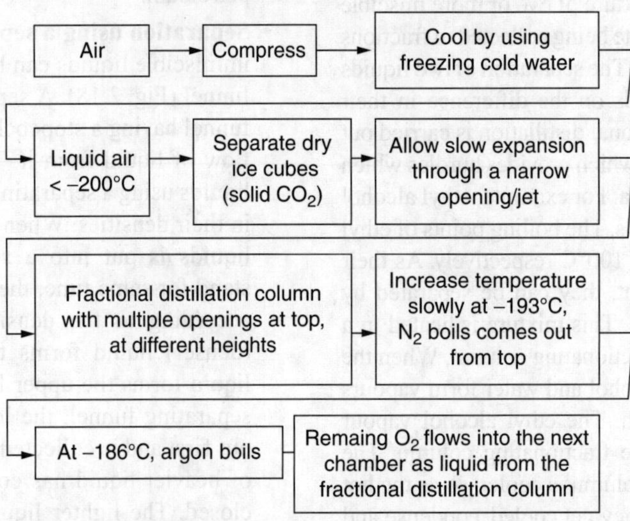

CHAPTER AT A GLANCE

- A **substance** is a type of matter that cannot be separated into other kinds of matter by any physical process. It is a pure form of matter.

- A **pure substance** comprises a single type of particle and is always homogeneous.

- An **element** is a substance that cannot be split into two or more simpler substances by the usual chemical methods of applying heat, light and electric energy.

- **Elements** can be divided into three types—non-metals, metalloids and metals.

- A **non-metal** is an element that is neither malleable nor ductile and cannot conduct electricity [with the exception graphite (C)]. All the non-metals are gases or solids, except bromine.

- The elements which show some properties of metals and some other properties of non-metals are known as **metalloids**.

- A **metal** is an element that is malleable and ductile and can conduct electricity. All the metals are solids except mercury, which is a liquid.

- The bonding or strong interaction that holds the metal atoms firmly together, due to forces of attraction between metal ions and the mobile electrons, is known as **metallic bonding**.

- A **compound** is a substance made up of two or more elements chemically combined in a fixed proportion by mass. It is formed as a result of chemical reaction/s, between the constituent elements.

- **Mixtures** are constituted by more than one kind of pure form of matter. That is, a mixture is a substance which consists of two or more elements or compounds not chemically combined together.

- Those mixtures in which the substances are completely mixed, are indistinguishable from one another and form a single phase are known as **homogeneous mixtures** (solutions). These can be of three types—Solid–liquid mixture (mixture of sugar in water), solid–solid mixture (alloys) and liquid–liquid mixture [rectified sprit (alcohol + gasoline)].

- The mixtures in which all the substances remain separate and one substance is spread throughout the other substance as small particles, droplets or bubbles are known as **heterogeneous mixtures**.

- A **solution** is a homogeneous mixture of two or more pure substances.

- A **suspension** is a heterogeneous mixture in which the small particles of a solid are spread throughout a liquid without dissolving in it.

- A **colloid** is a type of solution in which the size of the solute particles is intermediate between those in true solutions and those in suspensions. Colloidal solutions are heterogenous in nature and consist of two phases—the dispersed phase and the dispersion medium.

- Based on dispersed phase and dispersion medium, colloids are of eight types. **Lyophillic** colloids are solvent-loving while **lyophobic** colloids are solvent-hating.

- Colloid particles show a ceaseless random and swarming zig-zag motion in the dispersal medium known as **Brownian movement**.
- The heavier sol particles tend to settle down very slowly under the influence of gravity. This is called **sedimentation**.
- **Tyndall effect:** When a strong and converging beam of light is passed through a colloidal solution, its path becomes visible (bluish light) when viewed at right angles to the beam of light.
- The **concentration** of a solution is the amount of solute present in a given amount of solution (mass or volume) or, the amount of solute dissolved in a given mass or volume of solvent.
- The maximum amount of a solute that can be dissolved in 1 litre of a solution at a specified temperature is known as the **solubility** of that solute in that solvent. It can be given in g/L or mol/L.
- A change in which no new substances are formed but the physical form of the substance changes is known as a **physical change**.
- A change in which one or more substances change into new substance(s) is known as a **chemical change**. It cannot be reversed easily.
- **Separation of mixtures:** Separation makes it possible to study and use the individual components of a mixture.
- The process of **sublimation** can be used to separate those substances which sublime on heating from a mixture.
- A mixture of iron filings and sulfur powder can be separated by using a **magnet** as iron filings are attracted by a magnet while sulfur is not.
- **Winnowing** is used to separate lighter particles from heavier particles by blowing air/by the force of wind.
- The process of removing insoluble solids from a liquid using a filter paper is known as **filtration**.
- In the **sedimentation** process, heavier components of the mixture settle at the bottom due to gravity.
- **Decantation** follows sedimentation and it involves pouring the clear, upper liquid out of the container, without disturbing the sediment.
- The suspended particles of a substance in a liquid can be very rapidly separated by using the centrifugation process with the help of a machine called a **centrifuge**.
- **Evaporation** is used to separate a solid substance that has been dissolved in water or any other liquid solvent.
- The process of cooling a hot, concentrated solution of a substance to obtain crystals is called **crystallisation**.
- To separate colours in a dye, pigments from natural colours, drugs from blood, **chromatography** is used.
- **Distillation** is the process of heating a liquid to form vapour and then cooling the vapour to get back liquid.
- **Fractional distillation** is the process of separating two or more miscible liquids by distillation, the distillate being collected in fractions boiling at different temperatures.
- A mixture of two immiscible liquids can be separated by using a separating funnel.

PRACTICE QUESTIONS

Analyse Your Concepts (School Exam Based)

Fill in the Blanks

Instructions: Complete the following statements with an appropriate word/term to be filled in the blank spaces.

1. _______ is a pure substance.

2. An element is made up of only one kind of _______.

3. _______ is a metal which exists in the liquid state.

4. Sublimation of ammonium chloride is a _______ change.

5. Gases can be separated from air by _______ method.

6. Rusting of iron is a _______ change.

7. Brass is a mixture of _______ and _______.

8. Miscible liquids are separated by _______.

9. Immiscible liquids are separated by using a _______.

10. A solution having more of the solvent is a _______ solution.

11. A heterogeneous mixture of liquid and solid is conveniently separated by _______.

12. If a mixture contains iron filings as one of the constituents, it can be separated by using a _______.

13. Vinegar is a _______ solution and milk is a _______ solution.

14. The size of solute particles in a colloid is between _______ and _______ m.

15. The composition of a mixture is _______.

True or False

Instructions: Read the following statements and write if they are true or false.

1. Mercury, diamond and bromine are elements.

2. Sugar is a compound which contains the elements carbon, hydrogen, nitrogen and oxygen.

3. Wood is a mixture.

4. Fractional distillation is used when the substances differ substantially in their boiling points.

5. The separation of liquids by fractional distillation is based on the difference in their boiling points.

6. Solubility of a gas is directly proportional to temperature.

7. Chalk is a compound and gold is an element.

8. Gun powder is a compound.

9. Gas in gas is always a homogeneous mixture.

10. Blood is a homogeneous mixture.

11. Digestion of food is a physical change.

12. Colloids can show Tyndall effect.

13. Mercury is a liquid metal.

14. Filtered tea is a heterogenous mixture.

15. Pt and Au are noble metals.

Match the Following

Instructions: Each question contains statements given in two columns which have to be matched. Statements (a, b, c, d) in column I have to be matched with statements (p, q, r, s) in column II.

1.

Column I	Column II
(a) Hydragyrum	(p) Mixture
(b) Soap	(q) Pure substance
(c) Ozone	(r) Mercury
(d) Calcium	(s) Homonuclear molecule
(e) Urine	(t) Compound

Very Short Answer Type Questions

1. What is meant by a pure substance?

2. Name a method to check the purity of a liquid.

3. Give one example each of a homogeneous and a heterogeneous mixture.

4. Which of the two will scatter light—soap solution or sugar solution?

5. A solution contains 36 g of sugar dissolved in 364 g of water. What is the concentration of the sugar solution?

6. Name the process by which the coloured components can be obtained from blue ink?

7. Name the process used to separate a mixture of salt and ammonium chloride.

8. When we heat iron filings and sulfur till red hot, do we get a compound or a mixture?

9. Name the solutions which show the Tyndall effect.

10. Classify the following into elements and compounds:
    (a) H_2O (b) He (c) Cl_2 (d) CO_2
    (e) Co (f) Ar (g) NH_3

11. Name the apparatus used to separate oil from water.

12. Name the process which is used in milk dairies to separate cream from milk.

Short Answer Type Questions

1. What are the three general classes of matter? Give one example of each type.

2. Try segregating the things around you as pure substances or mixtures.
    (a) Wood (b) Coal (c) Milk (d) Sugar
    (e) Common salt (f) Soap (g) Soil (h) Rubber

3. Classify the substances given in the figure below into elements and compounds.

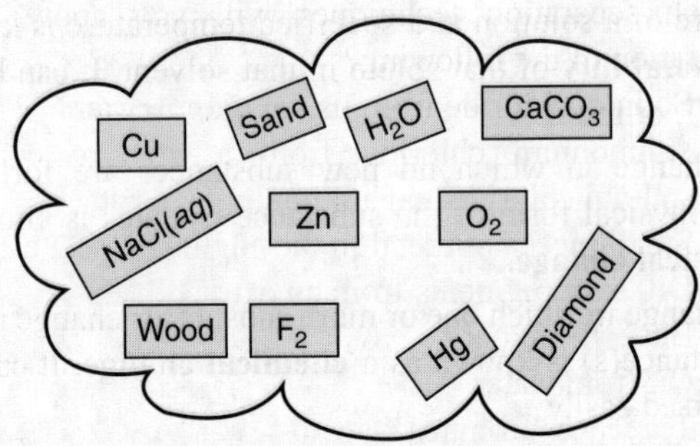

4. List the points of differences between homogeneous and heterogeneous mixtures.

5. What is the meaning of 'non-metals are brittle'?

6. Sucrose (sugar) crystals obtained from sugarcane and beetroot are mixed together. Will it be a pure substance or a mixture? Give reasons for the same.

7. Can we separate alcohol dissolved in water by using a separating funnel? If yes, then describe the procedure. If not, explain.

8. What types of mixtures are separated by the technique of crystallisation?

9. Define the following:
    (a) Sol (b) Aerosol
    (c) Emulsion (d) Foam

10. Define the terms solution, suspension and colloid. Make a comparison of the size of their particles.

11. How will you separate a mixture containing kerosene and petrol (difference in their boiling points is more than 25°C), which are miscible with each other?

12. Name the technique used to separate
    (a) Butter from curd (b) Salt from sea water
    (c) Camphor from salt

Long Answer Type Questions

1. Classify the following as chemical or physical changes
    (a) Cutting of trees
    (b) Melting of butter in a pan
    (c) Rusting of an almirah

(d) Boiling of water to form steam
(e) Passing of electric current through water and the water breaking down into hydrogen and oxygen gases
(f) Dissolving common salt in water
(g) Making a fruit salad with raw fruits
(h) Burning of paper and wood

2. Define the following terms:
   (a) Dispersed phase   (b) Dispersion medium
   (c) Brownian movement   (d) Solvent

3. Explain the following terms used in the separation of mixtures.
   (a) Filtration   (b) Crystallisation
   (c) Evaporation   (d) Sublimation

4. To make a saturated solution, 36 g of sodium chloride is dissolved in 100 g of water at 293 K. Find its concentration at this temperature.

5. Calculate the mass of sodium sulfate required to prepare a 20% (mass percent) solution in 100 g of water.

NCERT Corner

Exercises

1. Which separation techniques will you apply for the separation of the following?
   (a) Sodium chloride from its solution in water
   (b) Ammonium chloride from a mixture containing sodium chloride and ammonium chloride
   (c) Small pieces of metal in the engine oil of a car
   (d) Different pigments from an extract of flower petals
   (e) Butter from curd
   (f) Oil from water
   (g) Tea leaves from tea
   (h) Iron pins from sand
   (i) Wheat grains from husk
   (j) Fine mud particles suspended in water

2. Write the steps you would use for making tea. Use the words solution, solvent, solute, dissolve, soluble, insoluble, filtrate and residue.

3. Pragya tested the solubility of three different substances at different temperatures and collected the data as given below (results are given in the following table, as grams of substance dissolved in 100 grams of water to form a saturated solution).

Substance dissolved	Temperature in K and solubility				
	283	293	313	333	353
Potassium nitrate	21	32	62	106	167
Sodium chloride	36	36	36	37	37
Potassium chloride	35	35	40	46	54
Ammonium chloride	24	37	41	55	66

   (a) What mass of potassium nitrate would be needed to produce a saturated solution of potassium nitrate in 50 grams of water at 313 K?
   (b) Pragya makes a saturated solution of potassium chloride in water at 353 K and leaves the solution to cool at room temperature. What would she observe as the solution cools? Explain.
   (c) Find the solubility of each salt at 293 K. Which salt has the highest solubility at this temperature?
   (d) What is the effect of change of temperature on the solubility of a salt?

4. Explain the following giving examples.
   (a) Saturated solution   (b) Pure substance
   (c) Colloid   (d) Suspension

5. Classify each of the following as a homogeneous or heterogeneous mixture: soda water, wood, air, soil, vinegar, filtered tea.

6. How would you confirm that a colourless liquid given to you is pure water?

7. Which of the following materials fall in the category of a 'pure substance'?
   (a) Ice   (b) Milk   (c) Iron
   (d) Hydrochloric acid (e) Calcium oxide
   (f) Mercury   (g) Brick
   (h) Wood   (i) Air

8. Identify the solutions among the following mixtures.
   (a) Soil   (b) Sea water   (c) Air
   (d) Coal   (e) Soda water

9. Which of the following will show the Tyndall effect?
   (a) Salt solution   (b) Milk
   (c) Copper sulfate solution   (d) Starch solution

10. Classify the following into elements, compounds and mixtures.
    (a) Sodium   (b) Soil   (c) Sugar solution
    (d) Silver   (e) Calcium carbonate (f) Tin
    (g) Silicon   (h) Coal   (i) Air   (j) Soap
    (k) Methane   (l) Carbon dioxide   (m) Blood

11. Which of the following are chemical changes?
    (a) Growth of a plant   (b) Rusting of iron
    (c) Mixing of iron filings and sand
    (d) Cooking of food   (e) Digestion of food
    (f) Freezing of water   (g) Burning of a candle

Exemplar Problems

Short Answer Type Questions

1. Suggest separation technique(s) one would need to employ to separate the following mixtures.
   (a) Mercury and water
   (b) Potassium chloride and ammonium chloride
   (c) Common salt, water and sand
   (d) Kerosene, water and salt

2. Which of the tubes [(a) or (b)] will be more effective as a condenser in the distillation apparatus?

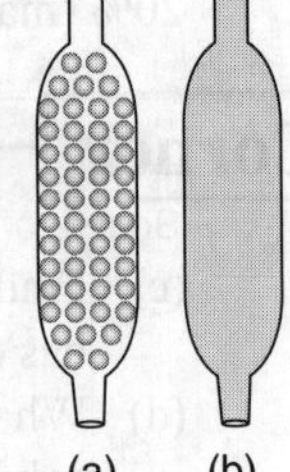

(a)   (b)

3. Salt can be recovered from its solution by evaporation. Suggest some other technique for the same.

4. 'Sea-water' can be classified as a homogeneous as well as a heterogeneous mixture. Comment.

5. While diluting a solution of salt in water, a student by mistake added acetone (boiling point 56°C). What technique can be employed to get back the acetone? Justify your choice.

6. What would you observe when,
   (a) a saturated solution of potassium chloride prepared at 60°C is allowed to cool to room temperature?
   (b) an aqueous sugar solution is heated to dryness?
   (c) a mixture of iron filings and sulfur powder is heated strongly?

7. Explain why particles of a colloidal solution do not settle down when left undisturbed, while they do in the case of a suspension.

8. Classify the following as physical or chemical properties.
   (a) The composition of a sample of steel is 98% iron, 1.5% carbon and 0.5% other elements.
   (b) Zinc dissolves in hydrochloric acid with the evolution of hydrogen gas.
   (c) Metallic sodium is soft enough to be cut with a knife.
   (d) Most metal oxides form alkalis on interacting with water.

9. Name the process associated with the following:
   (a) Dry ice is kept at room temperature and at 1 atmospheric pressure.
   (b) A drop of ink placed on the surface of water contained in a glass spreads throughout the water.
   (c) A potassium permanganate crystal is in a beaker and water is poured into the beaker with stirring.
   (d) An acetone bottle is left open and the bottle becomes empty.
   (e) Milk is churned to separate cream from it.
   (f) Settling of sand when a mixture of sand and water is left undisturbed for some time.
   (g) Fine beam of light entering through a small hole in a dark room, illuminates the particles in its path.

10. You are given two samples of water labelled A and B. Sample A boils at 100°C and sample B boils at 102°C. Which sample of water will not freeze at 0°C? Comment.

11. What are the favourable qualities given to gold when it is alloyed with copper or silver for the purpose of making ornaments?

12. An element is sonorous and highly ductile. Under which category would you classify this element? What other characteristics do you expect the element to possess?

13. Give an example each for the mixture having the following characteristics. Suggest a suitable method to separate the components of these mixtures.
    (a) A volatile and a non-volatile component.
    (b) Two volatile components with an appreciable difference in boiling points.
    (c) Two immiscible liquids.
    (d) One of the components changes directly from the solid to the gaseous state.
    (e) Two or more coloured constituents soluble in the same solvent.

14. Fill in the blanks
    (a) A colloid is a ________ mixture and its components can be separated by the technique known as ________.
    (b) Ice, water and water vapour look different and display different ________ properties but they are ________ the same.
    (c) A mixture of chloroform and water taken in a separating funnel is mixed and left undisturbed for some time. The upper layer in the separating funnel will be of ________ and the lower layer will be that of ________.
    (d) A mixture of two or more miscible liquids, for which the difference in the boiling points is less than 25 K can be separated by the process called ________.
    (e) When light is passed through water containing a few drops of milk, it shows a bluish tinge. This is due to the ________ of light by milk and the phenomenon is called ________. This indicates that milk is a ________ solution.

15. On heating, calcium carbonate gets converted into calcium oxide and carbon dioxide.
    (a) Is this a physical or a chemical change?
    (b) Can you prepare one acidic and one basic solution by using the products formed in the above process? If so, write the chemical equations involved.

16. Non-metals are usually poor conductors of heat and electricity. They are non-lustrous, non-sonorous, non-malleable and are coloured.
    (a) Name a lustrous non-metal.
    (b) Name a non-metal which exists as a liquid at room temperature.
    (c) The allotropic form of a non-metal is a good conductor of electricity. Name the allotrope.
    (d) Name a non-metal which is known to form the largest

number of compounds.
(e) Name a non-metal other than carbon which shows allotropy.
(f) Name a non-metal which is required for combustion.

Long Answer Type Questions

1. Fractional distillation is suitable for the separation of miscible liquids with a boiling point difference of about 25 K or less. What part of the fractional distillation apparatus makes it more efficient and possess an advantage over a simple distillation process? Explain using a diagram.

2. (a) Under which category of mixtures will you classify alloys and why?
   (b) A solution is always a liquid. Comment.
   (c) Can a solution be heterogeneous?

3. Iron filings and sulfur were mixed together and divided into two parts—A and S. Part A was heated strongly while Part S was not heated. Dilute hydrochloric acid was added to both the parts and evolution of gas was seen in both the cases. How will you identify the gases that are evolved?

4. A child wanted to separate the mixture of dyes constituting a sample of ink. He marked a line by the ink on the filter paper and placed the filter paper in a glass containing water as shown in the figure. The filter paper was removed when the water moved near the top of the filter paper.

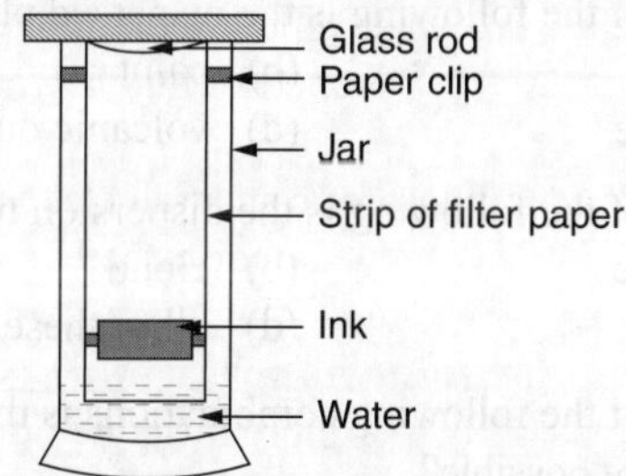

   (i) What would you expect to see, if the ink contains three different coloured components?
   (ii) Name the technique used by the child.
   (iii) Suggest one more application of this technique.

5. A group of students took an old shoe box and covered it with a black paper on all sides. They fixed a source of light (a torch) at one end of the box by making a hole in it and made another hole on the other side to view the light. They placed a milk sample contained in a beaker/tumbler in the box as shown in the figure. They were amazed to see that the milk taken in the tumbler was illuminated. They tried the same activity by taking a salt solution but found that light simply passed through it.

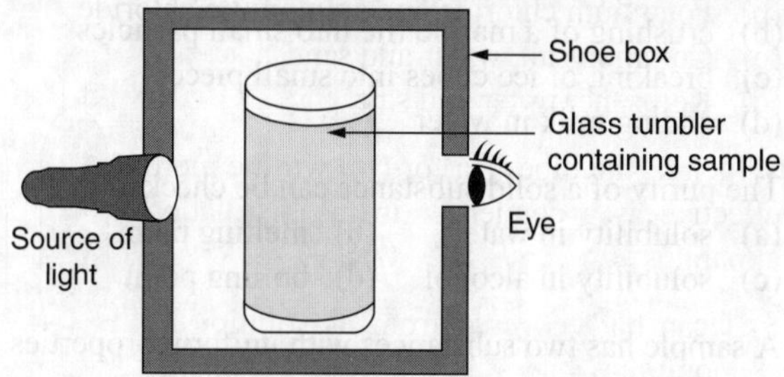

   (a) Explain why the milk sample was illuminated. Name the phenomenon involved.
   (b) The same results were not observed with a salt solution. Explain.
   (c) Can you suggest two more solutions which would show the same effect as shown by the milk solution?

6. During an experiment, the students were asked to prepare a 10% (Mass/Mass) solution of sugar in water. Ramesh dissolved 10 g of sugar in 100 g of water while Sarika prepared it by dissolving 10 g of sugar in water to make 100 g of the solution.
   (a) Are the two solutions of the same concentration?
   (b) Compare the mass% of the two solutions.

7. You are provided with a mixture containing sand, iron filings, ammonium chloride and sodium chloride. Describe the procedures you would use to separate these constituents from the mixture.

8. Arun prepared a 0.01% (by mass) solution of sodium chloride in water. Which of the following correctly represents the composition of the solutions?
   (a) 1.00 g of NaCl + 100 g of water
   (b) 0.11 g of NaCl + 100 g of water
   (c) 0.01 g of NaCl + 99.99 g of water
   (d) 0.10 g of NaCl + 99.90 g of water

COMPETITION WINDOW (OBJECTIVE TYPE)

[For NEET, JEE (Main and Advanced), NTSE, KVPY and Olympiads]

Topic-wise MCQs

Substances

1. Antimony and arsenic can be classified as
   (a) metalloids
   (b) non-metals
   (c) metals
   (d) cannot say

2. The most malleable metal here is
   (a) silver
   (b) iron
   (c) gold
   (d) aluminium

3. Which of the following is a compound?
   (a) graphite
   (b) stainless steel
   (c) brass
   (d) hydrogen bromide

4. In which of the following may the constituents be present in any ratio?
   (a) solution
   (b) colloid
   (c) mixture
   (d) compound

5. Two substances X and Y when brought together form a substance Z with the evolution of heat. The properties of Z

are entirely different from those of X and Y. The substance Z is
(a) a mixture (b) an element
(c) a compound (d) any of these

6. Which one of the following will result in the formation of a mixture?
(a) adding potassium metal to water
(b) crushing of a marble tile into small particles
(c) breaking of ice cubes into small pieces
(d) adding milk in water

7. The purity of a solid substance can be checked by its
(a) solubility in water (b) melting point
(c) solubility in alcohol (d) boiling point

8. A sample has two substances with uniform properties. The sample contains
(a) an element (b) a homogenous mixture
(c) a compound (d) a heterogeneous mixture

9. Which of the following is considered to be a pure substance?
(a) muddy water (b) milk of magnesia
(c) granite (d) sodium bromide

10. Compounds
(a) can be separated by their physical properties
(b) contain only one type of element
(c) are the same as mixtures
(d) are different kinds of atoms chemically combined with each other

Homogeneous and heterogeneous mixtures, solutions

11. Which of the following is an example of a heterogeneous substance?
(a) pieces of copper (b) candle
(c) bottled water (d) NaCl

12. Which of the following is a case of a homogeneous substance?
(a) M and M candy (b) muddy water
(c) granite (d) calcium sulfate

13. Which of the following is example of a homogenous mixture?
(a) alloy (b) blood
(c) milk (d) all of these

14. Assume air contains only nitrogen, oxygen and inert gases. Then air will be a
(a) heterogeneous mixture (b) colloid
(c) compound (d) homogeneous mixture

15. If addition of a solid in a liquid leads to a homogeneous mixture, then the solid will be
(a) partially soluble (b) insoluble
(c) soluble (d) all of these

16. Tincture of iodine has antiseptic properties. This solution is made by dissolving
(a) iodine in water (b) iodine in potassium iodide
(c) iodine in alcohol (d) iodine in vaseline

17. Which of the following reactions leads to the formation of a substance in the colloidal state?
(a) $2Mg + CO_2 \rightarrow MgO + C$
(b) $2HNO_3 + 3H_2S \rightarrow 3S + 4H_2O + 2NO$
(c) $Cu + HgCl_2 \rightarrow CuCl_2 + Hg$
(d) $Cu + CuCl_2 \rightarrow Cu_2Cl_2$

18. Solutions with low concentrations of solutes are known as
(a) dilute solutions (b) solvents
(c) concentrated solutions (d) none of these

Colloids

19. In the colloidal state, the particle size ranges from
(a) 10 to 1000 Å (b) 1 to 280 Å
(c) 1 to 10 Å (d) 20 to 50 Å

20. Substances whose solutions can readily diffuse through a parchment membrane are
(a) electrolytes (b) non-electrolytes
(c) colloids (d) crystalloids

21. Which of the following statements is true about a colloidal system?
(a) it can be made out of two gases
(b) it is electrically neutral as a whole
(c) it carries a net electric charge
(d) it consists of one phase only

22. In which of the following is the dispersed phase a liquid?
(a) fog (b) paint
(c) smoke (d) volcanic dust

23. In which of the following is the dispersion medium a gas
(a) smoke (b) cloud
(c) fog (d) all of these

24. In which of the following combinations is the formation of colloids not possible?
(a) gas in gas (b) gas in liquid
(c) solid in liquid (d) liquid in solid

25. Which of the following is a colloid?
(a) silicic acid (b) NaCl solution
(c) sugar solution (d) urea solution

26. Smoke is an example of a
(a) solid dispersed in a gas
(b) solid dispersed in a solid
(c) gas dispersed in a liquid
(d) gas dispersed in a solid

27. When the dispersed phase is a liquid and the dispersion medium is a gas, then the colloidal system is called a(an)
(a) jelly (b) cloud
(c) emulsion (d) smoke

28. When a colloidal solution is observed under an ultramicroscope, we can see the
(a) shape of the particles
(b) relative size
(c) light scattered by the colloidal particles
(d) size of the particles

29. Which of the following forms a colloidal solution in water?
    (a) starch   (b) barium nitrate
    (c) NaCl   (d) glucose

30. Tyndall effect will be observed in
    (a) vapour   (b) solution
    (c) sol   (d) precipitate

31. Small liquid droplets dispersed in another liquid is called
    (a) emulsion   (b) true solution
    (c) gel   (d) suspension

32. Butter is a colloid formed when
    (a) fat is dispersed in water
    (b) milk is dispersed in water
    (c) water is dispersed in fat
    (d) milk is dispersed in fat

Methods of separation

33. Salt can be obtained from sea water by
    (a) sublimation   (b) filtration
    (c) evaporation   (d) decantation

34. A mixture of $ZnCl_2$ and $PbCl_2$ can be separated by
    (a) sublimation   (b) fractional distillation
    (c) distillation   (d) crystallisation

35. Camphor can be purified by
    (a) sedimentation   (b) sublimation
    (c) distillation   (d) filtration

36. The process that can used to separate oil and water is
    (a) by using a separating funnel
    (b) chromatography
    (c) distillation
    (d) sublimation

37. Filtration can be used to separate
    (a) liquids from liquids   (b) liquids from gases
    (c) solids from solids   (d) liquids from solids

38. Which of the following is based on the difference in size?
    (a) fractional distillation   (b) sublimation
    (c) filtration   (d) solvent extraction

39. Fractional distillation is based on the difference in the _______ of two liquids.
    (a) solubility   (b) melting point
    (c) boiling point   (d) none of these

40. After evaporation, the residue left is
    (a) soluble solid   (b) solvent
    (c) Both (a) and (b)   (d) none of these

41. A mixture of water and C_2H_5OH can be separated to a certain extent by
    (a) fractional distillation   (b) centrifugation
    (c) sublimation   (d) evaporation

42. A mixture of sand and sulfur may best be separated by
    (a) fractional distillation
    (b) dissolving in CS_2 and filtering
    (c) fractional crystallisation from an aqueous solution
    (d) magnetic method

43. A mixture of sand and sulfur may best be separated by
    (a) fractional distillation
    (b) fractional crystallisation from an aqueous solution
    (c) magnetic method
    (d) dissolving in CS_2 and filtering

44. Distillation is a good separation technique for
    (a) solid alloys   (b) gases
    (c) solids   (d) liquids

45. Magnetism can be beneficial for separating
    (a) non-metallic solids and solids such as sulfur
    (b) non-magnetic solids from non-magnetic liquids
    (c) gases and non-metallic liquids
    (d) magnetic solids and solids like sulfur

Physical and chemical changes

46. Physical properties of a mixture
    (a) depend on the organisation of the substance
    (b) vary depending upon its components
    (c) vary with the amount of substance
    (d) depend on the volume of the substance

47. On adding a few drops of dilute sulfuric acid to blue vitriol (copper sulfate) solution, we get a blue coloured solution. When we drop a shaving blade into this blue solution, the blue colour changes to green after some time. What type of change is illustrated by this activity?
    (a) physical change   (b) chemical change
    (c) both (a) and (b)   (d) none of these

48. Which component of the mixture (Fe + S) reacts with dil. HCl and gives hydrogen gas?
    (a) iron   (b) sulfur
    (c) both   (d) none of these

49. Which of the following are physical changes?
    (i) melting of iron metal
    (ii) rusting of iron
    (iii) bending of an iron rod
    (iv) drawing a wire of iron metal
    (a) (i), (ii) and (iii)   (b) (i), (ii) and (iv)
    (c) (i), (iii) and (iv)   (d) (ii), (iii) and (iv)

50. Which of the following are chemical changes?
    (i) decaying of wood   (ii) burning of wood
    (iii) sawing of wood
    (iv) hammering of a nail into a piece of wood
    (a) (i) and (ii)   (b) (ii) and (iii)
    (c) (iii) and (iv)   (d) (i) and (iv)

Miscellaneous

1. Which of the following sets represents correct statements given here?
    (i) Metals are ductile and malleable
    (ii) Graphite can show electrical conductance
    (iii) Non-metals show conductance
    (iv) Metals are shiny while non-metals are dull (mostly)
    (a) (i), (iv)   (b) (i), (ii), (iv)
    (c) (ii), (iii), (iv)   (d) (i), (ii), (iii), (iv)

2. Which of the following set of element contains only non-metals?
   (a) H, C, Hg, Ge
   (b) Fe, Co, Ni, P
   (c) S, Cl, Ar, Br
   (d) Al, Ga, Si, P

3. Rusting of an article made up of iron is called
   (a) corrosion and it is a physical as well as chemical change
   (b) dissolution and it is a physical change
   (c) corrosion and it is a chemical change
   (d) dissolution and it is a chemical change

4. The most common solvent on earth is ________
   (a) turpentine
   (b) water
   (c) petrol
   (d) gasoline

5. A colloidal solution can be purified by
   (a) dialysis
   (b) Bredig's arc method
   (c) peptisation
   (d) coagulation

6. Which of the following statements is correct?
   (a) A homogeneous mixture must be uniform
   (b) A heterogeneous mixture must contain at least three elements
   (c) A mixture containing two compounds must be heterogeneous
   (d) A pure substance must contain only one type of atom

7. Which of the following is a lyophilic colloid?
   (a) sulfur sol
   (b) Fe_2O_2
   (c) gum
   (d) gold sol

8. Which of the following is a lyophobic colloid?
   (a) protein
   (b) gold sol
   (c) gelatine
   (d) starch

9. A mixture of sulfur and carbon disulfide is
   (a) heterogeneous and shows the Tyndall effect
   (b) homogeneous and shows the Tyndall effect
   (c) heterogeneous and does not show the Tyndall effect
   (d) homogeneous and does not show the Tyndall effect

10. Which of the following is not correct?
    (a) gemstone is an example of solid in solid type of colloid
    (b) cheese is an example of liquid in solid type of colloid
    (c) milk is an example of liquid in liquid type of colloid
    (d) whipped cream is an example of gas in gas type of colloid

11. Cloud or fog is an example of colloidal system of a
    (a) solid dispersed in a gas
    (b) solid dispersed in a liquid
    (c) liquid dispersed in a gas
    (d) gas dispersed in a gas

12. A mixture of common salt, sulfur, sand and iron filings is shaken with carbon disulfide and filtered through a filter paper. The filtrate is evaporated to dryness in a china dish. What will be left in the dish after evaporation?
    (a) common salt
    (b) sulfur
    (c) sand
    (d) iron filings

13. A mixture of ethanol and water can be separated by
    (a) fractional distillation
    (b) sublimation
    (c) filtration
    (d) decantation

14. A mixture of methyl alcohol and acetone can be separated by
    (a) steam distillation
    (b) fractional distillation
    (c) distillation under reduced pressure
    (d) distillation

15. Melting points can separate materials because
    (a) heat causes molecules to disintegrate
    (b) many substances fuse at the melting point
    (c) substances melt at different temperatures
    (d) molecules vibrate rapidly when heated

16. Which of the following will show Tyndall effect?
    (a) aqueous solution of soap above the critical micelle concentration
    (b) aqueous solution of soap below the critical micelle concentration
    (c) aqueous solution of sugar
    (d) aqueous solution of sodium chloride

17. Which of the following methods can be used for the separation of a liquid–liquid mixture?
    (a) separating funnel
    (b) solvent extraction
    (c) fractional distillation
    (d) both (a) and (c)

18. Which of the following sets of methods can be used for a mixture of a solid and solid type?
    (a) sublimation, filtration
    (b) evaporation, fractional distillation
    (c) magnetic separation, solvent extraction
    (d) sublimation, evaporation

19. Find the mole fraction of glucose in a solution having 36 g of glucose in 90 g of water.
    (a) 1.84%
    (b) 38.4%
    (c) 7.68%
    (d) 3.84%

20. A solution has 0.4 g of NaOH in 250 mL. Find its concentration.
    (a) 0.4 g/L
    (b) 0.8 g/L
    (c) 1.6 g/L
    (d) 3.2 g/L

Advanced and Olympiads

Single Choice

1. Which of the following statements are true for pure substances?
   (i) Pure substances contain only one kind of particle.
   (ii) Pure substances may be compounds or mixtures.
   (iii) Pure substances have the same composition throughout.
   (iv) Pure substances can be exemplified by all elements other than nickel.
   (a) (i) and (ii)
   (b) (i) and (iii)
   (c) (iii) and (iv)
   (d) (ii) and (iii)

2. Which of the following are homogeneous in nature?
   (i) ice   (ii) wood   (iii) soil   (iv) air
   (a) (i) and (iii)
   (b) (ii) and (iv)
   (c) (i) and (iv)
   (d) (iii) and (iv)

3. Two chemical species X and Y combine together to form a product P which contains both X and Y

$$X + Y \rightarrow P$$

X and Y cannot be broken down into simpler substances by simple chemical reactions. Which of the following concerning the species X, Y and P are correct?

  (i) P is a compound   (ii) X and Y are compounds
(iii) X and Y are elements   (iv) P has a fixed composition

(a) (i), (ii) and (iii)   (b) (i), (ii) and (iv)
(c) (ii), (iii) and (iv)   (d) (i), (iii) and (iv)

4. Two substances, A and B were made to react to form a third substance, A_2B according to the following reaction:

$$2A + B \rightarrow A_2B$$

Which of the following statements concerning this reaction are incorrect?

  (i) The product A_2B shows the properties of substances A and B.
 (ii) The product will always have a fixed composition.
(iii) The product so formed cannot be classified as a compound.
(iv) The product so formed is an element.

(a) (i), (ii) and (iii)   (b) (ii), (iii) and (iv)
(c) (i), (iii) and (iv)   (d) (iii) and (iv)

5. The teacher instructed three students X, Y and Z respectively, to prepare a 50% (mass by volume) solution of sodium hydroxide (NaOH). X dissolved 50 g of NaOH in 100 mL of water, Y dissolved 50 g of NaOH in 100 g of water while Z dissolved 50 g of NaOH in water to make 100 mL of solution. Which one of them has made the desired solution?

(a) X   (b) Y
(c) Z   (d) Both X and Y

6. Among the colloids cheese (C), milk (M) and smoke (S), the correct combination of the dispersed phase and dispersion medium, respectively is

(a) C: solid in liquid; M: solid in liquid; S: solid in gas
(b) C: solid in liquid; M: liquid in liquid; S: gas in solid
(c) C: liquid in solid; M: liquid in solid; S: solid in gas
(d) C: liquid in solid; M: liquid in liquid; S: solid in gas

7. In the separation of red ink from blue ink, a technique is used which is described below in the steps involved in the process. Which of the following set represents the proper sequence?

  (i) The blue and red ink form spots at a certain distance on the paper.
 (ii) Paper and solvent are taken as stationary and mobile phases, respectively.
(iii) Paper chromatography is used for this separation.
(iv) The paper is suspended in the closed jar by using a hook.
 (v) A narrow strip of paper with a line drawn is cut and a mixture of red and blue ink is placed on the line marked on the paper, with the help of capillary.

(a) (ii) (iv) (iii) (v) (i)   (b) (iii) (ii) (v) (iv) (i)
(c) (ii) (iii) (iv) (v) (i)   (d) (iii) (ii) (iv) (v) (i)

Multiple Choice

8. Which of the following is an example of a homogeneous substance?
   (a) Russian salad dressing
   (b) aspirin
   (c) sodium nitrate
   (d) mixture of sand and rice grains

9. Which of the following is/are the properties of a colloidal solution?
   (a) homogeneous   (b) scattering of light
   (c) electrophoresis   (d) Tyndall effect

10. Which of the following is an example(s) of alloy?
    (a) table salt   (b) gun metal
    (c) bronze   (d) brass

11. Which of the following are compound(s) here?
    (a) calcium sulfate   (b) sugar
    (c) diamond   (d) common salt

12. Which of the following is/are a mixture(s)?
    (a) metal amalgam   (b) solution
    (c) alloy   (d) phosphine

13. Which of the following are affected by temperature?
    (a) molarity   (b) molality
    (c) normality   (d) mole fraction

14. Which of the following is/are true for a compound?
    (a) Properties of a compound are entirely different from those of the elements present in it.
    (b) A compound contains different elements in a fixed ratio.
    (c) It is heterogeneous in nature.
    (d) Constituents of a compound cannot be separated by simple physical methods.

15. Which of the following are used for separation of solid–solid mixtures?
    (a) chromatography
    (b) winnowing
    (c) sublimation
    (d) solvent extraction

16. Which of the following statements are correct?
    (a) Presence of equal and similar charges on colloidal particles provides stability to the colloids.
    (b) Mixing two oppositely charged sols neutralises their charges and stabilises the colloid.
    (c) Brownian movement stabilises sols.
    (d) Any amount of dispersed liquid can be added to an emulsion without destabilising it.

Comprehension Type

Comprehension I: An element is a substance which cannot be split into simpler substances by using a chemical method. An element can be a non-metal, a metalloid or a metals. Most of the elements are metals and are solid in nature, with a few liquid elements also. Metals and non-metals differ greatly in physical properties like hardness, density, melting point and boiling point.

17. Most of the elements are present mainly as
   (a) non-metals in the liquid form
   (b) non-metals in the solid state
   (c) metals in the solid form
   (d) metals in the liquid state

18. Which of the following sets of elements contain liquid elements only?
   (a) Cl, Br, Hg   (b) Br, Hg, Ga
   (c) Br, Hg, He   (d) Br, Hg, Al

19. From the given elements select the elements with highest and lowest melting points—C, Fe, W, Hg, He, Na, Ca, Al
   (a) W, C   (b) Hg, W
   (c) W, Hg   (d) C, He

Comprehension II: A colloidal solution is a heterogeneous solution which contains particles of intermediate size, that is, a diameter between 1 and 100 nm. A colloid is not a substance but it is a state of a substance which depends upon the molecular size. Colloidal solutions are intermediate between true solutions and suspensions.

20. The size of the colloidal particles lies in the range
   (a) 10^{-5} cm–10^{-7} cm   (b) 10 nm–1000 nm
   (c) 1 nm–1000 nm   (d) 10 mμ–1000 mμ

21. The colloidal solution of a solid as the dispersed phase and a gas as the dispersion medium is called
   (a) an aerosol   (b) a solid foam
   (c) a sol   (d) a gel

22. The colloidal particles can pass through
   (a) filter paper but not through animal membrane
   (b) animal membrane but not through filter paper
   (c) filter paper as well as animal membrane
   (d) neither filter paper nor animal membrane

Matrix Matching

23. Match the following

Column I	Column II
(a) iron filings and common salt	(p) distillation
(b) salt and water	(q) magnetic separation
(c) benzene and toluene	(r) evaporation
(d) rice grains and alcohol	(s) filtration

24. Match the following

Column I	Column II
(a) burning diesel	(p) physical change
(b) cutting of wood	(q) reversible change
(c) burning of mgnesium ribbon	(r) chemical change
(d) melting of ice cream	(s) irreversible change

Integer Type

25. How many of the following elements are liquid in nature?
Cl, Br, I, Hg, Ga, Fr, Cs, Ca, Cu

26. How many of these elements are metalloid elements?
As, Br, Se, Sb, Al, B, Si, Hg, Ge

27. How many of the given substances are mixtures?
Cloud, iodised table salt, ice, glucose, brass, aerated drink, dry ice, milk, 22 carat gold, marsh gas, diesel, smoke

28. How many of the given substances are compounds?
Cloud, iodised table salt, steam, sucrose, steel, aerated drink, dry ice, milk, 22 carat gold, marsh gas, petrol, smoke

29. How many of the following compounds can undergo sublimation?
Naphthalene, iodine, potassium hydroxide, camphor, H_2SO_4, ammonium chloride, anthracene

30. Which of the following are not compounds?
   (a) chlorine gas   (b) potassium chloride
   (c) iron   (d) iron sulfide
   (e) aluminium   (f) iodine
   (g) carbon   (h) carbon monoxide
   (i) sulfur powder

31. How many of the given methods can be used for the separation of a mixture of a solid and a liquid?
Filtration, sedimentation, separating funnel, evaporation, distillation, solvent extraction, centrifugation, chromatography, fractional distillation

Answer Keys

Fill in the Blanks

1. element
2. atoms
3. mercury
4. physical
5. fractional distillation
6. chemical
7. zinc, copper
8. fractional distillation
9. separating funnel
10. dilute
11. filtration
12. magnet
13. true, colloidal
14. 10^{-7}, 10^{-9}
15. variable

True or False

1. T	2. F	3. T	4. F	5. T
6. F	7. F	8. F	9. T	10. T
11. F	12. T	13. T	14. F	15. T

Match the Following

1. (a)—(r); (b)—(t); (c)—(s); (d)—(q); (e)—(p)

Very Short Answer Type Questions

1. A substance having a single type of particle is known as a pure substance. For example, hydrogen and water.

2. By determination of the boiling point, the purity of a liquid can be checked.

3. Homogeneous mixture: salt in water, heterogeneous mixture: salt and sulfur.

4. Soap solution can scatter light.

5. Mass of sugar = 36 g

   Mass of water = 364 g

   So, Mass of solution = Mass of solute + Mass of solvent

   $$= 36 + 364 = 400 \text{ g}$$

   $$\text{Concentration of solution} = \frac{\text{Mass of solute}}{\text{Mass of solution}} \times 100$$

   $$\text{Concentration of solution} = \frac{36}{400} \times 100 = 9\% \text{ by mass}$$

6. By evaporation, the coloured components can be obtained from blue ink.

7. Sublimation is used to separate a mixture of salt and ammonium chloride.

8. When we heat iron filings and sulfur till red hot, we get a compound.

9. A colloidal solution shows Tyndall effect.

10. Elements: He, Ar, Cl_2 and Co, compounds: H_2O, CO_2 and NH_3.

11. A separating funnel is used to separate oil from water.

12. Centrifugation is used in milk dairies to separate cream from milk.

Short Answer Type Questions

1. See text part.

2. (a) Mixture (b) Mixture (c) Mixture
   (d) Pure substance (e) Pure substance
   (f) Compound/mixture (g) Mixture
   (h) Pure substance

3. Elements: Cu, Zn, F_2, O_2, diamond (C), Hg
   Compounds: $CaCO_3$, H_2O

4.

Homogeneous mixture	Heterogeneous mixture
1. Its constituent particles cannot be seen easily.	1. Its constituent particles can be seen easily.
2. There are no visible boundaries of separation in a homogeneous mixture.	2. Have visible boundaries of separation between the constituents.
3. Its constituents cannot be easily separated.	3. Its constituents can be separated by simple methods.
Examples: alloys, solution of salt in water.	Examples: mixture of sand and common salt, mixture of sand and water.

5. See text part.

6. The composition of the sugar (sucrose) will remain constant irrespective of the source of its preparation. Hence sugar or sucrose is a pure substance with fixed composition.

7. No, mixture of water and alcohol cannot be separated since both are miscible and they form a solution. Only immiscible liquids can be separated.

8. The crystallisation method can be used for the purification of those mixtures which contain insoluble and/or soluble impurities. Have crystalline nature. Cannot be separated by filtration as some impurities are soluble.

9. See text part.

10. See text part.

11. Simple distillation is the method to separate a mixture of kerosene and petrol (bp differs by more than 25°C).
    Method: In a distillation flask, a mixture of kerosene and petrol is taken as shown in Fig. 2.16. The mixture is heated slowly and the temperature is recorded with the help of a thermometer. Petrol (bp = 70°C to 1200°C) vaporises first and the temperature becomes constant for some time (till all petrol evaporates from the mixture). Vapours of petrol are condensed and collected in another container while the kerosene remains in the distillation flask. As soon as the temperature starts rising again, the heating is stopped and both the components are collected separately.

12. (i) By using the centrifugation method, butter can be separated from curd.
    (ii) By using the evaporation method, salt can be separated from sea water. Water vaporises on evaporation leaving behind the salt.
    (iii) Camphor from salt can be separated by the sublimation method. On subliming, camphor will be converted into vapour, leaving behind the salt.

Long Answer Type Questions

1. Physical changes: cutting of trees, melting of butter in a pan, boiling of water to form steam, dissolving common salt in water, making a fruit salad with raw fruits.
   Chemical changes: rusting of almirah, passing of electric current through water and the water breaking down into hydrogen and oxygen gases, burning of paper and wood.

2. See text part.

3. See text part.

4. Mass of sodium chloride (solute) = 36 g
   Mass of water (solvent) = 100 g
   We know that, Mass of solution = Mass of solute + Mass of solvent = 36 g + 100 g = 136 g
   Concentration (mass percentage) of the solution

   $$= \frac{\text{Mass of Solute}}{\text{Mass of Solution}} \times 100$$

   $$= \frac{36 \text{ g}}{136 \text{ g}} \times 100 = 26.47\%$$

5. Let the mass of sodium sulfate required be $= x$ g
   The mass of solution would be $= (x + 100)$ g $\times$ g of solute in $(x + 100)$ g of solution

$$20\% = \frac{x}{x+100} \times 100$$

$$20x + 2000 = 100x$$

$$80x = 2000$$

$$x = \frac{2000}{80} = 25 \text{ g}$$

NCERT Corner

Exercises

1. (a) Evaporation (b) Sublimation
   (c) Filtration (d) Chromatography.
   (e) Centrifugal machine or churning the curd by hand
   (f) Decantation (g) Filtration
   (h) Magnetic separation (i) Winnowing
   (j) Coagulation and decantation

2. Method of preparation of tea
   (i) Take some water (solvent) in a pan and heat it.
   (ii) Add some sugar (solute) and boil to dissolve the sugar completely. The obtained homogeneous mixture is called solution.
   (iii) Add tea leaves (or tea) in the solution and boil the mixture.
   (iv) Now add milk and boil again.
   (v) Filter the mixture through the tea strainer and collect the filtrate (tea) in a cup. The insoluble tea leaves are left behind as residue in the strainer.

3. (a) Mass of potassium nitrate needed to produce its saturated solution in 100 g of water at 313 K = 62 g
   Mass of potassium nitrate needed to produce its saturated solution in 50 g of water at

   $$313 \text{ K} = \frac{62}{100} \times 50 \text{ g} = 31 \text{ g}$$

   (b) Crystals of potassium chloride are formed. This happens as solubility of solid decreases when the temperature is decreased.
   (c) Solubility of each salt at 293 K

Potassium nitrate	32 g per 100 g water
Sodium chloride	36 g per 100 g water
Potassium chloride	35 g per 100 g water
Ammonium chloride	37 g per 100 g water

   Note: Solubility of a solid is that amount in grams which can be dissolved in 100 g of water (solvent) to make a saturated solution at a particular temperature. Ammonium chloride has the maximum solubility (37 g per 100 g of water) at 293 K.
   (d) Solubility of a (solid) salt decreases with a decrease in temperature while it increases with a rise in temperature.

4. (a) Saturated solution: A solution in which no more solute can be dissolved at a particular temperature is called a saturated solution. Example: when sugar is dissolved repeatedly in a given amount of water, a condition is reached at which further dissolution of sugar is not possible in that amount of water at room temperature.
   (b) Pure substance: A substance made up of a single type of particle (atoms and/or molecules) is called a pure substance. All elements and compounds are said to be pure. Example: water, sugar.
   (c) Colloid: A heterogeneous mixture in which the solute particle size is too small to be seen with the naked eye, but is big enough to scatter light is known as a colloid. There are two phases in a colloidal solution. Dispersed phase: solute particles are said to be the dispersed phase. Dispersion medium: the medium in which solute particles are spread is called the dispersion medium. Example: milk, clouds.
   (d) Suspension: A suspension is a heterogeneous mixture in which the solute particles do not dissolve but remain suspended throughout the bulk of the medium. The suspended particles are visible to the naked eye. Example: mixture of sand and water, muddy water.

5. Homogeneous mixtures: air, soda water, vinegar, filtered tea. Heterogeneous mixtures: Wood, soil.

6. If the given colourless liquid boils at 100°C sharp, it is pure water, otherwise it is not.

7. Ice, iron, calcium oxide, mercury are pure substances as they have a definite composition. Milk is a colloid, so it is a heterogeneous mixture. Hydrochloric acid is also a mixture of hydrogen chloride gas and water.

8. Sea water, air and soda water: homogeneous mixtures
   Coal, soil: heterogeneous solutions

9. Milk and starch solution will show the Tyndall effect, as both of these are colloids.

10. Elements: sodium, silver, tin and silicon
    Compounds: calcium carbonate, methane and carbon dioxide
    Mixtures: Soil, sugar solution, coal, air, soap and blood

11. Growth of a plant, rusting of iron, cooking of food, digestion of food, burning of a candle are chemical changes, because here the chemical composition of the substance changes.

Exemplar Problems

Short Answer Type Questions

1. (a) Mercury and water: Since there is a difference in the densities of mercury and water and both liquids are insoluble in each other, they can be separated using a separating funnel.
   (b) Potassium chloride and ammonium chloride can be separated by sublimation. Ammonium chloride being volatile will be converted into vapour. KCl does not sublime.
   (c) Filtration will separate the salt solution (salt and water) and sand. Evaporation of the salt solution will separate the salt from the solution.
   (d) A separating funnel can be used to separate kerosene and salt solution. Evaporation of salt solution will separate salt and water.

2. The tube (a) containing beads will be more effective as a condenser because the surface area of the tube increases.

3. Salt solution can be concentrated by heating to make a supersaturated solution. Crystallisation will occur when the solution is allowed to cool and salt will separate out from the solution.

4. Sea water can be classified as a homogeneous mixture because it contains salts dissolved in water. It can be classified as a heterogeneous mixture also, since it contains mud, sand and decayed parts of plants.

5. The mixture of acetone and salt solution in water can be separated by distillation since the difference in their boiling points is more than 25°C. Acetone will evaporate and get condensed first, leaving behind the salt solution.

6. (a) Crystals of potassium chloride will separate out.
   (b) On heating the sugar solution, water will evaporate first. Once the solution dries up, it will turn black and the sugar will get charred.
   (c) Iron sulfide is formed when a mixture of iron filings and sulfur is heated strongly.

7. The size of the colloidal particles in a colloidal solution is smaller than in a suspension. These particles are in a random motion hence do not settle down when left undisturbed. The particles of a suspension are larger and they tend to settle down under the effect of gravity.

8. (a) Physical property   (b) Chemical property
   (c) Physical property   (d) Chemical property

9. (a) Sublimation of dry ice (solid) to CO_2 (gas)
   (b) Diffusion of ink into water
   (c) Diffusion or dissolution of solid into liquid
   (d) Evaporation, diffusion of acetone in air
   (e) Centrifugation
   (f) Sedimentation
   (g) Tyndall effect: Scattering of light

10. Sample B which boils at 102°C contains impurities. It will not freeze at 0°C. There will be a depression in freezing point.

11. Pure gold is highly malleable and soft. When it is alloyed with copper or silver it becomes hard and strong and can be moulded into various shapes.

12. It is a metal. The element is expected to be lustrous, malleable and a good conductor of heat and electricity.

13. (a) Evaporation or distillation: Common salt solution
    (b) Distillation: Acetone–water mixture
    (c) Separation using separating funnel: Oil–water mixture
    (d) Sublimation: Mixture of common salt and ammonium chloride
    (e) Chromatography: Ink

14. (a) heterogeneous, centrifugation
    (b) physical, chemically
    (c) water, chloroform
    (d) fractional distillation
    (e) scattering, Tyndall effect, colloidal

15. $CaCO_3 \xrightarrow{\Delta} CaO + CO_2$
    (a) It is a chemical change in which new substances are formed.
    (b) Calcium oxide when dissolved in water, forms a basic solution.

$$CaO + H_2O \rightarrow Ca(OH)_2$$

Carbon dioxide when dissolved in water, forms an acidic solution.

$$CO_2 + H_2O \rightarrow H_2CO_3$$

16. (a) Iodine   (b) Bromine
    (c) Graphite   (d) Carbon
    (e) Sulfur, phosphorus (f) Oxygen

Long Answer Type Questions

1. In fractional distillation, a fractionating column is used which is packed with glass beads or small plates. This increases the surface area for the vapours to condense and they quickly lose energy when they come in contact with the beads or plates and can be quickly condensed. A longer column would increase the efficiency of the process.

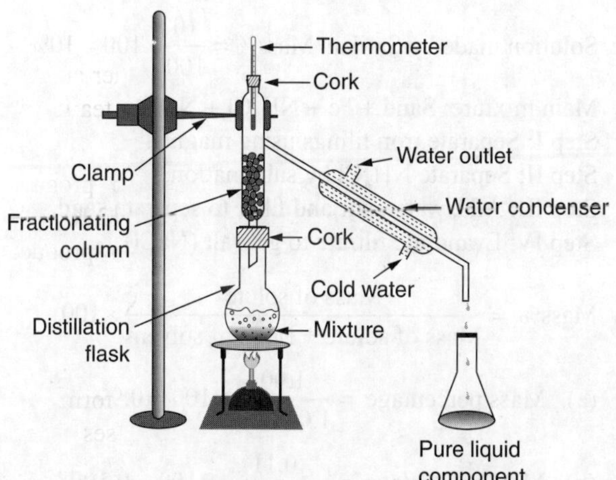

2. (a) Alloys are homogeneous mixtures because they have an uniform composition throughout.
   (b) No, a solution can be solid (alloys) or gaseous (air) also.
   (c) No, a solution is a homogeneous mixture.

3. Part A: Iron sulfide is formed which gives out hydrogen sulfide gas with HCl.

$$Fe + S \xrightarrow{\Delta} FeS$$

$$FeS + 2HCl \rightarrow FeCl_2 + H_2S$$

H_2S gas can be identified by its smell. It has a foul smell and it turns lead acetate solution black.

Part B: Fe and S will not react, when HCl is added to this mixture, only Fe will react with HCl to give out H_2 gas.

$$Fe + 2HCl \rightarrow FeCl_2 + H_2$$

Hydrogen gas burns with a pop sound and hence can be identified by bringing a burning matchstick near it.

4. (i) The components of the ink will travel with water and we will see three bands on the filter paper at various heights.

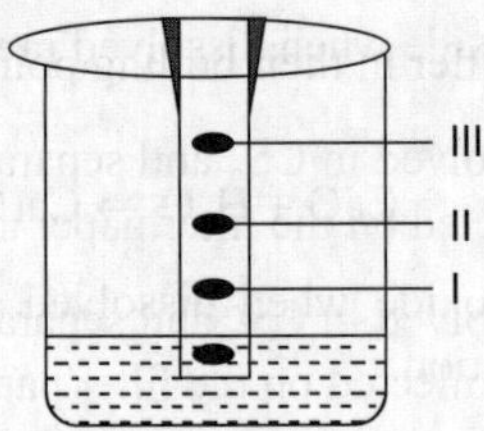

   (ii) The technique is called chromatography.
   (iii) Separation of pigments present in chlorophyll.

5. (a) Milk is a colloidal solution hence shows the Tyndall effect.
   (b) True solutions do not show the Tyndall effect because they do not scatter light.
   (c) Detergent solution, sulfur solution.

6. (a) No, Sarika has higher mass percentage.
   (b) Solution made by Ramesh,

$$\text{Mass \%} = \frac{10}{(10+100)} \times 100 = \frac{10}{110} \times 100 = 9.09\%$$

   Solution made by Sarika, $\text{Mass\%} = \frac{10}{100} \times 100 = 10\%$

7. Main mixture: Sand + Fe + NH_4Cl + NaCl
   Step I: Separate iron filings using magnet.
   Step II: Separate NH_4Cl by sublimation.
   Step III: Add, water, stir and filter to separate sand.
   Step IV: Evaporate filtrate to get salt (NaCl).

8. $\text{Mass\%} = \dfrac{\text{Mass of solute}}{\text{Mass of solute + Mass of solvent}} \times 100$

   (a) $\text{Mass percentage} = \dfrac{1.00}{1.0+100} \times 100 = 0.99$

   (b) $\text{Mass percentage} = \dfrac{0.11}{0.11+100} \times 100 = 0.1098$

   (c) $\text{Mass percentage} = \dfrac{0.01}{0.01+99.99} \times 100 = 0.01$

   (d) $\text{Mass percentage} = \dfrac{0.10}{0.10+99.90} \times 100 = 0.1$

Competition Window (Objective Type)
Topic-wise MCQs

1. (a)	**2.** (c)	**3.** (d)	**4.** (c)	**5.** (c)
6. (d)	**7.** (b)	**8.** (b)	**9.** (d)	**10.** (d)
11. (b)	**12.** (d)	**13.** (a)	**14.** (d)	**15.** (c)
16. (c)	**17.** (b)	**18.** (a)	**19.** (a)	**20.** (d)
21. (b)	**22.** (a)	**23.** (d)	**24.** (a)	**25.** (a)
26. (a)	**27.** (b)	**28.** (c)	**29.** (a)	**30.** (c)
31. (a)	**32.** (c)	**33.** (c)	**34.** (d)	**35.** (b)
36. (a)	**37.** (d)	**38.** (c)	**39.** (c)	**40.** (a)
41. (a)	**42.** (b)	**43.** (d)	**44.** (d)	**45.** (d)
46. (b)	**47.** (b)	**48.** (a)	**49.** (c)	**50.** (a)

Miscellaneous

1. (b)	**2.** (c)	**3.** (c)	**4.** (b)	**5.** (a)
6. (a)	**7.** (c)	**8.** (b)	**9.** (d)	**10.** (d)
11. (c)	**12.** (b)	**13.** (a)	**14.** (b)	**15.** (c)
16. (a)	**17.** (d)	**18.** (c)	**19.** (d)	**20.** (c)

Advanced and Olympiads
Single Choice

1. (b)	**2.** (c)	**3.** (d)	**4.** (c)	**5.** (c)
6. (d)	**7.** (b)			

Multiple Choice

8. (bc)	**9.** (bcd)	**10.** (bcd)	**11.** (abd)	**12.** (abc)
13. (ac)	**14.** (abd)	**15.** (bcd)	**16.** (ac)	

Comprehension Type

I. 17. (c)	**18.** (b)	**19.** (d)
II. 20. (c)	**21.** (a)	**22.** (a)

Matrix Matching
23. (a)—(q, s); (b)—(p, r); (c)—(p); (d)—(s)
24. (a)—(r, q); (b)—(p, q); (c)—(r, s); (d)—(p, s)

Integer Type

25. 5	**26.** 6	**27.** 7	**28.** 5	**29.** 5
30. 6	**31.** 6			

Hints and Solutions

Competition Window (Objective Type)
Topic-wise MCQs

3. Hydrogen bromide (HBr) is a compound of hydrogen and bromine. Stainless steel and brass are alloys whereas graphite is an allotropic form of the element carbon.

7. As every solid has a fixed melting point, it can be used.

8. A homogeneous mixture is a solution having a uniform composition and properties throughout the bulk of the solution.

9. Sodium bromide being a compound is a pure substance. Granite, muddy water and milk of magnesia all are mixtures.

11. Candle is a heterogeneous mixture of wax and threads. Copper is an element. Bottled water and table salt (NaCl) are compounds.

12. It is a compound.

13. Alloys are homogenous mixtures of metals.

15. A soluble solute leads to a homogenous mixture.

16. A solution of *tincture of iodine* is made by dissolving 2%–7% iodine in alcohol. Tincture solutions are characterised by the presence of alcohol since alcohol is a good solvent. Iodine does not dissolve in water. Tincture of iodine is used as an antiseptic.

17. $2HNO_3 + 3H_2S \rightarrow 3S + 4H_2O + 2NO$
${}_{(Sol)}$

19. The size of particles in the colloidal state ranges from 10–1000 Å.

Size of particles .... 10 Å .... 1000Å ....
True solution — particles in colloidal state — Suspension

20. Crystalloids can easily be obtained in the crystalline form and their solution can diffuse rapidly through a vegetable or animal membrane (such as a parchment membrane); for example, sugar, urea, salts, acids.

22. Fog: liquid in gas phase.

24. Gas in gas makes a homogeneous mixture.

25. Those substances which cannot pass through a membrane are termed as colloids. For example, silicic acid. Sugar solution, urea and NaCl can pass through a membrane.

26. Smoke consists of carbon particles (solid) dispersed in air (gas).

27. A cloud consists of fine droplets of water suspended in air.

28. Scattering of light from the surface of colloidal particles – Tyndall effect – is observed. When a colloidal solution is observed under an ultramicroscope.

29. Starch molecules have colloidal dimensions whereas NaCl, glucose and $Ba(NO_3)_2$ are crystalloids and are soluble in water.

30. Sol is a colloidal solution, so, it will show the Tyndall effect.

31. When the dispersed phase and the dispersion medium are both liquids, the colloidal system is called an emulsion. For example, milk, vanishing cream.

32. Butter is obtained from milk which itself is an emulsion of fat dispersed in water, the reverse is true for butter (water dispersed in fat).

33. Sea water is a solution of salt and water. During evaporation, water gets evaporated and the salt is left behind as a residue. So, evaporation can be used here.

35. Camphor sublimes, therefore it can be purified by sublimation.

36. The mixture of oil and water can be separated by using a separating funnel. As oil is less dense than water, it forms the upper layer.

38. Filtration is based on size difference.

39. Due to a difference in boiling point, different liquids have different tendencies to form vapour.

40. Soluble solid is left behind when the solvent evaporates.

41. Because both differ in their boiling points.

42. Sulfur gets dissolved in CS_2 and separates out as a filtrate. Sand gets collected on the filter paper as residue.

43. Sulfur gets dissolved in CS_2 and separates out as a filtrate. The sand gets collected on the filter paper as residue.

44. Distillation is a separation technique used for separation of miscible liquids having different boiling points.

45. Magnetism is useful for separation of magnetic and non-magnetic substances.

46. Physical properties of mixtures are the same as its individual components.

47. Copper sulfate (blue) reacts with iron (in shaving blade) to form iron sulfate (green).

49. Melting of iron metal, bending of an iron rod and drawing of an iron wire are physical changes since no new substances are formed during these changes. Only rusting of iron is a chemical change since a new substance, rust, is formed.

50. Decaying of wood and burning of wood are chemical changes since there is a change in the chemical composition of wood. Sawing of wood and hammering of a nail into a piece of wood are physical changes since there is no change in the composition of the wood during these changes.

Miscellaneous

1. Non-metals do not show conductance; however, graphite shows conductance.

2. S, Cl, Ar, Br are all non-metals.

3. Rusting of an article made up of iron is called corrosion. It is a chemical change because a new substance, hydrated iron oxide called rust, is formed.

$$Iron + Oxygen \xrightarrow{\ Water\ } Hydrated\ iron\ oxide\ (rust)$$

$$4Fe + 3O_2 \rightarrow 2Fe_2O_3.xH_2O$$
$${}_{(Rust)}$$

5. Colloidal solutions can be purified by dialysis. Dialysis may be defined as the process of separating a crystalloid from a colloid by diffusion or filtration through a fire membrane. The process of dialysis can be quickened by using hot water (hot dialysis) or by applying an electric field (electrolysis).

6. A homogeneous mixture must have uniform composition and properties.

9. Sulfur is soluble in carbon disulfide hence, a solution is formed when sulfur is mixed with carbon disulfide. The solution is homogeneous and does not show the Tyndall effect.

10. Whipped cream is an example of a gas in liquid type of colloid.

11. Fog is a colloidal system consisting water droplets dispersed in air.

12. Sulfur will be left behind. In the given mixture, only sulfur gets dissolved in carbon disulfide while rest do not dissolve.

13. A mixture of ethanol and water can be separated by fractional distillation as they have different boiling points.

14. A mixture of methyl alcohol and acetone can be separated by fractional distillation as there is only a small difference between the boiling points of methyl alcohol and acetone.

15. As different pure solid substances melt at different temperatures, the melting point method can be used to separate them.

18. A mixture of two solids can be separated by using sublimation and solvent extraction.

19. Moles of glucose $(n_2) = \dfrac{\text{Weight}}{\text{Molecular weight}} = \dfrac{36}{180} = 0.2$

Moles of water $(n_1) = \dfrac{\text{Weight}}{\text{Molecular weight}} = \dfrac{90}{18} = 5$

Total moles $= 0.2 + 5 = 5.2$

Mole fraction of glucose = Moles of glucose/Total moles
$$= 0.2/5.2 = 0.384$$
$$= 0.0384 \times 100 = 3.84\%$$

20. $C = \dfrac{w \text{ (g)}}{V \text{ (mL)}} \times 1000 = \dfrac{0.4}{250} \times 1000 = 1.6 \text{ g/L}$

Advanced and Olympiads

Single Choice

1. Pure substances are made up of only one kind of particle and they have the same composition throughout. Mixtures are not pure substances. Only elements and compounds are pure substances.

2. Ice and air are homogeneous in nature since they have the same composition throughout and there are no visible boundaries between the components.

3. $X + Y \rightarrow P$

X and Y are elements and hence, cannot be broken down into simpler substances. P is a compound, hence it has a fixed composition.

4. $2A + B \rightarrow A_2B$

The product A_2B is a newly formed compound. Hence, it does not show the properties of A and B. The product formed is a compound and not an element.

5. By definition, 50% mass by volume percent solution means 50 grams of a solute dissolved in 100 mL of solution. Student X dissolved 50 g of NaOH in 100 mL of water. So, the solution is dilute and it is not the desired solution. Student Y dissolved 50 g of NaOH in 150 g of solution so, it is not the desired solution. Z has made the desired solution by dissolving 50 g NaOH in water to make the volume of the solution 100 mL.

Mass by volume %
$$= \dfrac{\text{Mass of solute}}{\text{volume of solution}} \times 100 = \dfrac{50}{100} \times 100$$
$$= 50\% \text{ mass by volume}$$

Therefore, student Z made the desired solution.

6.

Colloid	Dispersed phase	Dispersion medium
Cheese (C)	Liquid	Solid
Milk (M)	Liquid	Liquid
Smoke (S)	Solid	Gas

7. The correct sequence of steps is as follows:
(iii) Paper chromatography is used for the separation.
(ii) Paper and solvent are taken as the stationary and mobile phases, respectively.
(v) A narrow strip of paper with a line drawn is cut. A mixture of red and blue ink with the help of a capillary is placed on the line marked on the paper.
(iv) The paper is suspended in the closed jar with the help of hook.
(i) The blue and red ink form spots at a certain distance on the paper.

Multiple Choice

9. As colloids are heterogeneous in nature.

10. Alloys are homogeneous mixtures of metals. Brass is an alloy of copper and zinc whereas bronze is an alloy of copper and tin.

11. Diamond is an allotrope of the element carbon while all the rest are compounds.

12. Ammonia is a compound whereas solution, alloy and amalgam are all mixtures.

13. Molality of a solution is the number of moles of a solute dissolved in 1000 g or 1 kg of the solvent. It is independent of the volume term. Therefore, it is not affected by the rise or fall of temperature.

Integer Type

25. Br, Hg, Ga, Fr, Cs are liquid elements.

26. As, Se, Sb, B, Si and Ge are metalloids.

27. The given mixtures are iodised table salt, brass, aerated drink, milk, 22 carat gold, diesel, smoke.

28. The given compounds are cloud, steam, sucrose, dry ice, marsh gas.

29. Naphthalene, iodine, camphor, ammonium chloride, anthracene undergo sublimation.

30. Chlorine gas, iron, aluminium, iodine, carbon and sulfur powder are not compounds.

31. Filtration, sedimentation, evaporation, distillation, centrifugation and chromatography can be used for the separation of a mixture of a solid and a liquid.

Atoms and Molecules

Introduction

Ancient Indian and Greek philosophers have always wondered about the unknown and unseen form of matter. The idea of divisibility of matter was considered long back in India, around 500 BC. An Indian philosopher Maharishi Kanad was one of the first persons to propose that matter (*padarth* in Hindi) is made up of very small particles called *parmanu* (John Dalton called these particles atoms). The word 'atom' means indivisible— something that cannot be divided further. Another Indian philosopher, Pakudha Katyayama, elaborated that the particles of matter (atoms) generally exist in a combined form and the combined forms of atoms are known as *molecules*. All matter is made up of small particles known as atoms and molecules. Different kinds of atoms and molecules have different properties due to which different kinds of matter also show different properties. By the end of the eighteenth century, scientists recognised the difference between elements and compounds and naturally became interested in finding out how and why elements combine and what happens when they combine together.

Laws of Chemical Combination

Antoine L Lavoisier laid the foundation for the chemical sciences by establishing two important laws of chemical combination. These two laws of chemical combination played a significant role in the development of Dalton's atomic theory of matter.

Law of Conservation of Mass

Law of conservation of mass was introduced by Lavoisier (father of chemistry) in 1774. According to this law of conservation of mass, matter can neither be created nor destroyed in a chemical reaction. The substances which combine together in a chemical reaction are called *reactants* whereas the new substances formed as a result of a chemical reaction are called *products*. The law of conservation of mass means that in a chemical reaction, the total mass of products is equal to the total mass of reactants and that is there is no change in mass during a chemical reaction.

For example,

(i) Lavoisier observed that on heating mercuric oxide, it produced free mercury and oxygen. The sum of the masses of mercury and oxygen was found to be equal to the mass of mercuric oxide.

Mercuric oxide $\rightarrow$ Mercury + Oxygen

100 g   92.6 g   7.4 g

Mass of the reactant = 100 g

Mass of the products = 92.6 + 7.4 g = 100.0 g = 100 g

Here, the total mass of the products formed is equal to the total mass of the reactants undergoing reaction, hence this data is in agreement with the law of conservation of mass.

(ii) A piece of ice (solid water) is taken in a small conical flask. It is well corked and weighed. The flask is now heated gently to melt the ice (solid) into water (liquid).

$$\text{Ice} \xrightarrow{\Delta} \text{Water}$$

The flask is again weighed. It is found that there is no change in the weight during this physical change also.

Law of Constant Proportions

Lavoisier, along with Joseph Proust and other scientists, observed that many compounds were composed of two or more elements and each such compound had the same elements in the same proportions, irrespective of where the compound came from and how it is prepared.

In a compound like water, the ratio of the mass of hydrogen to the mass of oxygen is always 1:8, whatever may be source (river, well) of water. If 18 g of water is decomposed, 2 g of hydrogen and 16 g of oxygen are always formed. Similarly in the case of ammonia, nitrogen and hydrogen are always present in the ratio 14:3 by mass, whatever be the method or the source from which it is obtained.

This led to the law of constant proportions which is also known as the *law of definite proportions*. According to this law, in a pure chemical substance or compound the elements are always present in definite proportions by mass.

For example, it is important to note that whatever sample of carbon dioxide is taken, it is observed that the ratio of carbon and oxygen is always same. That is, $12 : 32$ or $3 : 8$ in the methods given here.

(i) By burning carbon in oxygen: $C(s) + O_2(g) \rightarrow CO_2(g)\uparrow$

(ii) By heating sodium bicarbonate:

$$2NaHCO_3(s) \xrightarrow{\Delta} Na_2CO_3(s) + CO_2(g)\uparrow + H_2O(l)$$

(iii) By heating calcium carbonate:

$$CaCO_3(s) \xrightarrow{\Delta} CaO(s) + CO_2(g)\uparrow$$

(iv) By reacting calcium carbonate with hydrochloric acid:

$$CaCO_3(s) + 2HCl(aq) \rightarrow CaCl_2(aq) + CO_2(g)\uparrow + H_2O(l)$$

Law of Multiple Proportions

It was proposed by Dalton in 1804 and verified by Berzilius. According to this, different weights of an element that combine with a fixed weight of another element bear a simple numerical ratio. For example,

(i) CO, CO_2. Here, the weight of oxygen which combines with 12 g of carbon is in a $1 : 2$ ratio.

(ii) When nitrogen and oxygen combine, they form 5 stable oxides, N_2O, NO, N_2O_3, N_2O_4 and N_2O_5. In these oxides, the amounts of oxygen that react with 28 g of N_2 are in the ratio $16 : 32 : 48 : 64 : 80$, that is, $1 : 2 : 3 : 4 : 5$.

It does not hold good when the same compound is prepared from different isotopes of an element. For example, H_2O, D_2O.

Law of Reciprocal Proportions or Law of Equivalent Proportions

It was proposed by Richter in 1792 and verified by Star. According to this, when two different elements undergo combination with the same weight of a third element, the ratio in which they combine will either be the same or some simple multiple of the ratio in which they combine with each other. For example, CO_2, SO_2, CS_2.

This is also known as the *law of equivalent proportions* which states that, elements always combine in terms of their equivalent weight.

Law of Combining Volumes

It was proposed by Gay Lussac and it is applicable for gases. According to this, when gases react with each other they bear a simple whole number ratio with one another as well as the product under the conditions of same temperature and pressure.

For example, the volumes of hydrogen and chlorine which combine to form HCl is in the ratio $1 : 1 : 2$.

$$H_2(g) \quad + \quad Cl_2(g) \quad \rightarrow \quad 2HCl(g)$$

$$\text{1 unit vol.} \qquad \text{1 unit vol.} \qquad \text{2 unit vol.}$$

Ratio = $1 : 1 : 2$

The volumes of hydrogen and oxygen which combine to form water is in the ratio $2 : 1 : 2$.

$$2H_2(g) \quad + \quad O_2(g) \quad \rightarrow \quad 2H_2O(g)$$

$$\text{2 vol.} \qquad\qquad \text{vol.} \qquad\qquad \text{2 mol}$$

Avogadro's Law

It explains the law of combining volumes. According to this, under similar conditions of temperature and pressure equal volume of gases contain equal number of molecules. It is used in:

- deriving the molecular formula of a gas,
- determining the atomicity of a gas,
- deriving a relation, Molecular mass $(M) = 2 \times$ Vapour density (VD),
- Vapour density (VD) = Mass of same volume of gas at STP/ Mass of same volume of H_2 at STP,
- deriving the gram molecular volume, and
- making a clear distinction between atoms and molecules.

Avogadro number $(N_0$ or $N_a) = 6.023 \times 10^{23}$. Avogadro number of gas molecules occupies 22.4 litre or 22400 mL or cm^3 volume at STP. The number of molecules in 1 cm^3 of a gas at STP is equal to the Loschmidt number, that is, 2.68×10^{19}. The reciprocal of the Avogadro number is known as Avogram.

Dalton's Atomic Theory

In order to give appropriate explanations of these laws, British chemist John Dalton provided a basic theory about the nature of matter. Dalton picked up the idea of divisibility of matter and said that the smallest particles of matter are atoms. Dalton's atomic theory provided an explanation for the law of conservation of mass and the law of definite proportions.

Features of Dalton's Theory

The following are the features of of Dalton's theory.

(i) All matter is made of very tiny particles known as atoms.

(ii) Atoms are indivisible particles, which can neither be created nor destroyed in a chemical reaction.

(iii) Atoms are of various kinds. There are as many kinds of atoms as there are elements.

(iv) Atoms of a given element are identical in mass, size and chemical properties.

(v) Atoms of different elements have different mass, size and chemical properties.

(vi) Atoms combine in the ratio of small whole numbers to form compounds. The 'number' and 'kind' of atoms in a given compound is fixed. For example, in H_2O, the ratio of H : O is $2 : 1$.

(vii) Atoms of the same elements can combine in more than one ratio to form more than one compound. For example, SO_2 and SO_3.

Drawbacks of Dalton's Atomic Theory

Some of the drawbacks are given below.

(i) One major drawback is that atoms were thought to be indivisible. But, now we know that under special circumstances, atoms can be further divided into still smaller particles (electrons, protons and neutrons). So, atoms are themselves made up of these particles—electrons, protons and neutrons.

(ii) This theory says that all the atoms of an element have exactly the same mass but now it is known that atoms of the same element can have different masses also. For example, in the case of isotopes $_1^1H$, $_1^2H$, $_1^3H$.

(iii) This theory says that atoms of different elements have different masses but now it is known that even atoms of different elements can have the same mass. For example, in the case of isobars $_6^{14}C$, $_7^{14}N$.

(iv) It failed to explain why atoms of different elements differ from one other. That is, it cannot tell anything about internal structure of the atom.

(v) It could not explain how and why atoms of different elements combine with each other to form compound atoms or molecules. For example, NH_3, H_2S and so on.

(vi) It could not explain the nature of forces that hold different atoms together in a molecule.

Illustrations

1. Explain why/why not the law of constant compositions is true for all types of compounds.

Solution:

No, the law of constant composition is not true or valid for all types of compounds. It is valid only when the compounds are obtained from a single isotope. For example, carbon exists in two common isotopes, ^{12}C and ^{14}C. When it forms CO_2 from ^{12}C, the ratio of masses is $12 : 32 = 3 : 8$ but from ^{14}C, the ratio will be $14 : 32 = 7 : 16$ which is not the same as in the first case.

2. In a reaction, 5.3 g of sodium carbonate reacted with 6 g of ethanoic acid. The products were 2.2 g of carbon dioxide, 0.9 g water and 8.2 g of sodium ethanoate. Show that these observations are in agreement with the law of conservation of mass.

Sodium carbonate + ethanoic acid → sodium ethanoate + carbon + dioxide + water

Solution:

According to the law of conservation of mass, mass can neither be created nor destroyed during a chemical reaction. It means that the mass remains the same. So, we add the masses of the reactants on the LHS and add the masses of all the products on the RHS.

LHS = 5.3 g + 6 g = 11.3 g

RHS = 8.2 g + 2.2 g + 0.9 g = 11.3 g

LHS = RHS

Hence these observations are in agreement with the law of conservation of mass.

3. 1 litre of gas at STP weighs 2.053 g. Find its vapour density.

Solution:

22.4 L of the gas at STP will weigh = $2.053 \times 22.4 = 46$ g
Molecular mass = 46. So vapour density = $46/2 = 23$.

4. Element A and B form two different compounds. In the first compound, 0.324 g A is combined with 0.471 g B. In the second compound, 0.117 g A is combined with 0.509 g B. Show that these data illustrate the law of multiple proportions.

Solution:

In the first compound, 0.324 g of A combines with 0.471 g of B.

In the second compound, 0.117 g of A combines with 0.509 g of B.

Therefore, 0.324 g of A combines with the weight of

$$B = \frac{0.509 \times 0.324}{0.117} = 1.4095 \text{ g}$$

Now, the weight of B that combines with the same weight of A that is 0.324 g, is in the ratio of $0.471 : 1.4095$ or $1 : 3$. The ratio, being simple, illustrates the law of multiple proportions.

Atoms

The building blocks of all matter are *atoms*. An atom is the smallest particle of an element which may or may not exist independently but that can take part in a chemical reaction. Atoms of most elements are very reactive and do not exist in the free state. They exist in combination with the atoms of the same element or another element, as molecules. It means, the basic unit of matter in any state is considered as the *molecule* because a molecule is the smallest identity that can exist independently.

Atoms are extremely small, they are smaller than anything that we can imagine or compare with. Millions of atoms when stacked would make a layer barely as thick as this sheet of paper. The size of an atom is indicated by its radius which is called *atomic radius*. The atomic radius is measured in *nanometres* (nm).

$$1/10^9 \text{ m} = 1 \text{ nm}$$

$$1 \text{ m} = 10^9 \text{ nm}$$

KNOWLEDGE BOOSTER

We might think atoms are so insignificant in size that there is no need to bother about them. However, our entire world is made up of atoms. We may not be able to see them, but they are there and are constantly affecting whatever we do. They cannot be viewed using simple optical microscopes. However, through modern techniques such as scanning tunneling microscopy it is possible to produce magnified images of surfaces of elements showing atoms. we can now produce magnified images of surfaces of elements showing atoms.

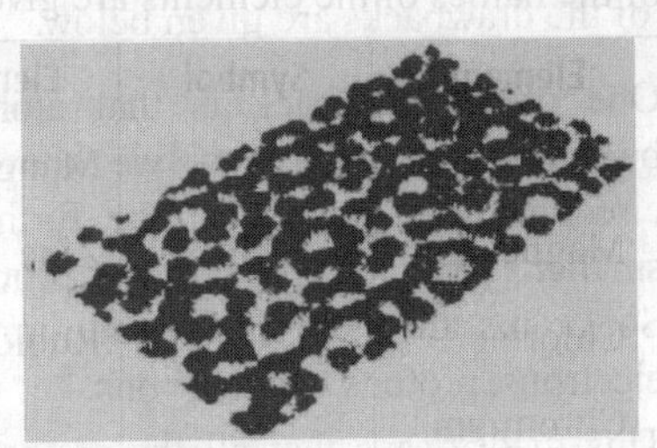

Fig. 3.1 An image of the surface of silicon

Relative Sizes

Radii (in m)	10^{-10}	10^{-9}	10^{-8}	10^{-4}	10^{-2}	10^{-1}
Example	Hydrogen atom	Water molecule	Hemoglobin molecule	Grain of sand	Ant	Watermelon

Symbols of Elements

Dalton was the first scientist to use symbols for elements. When he used a symbol for an element, he also meant a definite quantity of that element, that is, one atom of that element. Berzilius suggested that the symbols of elements can be made from one or two letters of the name of the element.

In the beginning, the names of elements were derived mainly from the names of the places where they were found for the first time. For example, the name copper was taken from Cyprus. Some names were also taken from some specific colours. For example, gold was taken from the Latin word meaning yellow.

Dalton's Symbols of Elements

Dalton was the first scientist to use symbols to represent elements. However, now it is just history.

Element	Dalton's symbol	Element	Dalton's symbol
Hydrogen	⊙	Iron	Ⓘ
Carbon	●	Copper	Ⓒ
Oxygen	○	Silver	Ⓢ
Phosphorus	⊘	Gold	◉
Sulfur	⊕	Lead	Ⓛ
Platinum	Ⓟ	Mercury	◉

Symbol

In order to represent the elements, instead of using the entire name, scientists use abbreviated names; these are known as symbols. Hence, symbol may be defined as the abbreviation used for the naming of an element.

Modern Method or IUPAC System of Symbols

IUPAC (International Union of Pure and Applied Chemistry) approves names of elements. The symbols of elements are generally either the first letter or the first two letters or the first and the third letters of the name of an element. The first letter of a symbol is always written as a capital letter (upper case) and the second letter as a small letter (lower case). For example,

  (i) hydrogen, H,

  (ii) aluminium, Al and not AL,

  (iii) cobalt, Co and not CO.

In the modern method, the symbols of elements can be classified as follows.

  (i) Some symbols derived from the first letter of the names of the elements are given below.

Element	Symbol	Element	Symbol
Hydrogen	H	Fluorine	F
Carbon	C	Phosphorus	P
Nitrogen	N	Sulfur	S
Oxygen	O	Iodine	I

  (ii) Some symbols derived from the first two letters of the names of the elements are given below.

Element	Symbol	Element	Symbol
Aluminium	Al	Silicon	Si
Barium	Ba	Argon	Ar
Lithium	Li	Calcium	Ca
Beryllium	Be	Nickel	Ni
Neon	Ne	Cobalt	Co

(iii) Some symbols derived from the first and the third letters of the names of the elements are given below.

Element	Symbol	Element	Symbol
Arsenic	As	Manganese	Mn
Magnesium	Mg	Zinc	Zn
Chlorine	Cl	Rubidium	Rb
Chromium	Cr		

(iv) Some symbols derived from the Latin or Greek names of the elements are given below.

Element	Latin name	Symbol
Copper	Cuprum	Cu
Potassium	Kalium	K
Iron	Ferrum	Fe
Gold	Aurum	Au
Sodium	Natrium	Na
Silver	Argentum	Ag
Mercury	Hydragyrum	Hg
Tin	Stannum	Sn
Lead	Plumbum	Pb
Antimony	Stibium	Sb

Significance of the Symbol of an Element

A symbol represents:

- Name of the element.
- One atom of the element.
- A definite mass of the element (equal to atomic mass expressed in grams).
- One mole of atoms of the element. That is, the symbol also represents 6.022×10^{23} atoms of the elements.

Atomic Mass

One of the most remarkable concepts of Dalton's atomic theory was the introduction of the atomic mass. According to this theory, each element had a characteristic atomic mass. This theory could explain the law of constant proportions so well that scientists were prompted to measure the atomic mass of an atom. As determining the mass of an individual atom was relatively a tough task, relative atomic masses were determined by using the laws of chemical combinations and the compounds formed. For example, consider a compound carbon monoxide (CO) formed from carbon and oxygen. It was observed experimentally that 12 g of carbon combines with 16 g of oxygen to form CO. That is, carbon combines with 4/3 times its mass of oxygen. Let us define the atomic mass unit [earlier abbreviated as *amu*, but now written as *u* (unified mass)] as equal to the mass of one carbon atom, then we would assign carbon an atomic mass of 1.0 u and oxygen an atomic mass of 1.33 u. However, it is more convenient to have these numbers as whole numbers or as near to a whole number as possible. While searching for various atomic mass units, scientists initially took 1/16 of the mass of an atom of naturally occurring oxygen as the unit. This was considered relevant due to two reasons.

- Oxygen reacted with a large number of elements and formed compounds.
- This atomic mass unit gave masses of most of the elements as whole numbers.

In 1961, the International Union of Chemists universally accepted the carbon-12 isotope as the standard reference for measuring atomic masses in terms of atomic mass units. One atomic mass unit is a mass unit equal to exactly one-twelfth $(1/12^{th})$ the mass of one atom of carbon-12. The relative atomic masses of all elements have been found with respect to an atom of carbon-12 (Table 3.1). That is atomic mass of an element may be defined as the average relative mass of an atom of the element as compared with mass of an atom of carbon (C-12 isotope) taken as 12 amu.

$$\text{Atomic mass} = \frac{\text{Mass of 1 atom of the element}}{\dfrac{1}{12} \text{ of the mass of an atom of carbon-12}}$$

For example, the atomic mass of magnesium is 24 u which indicates that one atom of magnesium is 24 times heavier than $\dfrac{1}{12}$ of a carbon 12 atom.

Table 3.1 Atomic mass of some elements

S. No.	Element	Symbol	Atomic mass
1	Hydrogen	H	1 u
2	Carbon	C	12 u
3	Nitrogen	N	14 u
4	Oxygen	O	16 u
5	Sodium	Na	23 u
6	Magnesium	Mg	24 u
7	Aluminium	Al	27 u
8	Phosphorus	P	31 u
9	Sulfur	S	32 u
10	Chlorine	Cl	35.5 u
11	Potassium	K	39 u
12	Calcium	Ca	40 u
13	Iron	Fe	56 u
14	Copper	Cu	63.5 u

Gram Atomic Mass

Gram atomic mass of an element is defined as that much quantity of the element whose mass expressed in grams is numerically equal to its atomic mass. To find the gram atomic mass we keep the numerical value the same as the atomic mass, but simply change the units from u to g. For example, the atomic mass of aluminium is 27 u. Its gram atomic mass is 27 g.

Gram Atomic Mass of Isotopes

Gram atomic mass of isotopes is expressed as average gram atomic mass using the relation given below.

$$(\text{Gram atomic mass})_{ave} = \frac{M_1 X_1 + M_2 X_2}{X_1 + X_2}$$

M_1 and M_2 are the relative masses of isotopes, and X_1 and X_2 are the relative % content.

Due to the presence of isotopes, the average atomic mass can be fractional. For example, in the case of chlorine, it is 35.5 g. Chlorine contains two types of atoms having relative masses 35 and 37 and their relative abundance is 3 : 1. In such cases, the atomic mass of the element is the average of the relative masses of the different isotopes of the element.

$$\text{Atomic mass of chlorine} = \frac{35 \times 3 + 37 \times 1}{4} = 35.5$$

Illustrations

1. Which of the following symbols of elements are correct?

   (i) Cobalt: CO (ii) Carbon: c   (iii) Aluminium: AL

  (iv) Helium: He (v) Sodium: So (vi) Iron: Fe

(vii) Gold: Gd

Solution: Here only helium and iron have been given correct symbols. The correct symbols of these elements are as follows:

   (i) Cobalt: Co (ii) Carbon: C   (iii) Aluminium: Al

  (iv) Helium: He (v) Sodium: Na (vi) Iron: Fe

(vii) Gold: Au

2. The mass of one steel screw is 4.11 g. Find the mass of one mole of these steel screws. Compare this value with the mass of the earth (5.98×10^{24} kg). Which one of the two is heavier and by how many times?

Solution:

Mass of 1 screw = 4.11 g

Mass of 1 mole of screws = $4.11 \times 6.022 \times 10^{23}$

$\qquad\qquad = 2.475 \times 10^{24}$ g $= 2.475 \times 10^{21}$ kg

$$\frac{\text{Mass of earth}}{\text{Mass of 1 mole screws}} = \frac{5.98 \times 10^{24} \text{ kg}}{2.475 \times 10^{21} \text{ kg}} = 2416$$

It shows that mass of the earth is 2416 times more than the mass of one mole of screws.

Molecules

In general, a molecule is a group of two or more atoms that are chemically bonded together or tightly held together by attractive forces. A molecule is defined as the smallest particle of an element or a compound that is capable of independent existence and shows all the properties of that substance. It means, a molecule is the smallest identity that can exist individually. For example, hydrogen exists as H_2 (dihydrogen) and not as H. Atoms of the same element or of different elements can join together to form molecules. For example, H_2, N_2 and NH_3.

KNOWLEDGE BOOSTER

How do atoms exist? Atoms of most of the elements are not able to exist independently. **Atoms usually exist in two ways**—molecules and ions. Atoms form molecules and ions. These molecules or ions aggregate in large numbers to form the matter that we can see, feel or touch.

Types of Molecules

There are two types of molecules.

(i) Molecules of elements: The molecules of an element are constituted by the same type of atoms (homoatomic). These may be as follows:

  (a) Monoatomic: Molecules of many elements like argon (Ar) and helium (He) are made up of only one atom of that element.

  (b) Polyatomic: Molecules of most of the non-metals are made of two or more atoms.

    • **Diatomic:** Such a molecule is formed by two atoms; for example, O_2, N_2, H_2, Cl_2.

    • **Triatomic:** Such a molecule is formed by three atoms; for example, O_3 (ozone).

    • **Tetraatomic:** Such a molecule is formed by four atoms; for example, P_4.

    • **Octaatomic:** Such a molecule is formed by eight atoms; for example, S_8.

(ii) Molecules of compounds: Atoms of different elements (heteroatomic) join together in definite proportions to form molecules of compounds. For example, hydrogen chloride is a compound. A molecule of hydrogen chloride (HCl) contains two different types of atoms: hydrogen (H) and chlorine (Cl) atoms (Table 3.2).

Table 3.2 Molecules of some compounds

Compound	Combining elements	Formula	Ratio by mass
Water	Hydrogen and oxygen	H_2O	1 : 8
Ammonia	Nitrogen and hydrogen	NH_3	14 : 3
Carbon dioxide	Carbon and oxygen	CO_2	3 : 8

Atomicity

The number of atoms constituting a molecule is known as its atomicity (Table 3.3).

Table 3.3 Atomicity of some common elements

Type of element	Name	Symbol	Atomicity
Non-metal	Helium	He	Monoatomic
	Argon	Ar	Monoatomic
	Neon	Ne	Monoatomic
	Hydrogen	H_2	Diatomic
	Chlorine	Cl_2	Diatomic
	Nitrogen	N_2	Diatomic
	Oxygen	O_2	Diatomic
	Phosphorus	P_4	Tetratomic
	Sulfur	S_8	Polyatomic
Metals	Sodium	Na	Monoatomic
	Iron	Fe	Monoatomic
	Aluminium	Al	Monoatomic
	Copper	Cu	Monoatomic

Ions

Compounds composed of metals and non-metals can contain charged species which are known as *ions*. An ion is a charged particle which can be negatively or positively charged. An ion is a positively or negatively charged atom (or group of atoms). It is formed by the loss or gain of electrons by an atom, so it contains an unequal number of electrons and protons.

Type of Ions

There are two types of ions—cations and anions.

Cation

A positively charged ion is known as a cation. It is formed by the loss of one or more electrons by an atom. The ions of all the metal elements are cations. For example, a sodium atom loses one electron to form a sodium ion, Na^+ (cation).

$$Na \xrightarrow{-1\ electron} Na^-$$

Sodium atom	Sodium ion
Protons = 11 (+ charge)	Protons = 11 (+ charge)
Electrons = 11 (− charge)	Electrons = 10 (− charge)
Overall charge = 0	Overall charge = 1+

Anion

A negatively charged ion is known as an anion. An anion is formed by the gain of one or more electrons by an atom. The ions of all the non-metal elements are anions. For example, a chlorine atom gains one electron to form a chloride ion, Cl^- (anion).

A normal or a neutral atom contains an equal number of protons and electrons but an anion contains more electrons than protons since it is formed by the addition of one or more electrons to an atom.

$$Cl \xrightarrow{+1\ electron} Cl^-$$

Chlorine atom	Chlorine ion
Protons = 17 (+ charge)	Protons = 17 (+ charge)
Electrons = 17 (− charge)	Electrons = 18 (− charge)
Overall charge = 0	Overall charge = 1−

Simple Ions

Such ions are formed from single atoms and are called *simple ions*. For example, the sodium ion Na^+, is a simple ion because it is formed from a single sodium atom, Na.

Compound Ions

Such ions are formed from groups of joined atoms and are called *compound ions*. For example, ammonium ion NH_4^+, is a compound ion which is made up of two types of atoms joined together—nitrogen and hydrogen atoms.

Ionic Compounds

The compounds which are made up of ions are called ionic compounds. In an ionic compound, the positively charged ions (cations) and negatively charged ions (anions) are held together by strong electrostatic forces of attraction. The forces that hold the ions together in an ionic compound are known as *ionic bonds* or *electrovalent bonds*. As an ionic compound consists of an equal number of positive ions and negative ions, the overall charge on an ionic compound is zero.

For example, sodium chloride (NaCl) is an ionic compound which is made up of an equal number of positively charged sodium ions (Na^+) and negatively charged chloride ions (Cl^-) (Table 3.4).

Table 3.4 Some ionic compounds

S. No.	Name	Formula	Ions present	S. No.	Name	Formula	Ions present
1	Sodium chloride	$NaCl$	Na^+ and Cl^-	5	Magnesium oxide	MgO	Mg^{2+} and O^{2-}
2	Potassium chloride	KCl	K^+ and Cl^-	6	Aluminium oxide	Al_2O_3	Al^{3+} and O^{2-}
3	Ammonium chloride	NH_4Cl	NH_4^+ and Cl^-	7	Sodium hydroxide	$NaOH$	Na^+ and OH^-
4	Magnesium chloride	$MgCl_2$	Mg^{2+} and Cl^-	8	Copper sulfate	$CuSO_4$	Cu^{2+} and SO_4^{2-}

Illustrations

1. Write the cations and anions present (if any) in the following compounds:

   (a) CH_3COONa (b) $NaCl$

   (c) H_3 (d) NH_4NO_3

   Solution:

   (a) CH_3COONa: CH_3COO^-, Na^+

   (b) $NaCl$: Na^+, Cl^-

   (c) H_2: It is a covalent compound; hence, no ions are present in it.

   (d) NH_4NO_3: NH_4^+, NO_3^-

2. Classify each of the following on the basis of their atomicity:

   (a) F_2 (b) NO_2 (c) N_2O (d) C_2H_6

   (e) P_4 (f) H_2O_2 (g) P_4O_{10} (h) O_3

   (i) HCl (j) CH_4 (k) He (l) Ag

   Solution:

   Monoatomic with atomicity (1): He, Ag

   Diatomic with atomicity (2): F_2, HCl

   Polyatomic with atomicity > 2

   Atomicity (3): NO_2, N_2O, O_3

   Atomicity (4): P_4, H_2O_2

   Atomicity (5): CH_4

   Atomicity (8): C_2H_6

   Atomicity (14): P_4O_{10}

Valency

The word *valency* means, the power to combine. That is, the combining capacity of an element is known as its *valency*. It can be measured in terms of the number of H-atoms or double the number of O-atoms that can combine with one atom of an element. As hydrogen has less combining capacity than any other element, its valency is taken as 1 as a standard value. For example, in CH_4 the valency of carbon is 4, in NH_3 the valency of nitrogen is 3 and in water (H_2O) the valency of oxygen is 2.

In the modern view, valency is equal to the number of electrons that an atom can share or lose or gain during a chemical reaction. The number of electrons present in the valence orbit of an atom are called valence electrons. In order to attain the octet state (8 valence electrons), every atom either loses or gains electrons or shares electrons.

Generally, elements having 1, 2, 3 valence electrons are metals and have a tendency to lose electrons and their valency is equal to the number of electrons lost. For Group 1, Group 2 and Group 13, the valency is 1, 2 and 3 respectively.

For example,
$$\underset{2,8,1}{Na} - e^- \rightarrow \underset{2,8}{Na^+} \quad \text{(monovalent)}$$

$$\underset{2,8,2}{Mg} - 2e^- \rightarrow \underset{2,8}{Mg^{2+}} \quad \text{(divalent)}$$

$$\underset{2,8,3}{Al} - 3e^- \rightarrow \underset{2,8}{Al^{3+}} \quad \text{(trivalent)}$$

Elements having 4 valence electrons are non-metals like C and Si, and they undergo sharing of electrons. Their common valency is 4. For example, in CH_4 and CCl_4 carbon has 4 valency. Elements with 5, 6 and 7 electrons in their valence orbit are non-metals and have a tendency to gain electrons to complete their octet.

For example,
$$\underset{(2,5)}{N} + 3e^- \rightarrow \underset{2,8}{N^{3-}} \quad \text{(trivalent)}$$

$$\underset{(2,6)}{O} + 2e^- \rightarrow \underset{2,8}{O^{2-}} \quad \text{(divalent)}$$

$$\underset{(2,8,7)}{Cl} + e^- \rightarrow \underset{2,8,8}{Cl^-} \quad \text{(trivalent)}$$

In general, valency = Number of valence electrons (up to 4). If the valence electrons are more than 4, the valency is given as, valency = valence electrons − 8.

Valency is not negative or positive it is just a number.

Number of electrons in outermost shell	1	2	3	4	5	6	7	8
Valency	1	2	3	4	3	2	1	0

Variable Valency

There are certain elements that exhibit more than one valency in their ions (or compounds). When an atom of an element loses not only its valence electron but also loses electrons from its penultimate orbit, variable valency is observed.

In case an element has two different positive valencies, then we use the suffix 'ous' for the lower valency state and the suffix 'ic' for the higher valency state. For example, iron can exist as Fe^{2+} (ferrous) or Fe^{3+} (ferric) in its compounds (Table 3.5).

Nowadays, in place of 'ous' or 'ic', we can write the name of the compound by writing the valency in roman numbers within brackets. For example, $SnCl_2$ (stannous chloride) can be written as tin(II) chloride and $SnCl_4$ (stannic chloride) can be written as tin(IV) chloride.

Table 3.5 Some basic ions exhibiting variable valency and their compounds

Name of ions	Formula	Formula of compound	Name of compound
Cuprous	Cu^+	Cu_2O	Cuprous oxide or copper(I) oxide
Cupric	Cu^{2+}	CuO	Cupric oxide or copper(II) oxide
Mercurous	Hg_2^{2+}	Hg_2Cl_2	Mercurous chloride or mercury(I) chloride
Mercuric	Hg^{2+}	$HgCl_2$	Mercuric chloride or mercury(II) chloride
Ferrous	Fe^{2+}	$FeCl_2$	Ferrous chloride or iron(II) chloride
Ferric	Fe^{3+}	$FeCl_3$	Ferric chloride or iron(III) chloride
Plumbous	Pb^{2+}	PbO	Plumbous oxide or lead(II) oxide
plumbic	Pb^{4+}	PbO_2	Plumbic oxide or lead(IV) oxide
Stannous	Sn^{2+}	SnO	Stannous oxide or tin(II) oxide
Stannic	Sn^{4+}	SnO_2	Stannic oxide or tin(IV) oxide
Aurous	Au^+	$AuCl$	Aurous chloride or gold(I) chloride
Auric	Au^{3+}	$AuCl_3$	Auric chloride or gold(III) chloride

Radicals

The molecule of any compound is made up of two parts, each of which is called a radical. A radical is an atom or a group of atoms having the same or different elements, that behaves like a single unit with a positive or negative charge. For example, in NaCl, Na^+ and Cl^- are radicals.

A radicals having only a single atom is called a simple radical while a radical having a group of two or more atoms (polyatomic) is called a compound radical. For example, Na^+ and Ca^{2+} are simple radicals while SO_4^{2-} and PO_4^{3-} are compound radicals.

Acidic and Basic Radicals

An acidic radical may be an atom or a group of atoms having a negative charge. That is, anions or electronegative radicals are acidic radicals. A basic radical may be an atom or group of atoms having a positive charge. That is, cations or electropositive radicals are basic radicals. For example, in ammonium sulfate $(NH_4)_2SO_4$, the acidic radical is SO_4^{2-} with a combining power of 2 and the basic radical is NH_4^+ with a combining power of 1.

Chemical Formulae

A compound is represented in the abbreviated form using a chemical formula. The chemical formula of a compound represents the composition of a molecule of the compound in terms of the symbols of the elements present in it. For example, water is a compound made up of 2 atoms of hydrogen element and 1 atom of oxygen, so the formula of water can be written as H_2O. In the formula H_2O, the subscript 2 indicates 2 atoms of hydrogen. In the formula of water, oxygen O is written without a subscript and it indicates 1 atom of oxygen.

Formulae of Elements

The chemical formula of an element is a statement of the composition of its molecule in which the symbol tells us the element and the subscript tells us how many atoms are present in one molecule.

For example, one molecule of hydrogen element contains two atoms of hydrogen, hence the formula of hydrogen is H_2.

Formulae of Compounds

The chemical formula of a compound is a statement of its composition in which the chemical symbols tell us which elements are present and the subscripts tell us how many atoms of each element are present in one molecule of the compound. For example, water is a compound whose molecule contains 2 atoms of hydrogen and 1 atom of oxygen. So, the formula of water is H_2O (Table 3.6).

Writing the Formula of a Compound

The chemical formulae of different compounds can be written easily but for this we need to learn the symbols and combining capacity (valency) of the elements. Valency can be used to find

Table 3.6 Formulae and valencies of common ions

Electropositive ions (basic radicals)							
Monovalent		**Bivalent**		**Trivalent**		**Tetravalent**	
Name	**Formula**	**Name**	**Formula**	**Name**	**Formula**	**Name**	**Formula**
Potassium	K^+	Calcium	Ca^{2+}	Aluminium	Al^{3+}	Manganese(II)	Mn^{4+}
Sodium	Na^+	Magnesium	Mg^{2+}	Chromium	Cr^{3+}	Tin(IV) or stannic	Sn^{4+}
Copper(I) or cuprous	Cu^+	Manganese(II)	Mn^{2+}	Iron(III) or ferric	Fe^{3+}	Lead(IV)	Pb^{4+}
Silver	Ag^+	Zinc	Zn^{2+}	Gold	Au^{3+}	Platinum	Pt^{4+}
Ammonium	NH_4^+	Iron(II) or ferrous	Fe^{2+}				
Hydrogen	H^+	Nickel	Ni^{2+}				
Mercurous	Hg^+	Cobalt	Co^{2+}				
		Tin(II) or stannous	Sn^{2+}				
		Cadmium	Cd^{2+}				
		Lead(II)	Pb^{2+}				
		Copper(II)	Cu^{2+}				

Electropositive ions (acidic radicals)							
Monovalent		**Bivalent**		**Trivalent**		**Tetravalent**	
Name	**Formula**	**Name**	**Formula**	**Name**	**Formula**	**Name**	**Formula**
Hydroxide	OH^-	Carbonate	CO_3^{2-}	Nitride	N^{3-}	Ferrocyanide	$Fe(CN)_6^{4-}$
Hydride	H^-	Chromate	CrO_4^{2-}	Phosphate	PO_4^{3-}	Carbide	C^{4-}
Fluoride	F^-	Dichromate	$Cr_2O_7^{2-}$	Phosphite	PO_3^{3-}	Silicate	SiO_4^{4-}
Chloride	Cl^-	Manganate	MnO_4^{2-}	Phosphide	P^{3-}		
Bromide	Br^-	Permanganate	MnO_4^-	Borate	BO_3^{3-}		
Iodide	I^-	Tetrathionate	$S_4O_6^{2-}$	Ferricyanide	$Fe(CN)_6^{3-}$		
Bicarbonate	HCO_3^-	Sulfide	S^{2-}	Arsenite	AsO_3^{3-}		
Bisulfite	HSO_3^-	Sulfite	SO_3^{2-}				
Bisulfate	HSO_4^-	Sulfate	SO_4^{2-}				
Chlorate	ClO_3^-	Oxide	O^{2-}				
Hypochlorite	ClO^-	Zincate	ZnO_2^{2-}				
Nitrite	NO_2^-	Thiosulfate	$S_2O_3^{2-}$				
Nitrate	NO_3^-	Oxalate	$C_2O_4^{2-}$				
Permanganate	MnO_4^-	Peroxide	O_2^{2-}				

out how the atoms of an element will combine with the atom(s) of another element to form a chemical compound.

(i) **To write the chemical formula for a binary atomic compound:** The rules that we have to follow while writing a chemical formula are as follows:

- We first write the symbols of all the elements which form the compound.
- Below the symbol of every element, we have to write down its valency. The valencies or charges on the ions must balance.
- Finally, we cross-over the valencies of the combining atoms. That is, with the first atom we write the valency of second atom (as a subscript); and with the second atom we write the valency of first atom (as subscript).

For example,

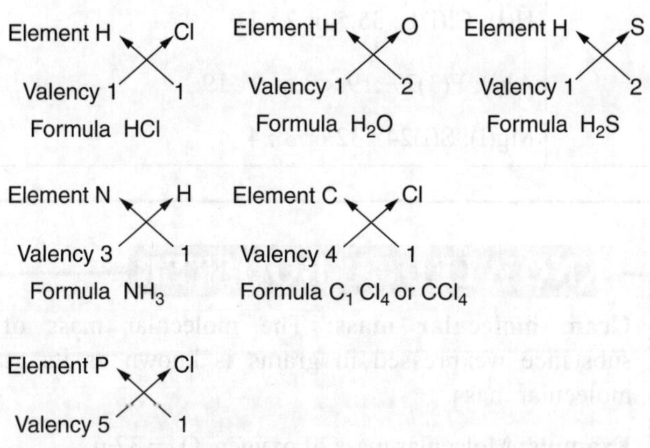

(ii) **To write the formula of a polyatomic compound:** In the case of ionic compounds follow these steps to write the formula.

Step 1: Write the symbols of formulae of the ions of the compound side by side with the positive ion on the left hand side and the negative ion on the right hand side.

Step 2: Enclose the polyatomic ion within brackets.

Step 3: Write the valency of each ion below its symbol.

Step 4: Reduce the valency numerals to a simple ratio by dividing with a common factor, if any.

Step 5: Cross the valencies. Do not write the charges positive or negative of the ions.

For example,

Formula of calcium nitrate.	Formula of calcium phosphate.
Step 1: Writing the formula of the ions: Ca^{2+} NO_3^-	**Step 1:** Writing the formula of the ions: Ca^{2+} PO_4^{3-}
Step 2: Ca^{2+} $(NO_3)^-$	**Step 2:** Ca^{2+} $(PO_4)^{3-}$
Step 3: Ca^{2+} $(NO_3)^-$ 　　　　2　　　1	**Step 3:** Ca^{2+} $(PO_4)^{3-}$ 　　　　2　　　3
Step 4: Not applicable, because the ratio is already simple.	**Step 4:** Not applicable, because the ratio is already simple.
Step 5: Ca^{2+} ✕ $(NO_3)^-$ 　　　　2　　　1	**Step 5:** Ca^{2+} ✕ $(PO_4)^3$ 　　　　2　　　3
Thus, the formula of calcium nitrate is $Ca(NO_3)_2$.	Thus, the formula of calcium phosphate is $Ca_3(PO_4)_2$.

Significance of the Formula of a Substance

The molecular formula of a compound has following significance.

- It represents the name of the substance.
- It represents one molecule of the substance.
- It gives the names of all the elements present in the molecule.
- It gives the number of atoms of each element present in one molecule.
- It represents a definite mass of the substance.
- It also represents one mole of molecules of the substance. That is, the formula also represents 6.022×10^{23} molecules of the substance.

<hr>

Illustrations

1. Write the molecular formulae for the following compounds:

- (a) Copper(II) bromide
- (b) Aluminium(III) nitrate
- (c) Calcium(II) phosphate
- (d) Iron(III) sulfide
- (e) Mercury(II) chloride
- (f) Magnesium(II) acetate

Solution:

- (a) Copper(II) bromide: $CuBr_2$
- (b) Aluminium(III) nitrate: $Al(NO_3)_3$
- (c) Calcium(II) phosphate: $Ca_3(PO_4)_2$
- (d) Iron(III) sulfide: Fe_2S_3
- (e) Mercury(II) chloride: $HgCl_2$
- (f) Magnesium(II) acetate: $Mg(CH_3COO)_2$

2. Write the molecular formulae of all the compounds that can be formed by the combination of the following ions.

$Cu^{2+}, Na^+, Fe^{3+}, Cl^-, SO_4^{2-}, PO_4^{3-}$

Solution: The cations are Cu^{2+}, Na^+, Fe^{3+} and Cl^-, SO_4^{2-}, PO_4^{3-} are the anions. The possible compounds are:

(i) $CuCl_2, CuSO_4, Cu_3(PO_4)_2$

(ii) $NaCl, Na_2SO_4, Na_3PO_4$

(iii) $FeCl_3, Fe_2(SO_4)_3, FePO_4$

3. Give the chemical formulae for the following compounds and compute the ratio by mass of the combining elements in each one of them. You may use Appendix III.

(a) Ammonia   (b) Carbon monoxide

(c) Hydrogen chloride   (d) Aluminium fluoride

(e) Magnesium sulfide

Solution:

	Chemical formula		Ratio by mass
(a)	Ammonia (NH_3)		N(l): H(3) 14:3
(b)	Carbon monoxide (CO)		0(1): 0(1)12 :16 or 3 : 4
(c)	Hydrogen chloride (HCl)		H(l): Cl(l)1 : 35.5 or 2 : 71
(d)	Aluminium fluoride (AlF_3)		Al(l): F(3)27 :19 $\times$ 3 or 9 : 19
(e)	Magnesium sulfide (MgS)		Mg(l): S(l)24 : 32 or 3 : 4

Molecular Mass and Formula Mass

Molecular Mass

As molecules are made up of two or more atoms of the same/different elements, the molecular mass may be calculated as the sum of the atomic masses of all the atoms in a molecule of that substance. The molecular mass of a substance is the sum of the atomic masses of all the atoms in a molecule of the substance. It is therefore the relative mass of a molecule expressed in *atomic mass units (u)*.

Relative Molecular Mass (RMM) expresses how many times a molecule of a substance is heavier than $\dfrac{1}{12}$ th of the mass of an atom of carbon (carbon-12).

Thus, Molecular mass

$$= \frac{\text{Mass of a molecule}}{\dfrac{1}{12}\text{th mass of a carbon atom (carbon} - 12)}$$

$$\text{Molecular mass} = 2 \times \text{vapour density}$$

For example, ammonia has the formula, NH_3. it consists of one atom of N and three atoms of H. The atomic mass of N and H are 14.0 and 1 respectively. Therefore,

$$\text{Molecular mass of } NH_3 = \text{At. mass of N} + 3 \times \text{At. mass of H}$$

$$= 14 + 3 \times 1 = 17\ u$$

KNOWLEDGE BOOSTER

Gram molecular mass: The molecular mass of a substance wexpressed in grams is known as its gram molecular mass.

Example: Molecular mass of oxygen, $O_2 = 32$ u

So, gram molecular mass of oxygen, $O_2 = 32$ grams

For example, sulfuric acid has the formula H_2SO_4. It consists of two H, one S and four O atoms. The atomic masses of H, S and O are 1, 32 and 16 respectively. Therefore,

Molecular mass of H_2SO_4

$$= (2 \times \text{at. m of H}) + (1 \times \text{at. m of S}) + (4 \times \text{at. m of O})$$

$$= (2 \times 1) + (1 \times 32) + (4 \times 16) = 98\ u$$

Formula Unit Mass

The formula unit mass of a substance is the sum of the atomic masses of all atoms in a formula unit of a compound. Formula unit mass is calculated in the same manner as we calculate the molecular mass. The only difference is that we use the word formula unit for those substances whose constituent particles are ions. For example, sodium chloride as discussed above, has a formula unit NaCl. Its formula unit mass can be calculated as $1 \times 23 + 1 \times 35.5 = 58.5$ u.

Illustrations

1. Calculate the relative molecular mass of (a) water (H_2O), (b) HNO_3 and (c) H_3PO_4.

   Solution:

   (a) Atomic mass of hydrogen = 1 u, oxygen = 16 u

   So, the molecular mass of water, which contains two atoms of hydrogen and one atom of oxygen

   $= 2 \times 1 + 1 \times 16 = 18$ u

   (b) The molecular mass of HNO_3 = atomic mass of H + atomic mass of N + 3 × atomic mass of O

   $= 1 + 14 + 48 = 63$ u

   (c) The molecular mass of H_3PO_4 = 3 × atomic mass of H + atomic mass of P + 4 × atomic mass of O

   $= 3 + 31 + 64 = 98$ u

2. Calculate the formula unit mass of (a) $CaCl_2$ (b) $CaCO_3$ and (c) Na_2SO_4.

   Solution:

   (a) Formula unit mass of $CaCl_2$ =

   Atomic mass of Ca + (2 × atomic mass of Cl) = 40 + 2 × 35.5 = 40 + 71 = 111 u

   (b) Formula unit mass of $CaCO_3$ = Atomic mass of Ca + Atomic mass of C + (3 × atomic mass of O) = 40 + 12 + 3 × 16 = 100 u

   (c) Formula unit mass of Na_2SO_4 = 2 × Atomic mass of Na + Atomic mass of S + (4 × atomic mass of O) = 2 × 23 + 32 + 4 × 16 = 142 u

3. Calculate the molar mass of the following substances.

   (a) Ethyne, C_2H_2

   (b) Sulfur molecule, S_8

   (c) Phosphorus molecule, P_4 (atomic mass of phosphorus = 31)

   (d) Hydrochloric acid, HCl

   (e) Nitric acid, HNO_3

   Solution:

   (a) Molar mass of C_2H_2 = (2 × Atomic mass of C) + (2 × Atomic mass of H)

   $= (2 \times 12) + (2 \times 1) = 26$ u

   (b) Molar mass of S_8 = (8 × Atomic mass of S) = 8 × 32 = 256 u

   (c) Molar mass of P_4 = 4 × Atomic mass of P = 4 × 31 = 124 u

   (d) Molar mass of HCl = Atomic mass of hydrogen + Atomic mass of Cl

   $= 1 + 35.5 = 36.5$ u

   (e) Molar mass of HNO_3 = Atomic mass of H + Atomic mass of N + (3 × Atomic mass of O)

   $= 1 + 14 + (3 \times 16) = 15 + 48 = 63$ u

Mole Concept

In order to understand the mole concept, let us consider the reaction of hydrogen and oxygen to form water.

$$2H_2 + O_2 \rightarrow 2H_2O$$

This reaction predicts that

(i) two molecules of hydrogen are combining with one molecule of oxygen to form two molecules of water, or

(ii) 4 u of hydrogen molecules are combining with 32 u of oxygen molecules to form 36 u of water molecules.

It means that the quantity of a substance can be characterised by its mass or the number of molecules. However, a chemical reaction equation indicates directly the number of atoms or molecules taking part in that reaction. Hence, it is more convenient to refer to the quantity of a substance in terms of the number of its molecules or atoms, rather than their masses. That's why a new unit *mole* was introduced around 1896 by Wilhelm Ostwald who derived the term from the Latin word 'moles' meaning a 'heap' or a 'pile'. A substance may be considered as a heap of atoms or molecules. One mole of any species (atoms, molecules, ions or particles) is that quantity in number having a mass equal to its atomic or molecular mass in grams. For example, 12 g carbon and 44 g of CO_2 both represent 1 mole.

The number of particles (atoms, molecules or ions) present in 1 mole of any substance is fixed, with a value of 6.022×10^{23}. This is an experimentally obtained value and this number is known as the *Avogadro constant* or *Avogadro number*. It is represented by N_0 or N_A.

1 mole of any species = 6.022×10^{23} in number

1 mole of atoms = 6.022×10^{23} atoms

1 mole of molecules = 6.022×10^{23} molecules

Methods to Find Moles

(i) Number of moles $(n) = \dfrac{\text{Mass of element } (w)}{\text{Atomic mass } (M)}$

(ii) Number of moles $(n) = \dfrac{\text{Mass of compound } (w)}{\text{Molecular mass (M)}}$

(iii) Number of moles (n)

$$= \frac{\text{Given number of atoms or entities } (N)}{\text{Avogadro number } (N_A)}$$

These relations can also be interchanged as follows:

Mass of element $(w) = n \times M$

Mass of molecule $(w) = n \times M$

Number of particles of element, $N = n \times N_A$

1 mole of molecules $= 6.022 \times 10^{23}$ molecules $=$ Gram molecular mass of molar mass

(iv) Number of moles $(n) = \dfrac{\text{Given volume } (V)}{\text{Volume at NTP (22.4 L)}}$

Molarity (M)

The number of moles or gram moles of solute in one litre of solution is known as *molarity*. It is representedby M. A solution having a molarity of one is known as a molar solution.

$$\text{Molarity } (M) = \frac{\text{Number of moles of solute } (n)}{\text{Volume of solution in litre } (V)}$$

$$= \frac{\text{Weight of substance } (w)}{\text{Molecular mass } (M) \times \text{Volume in litre } (V)}$$

Molality (m)

The number of moles or gram moles of solute in one kg of solvent is known as *molality*. It is represented by m. A solution having molality one is known as a molal solution.

$$\text{Molality } (m) = \frac{\text{Number of moles of solute } (n)}{\text{Weight of solvent in kg } (W)}$$

$$= \frac{\text{Weight of substance } (w)}{\text{Molecular mass } (M) \times \text{Weight of solvent in kg } (W)}$$

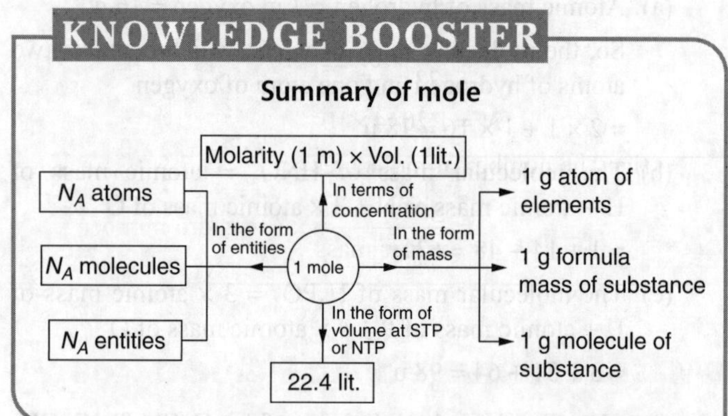

To find the total number of atoms, electrons and charge

Total number of molecules $n \times N_A$

Total number of atoms $= n \times N_A \times$ number of atoms in one molecule

Total number of electrons $= n \times N_A \times$ number of electrons in one molecule/ion

Total charge on any ion $= n \times N_A \times$ charge on one ion $\times 1.6 \times 10^{-19}$ coloumb

---

Illustrations

1. An ornament of silver contains 40 g of silver. Calculate the moles of silver present (atomic mass of silver = 180 u)

Solution: Moles of silver, $n = \dfrac{m}{M}$

Mass of silver, $m = 40$ g

Molar mass of silver, $M = 108$ g

$\therefore \quad n = \dfrac{40}{108} = 0.370$ mol

2. How many moles of CO_2 are present in 56.2 g?

Solution: Molecular mass of $CO_2 = 12 + 2 + 16 = 44$ u

Molar mass of CO_2 $(M) = 44$ g

Mass of CO_2 $(m) = 56.2$ g

Moles of CO_2 $n = \dfrac{m}{M} = \dfrac{56.2}{44} = 1.27$ mol

3. Calculate the mass of

(i) 0.6 moles of N_2 gas

(ii) 0.6 moles of N atoms

Solution:

(i) 0.6 moles of N_2 gas

Mass = Molar mass × Number of moles

$m = M \times n$

$M = 28$ g, $n = 0.6$

$\therefore \quad m = 28 \times 0.6 = 16.8$ g

(ii) Mass = Molar mass × Number of moles

$m = M \times n$

$n = 0.6$ mol, $M = 14$ g

$m = 14 \times 0.6 = 8.4$ g

4. Calculate the number of particles in each of the following:

(i) 4.6 g of Na atoms (number from mass) molecules (number of molecules from mass)

(ii) 16 g O_2

(iii) 0.2 mole of carbon atoms (number from given moles)

Solution:

(i) The number of atoms

$$= \frac{\text{given mass}}{\text{molar mass}} \times \text{Avogadro number}$$

$$N = \frac{m}{M} \times N_A$$

$$N = \frac{4.6}{23} \times 6.022 \times 10^{23}$$

$$N = 1.2044 \times 10^{23}$$

(ii) The number of molecules

$$= \frac{\text{given mass}}{\text{molar mass}} \times \text{Avogadro number}$$

$$N = \frac{m}{M} \times N_A$$

Atomic mass of oxygen = 16 u

So molar mass of O_2 molecules = $16 \times 2 = 32$ g

$$N = \frac{16}{32} \times 6.022 \times 10^{23} = 3.01 \times 10^{23}$$

(iii) The number of particles (atom) $n \times N_A$

$$= 0.2 \times 6.022 \times 10^{23} = 1.2044 \times 10^{23}$$

Mass Percentage of an Element from the Molecular Formula of a Compound

The molecular formula of a compound may be defined as the formula which specifies the number of atoms of various elements in the molecule of the compound. The percentage composition of a compound is the percentage by weight of each element present in it.

The mass percentage of each element is then calculated by the following formula:

Mass percentage of element

$$X = \frac{\text{Mass of element } X \text{ in one molecule}}{\text{Gram molecular mass}} \times 100$$

For example, the molecular formula of sucrose is $C_{12}H_{22}O_{11}$. This indicates that a molecule of sucrose contains 12 atoms of carbon, 22 atoms of hydrogen and 11 atoms of oxygen.

Illustrations

1. Find the percentage composition by mass of (a) formaldehyde (CH_2O) and (b) formic acid (CH_2O_2).

   Solution:

   (a) Molecular mass of formaldehyde, $CH_2O = 12 \times 1 + 1 \times 2 + 16 \times 1 = 30$

   Mass of one mole of formaldehyde = 30 g

   1 mol of CH_2O contains 1 mol (12 g) of carbon, 2 mol of hydrogen (2 g) and 1 mol of oxygen (16 g)

   $$\text{Percentage of carbon} = \frac{12 \text{ g}}{30 \text{ g}} \times 100 = 40.0\%$$

   $$\text{Percentage of hydrogen} = \frac{2 \text{ g}}{30 \text{ g}} \times 100 = 6.7\%$$

   $$\text{Percentage of oxygen} = \frac{16 \text{ g}}{30 \text{ g}} \times 100 = 53.3\%$$

   (b) Molecular mass of formic acid, $CH_2O_2 = 12 \times 1 + 1 \times 2 + 16 \times 2 = 46$

   Mass of one mol of formic acid = 46 g

   1 mol of CH_2O_2 contains 1 mol (12 g) of carbon. 2 mol of hydrogen (2 g) and 2 mol of oxygen (32 g)

   $$\text{Percentage of carbon} = \frac{12 \text{ g}}{46 \text{ g}} \times 100 = 26.08\%$$

   $$\text{Percentage of hydrogen} = \frac{2 \text{ g}}{46 \text{ g}} \times 100 = 4.34\%$$

   $$\text{Percentage of oxygen} = \frac{32 \text{ g}}{46 \text{ g}} \times 100 = 69.56\%$$

2. Find total number of (a) atoms, (b) electrons and (c) charge in 4.8 g SO_4^{2-}.

   Solution: First find total moles

   $$= \frac{w}{\text{Molecular weight}} = \frac{4.8}{96} = \frac{1}{20} = 0.05$$

   Total number of atoms $= n \times N_A \times$ number of atoms in one molecule

   $$= 0.05 \times 6.02 \times 10^{23} \times 5 = 3.01 \times 10^{22}$$

   Total number of $e^- = n \times N_A \times$ number of e^- in one molecule

   Total $e^- = 0.05 \times 6.02 \times 10^{23} \times (16 + 32 + 2)$

   $$= 0.05 \times 6.02 \times 10^{23} \times 50$$

   $$= 2.5 \times 6.02 \times 10^{23} = 1.505 \times 10^{24}$$

   Total charge $= n \times N_A \times$ charge on one ion $\times 1.6 \times 10^{-19}$ C

   $$= 0.05 \times 6.02 \times 10^{23} \times 2 \times 1.6 \times 10^{-19} \text{ C}$$

   $$= 0.05 \times 6.02 \times 2 \times 1.6 \times 10^{4} \text{ C}$$

   $$= 9.632 \times 10^{3} \text{ C}$$

Empirical Formula

The empirical formula of a compound is defined as the formula which gives the simplest whole number ratio of atoms of the various elements present in one molecule of the compound. For example, the empirical formula of the compound glucose ($C_6H_{12}O_6$), is CH_2O which shows that C, H and O are present in the simplest ratio of 1 : 2 : 1.

The empirical formula of the compound acetic acid ($C_2H_4O_2$ or CH_3COOH), is CH_2O which shows that C, H and O are present in the simplest ratio of 1 : 2 : 1.

Rules for Writing the Empirical Formula

The empirical formula can be determined by using the following steps:

(i) First divide the percentage of every element by its atomic mass to get the relative number of moles of various elements present in the compound.

(ii) Now divide the quotients obtained in the above step by the smallest of them so as to get a simple ratio of moles of various elements.

(iii) Now multiply the figures so obtained by a suitable integer, if needed, in order to obtain a whole number ratio (for example, if it is 1.5, multiply by 2).

(iv) Finally write down the symbols of the various elements side by side and put the above numbers as the subscripts to the lower right hand corner of each symbol. This will give the empirical formula of the compound.

Determination of the Molecular Formula

In order to get the molecular formula, first find n (any natural number like 1, 2, 3) and then multiply the empirical formula with n as follows:

$$n = \frac{\text{Molecular formula weight}}{\text{Empirical formula weight}}$$

$$\text{Molecular formula} = \text{Empirical formula} \times n$$

Illustrations

1. A substance, on analysis, gave the following composition: Na = 43.5%, C = 11.2%, O = 45.3%. Calculate its empirical formula [Atomic masses = Na = 23, C = 12, O = 16]

   Solution:

Element	Symbol	%	Atomic mass	Relative number of moles	Simple ratio of moles	Simplest whole no. ratio
Sodium	Na	43.5	23	$\frac{43.5}{23} = 1.89$	$\frac{1.89}{0.93} = 2$	2
Carbon	C	11.2	12	$\frac{11.2}{12} = 0.93$	$\frac{0.93}{0.93} = 1$	1
Oxygen	O	45.3	16	$\frac{45.3}{16} = 2.83$	$\frac{2.83}{0.94} = 3$	3

   Hence, the empirical formula is Na_2CO_3.

2. What is the simplest formula of the compound which has the following percentage composition: carbon 40%, hydrogen 6.66% and the rest is oxygen. If the vapour density is 30, calculate its molecular formula.

   Solution: Calculation of empirical formula

Element	%	Atomic mass	Relative number of moles	Simple ratio of moles	Simplest whole no. ratio
C	40	12	$\frac{40}{12} = 3.33$	$\frac{3.33}{3.33} = 1$	1
H	6.66	1	$\frac{6.66}{1} = 6.66$	$\frac{6.66}{3.33} = 2$	2
O	53.34	16	$\frac{53.34}{16} = 3.33$	$\frac{3.33}{3.33} = 1$	1

$\therefore$ Empirical formula is CH_2O.

Calculation of molecular formula:

Empirical formula mass $= 12 \times 1 + 1 \times 2 + 16 \times 1 = 30$

Molecular weight $= 2 \times$ vapour density $= 2 \times 30 = 60$

$$n = \frac{\text{Molecular mass}}{\text{Empirical formula mass}} = \frac{60}{30} = 2$$

Molecular formula $=$ Empirical formula $\times 2$

$$= CH_2O \times 2 = C_2H_4O_2$$

KNOWLEDGE BOOSTER

Limiting reagent: In a reaction having more than one reactant, we must identify the limiting reagent. The reagent which is finished early or present in less amount as per as the ratio required is known as the *limiting reagent*. For example, in the Haber's process of manufacturing NH_3,

$$N_2 + 3H_2 \rightleftharpoons 2NH_3$$

If we take 2 mol (56 g) of N_2 along with 3 mol (6 g) of H_2, we can find that just 1 mol (28 g) of N_2 is sufficient to react with 3 mol (6 g) of H_2. It means, 1 mol (28 g) N_2 is in excess and H_2 will be used up when the reaction will be complete. That is, H_2 is the limiting reagent.

CHAPTER AT A GLANCE

- **Law of conservation of mass:** Matter can neither be created nor destroyed in a chemical reaction.
- **Law of constant proportions:** In a pure chemical substance or compound the elements are always present in definite proportions by mass.
- **Law of multiple proportions**: Different weights of an element that combine with a fixed weight of another element bear a simple numerical ratio.
- **Law of reciprocal proportions:** When two different elements undergo combination with the same weight of a third element, the ratio in which they combine will either be the same or some simple multiple of the ratio in which they combine with each other.
- **Law of combining volumes**: When gases react with each other they bear a simple whole number ratio with one another as well as the product under conditions of same temperature and pressure.
- **Avogadro's law:** Under similar conditions of temperature and pressure, equal volumes of gases contain an equal number of molecules.
- The building blocks of all matter are **atoms**. An atom is the smallest particle of an element which may or may not exist independently but that can take part in a chemical reaction.
- The **atomic mass** of an element may be defined as the average relative mass of an atom of the element as compared with mass of an atom of carbon (C-12 isotope) taken as 12 amu.
- **Gram atomic mass** is defined as that much quantity of the element whose mass expressed in grams is numerically equal to its atomic mass.

- The number of atoms constituting a molecule is known as its **atomicity**.
- In the modern view, **valency** is equal to number of electrons that an atom can share or lose or gain during a chemical reaction. The number of electrons present in the valence orbit of an atom are called **valence electrons**.
- A **radical** is an atom or a group of atoms having the same or different elements that behaves like a single unit with a positive or negative charge.
- An **acidic** radical is an atom or group of atoms having a negative charge. Anions or electronegative radicals are acidic radicals. A **basic** radical is an atom or group of atoms having a positive charge. Cations or electropositive radicals are basic radicals.
- The **chemical formula** of a compound represents the composition of a molecule of the compound in terms of the symbols of the elements present in it.
- The **molecular mass** of a substance is the sum of the atomic masses of all the atoms in a molecule of that substance. It is therefore the relative mass of a molecule expressed in *atomic mass units (u)*.
- **One mole** of any species (atoms, molecules, ions or particles) is that quantity in number having a mass equal to its atomic or molecular mass in grams. For example, 12 g carbon and 44 g of CO_2 both represent 1 mole.
- The number of moles or gram moles of solute in one litre of solution is known as **molarity,** represented by M. A solution having a molarity of one is known as a **molar solution**.

- The number of moles or gram moles of solute in one kg of solvent is known as **molality,** represented by *m*. A solution having a molality one is known as a **molal solution**.
- The **molecular formula** of a compound may be defined as the formula which specifies the number of atoms of various elements in the molecule of the compound. Percentage composition of a compound is the percentage by weight of each element present in it.
- The **empirical formula** of a compound is defined as the formula which gives the simplest whole number ratio of atoms of the various elements present in one molecule of the compound.

$$\text{Molecular formula} = \text{Empirical formula} \times n$$

- The reagent which is finished early or present in less amount as per as the ratio required is known as the **limiting reagent**.

PRACTICE QUESTIONS

Analyse Your Concepts (School Exam Based)

Fill in the Blanks

Instructions: Complete the following statements with an appropriate word/term to be filled in the blank spaces.

1. In water, the proportion of hydrogen and oxygen is _______ by mass.

2. He, Ne are _______ molecules.

3. _______ is a pure substance which is made up of only one kind of atoms.

4. In CO_2, carbon and oxygen are in the weight ratio of _______.

5. The formula of oxide of Fe^{3+} and O^{2-} is _______.

6. The atomicity of ozone is _______.

7. N_2O_5 is known as nitrogen _______ oxide.

8. The formula of nitrous acid is _______.

9. 1 mole of oxygen atoms = _______ oxygen atoms.

10. Phosphate ion is _______ valent.

11. The ratio by mass of S and O in SO_2 is _______.

12. SO_2 and SO_3 are the examples for law of _______.

True or False

Instructions: Read the following statements and write your answer as true or false.

1. Two elements sometimes form more than one compound.

2. S_8 is a polyatomic element.

3. The smallest particle of a compound is element.

4. 1 amu is equal to 1.66056×10^{-24} g.

5. Mass of 6.022 atoms of an element is called atomic mass.

6. In a pure chemical compound, the elements are always present in a definite proportion by mass.

7. One mole of every substance has the same mass.

8. Ions always have the same number of protons.

9. One mole of NO_2 and SO_2 contain the same number of oxygen atoms.

10. Formula mass of CaO is 56 amu.

11. Mass of 1 mole of NO_2 is 46 g.

12. The mass of a hydrogen atom is equal to the mass of a carbon atom.

Match the Following

Instructions: Each question contains statements given in two columns which have to be matched. Statements (a, b, c, d) in column I have to be matched with statements (p, q, r, s) in column II.

1.

Column I	Column II
(a) Law of constant proportions	(q) Dalton
(b) Law of conservation of mass	(q) Richer
(c) Law of multiple proportions	(r) Lavoisier
(d) Law of reciprocal proportions	(s) Louist Proust

Very Short Answer Type Questions

1. Name the building block of all matter.

2. What are the symbols for copper and zinc?

3. Name two elements whose symbols are derived from Latin names. Give their symbols.

4. What is the mass of 1 mole of water?

5. What is meant by 1 mole of carbon atoms?

6. What is the molecular mass of H_2SO_4?

7. Sodium carbonate ($Na_2CO_3.10H_2O$) is an important industrial compound. Calculate its formula mass.

8. Which of the following are tetra-atomic molecules? CO_2, NH_3, SO_3, CH_3OH, CH_4, H_2O_2

9. Neon gas consists of single atoms. What mass of neon contains 6.022×10^{23} atoms?

10. What is the ratio by mass of nitrogen and hydrogen in ammonia?

11. Give three examples of any trivalent metal ions.

12. What is the chemical formula of silicon oxide?

13. Which of the following has larger mass?
    (i) A mole of ammonia (NH_3)
    (ii) A mole of methane (CH_4)

14. How many moles of helium are present in 108 g of He?

15. What is the molar mass of the sulfur molecule (S_8)?

Short Answer Type Questions

1. What are ionic and molecular compounds? Give examples.

2. Which postulate of Dalton's atomic theory can explain the law of definite proportions?

3. Write down the formulae for the following compounds.
   (a) calcium oxide
   (b) magnesium hydroxide
   (c) aluminium sulfate

4. An element A has a valency of 4. Write the formula of its:
   (a) chloride   (b) oxide
   (c) sulfate   (d) carbonate
   (e) nitrate

5. An element X shows valencies of 4 and 6. Write the formulae of its two oxides.

6. What is the simplest formula of the compound which has the following percentage composition: carbon 80%, hydrogen 20%. If the molecular mass is 30, calculate its molecular formula.

7. Which of the following represents a correct chemical formula? Name it.
   (a) CaCl   (b) $BiPO_4$
   (c) $NaSO_4$ (d) NaS

8. Give the formulae of the compounds formed from the following sets of elements:
   (a) calcium and fluorine
   (b) hydrogen and sulfur
   (c) nitrogen and hydrogen
   (d) carbon and chlorine
   (e) sodium and oxygen
   (f) carbon and oxygen

9. Write the formulae for the following and calculate the molecular mass for each one of them.
   (a) caustic potash
   (b) baking powder
   (c) limestone
   (d) caustic soda
   (e) ethanol
   (f) common salt

Long Answer Type Questions

1. Explain the following terms.
   (i) Atomic mass
   (ii) Molecular mass
   (iii) Mole
   (iv) Avogadro constant
   (v) Polyatomic ions

2. Calculate the molecular masses of H_2, O_2, Cl_2, CO_2, CH_4, C_2H_6, C_2H_4, NH_3, CH_3OH.

3. Raunak took 5 moles of carbon atoms in a container and Krish took 5 moles of sodium atoms in another container of the same weight.
   (a) Whose container is heavier?
   (b) Whose container has more number of atoms?

4. (a) In a chemical reaction, the sum of the masses of the reactants and products remains unchanged. This is called ______.
   (b) A group of atoms carrying a fixed charge on them is called ______.
   (c) The formula unit mass of $Ca_3(PO_4)_2$ is ______.
   (d) Formula of sodium carbonate is ______ and that of ammonium sulfate is ______.

NCERT Corner

In-Text Questions

1. Hydrogen and oxygen combine in the ratio of 1 : 8 by mass to form water. What mass of oxygen gas would be required to react completely with 3 g of hydrogen gas?

2. Which postulate of Dalton's atomic theory is the result of the law of conservation of mass?

3. Define the atomic mass unit.

4. Why is it not possible to see an atom with the naked eye?

5. Write down the formulae of:
   (i) sodium oxide
   (ii) ammonium chloride
   (iii) sodium sulfate
   (iv) magnesium hydroxide

6. Write down the names of compounds represented by the following formulae.
   (i) $Al_2(SO_4)_3$ (ii) $CaCl_2$
   (iii) K_2SO_4 (iv) KNO_3
   (v) $CaCO_3$

7. What is meant by the term chemical formula?

8. How many atoms are present in a:
   (i) H_2S molecule and
   (ii) PO_4^{3-} ion?

9. Calculate the formula unit masses of ZnO, Na_2O, K_2CO_3. Given atomic masses of Zn = 65 u, Na = 23 u, K = 39 u, C = 12 u and O = 16 u.

10. If one mole of carbon atoms weighs 12 grams, what is the mass (in grams) of 1 atom of carbon?

11. Which has more number of atoms, 100 grams of sodium or 100 grams of iron (given, atomic mass of Na = 23 u, Fe = 56 u)?

Textbook Exercises

1. A 0.24 g sample of compound of oxygen and boron was found by analysis to contain 0.096 g of boron and 0.144 g of oxygen. Calculate the percentage composition of the compound by weight.

2. When 3.0 g of carbon is burnt in 8.00 g oxygen, 11.00 g of carbon dioxide is produced. What mass of carbon dioxide will be formed when 3.00 g of carbon is burnt in 50.00 g of oxygen? Which law of chemical combination will govern your answer?

3. What are polyatomic ions? Give examples.

4. Write the chemical formulae of the following compounds.
   (i) Magnesium chloride
   (ii) Calcium oxide
   (iii) Copper nitrate
   (iv) Aluminium chloride
   (v) Calcium carbonate

5. Give the names of the elements present in the following compounds.
   (a) Quick lime
   (b) Hydrogen bromide
   (c) Baking powder
   (d) Potassium sulfate

6. What is the mass of:
   (a) 1 mole of nitrogen atoms?
   (b) 4 moles of aluminium atoms (atomic mass of aluminium = 27)?
   (c) 10 moles of sodium sulfite (Na_2SO_3)?

7. Convert into moles.
   (a) 12 g of oxygen gas
   (b) 20 g of water
   (c) 22 g of carbon dioxide

8. What is the mass of:
   (a) 0.2 mol of oxygen atoms?
   (b) 0.5 mol of water molecules?

9. Calculate the number of molecules of sulfur (S_8) present in 16 g of solid sulfur.

10. Calculate the number of aluminium ions present in 0.051 g of aluminium oxide.
    [**Hint:** The mass of an ion is the same as that of an atom of the same element. Atomic mass of Al = 27 u.]

Exemplar Problems

1. Verify by calculating that
   (a) 5 moles of CO_2 and 5 moles of H_2O do not have the same mass.
   (b) 240 g of calcium and 240 g magnesium elements have a mole ratio of 3:5.

2. Find the ratio by mass of the combining elements in the following compounds. (You may use Appendix III.)
   (a) $CaCO_3$　　　　　(b) $MgCl_2$
   (c) H_2SO_4　　　　　(d) C_2H_5OH
   (e) NH_3　　　　　　(f) $Ca(OH)_2$

3. The difference in the mass of 100 moles each of sodium atoms and sodium ions is 0.0548002 g. Compute the mass of an electron.

4. Cinnabar (HgS) is a prominent ore of mercury. How many grams of mercury are present in 225 g of pure HgS? Molar mass of Hg and S are 200.6 g mol^{-1} and 32 g mol^{-1} respectively.

5. A sample of vitamin C is known to contain 2.58×10^{24} oxygen atoms. How many moles of oxygen atoms are present in the sample?

6. Fill in the missing data in the table.

Property of species	H_2O	CO_2	Na atom	$MgCl_2$
No. of moles	2		–	0.5
No. of particles	–	3.01×10^{23}		–
Mass	36 g	–	115 g	–

7. What is the SI prefix for each of the following multiples and submultiples of a unit?
   (a) 10^3　　　　　　　(b) 10^{-1}
   (c) 10^{-2}　　　　　　(d) 10^{-6}
   (e) 10^{-9}　　　　　　(f) 10^{-12}

8. Express each of the following in kilograms
   (a) 5.84×10^{-3} mg　　(b) 58.34 g
   (c) 0.584 g　　　　　　(d) 5.873×10^{-21} g

9. Compute the difference in masses of 10^3 moles each of magnesium atoms and magnesium ions. (Mass of an electron = 9.1×10^{-31} kg)

10. Which has more number of atoms, 100 kg of N_2 or 100 g of NH_3?

11. A gold sample contains 90% of gold and the rest copper. How many atoms of gold are present in one gram of this sample of gold?

12. Compute the difference in masses of one mole each of aluminium atoms and one mole of its ions. (Mass of an electron is 9.1×10^{-28} g). Which one is heavier?

13. A silver ornament of mass m grams is polished with gold equivalent to 1% of the mass of silver. Compute the ratio of the number of atoms of gold and silver in the ornament.

14. A sample of ethane (C_2H_6) gas has the same mass as 1.5×10^{20} molecules of methane (CH_4). How many C_2H_6 molecules does the sample of gas contain?

15. In photosynthesis, 6 molecules of carbon dioxide combine with an equal number of water molecules through a complex series of reactions to give a molecule of glucose having a molecular formula $C_6H_{12}O_6$. How many grams of water would be required to produce 18 g of glucose? Compute the volume of water so consumed assuming the density of water to be 1 g cm^{-3}.

COMPETITION WINDOW (OBJECTIVE TYPE)

[For NEET, JEE (Main and Advanced), NTSE, KVPY and Olympiads]

Topic-wise MCQs

Laws of chemical combination and Dalton's theory

1. The law of definite proportions was given by
   - (a) Proust
   - (b) Michael Faraday
   - (c) John Dalton
   - (d) Humphry Davy

2. One mole of a gas occupies a volume of 22.4 L. This is derived from
   - (a) Gay-Lussac's law
   - (b) Berzelius hypothesis
   - (c) Dalton's law
   - (d) Avogadro's law

3. The percentage of copper and oxygen in samples of CuO obtained by different methods were found to be the same. This illustrates the law of
   - (a) multiple proportions
   - (b) reciprocal proportions
   - (c) constant proportions
   - (d) conservation of mass

4. For which of the following reactions, is Gay-Lussac's law not applicable?
   - (a) formation of CO_2 from its constituents
   - (b) formation of SO_3 from SO_2 and O_2
   - (c) formation of HCl from its constituents
   - (d) formation of H_2S from its constituents

5. The ratio of the number of molecules present in a given mass of oxygen and sulfur trioxide is
   - (a) 2 : 5
   - (b) 3 : 2
   - (c) 2 : 3
   - (d) 5 : 2

6. Which of the following reactions is associated with the same volume ratio as the reaction of formation of CO_2 from CO and O_2?
   - (a) synthesis of H_2S
   - (b) synthesis of Br
   - (c) formation of SO_3 from SO_2 and O_2
   - (d) synthesis of nitrous oxide

7. In the reaction $4NH_3(g) + 5O_2(g) \rightarrow 4NO(g) + 6H_2O(l)$, when 1 mole of ammonia and 1 mole of O_2 are made to react to completion,
   - (a) all the oxygen will be consumed
   - (b) all the ammonia will be consumed
   - (c) 1.0 mole of H_2O is produced
   - (d) 1.0 mole of NO will be produced

8. A change in the physical state can be brought about
   - (a) only when energy is given to the system
   - (b) only when energy is taken out from the system
   - (c) when energy is either given to, or taken out from the system
   - (d) without any energy change

9. Which one of the following pairs of compounds illustrate the law of multiple proportions?
   - (a) Na_2O and SrO
   - (b) $SnCl_2$ and $SnCl_4$
   - (c) H_2O and K_2O
   - (d) CaO and K_2O

10. The formation of SO_2 and SO_3 explains
    - (a) the law of definite properties
    - (b) Boyle's law
    - (c) the law of conservation of mass
    - (d) the law of multiple proportions

Atoms, ions, symbols and their representation

11. The chemical symbol for sodium is
    - (a) So
    - (b) Sd
    - (c) NA
    - (d) Na

12. The chemical symbol for Bismuth is
    - (a) Be
    - (b) Bi
    - (c) B
    - (d) Ba

13. A group of atoms chemically bonded together is a (an)
    - (a) salt
    - (b) element
    - (c) molecule
    - (d) ion

14. Adding electrons to an atom will result in a (an)
  (a) cation
  (b) salt
  (c) molecule
  (d) anion

15. The correct symbol for gold is
  (a) Ag
  (b) Si
  (c) Au
  (d) Al

16. The total number of atoms represented by the compound $CuSO_4.5H_2O$ is
  (a) 24
  (b) 8
  (c) 9
  (d) 21

17. The chemical symbol for nitrogen gas is
  (a) Ni
  (b) N_2
  (c) N^+
  (d) N

18. Which of the following symbols of elements is correct?
  (a) Cobalt: CO
  (b) Carbon: c
  (c) Aluminium: AL
  (d) Helium: He

19. How many of the following elements have been given correct symbol here:
  (a) Nickel: NI
  (b) Silver: Ag
  (c) Vanadium: V
  (d) Both (b) and (c)

20. To which of the following elements have symbols been given according to their Latin names?
  (a) Mercury: Hg
  (b) Lead: Pb
  (c) Potassium: K
  (d) All of these

Atomic mass, molecular mass and mass percentage

21. The mass of sodium in 5.85 g of sodium chloride is
  (a) 2.3 g
  (b) 5.02 g
  (c) 6.8 g
  (d) 2.51 g

22. The mass of one C atom is
  (a) 12 g
  (b) 1.99×10^{-23} g
  (c) 6.02×10^{23} g
  (d) 6 g

23. The percentage of hydrogen in H_2O is
  (a) 8.88
  (b) 4.44
  (c) 20.40
  (d) 11.12

24. The percentage by weight of O_2 in $CaSO_4$ (O = 16, S = 32, Ca = 40) is
  (a) 64
  (b) 28
  (c) 47
  (d) 32

25. Which among the following contain(s) 43.4% of sodium?
  (a) sodium carbonate
  (b) sodium bromide
  (c) sodium bicarbonate
  (d) sodium sulfate

26. Find the percentage of hydrogen in water given that the relative atomic masses of H = 1, O = 16.
  (a) 22.2%
  (b) 11.1%
  (c) 5.6%
  (d) 2.8%

27. Find the percentage mass of water in washing soda crystals ($Na_2CO_3.10H_2O$). Molecular mass of washing soda is 286 amu.
  (a) 15.73%
  (b) 31.45%
  (c) 62.9%
  (d) 6.29%

Valency and molecular formula

28. In MnO_2, the valency of Mn is
  (a) 4
  (b) 6
  (c) 2
  (d) 1

29. A molecular positive radical (M^+), combines with a chromate radical. Now, identify the formula of the compound formed.
  (a) $M_2Cr_2O_7$
  (b) $M(Cr_2O_7)_2$
  (c) $MCrO_4$
  (d) M_2CrO_4

30. The formula of magnesium dihydrogen phosphate is
  (a) $Ca(H_2PO_4)_2$
  (b) $Mg_3(PO_4)_2$
  (c) MgH_2PO_4
  (d) $Mg(HPO_4)_2$

31. In which of the following cases, is the empirical formula the same as the molecular formula?
  (a) C_2H_5COOH
  (b) glucose
  (c) sucrose
  (d) C_6H_6

32. If the formula of a metallic nitrate is $A(NO_3)_3$, then what will be the formula of the nitride of that metal?
  (a) AN_2
  (b) AN
  (c) A_2N
  (d) A_2N_3

33. The formula of a metal chloride is MCl_3. The formula of its phosphate is
  (a) $M(PO_4)_2$
  (b) MPO_4
  (c) M_2PO_4
  (d) M_3PO_4

34. In P_2O_3 and P_2O_5, the valencies of phosphorus are respectively
  (a) 2, 2
  (b) 3, 5
  (c) 5, 3
  (d) 6, 10

35. Which of the following metal atoms can show variable valency?
  (a) Ca
  (b) Fe
  (c) Cu
  (d) both (b) and (c)

36. Which of the following pairs are possible?
  (a) Hg^+, Hg^{2+}
  (b) Sn^{2+}, Sn^{4+}
  (c) Na^+, Na^{2+}
  (d) Both (a) and (b)

Mole concept

37. Which of the following has the largest number of particles?
  (a) 8 g of CH_4
  (b) 8.8 g of CO_2
  (c) 3.42 g of $C_{12}H_{22}O_{11}$
  (d) 4 g of H_2

38. Vapour density of SO_3 is ________.
  (a) 20
  (b) 40
  (c) 80
  (d) 60

39. The volume of CO_2 liberated at STP on burning 2.4 g of carbon in excess oxygen is ________.
  (a) 1.68 L
  (b) 4.48 L
  (c) 22.4 L
  (d) 6.72 L

40. The number of molecules present in 0.28 g of nitrogen is
  (a) 6.023×10^{22}
  (b) 6.023×10^{20}
  (c) 6.023×10^{21}
  (d) 6.023×10^{23}

41. How many grams are contained in 1 g atom of Fe?
  (a) 28 g
  (b) 10 g
  (c) 1 g
  (d) 56 g

42. The weight of H atoms present in 0.2 mol of C_2H_6 is
  (a) 1.2 g
  (b) 8 g
  (c) 0.2 g
  (d) 6 g

43. 2 mol of O atoms at NTP occupy a volume of
  (a) 2 L
  (b) 22.4 L
  (c) 11.2 L
  (d) 44.8 L

44. Which of the following contains the maximum number of molecules?
  (a) 1 g CO_2
  (b) 1 g N_2
  (c) 1 g H_2
  (d) 1 g CH_4

45. Calculate the total number of ions present in 5.85 g of sodium chloride.
  (a) 1.2044×10^{23}
  (b) 1.02044×10^{23}
  (c) 12.044×10^{23}
  (d) 0.6022×10^{23}

46. Which of the following represents 1 mol of quantity?
  (a) 16 g of oxygen atom
  (b) 48 g of ozone
  (c) 15 g of ethane
  (d) Both (a) and (b)

47. Find the molarity of 200 mL of a solution containing 3.65 g of HCl.
  (a) 0.25
  (b) 0.5
  (c) 0.75
  (d) 1

48. Find the molality of 9 g of glucose present in 90 g of water (molar mass of glucose = 180).
  (a) 0.055
  (b) 5.5
  (c) 0.55
  (d) 0.275

49. Find the total number of electrons present in 1.7 g of ammonia. (Given: $N_A = 6.022 \times 10^{23}$)
  (a) 6.022×10^{22}
  (b) 6.022×10^{23}
  (c) 6.022×10^{20}
  (d) 3.011×10^{23}

50. Find the total charge present in 3 g of carbonate (CO_3^{2-}). (Given: 1 unit charge = 1.6×10^{-19} C)
  (a) 963.52 C
  (b) 96352 C
  (c) 9635.2 C
  (d) 96500 C

Miscellaneous

1. 5.6 L of N_2 at NTP is equivalent to
  (a) $\dfrac{1}{4}$ mol
  (b) $\dfrac{1}{8}$ mol
  (c) 1 mol
  (d) $\dfrac{1}{2}$ mol

2. Which of the following would weigh the highest?
  (a) 0.2 mol of sucrose ($C_{12}H_{22}O_{11}$)
  (b) 2 mol of CO_2
  (c) 2 mol of $CaCO_3$
  (d) 10 mol of H_2O

3. Calculate the number of moles of magnesium present in a magnesium ribbon weighing 12 g. Molar atomic mass of magnesium is 24 g mol^{-1}.
  (a) 0.05
  (b) 0.25
  (c) 0.5
  (d) 1

4. The visible universe is estimated to contain 10^{22} stars. How many moles of stars are present in the visible universe?
  (a) 1.66
  (b) 0.166
  (c) 0.0166
  (d) 0.00166

5. Which of the statements is(are) true?
  (a) Vapour density of a gas is twice its molecular mass
  (b) Atomic masses of most elements are fractional
  (c) Law of constant composition is true for all types of compounds
  (d) Molar volume of a gas at standard conditions is 22.4 L

6. Which of the following has the maximum number of atoms?
  (a) 18 g of H_2O
  (b) 18 g of O_2
  (c) 18 g of CO_2
  (d) 18 g of CH_4

7. What is the fraction of the mass of water due to neutrons?
  (a) 9/4
  (b) 4/9
  (c) 2/9
  (d) 6/9

8. A solution is prepared by adding 4 g of a substance X to 18 g of water. Calculate the mass percent of the solute.
  (a) 181.8
  (b) 1.81
  (c) 9.9
  (d) 18.18

9. What is the percentage of aluminium in Al_2O_3? (Al = 27, O = 16)
  (a) 52.94
  (b) 26.47
  (c) 105.5
  (d) 5.94

10. Mass of one atom of oxygen is
  (a) $\dfrac{16}{6.023 \times 10^{23}} g$
  (b) $\dfrac{23}{6.023 \times 10^{23}} g$
  (c) $\dfrac{1}{6.023 \times 10^{23}} g$
  (d) 8 u

11. A gaseous mixture contains O_2 and N_2 in the ratio of 1 : 4 by weight. The ratio of their number of molecules is
  (a) 7 : 21
  (b) 1 : 8
  (c) 7 : 32
  (d) 32 : 7

12. What is the simplest formula of a compound that contains 0.25 g atom of silicon per 0.50 g atom of oxygen?
  (a) SiO
  (b) SiO_2
  (c) Si_2O_3
  (d) none of these

13. What is the empirical formula for a compound that contains 22% S and 78% F? (S = 32, F = 19).
    (a) SF_2
    (b) SF_4
    (c) SF_6
    (d) SF_3

14. 3.42 g of sucrose are dissolved in 18 g of water in a beaker. The number of oxygen atoms in the solution is
    (a) 6.68×10^{23}
    (b) 6.09×10^{22}
    (c) 6.022×10^{23}
    (d) 6.022×10^{21}

15. The empirical formula of a compound is CH_2. One mol of this compound has a mass of 42 grams. Its molecular formula is
    (a) C_3H_6
    (b) C_2H_4
    (c) C_3H_8
    (d) CH_2

16. Find the number of g atoms and weight of an element having 2×10^{23} atoms. The atomic mass of the element is 32.
    (a) 5.82 g
    (b) 10.63 g
    (c) 1.063 g
    (d) 15.83 g

17. The simplest formula of a compound containing 50% of an element P (atomic weight 10) and 50% of element Q (atomic weight 20) is
    (a) PQ
    (b) P_2Q
    (c) PQ_2
    (d) P_2Q_3

18. An organic compound contains 4% sulfur. Its minimum molecular weight is
    (a) 100
    (b) 400
    (c) 800
    (d) 200

19. The weight of 1×10^{22} molecules of $CuSO_4.5H_2O$ is
    (a) 4.14 g
    (b) 20.07 g
    (c) 6.42 g
    (d) 1.035 g

20. Assuming the atomic weight of a metal A to be 56, find the empirical formula of its oxide containing 70.00% of A.
    (a) A_2O
    (b) AO_2
    (c) A_2O_3
    (d) AO_3

Advanced and Olympiads

Single Choice

1. Two samples of lead oxide were separately reduced to metallic lead by heating in a current of hydrogen. The weight of lead from one oxide was half the weight of lead obtained from the other oxide. The data illustrates the
    (a) law of multiple proportions
    (b) law of equivalent proportions
    (c) law of reciprocal proportions
    (d) law of constant proportions

2. Among the following pairs of compounds, the one that illustrate the law of multiple proportions is
    (a) CS_2 and $FeSO_4$
    (b) CuO and Cu_2O
    (c) PH_3 and PCl_3
    (d) H_2S and SO_2

3. The simplest formula of a compound containing 50% of element P (atomic mass 10) and 50% of element Q (atomic mass 20) is
    (a) PQ
    (b) PQ_3
    (c) P_2Q
    (d) P_2Q_3

4. The empirical formula of a hydrocarbon containing 80% carbon and 20% hydrogen is
    (a) CH_3
    (b) CH_2
    (c) CH_6
    (d) CH

5. Which of the following correctly represents 360 g of water?
    (i) 2 mol of H_2O
    (ii) 20 mol of water
    (iii) 6.022×10^{23} molecules of water
    (iv) 1.2044×10^{25} molecules of water
    (a) (i)
    (b) (i) and (iv)
    (c) (ii) and (iii)
    (d) (ii) and (iv)

6. An organic compound consists of 6.023×10^{23} carbon atoms, 1.8069×10^{24} hydrogen atoms and 3.0115×10^{23} oxygen atoms. What is its simplest formula?
    (a) $C_2H_4O_2$
    (b) CH_3O
    (c) C_2H_6O
    (d) C_2H_4O

7. The vapour density of a chloride of an element is 39.5. The Ew of the elements is 3.82. The atomic weight of the element is
    (a) 3.82
    (b) 22.92
    (c) 15.28
    (d) 7.64

8. How many moles of methane are required to produce 22 g of CO_2 after combustion?
    (a) 1
    (b) 0.5
    (c) 1.5
    (d) 0.25

Multiple Choice

9. Which of the following statements are true about an atom?
    (a) Atoms are not able to exist independently.
    (b) Atoms are the basic units from which molecules and ions are formed.
    (c) Atoms are always neutral in nature.
    (d) Atoms aggregate in large numbers to form the matter that we can see, feel or touch.

10. Which of the following are the best examples of law of conservation of mass?
    (a) 2 g of hydrogen combines with 16 g of oxygen to form 18 g of water.
    (b) A sample of air increases in volume when heated at constant pressure but its mass remains unaltered.
    (c) When 6 g of carbon is heated in a vacuum there is no change in mass.
    (d) 12 g of carbon combines with 32 g of oxygen to form 44 g of CO_2.

11. Which of the following represents a polyatomic ion?
    (a) phosphate
    (b) nitride
    (c) sulfite
    (d) sulfate

12. Which of the following statements is/are correct?
    (a) An atom is the smallest indivisible particle of an element that can take part in a chemical change.
    (b) An atom is the smallest particle of matter according to Dalton's theory.
    (c) An atom is the smallest particle of an element.
    (d) An atom is always the radioactive emission.

13. Which of the following element(s) has a symbol consisting of two letters?
    (a) aluminium
    (b) lead
    (c) oxygen
    (d) tin

14. 8 g of O_2 has the same number of molecules as
    (a) 7 g CO
    (b) 7 g N_2
    (c) 11 g CO_2
    (d) 8 g SO_2

15. Which of the following pairs of substances illustrate(s) the law of multiple proportions?
    (a) CO and CO_2
    (b) Ca and $Ca(OH)_2$
    (c) SO_2 and SO_3
    (d) H_2O and D_2O

16. Which of the following set(s) of compounds illustrate(s) the law of reciprocal proportions?
    (a) PH_3, PCl_3, P_2O_3
    (b) CO_2, CH_4, H_2O
    (c) NH_3, N_2O_5, H_2O
    (d) HCl, HBr, HI

17. Which of the following statements are correct?
    (a) Richter proposed the law of reciprocal proportions.
    (b) Dalton proposed the law of multiple proportions.
    (c) Joseph Proust proposed the law of definite proportions.
    (d) A Lavosier is called the father of chemistry and proposed the law of conservation of mass.

18. Which of the following pair(s) of compounds illustrate the law of multiple proportions?
    (a) NO and N_2O_5
    (b) NO_2 and N_2O
    (c) MgO and $Mg(OH)_2$
    (d) SO_2 and SO_3

Comprehension Type

Comprehension I: A mole can be expressed in terms of weight, volume, number of entities, and so on. Mole is a very important quantity that can be used to express various concentration terms such as molarity, molality and to find the total number of entities such as molecules, atoms, electrons, total charges, and so on. The total number of entities can be found out by using the mole as follows.

Total number of molecules $n \times N_A$

Total number of atoms $= n \times N_A \times$ number of atoms in one molecule

Total number of electrons $= n \times N_A \times$ number of electrons in one molecule/ion

Total charge on any ion $= n \times N_A \times$ charge on one ion $\times 1.6 \times 10^{-19}$ Coloumb

19. A solution of HNO_3 has 0.63 g of HNO_3 present in 500 mL solution. The molarity of this solution is
    (a) 0.01 M
    (b) 0.02 M
    (c) 0.2 M
    (d) 0.1 M

20. A drop of water has a density of 1 g/mL and a volume of 3.6 mL. Find the total number of atoms present in it.
    (a) $0.3\ N_A$
    (b) $1.2\ N_A$
    (c) $0.6\ N_A$
    (d) $6\ N_A$

21. Find the total number of electrons present in 4 g of methane.
    (a) $5\ N_A$
    (b) $0.25\ N_A$
    (c) $0.5\ N_A$
    (d) $2.5\ N_A$

Comprehension II: In the modern view, valency is equal to the number of electrons that an atom can share or lose or gain during a chemical reaction. The number of electrons present in the valence orbit of an atom are called valence electrons. In order to attain the octet state (8 valence electrons) every atom should lose or gain or share electrons. There are certain elements that exhibit more than one valency in their ions (or compounds). When an atom of an element not only loses its valence electron but also loses electrons from its penultimate orbit, variable valency is observed.

22. In which of the following pair(s) can both the metals show variable valencies?
    (a) Na, Mg
    (b) Na, Pb
    (c) Sn, Pb
    (d) Zn, Pb

23. If mercury can exist in 1 and 2 valency states as Hg^+ and Hg^{2+}, the possible pair of compounds with Cl^- and O^{2-} may be
    (a) Hg_2Cl_2, $HgCl_2$
    (b) Hg_2O, HgO
    (c) HgO, HgO_2
    (d) Both (a) and (b)

Matrix Matching

24. Match the following

Column I	Column II
(a) CO_3^{2-}	(p) $PO_4^{3-} = 1P + 4\,(O) = 5$
(b) PQ_4^{3-}	(q) $CO = 1C + 1(O) = 2$
(c) P_2O_5	(r) $CO_3^{2-} = 1C + 3\,(O) = 4$
(d) CO	(s) $P_2O_5 = 2P + 5\,(O) = 7$

25. Match the following

Column I	Column II
(a) 2.24 L SO_3 at NTP	(p) 0.2 mol
(b) 1.7 g NH_3	(q) 0.1 mol
(c) 3.6 g H_2O	(r) Total atoms $= 0.4\ N_A$
(d) 6 g C_2H_6	(s) Total atoms $= 0.6\ N_A$
	(t) Total atoms $= 1.6\ N_A$

Integer Type

26. If the molecular mass of a compound X is n00 amu, and if its 3.0115×10^9 molecules weigh 1.0×10^{-12} g, find the value of n here.

27. The number of moles of NaOH present in 3 litre of 0.5 M NaOH is ________.

28. A compound was found to contain 55.2% xenon and 44.8% chlorine. If the empirical formula of the compound is $XeCl_x$, calculate the value of x.

29. The dot at the end of this sentence has a mass of about one microgram. Assuming that the black stuff is carbon, if the approximate atoms of carbon needed to make such a dot is $n \times 10^{16}$, find n.

30. If the molarity of NaOH in a solution prepared by dissolving 4 g in enough water to form 250 mL of the solution is 0.X mL, find X.

31. The weight of one atom of an element is 6.644×10^{-23} g. If 10^X g atom of element are present in 40 kg, find X.

32. Haemoglobin contains 0.25% iron by weight. The molecular weight of haemoglobin is 89600. Calculate the number of iron atoms per molecule of haemoglobin.

Answer Keys

Fill in the Blanks

1. 1 : 8
2. Monoatomic
3. Element
4. 3 : 8
5. Fe_2O_3
6. 3
7. Penta
8. HNO_2
9. 6.022×10^{23}
10. Tri
11. 1 : 1
12. Multiple proportions

True or False

1. T   **2.** T   **3.** F   **4.** T   **5.** F
6. T   **7.** F   **8.** F   **9.** T   **10.** T
11. T   **12.** F

Match the Following

1. (a)—(s); (b)—(r); (c)—(p); (d)—(q)

Very Short Answer Type Questions

1. Atoms are the building blocks of all matter.

2. The symbols of copper and zinc are Cu and Zn respectively.

3. (i) Na for sodium from natrium
(ii) Fe for iron from ferrum

4. The mass of 1 mole of water is 18 g.

5. 1 mole of carbon atoms is 12 g carbon.

6. The molecular mass of H_2SO_4 is 98 u (2 + 32 + 64 = 98 u).

7. The formula mass of $(Na_2CO_3.10H_2O)$ is 286 u as follows
Na = 23 u, C = 12 u, O = 16 u, H = 1 u
So, formula mass of $Na_2CO_3.10H_2O = 2 \times 23 + 12 + 48 + 10 \times 18 = 286$ u

8. NH_3, SO_3 and H_2O_2 are tetra-atomic molecules.

9. Number of atoms = 6.022×10^{23} = 1 mol of neon = 20 u of neon.

10. The ratio by mass of nitrogen and hydrogen in ammonia (NH_3) is 14 : 3.

11. Al^{+3}, Fe^{+3} and Cr^{+3} are trivalent metal ions (cations).

12. The chemical formula of silicon oxide SiO_2.

13. 1 mol of ammonia = 17 g
1 mol of methane = 16 g
So, 1 mol of ammonia has more mass than 1 mol of methane.

14. Number of moles of helium

$$= \frac{\text{weight in g}}{\text{atomic weight of helium}} = \frac{108}{4} = 27$$

15. The molar mass of sulfur molecule (S_8) is $32 \times 8 = 256$ g.

Short Answer Type Questions

1. An ionic compound is made up of ions. Or, it contains a cation which is a positive ion and an anion which is a negative ion. For example, potassium chloride is an ionic compound made up of K^+ and Cl^- ions.
A molecular compound is made up of molecules. For example, ammonia (NH_3), carbon dioxide (CO_2).

2. The following postulate of Dalton's atomic theory can explain the law of definite proportions. The relative number and kinds of atoms are constant in a given compound.

3. (a) CaO (b) $Mg(OH)_2$ (c) $Al_2(SO_4)_3$

4. (a) ACl_4 (b) AO_2 (c) $A(SO_4)_2$ (d) $A(CO_3)_2$ (e) $A(NO_3)_4$

5. Here X is a sulfur like element and its oxides are XO_2 and XO_3.

6. Calculation of empirical formula:

Element	%	Atomic mass	Relative number of moles	Simple ratio of moles	Simplest whole no. ratio
C	80	12	$\frac{80}{12} = 6.66$	$\frac{6.66}{6.66} = 1$	1
H	20	1	$\frac{20}{1} = 20$	$\frac{20}{6.66} = 3$	3

So the empirical formula is CH_3.
Calculation of molecular formula:
Empirical formula mass = $12 \times 1 + 1 \times 3 = 15$

$$n = \frac{\text{Molecular mass}}{\text{Empirical formula mass}} = \frac{30}{15} = 2$$

Molecular formula = Empirical formula × 2

$$= CH_3 \times 2 = C_2H_6$$

7. (a) The correct formula is $CaCl_2$ (valency of Ca = 2, valency of Cl = 1).

(b) $BiPO_4$ is correct because the valency of Bi = 3, valency of PO_4 = 3.

(c) The correct formula is Na_2SO_4 (valency of Na = 1, valency of SO_4 = 2).

The correct formula is Na_2S (valency of Na = 1, valency of sulfide = 2).

Hence, the correct formula is $BiPO_4$ and its name is bismuth phosphate.

8. The possible compounds and their formulae are as follows:

(a) Calcium fluoride: CaF_2

(b) Hydrogen sulfide: H_2S

(c) Ammonia: NH_3

(d) Carbon tetrachloride: CCl_4

(e) Sodium oxide: Na_2O

(f) Carbon monoxide: CO and carbon dioxide: CO_2

9. (a) Caustic potash, $KOH = (39 + 16 + 1) = 56$ g mol^{-1} = $(39 + 16 + 1) = 56$ mol^{-1}

(b) Baking powder, $NaHCO_3 = (23 + 1 + 12 + 48) = 84$ g mol^{-1}

(c) Limestone, $CaCO_3 = (40 + 12 + 48) = 100$ g mol^{-1}

(d) Caustic soda, $NaOH = (23 + 16 + 1) = 40$ g mol^{-1}

(e) Ethanol, $C_2H_5OH = (2 \times 12 + 6 + 16) = 46$ g mol^{-1}

(f) Common salt, $NaCl = (23 + 35.5) = 58.5$ g mol^{-1}

Long Answer Type Questions

1. See text part.

2. (i) Molecular mass of H_2 = Atomic mass of hydrogen × $2 = 1 \times 2 = 2$ u

(ii) Molecular mass of O_2 = Atomic mass of oxygen × $2 = 16 \times 2 = 32$ u

(iii) Molecular mass of Cl_2 = Atomic mass of chlorine × $2 = 35.5 \times 2 = 71$ u

(iv) Molecular mass of CO_2 = (Atomic mass of carbon × 1) + (Atomic mass of oxygen × 2)
$= 12 + (16 \times 2) = 12 + 32 = 44$ u

(v) Molecular mass of CH_4 = (Atomic mass of carbon × 1) + (Atomic mass of hydrogen × 4)
$= 12 + (1 \times 4) = 12 + 4 = 16$ u

(vi) Molecular mass of C_2H_6 = (Atomic mass of carbon × 2) + (Atomic mass of hydrogen × 6)
$= (12 \times 2) + (1 \times 6) = 24 + 6 = 30$ u

(vii) Molecular mass of C_2H_4 = (Atomic mass of carbon × 2) + (Atomic mass of hydrogen × 4)
$= (12 \times 2) + (1 \times 4) = 24 + 4 = 28$ u

(viii) Molecular mass of NH_3 = (Atomic mass of nitrogen × 1) + (Atomic mass of hydrogen × 3)
$= (14 \times 1) + (1 \times 3) = 14 + 3 = 17$ u

(ix) Molecular mass of CH_3OH = (Atomic mass of carbon × 1) + (Atomic mass of hydrogen × 3) + (Atomic mass of oxygen × 1) + (Atomic mass of hydrogen × 1)
$= 12 + 3 + 16 + 1 = 32$ u

3. (a) Mass of container containing 5 mol of C atoms = $5 \times 12 = 60$ g

Mass of container containing 5 mol of Na atoms = $5 \times 23 = 115$ g

The container that Krish has is heavier.

(b) Both containers have the same number of atoms as they contain same number of moles.

4. (a) Law of conservation of mass

(b) Polyatomic ion

(c) The formula unit mass of $Ca_3(PO_4)_2$

$$= 310 = (3 \times Ca) + (2 \times P) + (8 \times O)$$

$$= (3 \times 40) + (2 \times 31) + (8 \times 16)$$

$$= (120 + 62 + 128) = 310 \text{ u}$$

(d) Na_2CO_3, $(NH_4)_2SO_4$

NCERT Corner

In-Text Questions

1. According to the law of constant proportions, the composition of a compound is always fixed. By applying this, 1 g of hydrogen gas combines with oxygen = 8 g
So, 3 g of hydrogen gas will combine with oxygen = $8 \times 3 = 24$ g

2. The given postulate of Dalton's atomic theory is the result of the law of conservation of mass. Atoms are indivisible particles, which cannot be created or destroyed in a chemical reaction.

3. One atomic mass unit (amu) is a mass unit equal to exactly one twelfth (1/12th) the mass of one atom of carbon-12. 'The relative atomic masses of all the elements have been found with respect to an atom of carbon-12.

4. It is not possible to see it with the naked eye as an atom is extremely small in size. Generally, the radius of an atom is of the order of nanometres. For example, the atomic radius of the hydrogen atom is 10^{-10} m.

5. (i) The formula of sodium oxide is Na_2O.

(ii) The formula of ammonium chloride is NH_4Cl.

(iii) The formula of sodium sulfate is Na_2SO_4.

(iv) The formula of magnesium hydroxide is $Mg(OH)_2$.

6. (i) $Al_2(SO_4)_3$ is aluminium sulfate.

(ii) $CaCl_2$ is calcium chloride.

(iii) K_2SO_4 is potassium sulfate.

(iv) KNO_3 is potassium nitrate.

(v) $CaCO_3$ is calcium carbonate.

7. The chemical formula of a compound or element is the symbolic representation of its composition. It represents:
   (i) the number and kind of atoms present per molecule of the compound,
   (ii) molar mass of the compound,
   (iii) one mole of the compound.

8. In H_2S there are three atoms:
   (i) 2 atoms of hydrogen + 1 atom of sulfur = three (3) atoms
   In phosphate there are five atoms:
   (ii) 1 atom of phosphorus + 4 atoms of oxygen = five (5) atoms

9. (i) Formula unit mass of ZnO = 65 + 16 = 81 u
   (ii) Formula unit mass of Na_2O = $(23 \times 2) + (16 \times 1)$ = 46 + 16 = 62 u
   (iii) Formula unit mass of K_2CO_3
   $= (39 \times 2) + (12 \times 1) + (16 \times 3)$
   $= 78 + 12 + 48 = 138$ u

10. 1 mole of carbon atoms = 6.022×10^{23} atoms
    Molar atomic mass = 12 g
    6.022×10^{23} carbon atoms weigh = 12 g
    1 carbon atom weighs
    $$= \frac{\text{Weight of carbon}}{\text{number of carbon atom}} = \frac{12}{6.022 \times 10^{23}} = 1.99 \times 10^{-23} \text{ g}$$

11. Molar mass of sodium = 23 g
    1 mole atom = 6.022×10^{23} atoms
    23 g sodium contains = 6.022×10^{23} atoms
    1 g sodium contains = 6.022×10^{23} atoms
    100 g sodium contains = 6.022×10^{23} atoms = 2.618×10^{24} atoms
    By the above method or by formula we find number of atoms in 100 g Fe;
    Number of atoms of an element in a given mass
    $$= \frac{\text{Given mass}}{\text{Gram atomic mass 100 g}} \times \text{Avogadro's number}$$
    $$= \frac{100 \text{ g}}{56 \text{ g}} \times 6.022 \times 10^{23} = 1.075 \times 10^{24} \text{ atoms}$$
    Thus, 100 g of sodium has more number of atoms when compared to 100 g of iron.

Textbook Exercises

1. Mass of the compound = 0.24 g
   Mass of boron = 0.096 g
   Mass of oxygen = 0.144 g
   $$\text{Percentage of boron} = \frac{\text{Mass of boron}}{\text{Mass of compound}} \times 100$$
   $$= \frac{0.096 \text{ g}}{0.240 \text{ g}} \times 100$$

$$\text{Percentage of oxygen} = \frac{\text{Mass of oxygen}}{\text{Mass of compound}} \times 100$$
$$= \frac{0.144 \text{ g}}{0.240 \text{ g}} \times 100$$

Alternatively % of oxygen can be found as follows:
% of oxygen = 100 % of boron = 100 − 40 = 60%

2. First, we find the proportion of mass of carbon and oxygen in carbon dioxide.
   In CO_2, C : O = 12 : 32 or 3 : 8
   In other words, it means
   12.00 g carbon reacts with oxygen = 32.00 g
   $$3.00 \text{ g carbon will react with oxygen} = \frac{32}{12} \times 3 = 8 \text{ g}$$
   Hence 3.00 g of carbon will always react with 8.00 g of oxygen to give CO_2 (11 g), even if a large amount (50.00 g) of oxygen is present. This is governed by the law of constant proportions.

3. Polyatomic ions are the group of atoms which carry a fixed charge (either positive or negative) and behave as ions.
   For example, (i) ammonium ion (NH_4^+), (ii) sulfate ion (SO_4^{2-}), (iii) carbonate ion (CO_3^{2-}).

4. (i) Magnesium chloride: $MgCl_2$
   (ii) Calcium oxide: CaO
   (iii) Copper nitrate: $Cu(NO_3)_2$
   (iv) Aluminium chloride: $AlCl_3$
   (v) Calcium carbonate: $CaCO_3$

5.

Compounds	Elements
(a) Quick lime: calcium oxide, CaO	Calcium, oxygen
(b) Hydrogen bromide: HBr	Hydrogen, bromine
(c) Baking powder: sodium hydrogen carbonate, $NaHCO_3$	Sodium, hydrogen, carbon, oxygen
(d) Potassium sulfate: K_2SO_4	Potassium, sulfur, oxygen

6. (a) Molar mass of N atom = Atomic mass of N
   Mass of 1 mol of N atoms = 14 g
   (b) Mass of 1 mole Al atoms = 27 g
   Mass of 4 moles of Al atoms = $27 \times 4 = 108$ g
   (c) Mass of 1 mole of Na_2SO_3 = $(23 \times 2) + 32 + (16 \times 3)$ = 46 + 32 + 48 = 126 g
   Mass of 10 moles of Na_2SO_3 = $126 \times 10 = 1260$ g

7. Use the general formula mole $= \dfrac{\text{given weight}}{\text{molecular weight}}$
   (a) Molar mass of oxygen (O_2) = $16 \times 2 = 32$ g
   32 g oxygen gas = 1 mol
   $$12 \text{ g oxygen gas} = \frac{1 \times 12 \text{ g}}{32 \text{ g}} = 0.375 \text{ mol}$$

(b) Molar mass of water $(H_2O) = 2 + 16 = 18g$

18 g water = 1 mol

$$20 \text{ g water} = \frac{1 \times 20 \text{ g}}{18 \text{ g}} = 1.11 \text{ mol}$$

(c) 22 g of carbon dioxide (CO_2)

Molar mass of carbon dioxide $(CO_2) = 12 + 32 = 44g$

44 g CO_2 = 1 mol

$$22 \text{ CO}_2 = \frac{1 \times 22 \text{ g}}{44 \text{ g}} = 0.5 \text{ mol}$$

8. Use the relation mass or weight = mol $\times$ molecular weight

(a) Mass of 1 mol O atoms = 16 g

Mass of 0.2 mol O atoms = $16 \times 0.2 = 3.2$ g

(b) Mass of 1 mol of H_2O molecules = 18 g

Mass of 0.5 mol of H_2O molecules = $18 \times 0.5 = 9.0$ g

9. Molar mass of sulfur $(S_8) = 32 \times 8 = 256$ g

Number of S_8 molecules in 256 g of solid sulfur = 6.022×10^{23}

Number of S_8 molecules in 16 g of solid sulfur

$$= \frac{6.022 \times 10^{23} \times 16 \text{ g}}{256 \text{ g}} = 3.76 \times 10^{23}$$

10. Molar mass of $Al_2O_3 = (27 \times 2) + (16 \times 3) = 54 + 48 = 102$ g

$$\underset{\text{1 mol (102 g)}}{Al_2O_3} \rightleftharpoons \underset{\text{2 mol}}{2Al^{3+}} + 3O^{2-}$$

So 102 g Al_2O_3 contains Al^{3+} ions = $2 \times 6.022 \times 10^{23}$

Therefore, 0.051 g Al_2O_3 will contain

$$Al^{3+} \text{ ions} = \frac{2 \times 6.022 \times 10^{23}}{102} \times 0.051$$

$$= 6.022 \times 10^{20} \ Al^{3+} \text{ ions}$$

Exemplar Problems

1. (a) Mass of 1 mol CO_2 = 44 g

Mass of 5 mol of $CO_2 = 44 \times 5 = 200$ g

Mass of 5 mol of $H_2O = 18 \times 5 = 90$ g

(b) Molar mass of Ca = 40 g = 1 mol

$$\text{Number of mol in 240 g of Ca} = \frac{240}{40} = 6$$

Molar mass of Mg = 24 g = 1 mol

$$\text{Number of mol in 240 g of Mg} = \frac{240}{24} = 10$$

Mole ratio of Ca and Mg = 6 : 10 or 3 : 5

2. (a) $CaCO_3 = Ca(1) : C(1) : O(3) = 40 : 12 : 16 \times 3 = 40 : 12 : 48 = 10 : 3 : 12$

(b) $MgCl_2 = Mg(1) : Cl(2) = 24 : 35.5 \times 2 = 24 : 71$

(c) $H_2SO_4 = H(2) : S(1) : O(4) = 2 : 32 : 64 = 1 : 16 : 32$

(d) $C_2H_5OH = C(2) : H(5) : O(1) : H(1) = C(2) : H(6) : O(1) = 24 : 6 : 16 = 12 : 3 : 8$

(e) $NH_3 = N(1) : H(3) = 14 : 3$

(f) $Ca(OH)_2 = Ca(1) : O(2) : H(2) = 40 : 32 : 2 = 20 : 16 : 1.$

3. Number of electrons in Na atom = 11

Number of electrons in Na^+ = 10

For 1 mol of Na atom and Na^+ the difference in electrons = 1 mol

For 100 mol of Na atoms and Na ions the difference = 100 mol of electrons

Mass of 100 mol of electrons = 0.0548002 g

$$\text{Mass of 1 mol of electron} = \frac{0.0548002}{100} \text{ g}$$

Mass of 1 electron

$$= \frac{0.0548002}{100 \times 6.022 \times 10^{23}} = 9.1 \times 10^{-28} = 9.1 \times 10^{-31} \text{ kg}$$

4. Molar mass of HgS = 200.6 + 32 = 232.6 g

Mass of Hg in 232.6 g of HgS = 200.6 g

$$\text{Mass of Hg in 225 g of HgS} = \frac{200.6}{232.6} \times 225 = 194.04 \text{ g}$$

5. Number of oxygen atoms in the sample = 2.58×10^{24} = 2.58×10^{24}

6.022×10^{23} atoms of oxygen = 1 mol

2.58×10^{24} atoms of oxygen

$$= \frac{\text{Total atoms}}{\text{Atoms present in 1 mole}}$$

$$= \frac{2.58 \times 10^{24}}{6.022 \times 10^{23}} \text{ moles} = 4.28 \text{ moles}$$

6.

Property of species	H_2O	CO_2	Na atom	$MgCl_2$
No. of mol	2	0.5	5	0.5
No. of particles	1.2044×10^{24}	3.011×10^{23}	$5 \times 6.022 \times 10^{23}$ $= 3.011 \times 10^{24}$	$0.5 \times 6.022 \times 10^{23}$ $\times 3 = 9.033 \times 10^{23}$
Mass	36 g	22 g	115 g	47.5 g

7. (a) 10^3 = kilo   (b) 10^{-1} = deci   (c) 10^{-2} = centi

(d) 10^{-6} = nano   (f) 10^{-12} = pico

8. (a) 5.84×10^{-3} mg $= \dfrac{5.84 \times 10^{-3}}{10^3 \times 10^3} \text{ kg} = 5.84 \times 10^{-9} \text{ kg}$

(b) $58.34 = \dfrac{58.34}{10^3} = 5.834 \times 10^{-2} \text{ kg}$

(c) $0.584 \text{ g} = \dfrac{0.584}{10^3} \text{ kg} = 5.84 \times 10^{-4} \text{ kg}$

(d) $5.873 \times 10^{-21} \text{ g} = \dfrac{5.873 \times 10^{-21}}{10^3} = 5.873 \times 10^{-24} \text{ kg}$

9. Mg^{+2} ion = 10 electrons

Mg atom = 12 electrons

Difference between Mg^{2+} and $Mg = 2$ mol of electrons
The difference between 10^3 moles of Mg and 10^3 moles of $Mg^{2+} = 10^3 \times 2$ mol of electrons
Mass of $10^3 \times 2$ moles of electrons
$= 2 \times 10^3 \times 6.022 \times 10^{23} \times 9.1 \times 10^{-31}$ kg
$= 1.096 \times 10^{-3}$ kg

10. (i) 100 g of $N_2 = 100/28$ mol

Total number of molecules $= \dfrac{100}{28} \times 6.022 \times 10^{23}$ molecules

Total number of atoms $= \dfrac{2 \times 100}{28} \times 6.022 \times 10^{23}$ atoms
$= 43.01 \times 10^{23}$ atoms

(ii) 100 g of $NH_3 = \dfrac{100}{17}$ mol

Total number of molecules $= \dfrac{100}{17} \times 6.022 \times 10^{23}$

Total number of atoms $= \dfrac{100}{17} \times 6.022 \times 10^{23} \times 4$
atoms $= 141.69 \times 10^{23}$

Hence here, NH_3 will have more number of total atoms.

11. 1 g of gold sample contains $= \dfrac{90}{100} = 0.9$ g gold

Number of mol of gold

$= \dfrac{\text{Mass of gold}}{\text{Atomic mass of gold}} = \dfrac{0.9}{197} = 0.0046$

1 mol of gold contains $= 6.022 \times 10^{23}$ atoms

0.0046 mol of gold $= 6.022 \times 10^{23} \times 0.0046 = 2.77 \times 10^{21}$ atoms

12. Mass of 1 mol of aluminium atoms $= 27$ g mol^{-1}

$Al - 3e^- \rightarrow Al^{3+}$ (3 mol of electrons)
Mass of 3 mol of electrons

$= 3 \times (9.1 \times 10^{-28}) \times 6.022 \times 10^{23}$ g $= 0.00164$ g

Molar mass of $Al^{3+} = 27 - 0.00164$ g mol^{-1}

$\qquad = 26.9984$ g mol^{-1}

Difference $= 27 - 26.9984 = 0.0016$ g
1 mol of Al atoms is heavier than 1 mol of Al^{3+} ions.

13. Mass of silver $= m$ g

Mass of gold $= \dfrac{m}{100}$ g

Number of atoms of silver $= \dfrac{\text{Mass}}{\text{Atomic mass}} \times N_A$

$= \dfrac{m}{108} \times N_A$

Number of atoms of gold $= \dfrac{m}{100 \times 197} \times N_A$

Ratio of number of atoms of gold to silver $= Au : Ag$

$= \dfrac{m}{100 \times 197} \times N_A : \dfrac{m}{108} \times N_A$

$= 108 : 100 \times 197 = 108 : 19700 = 1 : 182.41$

14. Molar mass of $CH_4 = 12 + 4 = 16$ g mol^{-1}

1 mole of $CH_4 = 16$ g $= 6.022 \times 10^{23}$ molecules

Thus, 6.022×10^{23} molecules of CH_4 have a mass $= 16$ g

So, 1.5×10^{20} molecules of CH_4 will have mass

$= \dfrac{16}{6.022 \times 10^{23}} \times 1.5 \times 10^{20} = 4 \times 10^{-3}$ g

Molar mass of $C_2H_6 = 2 \times 12 + 6 = 30$ g mol^{-1}

So, 1 mol of $C_2H_6 = 30$ g $= 6.022 \times 10^{23}$ molecules

Thus, number of molecules in 30 g of $C_2H_6 = 6.022 \times 10^{23}$

So, number of molecules in 4×10^{-3} g of C_2H_6

$= \dfrac{6.022 \times 10^{23}}{30} \times 4 \times 10^{-3} = 0.803 \times 10^{20} = 8.03 \times 10^{19}$

15. $6CO_2 + 6H_2O \xrightarrow[\text{Sunlight}]{\text{Chlorophyll}} C_6H_{12}O_6 + 6O_2$

1 mol (180 g) of glucose needs 6 mol (6×8 g) of H_2O

1 g of glucose needs $= \dfrac{108}{180}$ g of H_2O

18 g glucose needs $= \dfrac{108}{180} \times 18 = 10.8$ g

Volume of water used = Mass/Density $= \dfrac{10.8}{1} = 10.8$ cm^3

Competition Window (Objective Type)
Topic-wise MCQs

1. (a)	**2.** (d)	**3.** (c)	**4.** (a)	**5.** (d)
6. (c)	**7.** (a)	**8.** (c)	**9.** (b)	**10.** (d)
11. (d)	**12.** (b)	**13.** (c)	**14.** (d)	**15.** (c)
16. (d)	**17.** (b)	**18.** (d)	**19.** (d)	**20.** (d)
21. (a)	**22.** (b)	**23.** (d)	**24.** (c)	**25.** (a)
26. (b)	**27.** (c)	**28.** (a)	**29.** (d)	**30.** (a)
31. (c)	**32.** (b)	**33.** (b)	**34.** (b)	**35.** (d)
36. (d)	**37.** (d)	**38.** (b)	**39.** (b)	**40.** (c)
41. (d)	**42.** (a)	**43.** (b)	**44.** (c)	**45.** (a)
46. (d)	**47.** (b)	**48.** (c)	**49.** (b)	**50.** (c)

Miscellaneous

1. (a)	**2.** (c)	**3.** (c)	**4.** (c)	**5.** (b)
6. (d)	**7.** (b)	**8.** (d)	**9.** (a)	**10.** (a)
11. (c)	**12.** (b)	**13.** (c)	**14.** (a)	**15.** (a)
16. (b)	**17.** (b)	**18.** (c)	**19.** (a)	**20.** (c)

Advanced and Olympiads
Single Choice

1. (a)	**2.** (b)	**3.** (c)	**4.** (a)	**5.** (d)
6. (c)	**7.** (d)	**8.** (b)		

Multiple Choice

9. (bcd)	**10.** (ad)	**11.** (acd)	**12.** (abc)	**13.** (ad)
14. (acd)	**15.** (ac)	**16.** (bc)	**17.** (abcd)	**18.** (abd)

Comprehension Type

  I. 19. (b)   20. (c)   21. (d)

 II. 22. (c)   23. (d)

Matrix Matching

24. (a)—(r); (b)—(p); (c)—(s); (d)—(q)

25. (a)—(q, r); (b)—(q, r); (c)—(p, s); (d)—(p, t)

Integer Type

26. 2   27. 2   28. 3   29. 5   30. 4

31. 3   32. 4

Hints and Solutions

Competition Window (Objective Type)

Topic-wise MCQs

3. Constant proportions according to which a pure chemical compound always contains the same elements combined together in the same definite proportion of weight.

4. The constituents of CO_2 are carbon and oxygen. Here, carbon is a solid and Gay-Lussac's law is applicable for only gaseous reactions.

5. 32 g of oxygen contains X molecules

   32 g of SO_3 contains $\frac{80}{32}$ X molecules

   Ratio of number of molecules present in 32 g of O_2 and

   $SO_3 = X : \frac{80}{32} X = 2.5 : 1 = 5 : 2$

6. $2CO + O_2 \rightarrow 2CO_2$. The volume ratio is 2 : 1 : 2

   2 vol : 1 vol : 2 vol

   Formation of SO_3 can be represented as $2SO_2 + O_2 \rightarrow 2SO_3$. This also possesses same volume ratio (2 : 1 : 2).

8. When energy is given to the system, the solid state changes to liquid. When energy is taken out from a liquid it changes to solid. For example, ice changes to water, and water to water vapour when heat energy is given. Water vapour or steam condenses to water and water freezes to ice when energy is decreased.

9. $SnCl_2 = 119 : 2 \times 35.5$

   $SnCl_4 = 119 : 4 \times 35.5$

   Chlorine ratio in both the compounds is $= 2 \times 35.5 : 4 \times 35.5 = 1 : 2$.

16. 1 atom of Cu + 1 atom of sulfur + 9 of O + 10 atoms of H. Hence, total number of atoms in this compound is 21.

17. Nitrogen gas exists as a diatomic molecule hence, its symbol is N_2.

18. The correct symbols of these elements are as follows:

    (a) Cobalt: Co   (b) Carbon: C

    (c) Aluminium: Al   (d) Helium: He

    Here, only helium is given correct the symbol.

19. The correct symbol of nickel is Ni.

21. 58.5 g of sodium chloride contains = 23 g of sodium

    So, 5.85 g of sodium chloride contains

    $= \frac{23}{58.5} \times 5.85 = 2.3$ g of sodium

22. The mass of 6.023×10^{23} atoms of carbon = 12 g

    So, the mass of 6.023×10^{23} atoms of carbon

    $= \frac{12}{6.023 \times 10^{23}}$ g $= 1.99 \times 10^{-23}$ g

23. Percentage of hydrogen

    $= \dfrac{\text{Mass of hydrogen in water}}{\text{Molar mass of water}} \times 100 = \dfrac{2}{18} \times 100 = 11.12\%$

24. Molar mass of $CaSO_4 = 40 + 32 + 16 \times 4 = 136$

    % of O_2 in $CaSO_4 = \dfrac{64}{136} \times 100 = 47\%$

26. Relative molecular mass of $H_2O = 1 \times 2 + 16 = 18$

    18 g of water contains 2 g of hydrogen

    So, 100 g of water contains $\dfrac{2}{18} \times 100 = 11.11$ g of hydrogen

    Hydrogen in water is 11.1%.

27. Relative molecular mass of $Na_2CO_3 \cdot 10H_2O$

    $= 23 \times 2 + 12 + 16 \times 3 + 10(2 + 16) = 106 + 180$

    $= 286$ amu

    286 g of washing soda contains = 180 g of water of crystallisation

    So, 100 g of washing soda contains

    $= \dfrac{180 \times 100}{286} = \dfrac{18000}{286} = 62.9$ g of H_2O

    The percentage of H_2O in $Na_2CO_3 \cdot 10H_2O = 62.9$.

32. $A(NO_3)_3 \Rightarrow$ valency of A is $+3$

    Metal nitride $A^{+3} N^{-3} \Rightarrow AN$

34. In P_2O_3 and P_2O_5, the valencies of phosphorus are 3, 5 respectively.

35. Calcium can have only 1 valency (2) while copper can have 2 valencies (1 and 2) and iron can have 2 valencies (2 and 3).

36. Sodium can have only Na^+ and not Na^{2+} hence this pair is not possible.

37. (a) 8 g of $CH_4 = \dfrac{8}{16} = \dfrac{1}{2}$ mol of $CH_4 = \dfrac{1}{2} \times 6.023 \times 10^{23}$

(b) 8.8 of $CO_2 = \dfrac{8.8}{44} = \dfrac{1}{5}$ mol of $CO_2 = \dfrac{1}{5} \times 6.023 \times 10^{23}$

(c) 3.42 g of $C_{12}H_{22}O_{11} = \dfrac{3.42}{342} = \dfrac{1}{100}$ mol of $C_{12}H_{22}O_{11}$

$$= \dfrac{1}{100} \times 6.023 \times 10^{23}$$

(d) 4 g of $H_2 = 2$ mol of $H_2 = 2 \times 6.023 \times 10^{23}$

40. 28 g nitrogen contains 6.023×10^{23} molecules

0.28 g of nitrogen contains 6.023×10^{21} molecules.

41. 1 g atom = 1 mol

42. $W_H = 0.2 \times 6 \times 1 = 1.2$ g

43. 2 mol O atom = 1 mol O_2

44. 1 g of $CO_2 = 1/44 \times 6.022 \times 10^{23}$ molecules $= 1.368 \times 10^{22}$ molecules

$$1 \text{ g of } N_2 = \dfrac{1}{28} \times 6.022 \times 10^{23} \text{ molecules}$$

$$= 2.15 \times 10^{22} \text{ molecules}$$

$$1 \text{ g of } H_2 = \dfrac{1}{2} \times 6.022 \times 10^{23} \text{ molecules}$$

$$= 3.011 \times 10^{23} \text{ molecules}$$

$$1 \text{ g of } CH_4 = \dfrac{1}{16} \times 6.022 \times 10^{23} \text{ molecules}$$

$$= 3.76 \times 10^{22} \text{ molecules}$$

45. 1 mol of NaCl = 23 + 35.5 = 58.5 g of NaCl

Number of mol in 5085 g of NaCl = 5085 / 58.5 = 0.1 mol

Each NaCl formula unit = $Na^+ + Cl^- = 2$ ions

Number of ions $= 0.2 \times 6.022 \times 10^{23} = 1.2044 \times 10^{23}$ ions

46. 16 g of oxygen atom $= \dfrac{16}{16} = 1$ mol

48 g of ozone $= \dfrac{48}{48} = 1$ mol

15 g of ethane $(C_2H_6) = \dfrac{15}{30} = \dfrac{1}{2}$ mol

47. Molarity $(M) = \dfrac{w}{\text{Molar mass} \times V} \times 1000$

$$= \dfrac{3.65}{36.5 \times 200} \times 1000 = 0.5$$

48. Molality $(m) = \dfrac{w}{\text{Molar mass} \times W} \times 1000$

$$= \dfrac{9}{180 \times 90} \times 1000 = \dfrac{1000}{1800} = 0.55$$

49. Total number of electrons

$$= \dfrac{w}{M} \times N_A \times \text{Number of electron in 1 molecule}$$

$$= \dfrac{1.7}{17} \times 6.022 \times 10^{23} \times 10 = 6.022 \times 10^{23}$$

50. Total charge on carbonate

$$= \dfrac{w}{M} \times N_A \times \text{Charge on carbonate in 1 molecule}$$

$$\times 1.6 \times 10^{-19} \text{ C}$$

$$= \dfrac{3}{60} \times 6.022 \times 10^{23} \times 2 \times 1.6 \times 10^{-19}$$

$$= 6.022 \times 1.6 \times 10^{3} = 9635.2 \text{ C}$$

Miscellaneous

1. $n = \dfrac{5.6}{22.4} = \dfrac{1}{4}$ mol

2. Weight = Number of mol × molar mass

0.2 mol of sucrose $(C_{12}H_{22}O_{11}) = 0.2 \times 342 = 68.4$ g

2 mol of $CO_2 = 2 \times 44 = 88$ g

2 mol of $CaCO_3 = 2 \times 100 = 200$ g

10 mol of $H_2O = 10 \times 18 = 180$ g

3. Atomic mass of Mg = 24 g mol^{-1}

24 g of Mg = 1 mol

12 of Mg = 12/24 = 0.5 mol

4. 1 mol of stars $= 6.022 \times 10^{23}$ stars

Number of mol of stars

$$= \dfrac{\text{Given stars}}{\text{Number of stars in 1 mole}}$$

$$= \dfrac{10^{22}}{6.023 \times 10^{23}} = 0.0166 \text{ mol}$$

6. Number of atoms

$$= \dfrac{\text{mass substance} \times (\text{Number of atoms in the molecule}) \times N_A}{\text{Molar mass}}$$

18 g of $H_2O = 1$ mol $= 1 \times 6.022 \times 10^{23}$

$$= 3 \times 6.022 \times 10^{23} \text{ atoms} = 1.8066 \times 10^{23} \text{ atoms}$$

18 g of $O_2 = \dfrac{18 \times 2}{32} \times 6.022 \times 10^{23} = 6.77 \times 10^{23}$ atoms

18 g of $CO_2 = \dfrac{18 \times 3}{44} \times 6.022 \times 10^{23} = 7.39 \times 10^{23}$ atoms

18 g of $CH_4 = \dfrac{18 \times 5}{16} \times 6.022 \times 10^{23} = 3.38 \times 10^{24}$ atoms

7. Mass of 1 neutron $= \dfrac{1}{N_A}$ g

Mass of 1 mol of $H_2O = 18$

Mass of one molecule of water $= \dfrac{18}{N_A}$ g

Number of neutrons in one atom of $H = 0$

Number of neutrons in one atom of $O = 8$

Mass of 8 neutrons (in H_2O) $= \dfrac{8}{N_A}$

Fraction of mass of water due to neutrons

$= \dfrac{8}{N_A} \Big/ \dfrac{18}{N_A}$ or $\dfrac{8}{18} = \dfrac{4}{9}$

8. Mass percent $= \dfrac{\text{Mass of solute}}{\text{Mass of solution}} \times 100$

$= \dfrac{4\text{ g}}{4\text{ g}+18\text{ g}} \times 100 = \dfrac{400}{22} = 18.18$

9. Mw of $Al_2O_3 = 27 \times 2 + 16 \times 3 = 102$ g

% of Al $= \dfrac{54 \times 100}{102} = 52.94\%$

10. Mass of one atom of oxygen = Atomic mass/N

$= \dfrac{16}{6.023 \times 10^{23}}$ g

11. Moles or molecules of $O_2 : N_2 = \dfrac{1}{32} : \dfrac{4}{28}$

$= \dfrac{1}{32} : \dfrac{1}{7} = \dfrac{7}{32} : 1 = 7 : 32$

12. 0.25 g atom or mol of Si = 0.5 mol of oxygen

1 mol of Si $= \dfrac{0.5}{0.25} = 2$ mol of oxygen

Formula: SiO_2

13.

Element	Moles	Least ratio
S	$\dfrac{22}{32} = 0.6875$	$\dfrac{0.6875}{0.6875} = 1$
F	$\dfrac{78}{19} = 4.10$	$\dfrac{4.10}{0.6875} = 5.97 \approx 6$

Formula $= SF_6$

14. Number of mol of sucrose = 3.42/342 = 0.01 mol

1 mol of sucrose $(C_{12}H_{22}O_{11}) = 11 \times N_A$ atoms of O

0.01 mol of $C_{12}H_{22}O_{11} = 0.01 \times 11 \times N_A = 0.11 \times N_A$ atoms of O

Number of mol of water $= \dfrac{18}{8} = 1$ mol

1 mol of water contains $= 1 \times N_A$ atoms of O

Total number of oxygen atoms

$= 0.11\, N_A + 1.0\, N_A = 1.11\, N_A$

Number of oxygen atoms in solution $= 1.11\, N_A$

$= 1.11 \times 6.022 \times 10^{23} = 6.68 \times 10^{23}$

15. Empirical formula of compound $= CH_2$

Molecular mass of the compound = 42

So, $n = 42/14 = 3$

So molecular formula $= C_3H_6$

16. N atoms have 1 g atom

So 2×10^{23} atoms $= \dfrac{2 \times 10^{23}}{6.023 \times 10^{23}} = 0.33$ g atoms

Hence N atoms of element weight 32 g

So 2×10^{23} atoms of element weigh

$= \dfrac{32 \times 2 \times 10^{23}}{6.023 \times 10^{23}} = 10.63$ g

17. P: 50%   Q: 50%

$P : Q = 5 : 2.5 = 2 : 1$

Hence, P_2Q

18. 4 g sulfur is in 100 g compound, hence 32 g sulfur is in

$= \left(\dfrac{100}{4} \times 32 \right) = 80$ g compound

19. Weight of 6.023×10^{23} (Avogadro's number)

= Mw of $CuSO_4.5H_2O = 249$ g

= 1 mol of $CuSO_4.5H_2O$

Weight of 1×10^{22} molecules of $CuSO_4.5H_2O$

$= \dfrac{249 \times 1 \times 10^{22}}{6.023 \times 10^{23}} = 4.14$ g

20. Percentage of metal A = 70.00 and that oxygen = 30.00.

Element	%	Atomic ratio	Least ratio	Whole number ratio
Metal	70.00	$\dfrac{70}{56} = 1.25$	$\dfrac{1.25}{1.25} = 1$	2
Oxygen	30.00	$\dfrac{30}{16} = 1.875$	$\dfrac{1.875}{1.25} = 1.5$	3

Hence, empirical formula $= A_2O_3$

Advanced and Olympiads
Single Choice

2. In CuO and Cu_2O the O : Cu is 1 : 1 and 1 : 2 respectively. This is the law of multiple proportions.

3. 50% of P (atomic mass 10), 50% of Q (atomic mass 20).

Relative number of atoms of $P = \dfrac{50}{10} = 5$ and $Q = \dfrac{50}{20} = 2.5$

Simple ratio 2 : 1. Formula P_2Y.

4.

Element	%	Atomic mass	Relative number of atoms	Simple ratio of atoms
C	80	12	$\dfrac{80}{12} = 6.66$	$\dfrac{6.66}{6.66} = 1$
H	20	1	$\dfrac{20}{1} = 20.0$	$\dfrac{20.0}{6.66} = 3$

So empirical formula is CH_3.

5. (i) 1 mol of water = 18 g
    2 mol of water = 2×18 g = 36 g
(ii) 20 mol of water = $18 \times 20 = 360$ g
(iii) 6.022×10^{23} molecules of water = 1 mol = 18 g
(iv) 1 mol of water = 6.022×10^{23} molecules
6.22×10^{23} molecules = 1 mol = 1.2044×10^{25} molecules

$$= \dfrac{1}{6.022 \times 10^{23}} \times 1.2044 \times 10^{25} = 20 \text{ mol} = 20 \times 18 = 360 \text{ g}$$

6. Moles of $C = \dfrac{6.023 \times 10^{23}}{6.023 \times 10^{23}} = 1$

Moles of $H = \dfrac{1.8069 \times 10^{24}}{6.023 \times 10^{23}} = 3$

Moles of $O = \dfrac{3.0115 \times 10^{23}}{6.023 \times 10^{23}} = 0.5$

$C : H : O = 1 : 3 : 0.5 = 2 : 6 : 1$
Empirical formula = C_2H_6O

7. Mw of metal chloride $(MCl_n) = M + n \times 35.5$
$$= 2 \times VD = 2 \times 39.5 = 79.0$$

$Mw = Ew \times n$
Valency of metal $= n$
Atomic weight of element $= Ew \times n$
$MCl_n = 3.82 \times n \times n \times 35.5 = 79$

$n = \dfrac{79}{3.82 \times 35.5} \approx 2$

Atomic weight $= Ew \times n = 3.82 \times 2 = 7.64$

8. According to equation:

$$CH_4(g) + 2O_2(g) \to CO_2(g) + 2H_2O(g)$$
$$\underset{16 \text{ g}}{} \qquad\qquad\qquad \underset{44 \text{ g}}{}$$

44 g of CO_2 is obtained from 16 g of $CH_4(g)$
So, 22 g of CO_2 is obtained from 8 g of $CH_4(g)$
Mol of $CH_4(g) = \dfrac{8 \text{ g}}{16 \text{ g}} = 0.5$ mol of $CO_2(g)$

Multiple Choice

9. Atoms of inert gases exist in monoatomic or independent form.

11. Sulfite $\left(SO_3^{2-}\right)$, sulfate $\left(SO_4^{2-}\right)$ and phosphate $\left(PO_4^{3-}\right)$ ions are polyatomic as they contain more than one ion. N^{3-} is monoatomic.

13. Symbol of tin, aluminium and lead are Sn, Al and Pb respectively.

14. $8 \text{ g O}_2 = \dfrac{8}{32}$ mol = 0.25 mol

$7 \text{ g CO} = \dfrac{7}{28} = 0.25$ mol, $7 \text{ g N}_2 = \dfrac{7}{28} = 0.25$ mol

$8 \text{ g SO}_2 = \dfrac{8}{64} = 0.125$ g mol

$11 \text{ g CO}_2 = \dfrac{11}{44} = 0.25$ mol

Equal moles contain equal number of molecules.

15. Here the amount of oxygen which combines with fixed amount of C and S in their oxides will be in a simple whole number ratio.

Comprehension Type

19. Molarity $(M) = \dfrac{w}{M \times V} \times 1000 = \dfrac{0.63}{63 \times 500} \times 1000 = 0.02$

20. Weight of water $= \dfrac{\text{Volume}}{\text{Density}} = \dfrac{3.6}{1} = 3.6$ g

Moles of water $= \dfrac{3.6}{18} = 0.2$

Total number of atoms $n \times N_A \times$ number of atoms in 1 molecule
$= 0.2 \times N_A \times 3 = 0.6 \, N_A$

21. Moles of methane $= \dfrac{4}{16} = \dfrac{1}{4} = 0.25$

Total electrons $n \times N_A \times$ number of electrons in 1 molecule
$= 0.25 \times N_A \times 10 = 2.5 \, N_A$

22. Both Sn and Pb can show valencies 2 and 4. The valencies of Na, Mg, Zn are fixed that is 1, 2, 2 respectively.

23. Hg cannot form HgO_2 while the rest other are possible.

Integer Type

26. 3.0115×10^9 molecules of $n = 10^{-12}$ g

6.023×10^{23} molecules of $n = \dfrac{10^{-12} \times 6.023 \times 10^{23}}{3.0115 \times 10^9}$

$= 200 \text{ g} = 200$ amu

So, $n = 2$

27. Given $V = 2$ L, molarity $= 0.5$ M, mol $= ?$

$$\text{Molarity} = \frac{\text{Number of moles of solute}}{V \text{ of solution in } L}$$

or $0.5 = \dfrac{\text{Moles}}{4}$

So, mol $= 4 \times 0.5 = 2$

28. Moles Xe $= \dfrac{55.2}{131 \text{ g/mole}} = 0.42$ mol

Moles Cl $= \dfrac{44.8 \text{ g}}{35.5 \text{ g/mole}} = 1.26$ mol

$\dfrac{\text{Moles Cl}}{\text{Moles Xe}} = \dfrac{1.26}{0.42} = 3$

So, the empirical formula is $XeCl_3$

Hence, the value of x is 3.

29. 1 microgram $= 1\ \mu g = 10^{-6}$ g

1 mol of C $= 12$ g $= 6.023 \times 10^{23}$ atoms

12 g of C $= 6.023 \times 10^{23}$ atoms

10^{-6} g of C $= \dfrac{6.023 \times 10^{23} \times 10^{-6}}{12}$

$= 5 \times 10^{16}$ atoms of C

So, $n = 5$

30. Molar mass (Mw_2) of solute (NaOH)

$= (23 + 16 + 1)\text{g} = 40$ g

Use the formula of molarity (M)

$$= \frac{W_2 \times 1000}{Mw_2 \times V_{\text{sol}}(\text{in L})} = \frac{4.0 \times 1000}{40 \times 250}$$

$= 0.4$ mol L$^{-1} = 0.4$ M

So, $X = 4$

31. Weight of 1 atom of element $= 6.644 \times 10^{-23}$ g

So, weight of N atoms of element

$= 6.644 \times 10^{-23} \times 6.023 \times 10^{23} = 40$

Hence 40 g weight of element has 1 g atom

So 40×10^{-23} g weight of element

$= \dfrac{40 \times 10^3}{40} = 10^3$ g atom

So, $X = 3$

32. 100 g haemoglobin has $= 0.25$ g Fe

So, 89600 g haemoglobin has $= \dfrac{0.25 \times 89600}{100} = 224$ g Fe

So, 1 mol or N molecules of haemoglobin have

$= \dfrac{224}{56}$ g atoms Fe

$= 4$ g atom Fe

Hence, 1 molecule of haemoglobin has 4 atoms of Fe

Chemical Reactions and Equations

LEARNING OBJECTIVES

After studying this unit, you will be able to understand:

- Chemical reactions and their characteristics, reactants, products
- How to define a chemical equation and to write it
- Combination reactions, decomposition reactions, displacement reactions, double displacement reactions, oxidation and reduction reactions, and differentiate between them

Introduction

The changes occurring in a substance involve physical and chemical changes. In a physical change, the form of the matter changes but its chemical identity remains the same. A chemical change occurs when a substance combines with another to form a new substance and is called chemical synthesis; alternatively, chemical decomposition into two or more different substances can occur. These processes are called chemical reactions and are in general not reversible, except by further chemical reactions. For example, the reaction between sodium and water to produce sodium hydroxide and hydrogen is a chemical change.

Chemical and physical changes

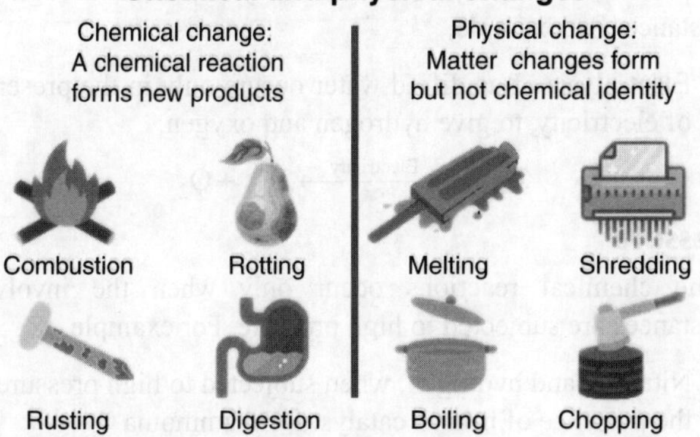

Chemical Reactions

Chemical reactions are the processes by which new substances with new properties are formed. Chemical reactions involve chemical changes. During a chemical reaction, a rearrangement of atoms takes place between the reacting species to form new species having entirely different properties. Chemical reactions involve breaking of old chemical bonds which exist between the atoms of the reacting substances, and then forming of new chemical bonds between the rearranged atoms of the new substances. During a chemical reaction, atoms of one element do not change into those of another element; however, a rearrangement of atoms takes place in a chemical reaction. Let us now discuss reactants and products of a chemical reaction.

Reactants

The substances which take part in a chemical reaction are known as *reactants*.

For example, in the breaking up of calcium carbonate into calcium oxide and carbon dioxide, calcium carbonate is the reactant. Similarly, sodium and water are the reactants when they react.

$$CaCO_3 \rightarrow CaO + CO_2$$
Calcium carbonate   Calcium oxide   Carbon dioxide

Products

The new substances produced as a result of a chemical reaction are known as *products*.

For example, hydrogen and sodium hydroxide are the products of the reaction between sodium and water.

$$Na + H_2O \rightarrow NaOH + H_2$$
Sodium   Water   Sodium hydroxide   Hydrogen

When a magnesium ribbon is heated, it burns in air with a dazzling white flame to form a white powder called magnesium oxide.

$$Magnesium + Oxygen \xrightarrow{Heat} Magnesium\ oxide$$
(As ribbon)   (From air)   (White powder)

The burning of magnesium in air to form magnesium oxide is an example of a chemical reaction.

KNOWLEDGE BOOSTER

The magnesium ribbon used here usually has a coating of magnesium oxide on its surface, which is formed by the slow action of oxygen of air on it. So, before burning in air, the magnesium ribbon is cleaned by rubbing with a sand paper. This is done to remove the protective layer of magnesium oxide from the surface of the magnesium ribbon so that it may readily combine with the oxygen of air (on heating). The dazzling (very bright) white light given out during the burning of a magnesium ribbon is harmful to the eyes. So, the magnesium ribbon should be burned by keeping it as far as possible from the eyes.

A larger number of chemical reactions keep on occurring in our daily life. For example, souring of milk (when left at room temperature during summer), formation of curd from milk, digestion of food in our body, process of respiration, rusting of iron (when left exposed to humid atmosphere), burning of fuels (like wood, coal, kerosene, petrol and LPG), burning of candle wax, ripening of fruits, and so on, are all chemical changes which involve chemical reactions.

Conditions Needed for a Chemical Change

A chemical change or chemical reaction can take place when particles undergo collisions and collisions occur when reactants are in close contact or when energy is supplied. Hence, one or more of the following conditions are necessary for a chemical change to occur.

Mixing

In some cases, a chemical reaction takes place when two substances are mixed (for contact) in their solid states. For example, lead nitrate (white) and potassium iodide (white) react to form lead iodide (yellow).

$$Pb(NO_3)_2(s) + 2KI(s) \rightarrow 2KNO_3(s) + PbI_2(s)$$

Solution

In a few cases, a chemical reaction takes place when the substances are mixed in either molten or aqueous state. For example, sodium chloride and silver nitrate react in the solution state to form a precipitate of silver chloride and sodium nitrate.

$$NaCl(aq) + AgNO_3(aq) \rightarrow AgCl\downarrow + NaNO_3(aq)$$
$$\text{(White ppt.)}$$

Heat

Some chemical reactions occur only on heating. For example,

- Copper carbonate decomposes into copper oxide and carbon dioxide on heating.

$$CuCO_3(s) \xrightarrow{\Delta} CuO(s) + CO_2(g)$$

- Lead nitrate decomposes on heating leaving a yellow residue of lead monoxide, a brown gas (nitrogen dioxide) and a colourless gas (oxygen).

$$2Pb(NO_3)_2 \xrightarrow{\Delta} 2PbO + 4NO_2 + O_2$$

Light

Some chemical reactions can take place by the action of light. These are called *photochemical reactions* or *photolysis* reactions. Here, molecules of the reactants absorb light energy to get activated and then react quickly. For example,

- **Photosynthesis:** Plants form glucose from carbon dioxide and water in the presence of light.

$$6CO_2 + 12H_2O \xrightarrow{\text{Light}} \underset{\text{(Glucose)}}{C_6H_{12}O_6} + 6H_2O$$

- Hydrogen and chlorine react in the presence of sunlight to give HCl.

$$H_2 + Cl_2 \xrightarrow{\text{Sunlight}} 2HCl$$

- Solutions of silver nitrate and hydrogen peroxide are kept in brown bottles in the laboratory because they decompose in the presence of light. Silver nitrate solution becomes black as it decomposes to silver, nitrogen dioxide and oxygen while hydrogen peroxide changes to water and the colourless, odourless oxygen is gas evolved.

$$2AgNO_3 \xrightarrow{\text{Sunlight}} 2Ag + 2NO_2 + O_2$$
$$2H_2O_2 \xrightarrow{\text{Sunlight}} 2H_2O + O_2$$

Electricity

Some chemical reactions, such as the decomposition of certain compounds, occur only when electricity is passed through the substance. For example,

- Electrolysis of acidified water occurs only in the presence of electricity, to give hydrogen and oxygen.

$$2H_2O \xrightarrow{\text{Electricity}} 2H_2 + O_2$$

Pressure

Some chemical reactions occur only when the involved substances are subjected to high pressure. For example,

- Nitrogen and hydrogen, when subjected to high pressure in the presence of iron as catalyst form ammonia.

$$N_2 + 3H_2 \underset{}{\overset{\text{Above 200 atm}}{\rightleftharpoons}} 2NH_3$$

- Mercuric chloride and potassium iodide when crushed together in a mortar, give a scarlet-coloured substance called mercuric iodide.

$$HgCl_2 + 2KI \rightarrow HgI_2 + KCl$$

Catalyst

Some chemical reactions need a catalyst to accelerate or decelerate the rate at which they occur. The catalysts themselves do not take part in the reaction and so remain unchanged in mass and composition after the reaction. For example,

- Potassium chlorate decomposes only at 700°C and even then, the rate of release of oxygen is very slow. However, when it is heated in the presence of manganese dioxide (catalyst), decomposition begins at a much lower temperature, 300°C, and manganese dioxide remains unaffected.

$$2KClO_3 \xrightarrow{MnO_2} 2KCl + 3O_2$$

- Ammonia reacts with oxygen to produce nitric oxide and water vapour in the presence of platinum (catalyst).

$$4NH_3 + 5O_2 \xrightarrow[800°C]{Pt} 4NO + 6H_2O$$

KNOWLEDGE BOOSTER

(i) Positive catalyst: A positive catalyst accelerates the rate of a reaction and is known as an *accelerator*. For example, the rate of decomposition of hydrogen peroxide gets increased in the presence of manganese dioxide.

$$2H_2O \xrightarrow{MnO_2} 2H_2O + O_2$$

(ii) Negative catalyst: A negative catalyst is employed to retard the rate of a reaction. It is known as an *inhibitor*. For example,
- Phosphoric acid retards the rate of decomposition of hydrogen peroxide.
- The rate of oxidation of chloroform decreases in the presence of ethyl alcohol.

Characteristics or Tests of Chemical Reactions

The conversion of reactants into products in a chemical reaction is often accompanied by some features (changes) which can be observed easily. These easily observable features which take place as a result of chemical reactions are called *characteristics* of chemical reactions. The important characteristics of chemical reactions are:

- change in state,
- formation of a precipitate,
- change in temperature,
- change in colour, and
- evolution of a gas.

Change in State

Many chemical reactions are characterised by a change in state—the physical state changes from solid to liquid and gas. For example, the combustion reaction of candle wax is characterised by a change in state from solid to liquid and gas (because wax is a solid, combustion of wax leads to a liquid at room temperature, carbon dioxide produced by the combustion of wax is a gas).

Formation of a Precipitate

Some chemical reactions are characterised by the formation of a precipitate. A precipitate is a solid product which separates out from the solution during a chemical reaction. For example, when potassium iodide solution is added to a solution of lead nitrate, a yellow precipitate of lead iodide is formed.

Let us take another example of a chemical reaction in which a precipitate is formed. When dilute sulfuric acid is added to barium chloride solution taken in a test tube, a white precipitate of barium sulfate is formed. Thus, the chemical reaction between sulfuric acid and barium chloride solution is characterised by the formation of a white precipitate of barium sulfate.

Change in Temperature

Many chemical reactions are characterised by a change in temperature. For example, when quick lime reacts with water, slaked lime is formed and a huge amount of heat energy is produced. This heat raises the temperature due to which the reaction mixture becomes hot. This is an exothermic reaction (heat producing reaction).

When barium hydroxide [$Ba(OH)_2$] is added to ammonium chloride (NH_4Cl) taken in a test tube and mixed with a glass rod, barium chloride, ammonia and water are formed. A lot of heat energy is absorbed during this reaction due to which the temperature of the reaction mixture falls and the bottom of the test tube becomes very cold. This is a case of an endothermic reaction (heat absorbing reaction).

Change in Colour

Some chemical reactions are characterised by a change in colour. For example, when citric acid reacts with potassium permanganate solution, the purple colour of potassium permanganate solution disappears (it becomes colourless). When sulfur dioxide gas is passed through acidified potassium dichromate solution, the orange colour of potassium dichromate solution changes to green.

Evolution of a Gas

Many chemical reactions are characterised by the evolution of a gas. For example, when zinc granules react with dilute sulfuric acid, bubbles of hydrogen gas are produced. So, the chemical reaction between zinc and dilute sulfuric acid is characterised by the evolution of hydrogen gas.

When magnesium reacts with a dilute acid (like dilute hydrochloric acid or dilute sulfuric acid), hydrogen gas is evolved. So, the chemical reaction between sodium carbonate and dilute hydrochloric acid is characterised by the evolution of carbon dioxide gas.

Before we go further, we should know the meaning of the term precipitate. A precipitate can be formed by mixing an aqueous solution (water solution) of reactants when one of the products is insoluble in water. A precipitate can also be formed by passing a gas into an aqueous solution of a substance (like passing carbon dioxide gas into lime water).

KNOWLEDGE BOOSTER

Rearrangement and isomerisation reactions

As we know, atoms in a molecule are held together by a force of attraction called a *bond*. The molecules do not participate directly in a chemical reaction. First, they break down into atoms and these atoms then take part in the reaction. New bonds are formed between the atoms to form the products. That is, rearrangement or regrouping of atoms takes place in various ways to give products. For example, when ammonium cyanate is heated, different bonds in ammonium cyanate molecules are broken and new bonds are formed to produce urea.

$$\underset{\text{Ammonium cyanate}}{H_4N-N=C=O} \rightarrow \underset{\text{Urea}}{H_2N-\overset{\displaystyle O}{\overset{\|}{C}}-NH_2}$$

Here, you can see that the molecular formulae of both ammonium cyanate and urea are the same, but their properties are quite different as they are two different compounds. Such compounds are known as isomers of each other and the reactions that give such isomers are called *isomerisation reactions*.

Illustrations

1. Why should a magnesium ribbon be cleaned before burning in air?

Solution: Magnesium ribbon is cleaned or rubbed with a sand paper before burning to remove the layer of magnesium oxide formed on its surface (due to exposure to air). This layer interferes in the further reaction of the pure metal.

2. Which one is a chemical change—rusting of iron or melting of iron?

Solution: Rusting of iron is a chemical change.

3. Why is respiration considered as an exothermic reaction?

Solution: As respiration results in the oxidation of glucose to produce heat energy, it is an exothermic reaction.

Chemical Equations

All chemical reactions are represented by chemical equations. The method of representing a chemical reaction with the help of symbols and formulae of the substances involved in it is called a chemical equation. For example, zinc metal reacts with dilute sulfuric acid to give zinc sulfate and hydrogen gas. This is known as the *word equation*.

KNOWLEDGE BOOSTER

The amount of product actually obtained in a reaction is known as actual yield.

Balancing of a chemical equation refers to establishing a mathematical relationship between the quantity of reactants and products. These quantities are expressed as grams or moles.

By putting the symbols and formulae of all substances in the above word equation, we get the following chemical equation:

$$\underset{\text{Reactants}}{\underbrace{Zn+H_2SO_4}} \rightarrow \underset{\text{Products}}{\underbrace{ZnSO_4+H_2}}$$

The substances which combine or react are called reactants. For example, zinc and sulfuric acid are the reactants. The reactants are always written on the left hand side in an equation with a plus sign (+) between them.

The new substances produced in a reaction are called products. Zinc sulfate and hydrogen are the products in the case. The products are always written on the right hand side in an equation with a plus sign (+) between them.

When potassium nitrate is heated, it gives potassium nitrite and oxygen. This reaction may be represented in the form of a chemical equation as follows.

$$\underset{\text{Potassium nitrate}}{KNO_3} \rightarrow \underset{\text{Potassium nitrite}}{KNO_2} + \underset{\text{Oxygen}}{O_2}$$

Energy Change in Chemical Reactions

Every substance has a certain amount of stored energy in the form of potential energy and it is defined as its chemical energy. A chemical reaction involves a change of chemical energy, that is, there is a difference between the energies of reactants and products. This energy can be in the form of heat, light, sound and electricity. A chemical reaction involves the breaking and formation of bonds so energy change is definite. Depending upon the energy change, reactions are of two types.

Exothermic	Endothermic
A chemical reaction in which heat is evolved during the reaction is known as an exothermic reaction.	A chemical reaction in which heat is absorbed during the reaction is known as an endothermic reaction.
The total energy of the reactants is more than that of the products.	The total energy of the reactants is less than that of the products.
It causes a rise in temperature.	It causes a decrease in temperature.
Example, (i) Any combustion process: $C + O_2 \xrightarrow{\Delta} CO_2 + Heat$ $CH_4 + 2O_2 \xrightarrow{\Delta} CO_2 + H_2O + Heat$ (ii) Any neutralisation reaction: $NaOH + HCl \longrightarrow NaCl + H_2O + Heat$ (iii) Addition of quick lime in water: $CaO + H_2O \longrightarrow Ca(OH)_2 + Heat$ It is called slaking of lime and is accompanied by a hissing sound. (iv) Respiration: Combustion of glucose with O_2 in the cells of the body is an exothermic reaction and it provides energy. $C_6H_{12}O_6 + 6O_2 \longrightarrow 6CO_2 + 6H_2O + Energy$	Example, (i) Formation of carbon disulfide, nitric oxide $C + 2S \xrightarrow{\Delta} CS_2$ $N_2 + O_2 \xrightarrow[3000°C]{\Delta} 2NO$ (ii) Decomposition reaction: $CaCO_3 \xrightarrow[1000°C]{\Delta} CaO(s) + CO_2(g)$ $2KClO_3 \xrightarrow{\Delta} 2KCl + 3O_2$

Types of Chemical Reactions

Chemical reactions are mainly of these types:

- combination reactions,
- decomposition reactions,
- displacement reactions,
- double displacement reactions, and
- oxidation and reduction reactions.

Combination Reactions

The reactions in which two or more substances combine to form a single compound, are known as combination reactions. In a combination reaction, two or more elements can combine to form a compound, two or more compounds can combine to form a new compound, or an element and a compound can combine to form a new compound. For example,

(i) Magnesium and oxygen combine when heated, to form magnesium oxide.

$$2\underset{\text{Magnesium}}{Mg(s)} + \underset{\text{Oxygen}}{O_2(g)} \xrightarrow{\text{Combination}} \underset{\text{Magnesium oxide}}{2MgO(s)}$$

(ii) Hydrogen burns in oxygen to form water.

$$2\underset{\text{Hydrogen}}{H_2(g)} + \underset{\text{Oxygen}}{O_2(g)} \xrightarrow{\text{Combination}} \underset{\text{Water}}{2H_2O(l)}$$

(iii) Carbon (coal) burns in air to form carbon dioxide.

$$\underset{\text{Carbon (coal)}}{C(s)} + \underset{\text{Oxygen (From air)}}{O_2(g)} \xrightarrow{\text{Combination}} \underset{\text{Carbon dioxide}}{CO_2(g)}$$

(iv) Hydrogen combines with chlorine to give hydrogen chloride.

$$\underset{\text{Hydrogen}}{H_2(g)} + \underset{\text{Chloride}}{Cl_2(g)} \xrightarrow{\text{Combination}} \underset{\text{Hydrogen chloride}}{2HCl(g)}$$

(v) Sodium metal burns in chlorine to give sodium chloride.

$$2\underset{\text{Sodium}}{Na(s)} + \underset{\text{Chlorine}}{Cl_2(g)} \xrightarrow{\text{Combination}} \underset{\text{Sodium chloride}}{2NaCl(s)}$$

(vi) When iron powder is heated with sulfur, iron sulfide is formed.

$$\underset{\text{Iron}}{Fe(s)} + \underset{\text{Sulphur}}{S(s)} \xrightarrow{\text{Combination}} \underset{\text{Iron sulphide}}{FeS(s)}$$

(vii) Calcium oxide (lime or quicklime) reacts vigorously with water to form calcium hydroxide (slaked lime).

$$\underset{\substack{\text{Calcium oxide}\\\text{(lime or quicklime)}}}{CaO(s)} + \underset{\text{Water}}{H_2O(l)} \xrightarrow{\text{Combination}} \underset{\substack{\text{Calcium hydroxide}\\\text{(Slaked lime)}}}{Ca(OH)_2(s)}$$

(viii) Ammonia reacts with hydrogen chloride to form ammonium chloride.

$$\underset{\text{Ammonia}}{NH_3(g)} + \underset{\text{Hydrogen chloride}}{HCl(g)} \xrightarrow{\text{Combination}} \underset{\text{Ammonium chloride}}{NH_4Cl(s)}$$

Illustrations

1. Write balanced chemical equations for the following reactions:

   (i) Sodium metal reacts with water to form sodium hydroxide and hydrogen gas.

   (ii) Magnesium reacts with dil. HCl.

   Solution: The reactions for these equations are

   (i) $2Na(s) + 2H_2O(l) \rightarrow 2NaOH(aq) + H_2(g)$
   $\quad$ Sodium $\quad$ Water $\quad$ Sodium hydroxide $\quad$ Hydrogen

   (ii) $Mg + 2HCl \rightarrow MgCl_2 + H_2$

2. Write the chemical equations for the reactions taking place when:

   (i) Iron reacts with steam

   (ii) Copper is heated in air

Solution:

(i) $3Fe(s) + 4H_2O(g) \rightarrow Fe_3O_4(s) + 4H_2(g)$

(ii) $2Cu + O_2 \xrightarrow{\text{Heat}} 2CuO(s)$

3. Write the balanced chemical equations for the reactions that take place during respiration. Identify the type of combination reaction that takes place during this process and justify the name. Give one more example of this type of reaction.

   Solution: Reaction taking place during respiration:

   $$C_6H_{12}O_6 + 6O_2 \rightarrow 6CO_2 + 6H_2O + \text{Heat}$$

   An exothermic combination reaction takes place during this process due to the evolution of heat. Combustion of methane is another example of such a reaction.

   $$CH_4(g) + 2O_2(g) \rightarrow CO_2(g) + 2H_2O$$

Decomposition Reactions

The reactions in which a compound splits up into two or more simpler substances is called a decomposition reaction. Decomposition reactions are carried out by applying heat, light or electricity (Tables 4.1, 4.2). Heat, light or electricity provides energy which breaks a compound into two or more simpler compounds. A decomposition reaction is the opposite of a combination reaction. Here are some examples of decomposition reactions.

(i) When calcium carbonate is heated, it decomposes to give calcium oxide and carbon dioxide.

$$CaCO_3(s) \xrightarrow[\text{(Decomposition)}]{\text{Heat}} CaO(s) + CO_2(g)$$

Calcium carbonate (Limestone) $\quad$ Calcium oxide (Lime) $\quad$ Carbon dioxide

When a decomposition reaction is carried out by heating, it is known as thermal decomposition (thermal means, relating to heat). The decomposition of calcium carbonate into calcium oxide and carbon dioxide is an example of thermal decomposition (because it is carried out by heating).

(ii) When potassium chlorate is heated in the presence of manganese dioxide catalyst, it decomposes to give potassium chloride and oxygen.

$$2KClO_3(s) \xrightarrow[\text{(Decomposition)}]{\text{Heat}} 2KCl(s) + 3O_2(g)$$

Potassium chlorate $\quad$ Potassium chloride $\quad$ Oxygen

(iii) When lead nitrate is heated strongly, it breaks down to form simpler substances like lead monoxide, nitrogen dioxide and oxygen.

$$2Pb(NO_3)_2(s) \xrightarrow[\text{(Decomposition)}]{\text{Heat}}$$

Lead nitrate (Colourless)

$$2PbO(s) + 4NO_2(g) + O_2(g)$$

Lead monoxide (Yellow) $\quad$ Nitrogen dioxide (Brown fumes) $\quad$ Oxygen

(iv) When electric current is passed through acidified water, it decomposes to give hydrogen gas and oxygen gas. This reaction can be represented as:

$$2H_2O(l) \xrightarrow[\text{(Decomposition)}]{\text{Electricity}} 2H_2(g) + O_2(g)$$

Water $\quad$ Hydrogen $\quad$ Oxygen

(v) When electric current is passed through molten sodium chloride, it decomposes into sodium metal and chlorine gas.

$$2NaCl(l) \xrightarrow[\text{(Decomposition)}]{\text{Electricity}} 2Na(s) + Cl_2(g)$$

Sodium chloride (Molten) $\quad$ Sodium metal $\quad$ Chlorine gas

(vi) When electric current is passed through molten aluminium oxide, it decomposes into aluminium metal and oxygen gas.

$$2Al_2O_3(l) \xrightarrow[\text{(Decomposition)}]{\text{Electricity}} 4Al(l) + 3O_2(g)$$

Aluminium oxide (Molten) $\quad$ Aluminium metal

Table 4.1 Metal hydroxides and metal nitrides

Metal hydroxides		Metal nitrates	
Ka Na	Metal hydroxides are stable to heat	K Na	On heating they melt and decompose to give metal nitrite and oxygen $2KNO_3 \rightarrow 2KNO_2 + O_2$

(Continued)

Metal hydroxides		Metal nitrates	
Ca Mg Al	Decompose on heating to form metal oxide and water vapour	Ca Mg Al	Decompose on heating to form metal oxide, nitrogen dioxide and oxygen
Zn Fe Pb Cu	$Ca(OH)_2 \rightarrow CaO + H_2O$ $Zn(OH)_2 \rightarrow ZnO + H_2O$ $Pb(OH)_2 \rightarrow PbO + H_2O$ $Cu(OH)_2 \rightarrow CuO + H_2O$	Zn Fe Pb Cu	$2Ca(NO_3)_2 \rightarrow 2CaO + 4NO_2 + O_2$ $2Zn(NO_3)_2 \rightarrow 2ZnO + 4NO_2 + O_2$ $2Pb(NO_3)_2 \rightarrow 2PbO + 4NO_2 + O_2$ $2Cu(NO_3)_2 \rightarrow 2CuO + 4NO_2 + O_2$
Hg Ag	Yields metal, oxygen and water vapour $4AgOH \rightarrow 4Ag + O_2 + 2H_2O$	Hg Ag	Forms metal, nitrogen dioxide and oxygen $2AgNO_3 \xrightarrow{\Delta} 2Ag + 2NO_2 + O_2$

Table 4.2 Metal carbonates and metal bicarbonates (metal hydrogen carbonate)

Metal carbonates		Metal nitrates	
Ca, Mg Al, Zn, Fe, Pb Cu	Decompose on heating with decreasing vigour to form metal oxide and carbon dioxide $MCO_3 \xrightarrow{\Delta} MO + CO_2$ Here M = Ca, Mg, Zn, Pb, Cu	Na K	Metal bicarbonates or metal hydrogen carbonates decompose to give metal carbonates, water vapour and carbon dioxide $2MHCO_3 \xrightarrow{\Delta} M_2CO_3 + H_2O + CO_2$ Sodium hydrogen carbonate Sodium carbonate
Na K	Stable to heat and soluble in water	Ca Mg	$M(HCO_3)_2 \xrightarrow{\Delta} MCO_3 + H_2O + CO_2$ Calcium hydrogen carbonate Calcium carbonate M – Ca, Mg
Ag	$2Ag_2CO_3 \rightarrow 4Ag + O_2 + 2CO_2$		**Metal sulfate**
Hg	Decompose on heating and forms metal, oxygen and carbon dioxide	Zn, Cu, Ca, Mg, Pb	Metal sulfates on heating give metal oxides and SO_3 $MSO_4 \xrightarrow{\Delta} MO + SO_3$

Illustrations

1. Why are decomposition reactions called the opposite of combination reactions? Write equations for these reactions.

   Solution: A decomposition reaction is one in which a compound is broken down into simpler elements or compounds. For example,

   $$CaCO_3(s) \xrightarrow{\Delta} CaO(s) + CO_2(g)$$
   Calcium carbonate

   It is just the opposite of a combination reaction in which two or more reactants combine to form a single product.

   For example,

   $$CaSO_4(s) \xrightarrow{\Delta} CaO(s) + SO_3(g)$$
   Calcium sulphate

2. Why do we store silver chloride in dark-coloured bottles?

   Solution: We store silver chloride in dark-coloured bottles to avoid its decomposition in the sunlight.

$$CaO(s) + H_2O(l) \rightarrow Ca(OH)_2(aq)$$
Calcium hydroxide

3. Write balanced chemical equations for the following reactions:

   (i) Silver bromide on exposure to sunlight decomposes into silver and bromine.

   (ii) What is the colour of ferrous sulfate crystals? How does this colour change after heating?

   Solution: The reactions for these equations are

   (i) $2AgBr(s) \xrightarrow{\text{Sunlight}} 2Ag(s) + Br_2(g)$
   Silver bromide   Silver   Bromine

   (ii) The colour of ferrous sulfate crystals is pale green which changes to reddish-brown on heating due to the formation of iron (III) oxide.

   $2FeSO_4(s) \xrightarrow{\text{Heat}} Fe_2O_3(s) + SO_2(g) + SO_3(g)$
   Iron sulphate (Pale green)   Iron (III) oxide (Reddish-brown)

4. The following diagram displays a chemical reaction. Observe carefully and answer the following questions:

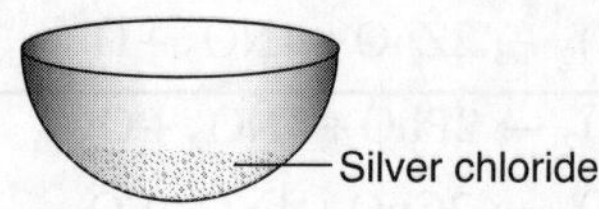

(i) Identify the type of chemical reaction that will take place and define it. How will the colour of the salt change?

(ii) Write the chemical equation of the reaction that takes place.

(iii) Mention one commercial use of this salt.

Solution:

(i) The chemical reaction is a photochemical decomposition reaction. It is a reaction in which a compound breaks down into simpler substances in the presence of light. The colour of the salt will change from white to grey.

(ii) Chemical equation:

$$2AgCl(s) \xrightarrow{\text{Sunlight}} 2Ag(s) + Cl_2(g)$$

(iii) Silver chloride ($AgCl$) is used commercially in photography.

Displacement Reactions

A reaction in which one element takes the place of another element in a compound is called a *displacement reaction*. In general, a more reactive element displaces a less reactive element from its compound.

> ## KNOWLEDGE BOOSTER
>
> Some metals are so reactive that they will displace hydrogen from water. This reaction can also be termed as a *hydrolysis reaction*.
>
> $$2K + 2H_2O \rightarrow 2KOH + H_2$$

Activity (reactivity) of elements

Metals		
Potassium	K	Most active metal
Sodium	Na	
Calcium	Ca	
Magnesium	Mg	
Aluminium	Al	
Zinc	Zn	
Iron	Fe	
Lead	Pb	
[Hydrogen]	[H]	
Copper	Cu	
Mercury	Hg	
Silver	Ag	
Gold	Au	
Platinum	Pt	Least active metal

Non-metals		
Fluorine	F	Most active
Chlorine	Cl	
Bromine	Br	
Iodine	I	Least active

Reactivity in decreasing order

> ### KNOWLEDGE BOOSTER
>
> A more reactive element displaces the less reactive element from its salt solution.
>
> For example, $2KBr + Cl_2 \rightarrow 2KCl + Br_2$

Here are few examples.

(i) When a strip of zinc metal is placed in copper sulfate solution, then zinc sulfate solution and copper are obtained.

$$CuSO_4(aq) + Zn(s) \rightarrow ZnSO_4(aq) + Cu(s)$$

Copper sulphate (Blue solution)   Zinc (Silvery-white)   Zinc sulphate (Colourless solution)   Copper (Red-brown)

Here, a displacement reaction takes place because zinc is more reactive than copper.

(ii) When a piece of magnesium metal is placed in copper sulfate solution, then magnesium sulfate solution and copper metal are formed.

$$CuSO_4(aq) + Mg(s) \rightarrow MgSO_4(aq) + Cu(s)$$

Copper sulphate (Blue solution)   Magnesium (Silvery white)   Magnesium sulphate (Colourless solution)   Copper (Red-brown)

(iii) When a piece of iron metal (say, an iron nail) is placed in copper sulfate solution, then iron sulfate solution and copper metal are formed.

$$CuSO_4(aq) + Fe(s) \rightarrow FeSO_4(aq) + Cu(s)$$

Copper sulphate (Blue solution)   Iron (Grey)   Iron sulphate (Greenish solution)   Copper (Red-brown)

(iv) When a strip of lead metal is placed in a solution of copper chloride, then lead chloride solution and copper metal are formed.

$$CuCl_2(aq) + Pb(s) \rightarrow PbCl_2(aq) + Cu(s)$$

Copper chloride (Green solution)   Lead (Bluish grey)   Lead chloride (Colourless solution)   Copper (Red-brown)

(v) When a copper strip is placed in a solution of silver nitrate, then copper nitrate solution and silver metal are formed.

$$2AgNO_3(aq) + Cu(s) \rightarrow Cu(NO_3)_2(aq) + 2Ag(s)$$

Silver nitrate (Colourless solution)   Copper (Red-brown)   Copper nitrate (Blue solution)   Silver (Greyish white)

This displacement reaction occurs because copper is more reactive than silver. A shining greyish white deposit of silver is formed on the copper strip and the solution becomes blue due to the formation of copper nitrate.

> ## KNOWLEDGE BOOSTER
>
> Note that the reaction between iron and copper sulfate takes a much longer time than magnesium and copper because iron and copper are closer together in the reactivity series than magnesium and copper.

(vi) Chlorine gas reacts with potassium iodide solution to give potassium chloride and iodine.

$$Cl_2(g) + 2KI(aq) \rightarrow 2KCl(aq) + I_2(g)$$
Chlorine Potassium iodide Potassium chloride Iodine

This reaction occurs as chlorine is more reactive than iodine.

(vii) When iron(III) oxide is heated with aluminium powder, then aluminium oxide and iron metal are formed.

$$Fe_2O_3(s) + 2Al(s) \rightarrow Al_2O_3(s) + 2Fe(l)$$
Iron (III) oxide Aluminium Aluminium oxide Iron (molten)
(Ferric oxide)

Illustration

1. Why does the colour of copper sulfate solution change when an iron nail is dipped in it? Write two observations.

Solution: The blue colour of copper sulfate solution changes due to the displacement reaction taking place. Iron displaces copper from copper sulfate solution to give the following observations:

- Pale green coloured solution of $FeSO_4$ is formed.
- Reddish-brown copper metal gets deposited.

$$Fe(s) + CuSO_4(aq) \rightarrow FeSO_4(aq) + Cu(s)$$

Double Displacement Reactions

The reactions in which two compounds react by an exchange of ions to form two new compounds is known as a double displacement reaction. Such a reaction usually occurs in solution and one of the products, being insoluble, precipitates out (separates as a solid). Some of the examples of double displacement reactions are given below.

(i) When silver nitrate solution is added to sodium, a white precipitate of silver chloride is formed along with sodium nitrate solution.

$$AgNO_3(aq) + NaCl(aq) \rightarrow AgCl(s) + NaNO_3(aq)$$
Silver nitrate Sodium chloride Silver chloride Sodium nitrate
(White ppt.)

(ii) When barium chloride solution is added to sodium sulfate solution, a white precipitate of barium sulfate is formed along with sodium chloride solution.

$$BaCl_2(aq) + Na_2SO_4(aq) \rightarrow BaSO_4(s) + 2NaCl(aq)$$
Barium chloride Sodium sulphate Barium sulphate Sodium chloride
(white ppt.)

(iii) If barium chloride solution is added to copper sulfate solution, a white precipitate of barium sulfate is produced along with copper chloride.

$$BaCl_2(aq) + CuSO_4(aq) \rightarrow BaSO_4(s) + CuCl_2(aq)$$
Barium chloride Copper sulphate Barium sulphate Copper chloride
(White ppt.)

(iv) When hydrogen sulfide gas is passed through copper sulfate solution, a black precipitate of copper sulfide is formed along with sulfuric acid solution.

$$CuSO_4(aq) + H_2S(g) \rightarrow CuS(s) + H_2SO_4(aq)$$
Copper sulphate Hydrogen sulphide Copper sulphide Sulphuric acid
(Black ppt.)

(v) When ammonium hydroxide solution is added to aluminium chloride solution, a white precipitate of aluminium hydroxide is formed along with ammonium chloride solution.

$$AlCl_3(aq) + 3NH_4OH(aq) \rightarrow Al(OH)_3(s) + 3NH_4Cl(aq)$$
Aluminium Ammonium Aluminium hydroxide Ammonium
chloride hydroxide (white ppt.) chloride

(vi) When potassium iodide solution is added to lead nitrate solution, a yellow precipitate of lead iodide is produced along with potassium nitrate solution.

$$Pb(NO_3)_2(aq) + 2KI(aq) \rightarrow PbI_2(s) + 2KNO_3(aq)$$
Lead nitrate Potassium iodide Lead iodide Potassium nitrate
(Yellow ppt.)

Types of Double Displacement Reactions

Double displacement reactions can be further classified as precipitation, gas formation and acid–base neutralisation reactions.

(i) Precipitation reaction: A precipitation reaction is possible when two solutions are mixed and a solid separates from the solution. The solid part that forms and separates from the solution is called the *precipitate*. For example,

- $Pb(NO_3)_2(aq) + 2KI(aq) \rightarrow PbI_2(s) \downarrow + 2KNO_3(aq)$
- $BaCl_2(aq) + Na_2SO_4(aq) \rightarrow BaSO_4(s) + 2NaCl(aq)$
  white ppt.
- $CuSO_4(aq) + H_2S(g) \rightarrow CuS(s) + H_2SO_4(aq)$
  black ppt.

(ii) Gas-formation reaction: A double displacement reaction is possible when an insoluble gas is formed. All gases are soluble in water to some extent, but only a few gases [$HCl(g)$ and NH_3 (g)] are highly soluble. For example, many sulfide salts react with acids to form gaseous hydrogen sulfide:

$$ZnS(s) + 2HCl(aq) \rightarrow ZnCl_2(aq) + H_2S(g)$$

(iii) Neutralisation reaction: A neutralisation is a special kind of double displacement reaction. In an acid–base reaction, an acid reacts with an equal quantity of a base. The acid–base reaction results in the formation of salt and water.

- Neutralisation of a soluble base (alkali) with an acid.

$$HCl(aq) + NaOH(aq) \rightarrow NaCl(aq) + H_2O(l)$$

$$2NaOH + H_2SO_4 \rightarrow Na_2SO_4 + 2H_2O$$
(Alkali) (Acid) (Salt)

- Neutralisation of an insoluble base with an acid.

$$PbO + 2HNO_3 \rightarrow Pb(NO_3)_2 + H_2O$$
$$\text{(Base)} \quad \text{(Acid)} \qquad \text{(Salt)}$$

$$Cu(OH)_2 + 2H_2SO_4 \rightarrow CuSO_4 + 2H_2O$$
$$\text{(Base)} \qquad \text{(Acid)} \qquad \text{(Salt)}$$

Uses of Neutralisation

- When someone is stung by a bee or other insects, formic acid enters the skin and causes pain which can be relieved by rubbing this spot with baking soda or slaked lime (basic in nature).

- When the stomach glands secrete excess hydrochloric acid, acidity occurs and to get relief from it, a solution of sodium hydrogen carbonate or magnesium hydroxide (milk of magnesia) is used.

- Excess of acid in soil is not suitable for growing of certain crops so it has to be neutralised. This is possible by adding slaked lime, $Ca(OH)_2$.

- Acidic stains on clothes can be removed by using ammonia solution.

Illustrations

1. Translate the following statement into a chemical equation and then balance it. Barium chloride reacts with aluminium sulfate to give aluminium chloride and a precipitate of barium sulfate. State how this reaction can be classified (two types).

   Solution: Chemical Equation:

   $$3BaCl_2(aq) + Al_2(SO_4)_3(aq) \rightarrow 3BaSO_4(s) + 2AlCl_3(aq)$$
   $$\text{Barium} \qquad \text{Aluminium} \qquad \text{Barium} \qquad \text{Aluminium}$$
   $$\text{chloride} \qquad \text{sulphate} \qquad \text{sulphate} \qquad \text{chloride}$$

   This reaction can be classified as a double displacement reaction and a precipitation reaction.

2. What is observed when a solution of potassium iodide is added to a solution of lead nitrate? Name the type of reaction. Write a balanced chemical equation to represent the above chemical reaction.

   Solution: A yellow precipitate of lead iodide is formed. It is a double displacement reaction and also a precipitation reaction.

   $$Pb(NO_3)_2(aq) + 2KI(aq) \rightarrow PbI_2(s) + 2KNO_3(aq)$$

3. A solution of potassium chloride when mixed with a solution of silver nitrate, forms an insoluble white substance. Write the chemical reaction involved and also mention the type of chemical reaction.

   Solution:

   $$KCl(aq) + AgNO_3(aq) \rightarrow \underset{\substack{\text{Silver chloride} \\ \text{(white ppt.)}}}{AgCl(s)} + \underset{\text{Potassium nitrate}}{KNO_3(aq)}$$

   The reaction is a double displacement reaction. It is also called a precipitation reaction due to the formation of the white precipitate of AgCl.

4. Write balanced chemical equations from the following word equations. Also include the physical state of each element or compound involved in the reaction.

   (i) An aqueous sodium chloride solution reacts with an aqueous silver nitrate solution to give a silver chloride precipitate (solid) and a sodium nitrate solution.

   (ii) An aqueous phosphoric acid solution reacts with an aqueous calcium hydroxide solution to give water and solid calcium phosphate.

   Solution:

   (i) $NaCl(aq) + AgNO_3(aq) \rightarrow AgCl(s) + NaNO_3(aq)$

   (ii) $3H_3PO_4(aq) + 3Ca(OH)_2(aq) \rightarrow$
   $$Ca_3(PO_4)_2(s) + 6H_2O(l)$$

Oxidation and Reduction Reactions

The earlier concept of oxidation and reduction is based on the addition or removal of oxygen or hydrogen. So, in terms of oxygen or hydrogen, oxidation and reduction reactions can be defined as follows.

Oxidation

It is defined as:

Addition of oxygen: The addition of oxygen to a substance is called oxidation. For example,

$$2Mg + O_2 \rightarrow 2MgO$$

Removal of hydrogen: The removal of hydrogen from a substance is also called oxidation. For example,

$$4HCl + MnO_2 \rightarrow MnCl_2 + Cl_2 + 2H_2O$$

Oxidation also involves following processes:

(i) Addition of an electronegative element or increase in its ratio.

$$2SnCl_2 + Cl_2 \rightarrow 2SnCl_4$$

(ii) Removal of an electropositive element or decrease in its ratio.

$$Hg_2Cl_2 \rightarrow HgCl_2 + Hg$$

(iii) Loss of electrons by an atom, ion or molecule—de-electronation.
$$Na - 1e^- \rightarrow Na^+$$
$$Fe^{2+} - 1e^- \rightarrow Fe^{3+}$$

(iv) Increase in oxidation number or positive charge on metal atom.
$$Fe^{2+} \rightarrow Fe^{3+}$$
Rusting of iron and combustion of fuels are examples of oxidation.

Reduction

It is defined as:

Removal of oxygen: The removal of oxygen from a substance is also called reduction.

$$PbO + C \rightarrow Pb + CO$$

Addition of hydrogen: The addition of hydrogen to a substance is called reduction.

$$H_2 + Cl_2 \rightarrow 2HCl$$

Reduction also involves following processes:

(i) Removal of an electronegative element or decrease in its ratio.
$$2FeCl_3 \rightarrow 2FeCl_2 + Cl_2$$

(ii) Addition of an electropositive element or increase in its ratio
$$2HgCl_2 \rightarrow Hg_2Cl_2 + Cl_2$$

(iii) Gain of electrons by atom or ion—electronation.
$$Cu^{2+} + e^- \rightarrow Cu^+$$

(iv) Decrease in the oxidation number of the metal atom.
$$Cu^{2+} \rightarrow Cu^+$$

Oxidising Agent

It oxidises the other substance and in the process itself gets reduced.

(i) The substance which gives oxygen for oxidation is called an oxidising agent.

(ii) The substance which removes hydrogen is also called an oxidising agent. For example,
$$\underset{\text{Oxidising agent}}{PbS + 4O_3} \rightarrow PbSO_4 + 4O_2$$

Some Common Oxidants

(i) Molecules of most electronegative elements. For example, O_3, O_2, F_2, Cl_2.

(ii) Elements in their highest oxidation states in compounds. For example,

$$\overset{+7}{K}MnO_4, \ \overset{+6}{K_2}Cr_2O_7, \ \overset{+6}{H_2}SO_4, \ \overset{+5}{H}NO_3, \ \overset{+6}{S}O_3, \ \overset{+4}{C}O_2, \ \overset{+3}{Fe}Cl_3$$
Some oxides like MgO, CaO, CrO_3, H_2O_2.

Reducing Agent

It reduces the other substance and in the process itself gets oxidised.

(i) The substance which gives hydrogen for reduction is called a reducing agent.

(ii) The substance which removes oxygen is also called a reducing agent.

Some Examples of Reductants

(i) Metals from group 1 (Li, Na) and group 2 (Mg, Ca).

(ii) Non-metals such as C, S, P, H.

(iii) HX such as HI, HBr, HCl.

(iv) Metal hydrides (MH) such as LiH, CaH_2.

(v) Molecules or compounds in which element has the lowest oxidation state (e^- needed to attain the octet). For example,
$$\overset{+1}{Hg_2}Cl_2, \ \overset{-2}{H_2}S, \ \overset{+1}{Cu_2}O, \ \overset{+2}{Sn}Cl_2, \ \overset{+2}{Fe}Cl_2$$

- Some organic acids such as $HCOOH, (COOH)_2$.

Redox Reactions

The oxidation and reduction reactions are also called *redox reactions* (the term *red* stands for reduction and *ox* stands for oxidation). Here, oxidation and reduction take place simultaneously, that is, one species gets oxidised while another one gets reduced. For example,

$$A_1 + A_2^{n+} \rightarrow A_1^{n+} + A_2$$

Here A_1 is oxidised into A_1^{n+} while A_2^{n+} is reduced into A_2.

(i) $\overset{+3}{2Fe}Cl_3 + \overset{+2}{Sn}Cl_2 \rightarrow 2FeCl_2 + \overset{+4}{Sn}Cl_4$

Here $SnCl_2$ is oxidised into $SnCl_4$ while $FeCl_3$ is reduced into $FeCl_2$.

(ii) When copper oxide is heated with hydrogen, copper metal and water are formed.

$$\underset{\text{Copper oxide}}{CuO} + \underset{\text{Hydrogen}}{H_2} \xrightarrow{\text{Heat}} \underset{\text{Copper}}{Cu} + \underset{\text{Water}}{H_2O}$$

Here, copper oxide is being reduced to copper while hydrogen is being oxidised to water.

Substance oxidised:	H_2
Substance reduced:	CuO
Oxidising agent:	CuO
Reducing agent:	H_2

(iii) When hydrogen sulfide reacts with chlorine, sulfur and hydrogen chloride are formed.

$$H_2S + Cl_2 \rightarrow S + 2HCl$$
Hydrogen sulphide Chlorine Sulphur Hydrogen chloride

Here hydrogen sulfide is being oxidised to sulfur while chlorine is being reduced to hydrogen chloride.

The oxidation–reduction reaction between hydrogen sulfide and chlorine can be shown more clearly as follows:

Removal of hydrogen: Oxidation

$$H_2S + Cl_2 \rightarrow S + 2HCl$$

Addition of hydrogen: Reduction

Here chlorine is the oxidising agent while hydrogen sulfide is the reducing agent.

(iv) When manganese dioxide reacts with hydrochloric acid, then manganese dichloride, chlorine and water are formed.

$$MnO_2 + 4HCl \rightarrow MnCl_2 + Cl_2 + 2H_2O$$
Manganese dioxide Hydrochloric acid Manganese dichloride Chlorine

Here MnO_2 loses oxygen to form $MnCl_2$, so manganese dioxide (MnO_2) is reduced to manganese dichloride ($MnCl_2$). On the other hand, HCl loses hydrogen to form Cl_2, so hydrochloric acid (HCl) is oxidised to chlorine (Cl_2). In this reaction, manganese dioxide (MnO_2) is the oxidising agent whereas hydrochloric acid (HCl) is the reducing agent.

Illustration

1. What is a redox reaction? Identify the substance oxidised and the substance reduced in the following reactions.

(i) $2PbO + C \rightarrow 2Pb + CO_2$

(ii) $2MnO_2 + 4HCl \rightarrow MnCl_2 + 2H_2O + Cl_2$

Solution: The reaction in which oxidation and reduction take place simultaneously is called a redox reaction.

(i) Substance oxidised: C

Substance reduced: PbO

(ii) Substance oxidised: HCl

Substance reduced: MnO_2

Photochemical and Electrochemical Reactions

Photochemical	Electrochemical
A photochemical reaction occurs by the absorption of light energy	An electrochemical reaction occurs by the absorption of electrical energy
Examples:	**Examples:**
$6CO_2 + 12H_2O \xrightarrow[\text{sunlight}]{\text{chlorophyll}}$ $C_6H_{12}O_6 + 6H_2O + 6O_2$ $2AgNO_3 \xrightarrow{\text{sunlight}}$ $2Ag + 2NO_2 + O_2$	$KCl \xrightarrow[\text{fused}]{\text{electric current}} K^+ + Cl^-$ $2H_2O \xrightarrow[\text{acidulated}]{\text{electric current}} 2H_2 + O_2$

KNOWLEDGE BOOSTER

Positive Catalyst or Accelerator

A positive catalyst is a substance that increases the rate of a reaction without undergoing any change in its mass and composition. A catalyst promoter increases the power of catalyst. For example,

- On a commercial scale, ammonia is manufactured by Haber's process in which iron fillings are used as catalyst and Mo as catalyst promoter.

$$N_2(g) + 3H_2(g) \rightleftharpoons 2NH_3(g)$$

- Ostwald process for manufacturing of nitric acid involves the use of platinum gauze as a catalyst.

$$2NO + O_2 \rightarrow 2NO_2$$
$$4NO_2(g) + O_2(g) + 2H_2O(l) \rightarrow 4HNO_3(aq)$$

- The contact process for manufacturing sulfuric acid involves the use of V_2O_5 as a catalyst.

$$2SO_2(g) + O_2(g) \rightleftharpoons 2SO_3(g)$$
$$H_2SO_4(l) + SO_3(g) \rightarrow H_2S_2O_7(l)$$
$$\text{(Conc)}$$
$$H_2S_2O_7(l) + H_2O(l) \rightarrow 2H_2SO_4(l)$$

- In the decomposition of potassium chlorate and H_2O_2, MnO_2 is used as a catalyst.

Negative Catalyst or Inhibitor

A negative catalyst is used to retard or decrease the rate of a reaction. For example,

- H_3PO_4 retards the decomposition of H_2O_2.

$$H_2O_2 \xrightarrow{H_3PO_4} 2H_2O + O_2$$

- Ethyl alcohol retards the rate of oxidation of chloroform ($CHCl_3$).

Enzyme or Biochemical Catalyst

Enzymes are used as positive catalysts in various biochemical reactions. For example, during digestion of complex food, during fermentation of starch and sucrose, and so on.

PRACTICE QUESTIONS

Analyse Your Concepts (School Exam Based)

Fill in the Blanks

Instructions: Complete the following statements with an appropriate word/term to be filled in the blank spaces.

1. A reaction in which heat is given out along with the products is called an _____________ reaction.

2. A reaction in which energy is absorbed is known as an _____________ reaction.

3. Two different atoms or groups of atoms (ions) are exchanged during _____________ reactions.

4. The uptake of oxygen by a substance is known as _____________.

5. The new substances produced in a reaction are known as _____________.

6. A reaction that takes place in the presence of sunlight is known as _____________ reaction.

7. An inhibitor _____________ rate of reaction.

8. AgBr decomposes in light, so it is used in _____________.

9. Calcium carbonate on heating gives CaO and CO_2. It is a case of _____________ reaction.

10. Potassium on reaction with water displaces hydrogen from water. This reaction is called _____________ reaction.

11. Magnesium reacts more _____________ than iron with copper sulfate.

True or False

Instructions: Read the following statements and write your answer as true or false.

1. Reactants and products are written on the left and right hand sides respectively in a reaction.

2. Oxidation is the loss of electrons from a species.

3. Reduction is the gain of electrons by a species.

4. Rusting is a double decomposition type reaction.

5. Formation of HCl in light is called photolysis.

6. Oxidation involves electronation.

7. The amount of product actually formed in a reaction is called actual yield.

8. Respiration is a redox reaction.

9. $PbS + 4H_2O_2 \rightarrow PbSO_4 + 4H_2O$. Here H_2O_2 acts as an oxidant.

10. A neutralisation reaction is always exothermic in nature.

11. Enzymes are examples of positive catalysts.

12. Reaction between liquid O_2 and liquid H_2 to form water is a case of exothermic reaction.

13. The number of atoms of each element is conserved during a chemical reaction.

Match the Following

Instructions: Each question contains statements given in two columns which have to be matched. Statements (a, b, c, d) in column I have to be matched with statements (p, q, r, s) in column II.

1.

Column I	Column II
(a) $C + O_2 \rightarrow CO_2$	(p) Displacement
(b) $AgBr \xrightarrow{light} Ag + Br$	(q) Combination
(c) $Zn + CuCl_2 \rightarrow ZnCl_2 + Cu$	(r) Decomposition
(d) $CH_3CH_2OH \xrightarrow{Cu} CH_3CHO + H_2$	(s) Oxidation

Very Short Answer Type Questions

1. Write the chemical equation involved when calcium metal reacts with aqueous hydrochloric acid to produce a solution of magnesium chloride and hydrogen gas.

2. What happens when sodium reacts with water?

3. Write the reaction of potassium chlorate ($KClO_3$) on heating.

4. Why should a magnesium ribbon may be cleaned before burning it in air?

5. Complete and balance the following equations:
   (a) $KOH + \text{.................} \rightarrow K_2SO_4 + H_2O$
   (b) $Mg(OH)_2 + \text{................} \rightarrow MgCO_3 + H_2O$

6. Why is silver nitrate solution kept in coloured bottles?

7. Why does iron turn aqueous copper sulfate into a green solution?

8. Write the reaction when hydrogen peroxide is exposed to sunlight.

9. Define chemical energy.

10. Write the reaction involved in photosynthesis.

11. Write an example of electrochemical reaction.

12. Write the example of an endothermic reaction.

Short Answer Type Questions

1. What type of chemical reactions take place when
   (a) limestone is heated?
   (b) a magnesium wire is burnt in air?
   (c) electricity is passed through water?
   (d) ammonia and hydrogen chloride are mixed?
   (e) silver bromide is exposed to sunlight?

2. During electrolysis of water, why is the amount of gas collected in one of the test tubes double of the amount collected in the other? Name this gas.

3. (a) Write a balanced chemical equation for the process of photosynthesis.
   (b) When do desert plants take up carbon dioxide and perform photosynthesis?

4. Consider the following chemical reaction:

   $$X + \text{Barium chloride} \rightarrow \underset{\text{White ppt.}}{Y} + \text{Sodium chloride}$$

   (a) Identify X and Y.
   (b) What type of reaction is it?

5. Name the reducing agent in the following reaction:

   $$3MnO_2 + 4Al \rightarrow 3Mn + 2Al_2O_3$$

   State which is more reactive, Mn or Al, and why.

6. State the type of chemical reactions with chemical equations that take place when ammonia and hydrogen chloride gases are mixed.

7. What happens when water is added to quick lime? Write the chemical equation.

8. Write the chemical equation of the reaction between lead nitrate and potassium iodide.

9. Name the type of chemical reaction represented by the following equations:
   (a) $CaO + H_2O \rightarrow Ca(OH)_2$
   (b) $3BaCl_2 + Al_2(SO_4)_3 \rightarrow 2AlCl_3 + 3BaSO_4$
   (c) $2FeSO_4 \rightarrow Fe_2O_3 + SO_2 + SO_3$

10. Define the term decomposition reaction. Give one example each of thermal decomposition and electrolytic decomposition.

11. Silver chloride turns grey when exposed to sunlight. Give a chemical equation to explain it.

Long Answer Type Questions

1. Discuss the importance of the decomposition reaction in the metal industry with three points.

2. (i) Define a balanced chemical equation. Why should an equation be balanced?

   (ii) Write the balanced chemical equations for the following reactions.
   (a) Phosphorus burns in the presence of chlorine to form phosphorus pentachloride.
   (b) Burning of natural gas.
   (c) The process of respiration.

3. (i) Write one example for each of the following decomposition reaction carried out with the help of:
   (a) Electricity   (b) Heat   (c) Light

   Or

   Decomposition reactions require energy either in the form of heat, light or electricity for breaking down the reactants. Write one equation each for decomposition reactions where energy is supplied in the form of heat, light and electricity.

   (ii) Which of the following statements is correct?
   (a) Copper can displace silver from silver nitrate.
   (b) Silver can displace copper from copper sulfate solution.

4. What is observed when a solution of sodium sulfate is added to a solution of barium chloride taken in a test tube? Write the equation for the chemical reaction involved and name the type of reaction.

5. 2 g of ferrous sulfate crystals are heated in a dry boiling tube.
   (a) List any two observations.
   (b) Name the type of chemical reaction taking place.
   (c) Write a balanced chemical equation for the reaction and name the products formed.

   Or

   You might have noted that when copper powder is heated in a china dish, the reddish-brown surface of copper powder becomes coated with a black substance.
   (a) Why has this black substance formed?
   (b) What is this black substance?
   (c) Write the chemical equation of the reaction that takes place.
   (d) How can the black coating on the surface be turned reddish-brown?

6. In an industrial process used for the manufacture of sodium hydroxide, a gas A is formed as a byproduct. The gas A reacts with lime water to give a compound B which is used as a bleaching agent in the chemical industry. Identify A and B. Also give the chemical equations of the reactions involved.

7. What would you observe on adding zinc granules to freshly prepared ferrous sulfate solution? Give reasons for your answer.

COMPETITION WINDOW (OBJECTIVE TYPE)

[For NEET, JEE (Main and Advanced), NTSE, KVPY and Olympiads]

Topic-wise MCQs

Chemical reactions and chemical equations

1. A chemical equation is balanced to satisfy one of the following laws in chemical reactions. This law is called:
   (a) law of conservation of magnetism
   (b) law of conservation of mass
   (c) law of conservation of motion
   (d) law of conservation of momentum

2. Which of the following is not a physical change?
   (a) boiling of water to give water vapour
   (b) melting of ice to give water
   (c) dissolution of salt in water
   (d) combustion of Liquefied Petroleum Gas (LPG)

3. In which of the following chemical equations do the abbreviations represent the correct states of the reactants and products involved at reaction temperature?
   (a) $2H_2(l) + O_2(l) \rightarrow 2H_2O(g)$
   (b) $2H_2(g) + O_2(l) \rightarrow 2H_2O(l)$
   (c) $2H_2(g) + O_2(g) \rightarrow 2H_2O(l)$
   (d) $2H_2(g) + O_2(g) \rightarrow 2H_2O(g)$

4. One of the following does not happen during a chemical reaction. This is
   (a) formation of new substances with completely different properties
   (b) breaking of old chemical bonds and formation of new chemical bonds
   (c) atoms of one element change into those of another element to give new products
   (d) rearrangement of atoms occurs to form new products

5. In the equation, $NaOH + HNO_3 \rightarrow NaNO_3 + H_2O$, nitric acid behaves as
   (a) a dehydrating agent   (b) a nitrating agent
   (c) an acid   (d) an oxidising agent

6. Electrolysis of water is a decomposition reaction. The mole ratio of hydrogen and oxygen gases liberated during the electrolysis of water is
   (a) $1:1$ (b) $2:1$
   (c) $4:1$ (d) $1:2$

7. Which one of the following processes involve chemical reactions?
   (a) storing of oxygen gas under pressure in a gas cylinder
   (b) liquefaction of air
   (c) keeping petrol in a china dish in the open
   (d) heating copper wire in the presence of air at a high temperature

8. Which of the following is not a characteristic of a chemical change?
   (a) it is irreversible
   (b) no net energy change is involved
   (c) a new substance is formed
   (d) it involves absorption or liberation of energy

9. Thermal decomposition of sodium carbonate will produce
   (a) carbon dioxide
   (b) oxygen
   (c) sodium hydroxide
   (d) no other product

Types of chemical reactions

10. Which of the following is not an endothermic reaction?
    (a) $CaCO_3 \rightarrow CaO + CO_2$
    (b) $C_6H_{12}O_6 + 6O_2 \rightarrow 6CO_2 + 6H_2O$
    (c) $6CO_2 + 6H_2O \rightarrow C_6H_{12}O_6 + 6O_2$
    (d) $2H_2O \rightarrow 2H_2 + O_2$

11. Consider the reaction
    $$NaBr(aq) + AgNO_3(aq) \rightarrow NaNO_3(aq) + AgBr(s)$$
    (a) double displacement reaction
    (b) displacement reaction
    (c) decomposition reaction
    (d) combination reaction

12. One of the following is an exothermic reaction. This is
    (a) process of photosynthesis
    (b) electrolysis of water
    (c) conversion of limestone into quicklime
    (d) process of respiration

13. The following reaction is an example of a
    $$4NH_3(g) + 5O_2(g) \rightarrow 4NO(g) + 6H_2O(g)$$
    (i) displacement reaction
    (ii) combination reaction
    (iii) redox reaction
    (iv) neutralisation reaction
    (a) (i) and (iv)   (b) (ii) and (iii)
    (c) (i) and (iii)   (d) (iii) and (iv)

14. Barium chloride on reacting with ammonium sulfate forms barium sulfate and ammonium chloride. Which of the following correctly represents the type of the reaction involved?
    (i) displacement reaction
    (ii) precipitation reaction
    (iii) combination reaction
    (iv) double displacement reaction
    (a) (i) only   (b) (ii) only
    (c) (iv) only   (d) (ii) and (iv)

15. Which among the following is(are) double displacement reaction(s)?
    (i) $Pb + CuCl_2 \rightarrow PbCl_2 + Cu$
    (ii) $Na_2SO_4 + BaCl_2 \rightarrow BaSO_4 + 2NaCl$
    (iii) $C + O_2 \rightarrow CO_2$
    (iv) $CH_4 + 2O_2 \rightarrow CO_2 + 2H_2O$
    (a) (i) and (iv)   (b) (ii) only
    (c) (i) and (ii)   (d) (iii) and (iv)

16. Which of the following is(are) an endothermic process(es)?
    (i) dilution of sulfuric acid
    (ii) sublimation of dry ice
    (iii) condensation of water vapours
    (iv) evaporation of water
    (a) (i) and (iii)   (b) (ii) only
    (c) (iii) only   (d) (ii) and (iv)

17. In the double displacement reaction between aqueous potassium iodide and aqueous lead nitrate, a yellow precipitate of lead iodide is formed. If lead nitrate is not available while performing the activity which of the following can be used in place of lead nitrate?
    (a) lead sulfate (insoluble)   (b) lead acetate
    (c) ammonium nitrate   (d) potassium sulfate

18. The following reaction is used for the preparation of oxygen gas in the laboratory

    $$2KClO_3(s) \xrightarrow[\text{Catalyst}]{\text{Heat}} 2KCl(s) + 3O_2(g)$$

    Which of the following statement(s) is(are) correct about the reaction?
    (a) It is a decomposition reaction and endothermic in nature.
    (b) It is a combination reaction.
    (c) It is a decomposition reaction and is accompanied by the release of heat.
    (d) It is a photochemical decomposition reaction and is exothermic in nature.

19. Which of the following reactions involves the combination of two elements?
    (a) $CaO + CO_2 \rightarrow CaCO_3$
    (b) $4K + O_2 \rightarrow 2K_2O$
    (c) $NH_3 + HCl \rightarrow NH_4Cl$
    (d) $SO_2 + \frac{1}{2}O_2 \rightarrow SO_3$

Reaction based

20. A white precipitate will be formed if we add common salt solution to:
    (a) $Ba(NO_3)_2$ solution   (b) KNO_3 solution
    (c) $AgNO_3$ solution   (d) $Mg(NO_3)_2$ solution

21. What happens when a copper rod is dipped in iron sulfate solution?
    (a) reaction is exothermic
    (b) copper displaces iron
    (c) no reaction takes place
    (d) blue colour of copper sulfate solution is obtained

22. An acid that can decolourise purple coloured potassium permanganate solution is
    (a) carbonic acid   (b) hydrochloric acid
    (c) sulfuric acid   (d) citric acid

23. When copper powder is heated it gets coated with
    (a) red copper oxide
    (b) black copper oxide
    (c) reduction
    (d) none of these

24. A student added dilute HCl to a test tube containing zinc granules and made the following observation:
    (a) the solution remained colourless
    (b) the solution becomes green in colour
    (c) the zinc surface became dull and black
    (d) a gas evolved which burnt with a pop sound

25. When copper turnings are added to silver nitrate solution, a blue-coloured solution is formed after some time. It is because copper:
    (a) is oxidised to Cu^{2+}
    (b) is reduced to Cu^{2+}
    (c) forms a blue-coloured complex with $AgNO_3$
    (d) displaces silver from the solution

26. A dilute solution of sodium carbonate was added to two test tubes—one containing dil HCl (x) and the other containing dilute NaOH (y). The correct observation was
    (a) a colourless gas liberated in test tube x
    (b) a colourless gas liberated in test tube y
    (c) a brown coloured gas liberated in test tube x
    (d) a brown coloured gas liberated in test tube y

Oxidation and reduction

27. The removal of oxygen from a substance is known as
    (a) corrosion   (b) oxidation
    (c) rancidity   (d) reduction

28. The process of respiration is:
    (a) a combination reaction which is endothermic
    (b) an oxidation reaction which is exothermic
    (c) an oxidation reaction which is endothermic
    (d) a reduction reaction which is exothermic

29. The chemical reaction involved in the corrosion of iron metal is that of
    (a) oxidation as well as combination
    (b) reduction as well as displacement
    (c) oxidation as well as displacement
    (d) reduction as well as combination

30. Hydrogen sulfide (H_2S) is a strong reducing agent. Which of the following reactions shows its reducing action?
    (a) $CuSO_4 + H_2S \rightarrow CuS + H_2SO_4$
    (b) $Cd(NO_3)_2 + H_2S \rightarrow CdS + 2HNO_3$
    (c) $Pb(NO_3)_2 + H_2S \rightarrow PbS + 2CH_3COOH$
    (d) $2FeCl_3 + H_2S \rightarrow 2FeCl_2 + 2HCl + S$

31. In which of the following reactions does H_2O_2 act as a reducing agent?

(a) $Ag_2O + H_2O_2 \rightarrow 2Ag + H_2O + O_2$

(b) $PbS + 4H_2O_2 \rightarrow PbSO_4 + 4H_2O$

(c) $2FeSO_4 + H_2SO_4 + H_2O_2 \rightarrow Fe_2(SO_4)_3 + 2H_2O$

(d) $2KI + H_2O_2 \rightarrow 2KOH + I_2$

32. When the gases sulfur dioxide and hydrogen sulfide mix in the presence of water, the reaction is $SO_2 + 2H_2S \rightarrow 2H_2O + 3S$. Here hydrogen sulfide is acting as:
(a) an oxidising agent   (b) a reducing agent
(c) a catalyst   (d) a dehydrating agent

33. Which of the following is not a decomposition reaction?
(a) $N_2 + 3H_2 \rightarrow 2NH_3$

(b) $2H_2O \xrightarrow{\text{Electrolysis}} H_2 + O_2$

(c) $CaCO_3 \xrightarrow{\text{Heat}} CaO + CO_2$

(d) $2HgO \xrightarrow{\text{Heat}} 2Hg + O_2$

34. The following reaction describes the rusting of iron $4Fe + 3O_2 \rightarrow Fe_3O_3$. Which one of the following statements is correct?
(a) metallic iron is oxidised to Fe_2O_3
(b) this is an example of an oxidation reaction
(c) Fe is reduced to Fe_2O_3
(d) both (a) and (b)

Miscellaneous

1. The chemical reaction between quicklime and water is characterised by:
(a) formation of slaked lime precipitate
(b) evolution of hydrogen gas
(c) change in the colour of the product
(d) change in the temperature of mixture

2. A reaction of the type $PQ + RS \rightarrow PS + RQ$ involves,
(a) no chemical change
(b) decomposition of PQ and RS
(c) exchange of ions of PQ and RS
(d) combination of PQ and RS

3. In which of the following reactions are gaseous species involved?
(a) decomposition of calcium carbonate
(b) reaction of zinc with dilute H_2SO_4
(c) decomposition of $KClO_3$
(d) all of these

4. Consider the following equation of the chemical reaction of a metal A:
$$4A + 3O_2 \rightarrow 2A_2O_3$$
This equation represents:
(a) combination reaction as well as oxidation reaction
(b) oxidation reaction as well as displacement reaction
(c) decomposition reaction as well as oxidation reaction
(d) combination reaction as well as reduction reaction

5. Which of the following can be decomposed by the action of light?
(a) $PbCl_2$ (b) KCl
(c) CuCl   (d) AgCl

6. A dilute ferrous sulfate solution was gradually added to a beaker containing acidified permanganate solution. The light purple colour of the solution fades and finally disappears. Which of the following is the correct explanation for the observation?
(a) $KMnO_4$ is an oxidising agent, it oxidises $FeSO_4$
(b) $FeSO_4$ acts as an oxidising agent and oxidises $KMnO_4$
(c) the colour disappears due to dilution; no reaction is involved
(d) $KMnO_4$ is an unstable compound and decomposes in the presence of $FeSO_4$ to a colourless compound.

7. Which of the following are examples of photochemical reactions?
(a) formation of glucose in photosynthesis
(b) decomposition of silver nitrate
(c) formation of nitric oxide (NO) from N_2 and O_2
(d) both (a) and (b)

8. Which of the following are exothermic processes?
 (i) reaction of water with quick lime
 (ii) dilution of an acid
 (iii) evaporation of water
 (iv) sublimation of camphor (crystals)
(a) (i) and (ii)   (b) (ii) and (iii)
(c) (i) and (iv)   (d) (iii) and (iv)

9. Which of the following process is not endothermic?
(a) formation of CS_2 from carbon and sulfur
(b) formation of NO from N_2 and O_2
(c) decomposition of calcium carbonate
(d) formation of ammonia from nitrogen and hydrogen

10. Which among the following statement(s) is(are) true? On exposure of silver chloride to sunlight for a long duration, it turns grey due to
 (i) the formation of silver by decomposition of silver chloride
 (ii) sublimation of silver chloride
 (iii) decomposition of chlorine gas from silver chloride
 (iv) oxidation of silver chloride
(a) (i) only   (b) (i) and (iii)
(c) (ii) and (iii)   (d) (iv) only

11. What will be the value of x, y and z in the following equation?
$$H_2C_2O_4 + xH_2O_2 \rightarrow yCO_2 + zH_2O$$
(a) $x = 1, y = 2, z = 2$ (b) $x = 2, y = 2, z = 2$
(c) $x = 1, y = 2, z = 4$ (d) $x = 2, y = 2, z = 1$

12. Solid calcium oxide reacts vigorously with water to form calcium hydroxide accompanied by the liberation of heat. This process is called slaking of lime. Calcium hydroxide dissolves in water to form its solution called lime water.

Which among the following is (are) true about slaking of lime and the solution formed?

  (i) it is an endothermic reaction

  (ii) it is an exothermic reaction

  (iii) the pH of the resulting solution will be more than seven

  (iv) the pH of the resulting solution will be less than seven

(a) (i) and (ii)   (b) (ii) and (iii)

(c) (i) and (iv)   (d) (iii) and (iv)

13. Which of the following are the characteristics of a chemical reaction?

  (i) evolution of a gas

  (ii) change in physical state

  (iii) formation of a precipitate

  (iv) constant temperature

(a) (i), (ii) and (iii)   (b) (ii) and (iii)

(c) (i) and (iv)   (d) (i), (ii), (iii) and (iv)

14. Which of the following gases can be used for storing a fresh sample of an oil for a long time?

(a) carbon dioxide or oxygen

(b) nitrogen or oxygen

(c) carbon dioxide or helium

(d) helium or nitrogen

15. $x\text{Fe} + y\text{H}_2\text{O} \rightarrow \text{Fe}_3\text{O}_4 + \text{H}_2$

What is the value of x and y here respectively?

(a) 3, 3   (b) 4, 4

(c) 3, 4   (d) 3, 2

16. Which of the following are combination reactions?

  (i) $2\text{KClO}_3 \xrightarrow{\text{Heat}} 2\text{KCl} + 3\text{O}_2$

  (ii) $\text{MgO} + \text{H}_2\text{O} \rightarrow \text{Mg(OH)}_2$

  (iii) $4\text{Al} + 3\text{O}_2 \rightarrow 2\text{Al}_2\text{O}_3$

  (iv) $\text{Zn} + \text{FeSO}_4 \rightarrow \text{ZnSO}_4 + \text{Fe}$

(a) (i) and (iii)   (b) (iii) and (iv)

(c) (ii) and (iv)   (d) (ii) and (iii)

17. When ferrous sulfate is heated strongly, decomposition occurs to give ferric oxide, SO_2 and SO_3. During this reaction, the colour changes from reactant to product is

(a) brown to green   (b) green to brown

(c) yellow to green   (d) green to blue

18. A shining brown coloured element M on heating in air becomes black in colour. This element M and the black-coloured compound are respectively

(a) Zn, ZnO   (b) Cu, CuO

(c) Ni, NiO   (d) Fe, FeO

19. The vapour density of fully dissociated ammonium chloride would be

(a) slightly less than half of that of ammonium chloride

(b) half of that of ammonium chloride

(c) double of that of ammonium chloride

(d) none of these

20. The pressure inside a closed glass vessel in which hydrogen iodide is heated would

(a) increase with temperature

(b) decreases with temperature

(c) remain constant

(d) first decrease and then increase

Advanced and Olympiads

Single Choice

1. Select the set showing the correct statements.

  (i) Mixing, heat, light, electricity, pressure are needed for a reaction.

  (ii) A reaction is characterised by change of state, change of colour and so on.

  (iii) In an exothermic reaction, heat is evolved.

  (iv) In an endothermic reaction, heat is evolved.

  (v) In an endothermic reaction, heat is absorbed.

(a) (i), (ii), (iii), (iv)   (b) (i), (ii), (iii), (v)

(c) (ii), (iii), (iv), (v)   (d) (i), (ii), (iii)

2. In a balanced equation, $\text{H}_2\text{SO}_4 + a\text{HI} \rightarrow \text{H}_2\text{S} + b\text{I}_2 + c\text{H}_2\text{O}$, the values of a, b, c are

(a) $a = 3, b = 5, c = 2$ (b) $a = 4, b = 8, c = 5$

(c) $a = 8, b = 4, c = 4$ (d) $a = 5, b = 3, c = 4$

3. Which of the following statements about the given reaction are correct?

$$3\text{Fe}(s) + 4\text{H}_2\text{O}(g) \rightarrow \text{Fe}_3\text{O}_4(s) + 4\text{H}_2(g)$$

  (i) iron metal is getting oxidised

  (ii) water is getting reduced

  (iii) water is acting as reducing agent

  (iv) water is acting as oxidising agent

(a) (i), (ii) and (iii)   (b) (iii) and (iv)

(c) (i), (ii) and (iv)   (d) (ii) and (iv)

4. Which of the following are examples of an electrochemical reaction?

(a) dissociation of fused potassium chloride

(b) combustion of glucose

(c) dissociation of acidulated water

(d) both (a) and (c)

5. Three beakers labelled as A, B and C each containing 25 mL of water were taken. A small amount of NaOH, anhydrous CuSO_4 and NaCl were added to the beakers A, B and C respectively. It was observed that there was an increase in the temperature of the solutions contained in beakers A and B, whereas in the case of beaker C, the temperature of the solution falls. Which one of the following statement(s) is(are) correct?

  (i) In beakers A and B, an exothermic process has occurred.

  (ii) In beakers A and B, an endothermic process has occurred.

(iii) In beaker C, an exothermic process has occurred.

(iv) In beaker C, an endothermic process has occurred.

(a) (i) only (b) (ii) only

(c) (i) and (iv) (d) (ii) and (iii)

Multiple Choice

6. In the reaction $PbO + C \rightarrow Pb + CO$

(a) C acts as an oxidising agent

(b) PbO is reduced

(c) this reaction does not represent a redox reaction

(d) C acts as a reducing agent

7. Which of the following involves a chemical reaction?

(a) process of respiration

(b) digestion of food in our body

(c) melting of candle wax on heating

(d) burning of candle wax when heated

8. Which of the following is(are) an example of a displacement reaction?

(a) $CaCO_3 \rightarrow CaO + CO_2$

(b) $CaO + CO_2 \rightarrow CaCO_3$

(c) $CuSO_4 + H_2S \rightarrow CuS + H_2SO_4$

(d) $Cu + 2AgNO_3 \rightarrow CuNO_3 + 2Ag$

9. $2Mg + O_2 \rightarrow 2MgO$. This reaction is an example of a/an

(a) displacement reaction (b) decomposition reaction

(c) oxidation reaction (d) combination reaction

10. Double decomposition reactions are

(a) $Mg + CuSO_4 \rightarrow MgSO_4 + Cu$

(b) $BaCl_2 + H_2SO_4 \rightarrow BaSO_4 + 2HCl$

(c) $Na_2S + 2HCl \rightarrow 2NaCl + H_2S$

(d) $KOH + HNO_3 \rightarrow H_2O + KNO_3$

11. Which of the following reactions is/are not redox reactions?

(a) $2S_2O_7^{2-} + 2H_2O \rightarrow 4SO_4^{2-} + 4H^+$

(b) $BaCl_2 + MgSO_4 \rightarrow BaSO_4 + MgCl_2$

(c) $Ca(OH)_2 + 2HCl \rightarrow CaCl_2 + 2H_2O$

(d) $Cu_2S + 2FeO \rightarrow 2Cu + 2Fe + SO_2$

12. Which of the following is/are not redox reactions?

(a) $CH_3COOH + C_2H_5OH \rightarrow CH_3COOC_2H_5 + H_2O$

(b) $H_2SO_4 + 2NaOH \rightarrow Na_2SO_4 + 2H_2O$

(c) $BaCl_2 + H_2SO_4 \rightarrow BaSO_4 + 2HCl$

(d) $2FeCl_3 + SnCl_2 \rightarrow 2FeCl_2 + SnCl_4$

Comprehension Type

Comprehension I: Certain substances such as hydroiodic acid and ammonium chloride are found to split up into simpler molecules on heating, which reunite on cooling to give the original substances. This type of process is referred to as thermal dissociation and differs from simple decomposition in being reversible. The vapour density of a gas, which decreases as dissociation proceeds, serves as a measure of the degree of dissociation at a particular temperature.

13. Thermal dissociation is a process in which

(a) an acid is neutralised by a base

(b) a compound decomposes on heating

(c) a compound ionises on heating

(d) the compound decomposes to simpler molecules which reunite

14. Which one of the following reactions involves thermal dissociation?

(a) $H_2 + Cl_2 \rightarrow HCl$

(b) $PCl_3 + Cl_2 \rightarrow PCl_5$

(c) $2Mg + O_2 \rightarrow 2MgO$

(d) $2Pb(NO_3)_2 \rightarrow 2PbO + 4NO_2 + O_2$

15. During the thermal dissociation of a gas, the vapour density

(a) increases

(b) decreases

(c) remains constant

(d) increases in some cases and decreases in others

Matrix Matching

16. Match the following

Column I	Column II
(a) Displacement reaction	(p) $AgCl(s) \xrightarrow{\text{Sunlight}} 2Ag(s) + Cl_2(g)$
(b) Double displacement reaction	(q) $Zn(s) + CuSO_4(aq) \rightarrow ZnSO_4(aq) + Cu(s)$
(c) Thermal decomposition reaction	(r) $CaCO_3(s) \xrightarrow{\text{Heat}} CaO(s) + CO_2(g)$
(d) Photolytic decomposition reaction	(s) $AgNO_3(aq) + NaCl(aq) \rightarrow AgCl(s) + NaNO_3(aq)$
	(t) $Pb(NO_3)_2(aq) + 2KI(aq) \rightarrow PbI_2(s) + 2KNO_3(aq)$

17. Match the following

Column I	Column II
(a) Addition reaction involving the combination of two compounds	(p) $C(s) + O_2(g) \rightarrow CO_2(g)$
(b) Reaction involving the combination between an element and a compound	(q) $Na_2SO_4(aq) + BaCl_2(aq) \rightarrow BaSO_4(s) + 2NaCl(aq)$

(Continued)

Column I	Column II
(c) Reaction in which a white precipitate is formed	(r) $Pb(NO_3)_2(s) \xrightarrow{\text{Heat}} 2PbO(s) + 4NO_2(g) + O_2(g)$
(d) Reaction in which brown fumes are evolved	(s) $CaO(s) + H_2O(l) \rightarrow Ca(OH)_2(aq)$

Integer Type

18. How many of these metals are more reactive than hydrogen?
    K, Na, Ca, Mg, Al, Cu, Hg, Ag, Fe

19. Consider the following reaction

$$PbS + xH_2O_2 \rightarrow PbSO_4 + 4H_2O$$

What is the value of the coefficient 'x' in the above equation?

20. In the reaction, $FeCl_3 + SnCl_2 \rightarrow FeCl_2 + SnCl_4$, the number of electrons transferred from one reactant to another for complete reaction is _____.

21. How many of the following compounds can behave both like an oxidant and a reductant CO_2, CO, HNO_3, H_2S, HNO_2, H_2O_2, SO_2, SO_3, H_2SO_4.

22. How many of the following may be the important characteristics of chemical reactions? (i) Change in state, (ii) formation of a precipitate, (iii) change in temperature, (iv) change in colour, (v) evolution of a gas, and (vi) formation of gaseous products only.

23. Metal nitrates of which of these metals on heating gives metal nitrite and oxygen?
    Na, K, Ca, Mg, Al, Ag, Zn, Fe, Pb

Answer Keys

Fill in the Blanks

1. exothermic
2. endothermic
3. double displacement
4. oxidation
5. products
6. photochemical
7. decreases
8. photography
9. thermal decomposition
10. hydrolysis
11. fast

True or False

1. T   2. T   3. T   4. F   5. T
6. F   7. T   8. T   9. T   10. T
11. T   12. T   13. T

Matching the Following

1. (a)—(q); (b)—(r); (c)—(p); (d)—(s)

Very Short Answer Type Questions

1. $Ca(s) + 2HCl(aq) \rightarrow CaCl_2(aq) + H_2(g)$

2. Sodium being more reactive than hydrogen, displaces hydrogen from water.

$$2Na(s) + H_2O(l) \rightarrow 2NaOH(aq) + H_2(g)$$

3. $2KClO_3(s) \rightarrow 2KCl(s) + 3O_2(g)$

4. Magnesium ribbon should be cleaned so that the oxide layer over it is removed and magnesium metal is available for the reaction.

5. (a) $2KOH + H_2SO_4 \rightarrow K_2SO_4 + 2H_2O$
   (b) $Mg(OH)_2 + CO_2 \rightarrow MgCO_3 + H_2O$

6. Silver nitrate solution is kept in a coloured bottle to avoid its decomposition in presence of sunlight.

7. Iron reacts with copper sulfate solution to give a green solution of ferrous sulfate.

$$Fe + CuSO_4 \rightarrow \underset{\text{Green}}{FeSO_4} + Cu$$

8. In the presence of sunlight, it decomposes into water and oxygen as follows:

$$2H_2O_2 \rightarrow 2H_2O + O_2$$

9. Every substance has a fixed amount of stored energy, which is in the form of potential energy and is referred to as its chemical energy.

10. Photosynthesis:

$$6CO_2 + 12H_2O \xrightarrow[\text{Sunlight}]{\text{Chlorophyll}} C_6H_{12}O_6 + 6H_2O + 6O_2$$

11. Acidified water breaks into hydrogen and oxygen when electric current is passed through it.

$$2H_2O \xrightarrow[\text{current}]{\text{Electric}} 2H_2 + O_2$$

12. Here is an example of an endothermic reaction:

$$N_2 + O_2 \xrightarrow[3000°C]{\Delta} 2NO$$

Short Answer Type Questions

1. (a) Decomposition (b) Combination (c) Decomposition (d) Combination (e) Decomposition

2. It is because water is formed when hydrogen and oxygen combine in a ratio of 2 : 1 by volume. The gas with double the volume is hydrogen.

3. (i) $6CO_2(g) + 6H_2O(l) \xrightarrow[\text{Glucose}]{\text{Sunlight}} C_6H_{12}O_6(s) + 6O_2(g)$

   (ii) In desert plants, stomata open during the night and take carbon dioxide that gets stored in the vacuoles. This stored CO_2 is utilised in the presence of sunlight during day time to perform photosynthesis.

4. (i) X is Na_2SO_4 and Y is $BaSO_4$

(ii) $\underset{\text{Sodium sulphate}}{Na_2SO_4(aq)} + BaCl_2(aq) \rightarrow \underset{\text{White ppt.}}{BaSO_4} \downarrow +2NaCl(aq)$

The reaction is a double displacement reaction. It is also called a precipitation reaction.

5. Al is a reducing agent. Al is more reactive than Mn because Al displaces Mn from its oxide.

6. Ammonia and hydrogen chloride gases are mixed to form ammonium chloride.

$$NH_3 + HCl \rightarrow NH_4Cl$$

The reaction is a combination reaction.

7. Slaked lime is formed with a hissing sound and a large amount of heat is evolved.

$$CaO + H_2O \rightarrow Ca(OH)_2 + Heat$$

8. $\underset{\text{Lead nitrate}}{Pb(NO_3)_2(aq)} + 2KI(aq) \rightarrow \underset{\substack{\text{Lead iodide} \\ \text{(Yellow ppt.)}}}{PbI_2(s)} + 2KNO_3(aq)$

Yellow precipitate of PbI_2 is formed.

9. The type of chemical reactions are as follows:
 (i) Combination reaction
 (ii) Double displacement reaction
 (iii) Decomposition reaction

10. Decomposition reaction: A reaction in which a compound is broken down into simpler elements or simple compounds is known as a decomposition reaction.

Thermal decomposition:

$$CaCO_3(s) \xrightarrow{\Delta} CaO(s) + CO_2(g)$$

Electrolytic decomposition:

$$2H_2O(l) \xrightarrow{\text{Electricity}} 2H_2(g) + O_2(g)$$

11. A large amount of energy of sunlight makes silver chloride (AgCl) decompose into silver (Ag^+) and chloride (Cl^-) ions. The metallic silver appears as small flakes which are essentially black. These tiny black spots evenly distributed in the mass of white AgCl makes it look grey.

$$2AgCl(s) \rightarrow 2Ag(s) + Cl_2(g)$$

Long Answer Type Questions

2. (i) A balanced chemical equation consists of an equal number of atoms of different elements in the reactants and products. According to the law of conservation of mass, matter can neither be created nor be destroyed in a chemical reaction. To follow the law, an equation should be balanced.

(ii) (a) $P_4(s) + 10Cl_2(g) \rightarrow 4PCl_5(s)$

(b) $CH_4(g) + 2O_2(g) \rightarrow CO_2(g) + 2H_2O(l) +$ Heat energy

(c) $C_6H_{12}O_6(aq) + 6O_2(g) \rightarrow$
$6CO_2(aq) + 6H_2O(l) +$ Heat energy

3. (a) Molten NaCl is electrolytically decomposed to sodium metal.

(b) Aluminium metal is obtained by the electric decomposition of bauxite ore mixed with cryolite.

(c) Carbonate ores are thermally decomposed to give the metal oxide which on reduction gives the metal.

Or (i) (a) Electricity (electrochemical decomposition)

$$2H_2O \xrightarrow{\text{Electricity}} 2H_2 + O_2$$

(b) Heat (thermal decomposition)

$$2FeSO_4 \xrightarrow{\text{Heat}} Fe_2O_3 + SO_2 + SO_3$$

(c) Light (photochemical decomposition)

$$2AgBr \xrightarrow{\text{Sunlight}} 2Ag + Br_2$$

(ii) The correct statement is
(a) Copper can displace silver from silver nitrate as copper is more reactive than silver.

$$Cu + 2AgNO_3(aq) \rightarrow Cu(NO_3)_2(aq) + 2Ag(s)$$

4. When a solution of sodium sulfate is added to a solution of barium chloride taken in a test tube, a white precipitate of barium sulfate is formed.

Chemical reaction involved:

$$\underset{\text{White ppt.}}{BaCl_2(aq) + Na_2SO_4(aq) \rightarrow BaSO_4} \downarrow +2NaCl(aq)$$

Type of reaction: This reaction is double displacement reaction.

5. (a) Observations:
The colour changes from green to white.
Formation of reddish-brown ferric oxide (Fe_2O_3) and evolution of SO_2/SO_3 gas.

(b) Decomposition reaction

(c) $2FeSO_4 \xrightarrow{\text{Heat}} \underset{\text{Ferric oxide}}{Fe_2O_3} + \underset{\text{Sulphuric oxide}}{SO_2} + \underset{\text{Sulphuric trioxide}}{SO_3}$

Or

(a) When copper is heated in air, oxidation takes place

(b) CuO/Copper oxide

(c) $2Cu + O_2 \rightarrow 2CuO$

(d) On passing hydrogen gas over the heated material

6. A ----- Cl_2 (Chlorine gas)
B ------ $CaOCl_2$ (Calcium oxychloride)
$2NaCl + 2H_2O$ ------ $2NaOH + Cl_2 + H_2$
$Cl_2 + Ca(OH)_2$ ----- $CaOCl_2 + H_2O$

7. The solution turns
 (i) green to colourless
 (ii) black coating is formed on zinc.
Reason: Zinc is more reactive than iron so it displaces the iron from its salt solution.

Competition Window (Objective Type)

Topic-wise MCQs

1. (b)	**2.** (d)	**3.** (d)	**4.** (c)	**5.** (c)
6. (b)	**7.** (d)	**8.** (b)	**9.** (d)	**10.** (b)
11. (a)	**12.** (d)	**13.** (c)	**14.** (d)	**15.** (b)
16. (d)	**17.** (b)	**18.** (a)	**19.** (b)	**20.** (c)
21. (c)	**22.** (d)	**23.** (b)	**24.** (d)	**25.** (c)
26. (a)	**27.** (d)	**28.** (b)	**29.** (a)	**30.** (d)
31. (a)	**32.** (b)	**33.** (a)	**34.** (d)	

Miscellaneous

1. (d)	**2.** (c)	**3.** (d)	**4.** (a)	**5.** (d)
6. (a)	**7.** (d)	**8.** (a)	**9.** (d)	**10.** (a)
11. (c)	**12.** (b)	**13.** (a)	**14.** (d)	**15.** (c)
16. (d)	**17.** (b)	**18.** (b)	**19.** (b)	**20.** (c)

Advanced and Olympiads

Single Choice

1. (b)	**2.** (c)	**3.** (c)	**4.** (d)	**5.** (c)

Multiple Choice

6. (ab)	**7.** (ab)	**8.** (cd)	**9.** (cd)	**10.** (bcd)
11. (abc)	**12.** (abc)			

Comprehension Type

13. (b)	**14.** (d)	**15.** (b)

Matrix Matching

16. (a)—(q); (b)—(st); (c)—(r); (d)—(p)

17. (a)—(s); (b)—(p); (c)—(q); (d)—(r)

Integer Type

18. 6	**19.** 4	**20.** 2	**21.** 4	**22.** 5
23. 2				

Hints and Solutions

Competition Window (Objective Type)

Topic Wise

5. The reaction represents a neutralisation reaction in which the base (NaOH) reacts with an acid (HNO_3) to form salt ($NaNO_3$) and water (H_2O).

8. Heat change is possible during a reaction.

9. Sodium carbonate is thermally stable and does not decompose normally.

17. Lead sulfate being insoluble, will not dissociate into Pb^{2+} ions.

19. Except (b) all other reactions involve compounds.

21. Iron is more reactive than copper, hence Cu will not displace iron from iron sulfate, hence no reaction will take place.

23. $2Cu + O_2 \xrightarrow{\Delta} 2CuO$ (Black)

24. $Zn + 2HCl \rightarrow ZnCl_2 + H_2$

Hydrogen gas burns with a pop.

25. Cu is more reactive than Ag.

$Cu + 2AgNO_3 \rightarrow CuNO_3 + 2Ag$ reaction occurs.

26. $Na_2CO_3 + 2HCl \rightarrow 2NaCl + H_2O + CO_2$

$Na_2CO_3 + NaOH \rightarrow$ No reaction

30. $FeCl_3 + H_2S \rightarrow FeCl_2 + HCl + S$

(with Reduction shown above $FeCl_3 \to FeCl_2$ and Oxidation shown below $H_2S \to S$)

In the given reaction H_2S undergoes oxidation and hence behaves as a reducing agent.

31. In (a), (b) and (c) H_2O_2 acts as the oxidising agent.

Miscellaneous

3. Decomposition of calcium carbonate and $KClO_3$ gives CO_2 and O_2 respectively. Zinc on reaction with H_2SO_4 gives H_2. For reactions see the text part.

7. Formation of glucose in photosynthesis and decomposition of silver nitrate are examples of photochemical reactions while formation of nitric oxide is a case of an endothermic reaction.

$$N_2 + O_2 \xrightarrow{\Delta} 2NO$$

9. Here reactions in option (a), (b) and (c) are endothermic while formation of ammonia is an exothermic reaction. (See the text part.)

15. Here the balanced equation is $3Fe + 4H_2O \rightarrow Fe_3O_4 + H_2$

So $x = 3$, $y = 4$

17. $2FeSO_4(s)$ (Ferrous sulphate (Green colour)) $\xrightarrow[\text{(Decomposition)}]{\text{Heat}}$

$Fe_2O_3(s)$ (Ferric oxide (Brown colour)) $+ SO_2(g) + SO_3(g)$

18. $Cu + \dfrac{1}{2}O_2 \xrightarrow{\Delta} CuO$ (Brown → Black)

19. Ammonium chloride on dissociation gives ammonia and HCl.

$$NH_4Cl \rightarrow NH_3 + HCl$$

Molecular weight = 2 × vapour density

The molecular weights of NH_4Cl, NH_3 and HCl are 53.5 g/mol, 17 g/mol and 36.5 g/mol respectively. The average molecular weight of fully dissociated NH_4Cl will be $\dfrac{17 + 36.5}{2} = 26.75$ g/mol which is one half of 53.5 g/mol (the molecular weight of pure undissociated NH_4Cl).

20. $2HI \rightarrow H_2 + I_2$

As the volume is constant after dissociation, using Boyle's law, the pressure will also be constant.

As $PV = $ Constant

Advanced and Olympiads

Single Choice

1. Heat is absorbed in an endothermic reaction.

2. The value of a, b, c are 8, 4, 4 respectively hence the reaction is

$$H_2SO_4 + 8HI \rightarrow H_2S + 4I_2 + 4H_2O$$

3. The substance which oxidises the other substances in a chemical reaction is known as an oxidising agent. Likewise, the substance which reduces the other substance in a chemical reaction is known as a reducing agent.

4. Dissociation of fused potassium chloride and dissociation of acidified water are examples of electrochemical reactions while combustion of glucose is a case of exothermic reaction and is known as respiration.

Multiple Choice

6.

$$\underset{\text{Oxidation}}{\overset{\text{Reduction}}{PbO + C \longrightarrow Pb + CO}}$$

Since C oxidises itself and reduces PbO, it acts as a reducing agent.

8. (a) is combination and (b) is decombination.

11. In redox reactions, oxidation and reduction take place side by side.

$$Cu_2S + 2FeO \rightarrow 2Cu + 2Fe + SO_2$$

Oxidation number of Cu changes from +1 to 0 (reduction) and oxidation number of S changes from -2 to +4 (oxidation). Hence (d) is a redox reaction.

12. The reaction in which both oxidation and reduction take place simultaneously is called a redox reaction.

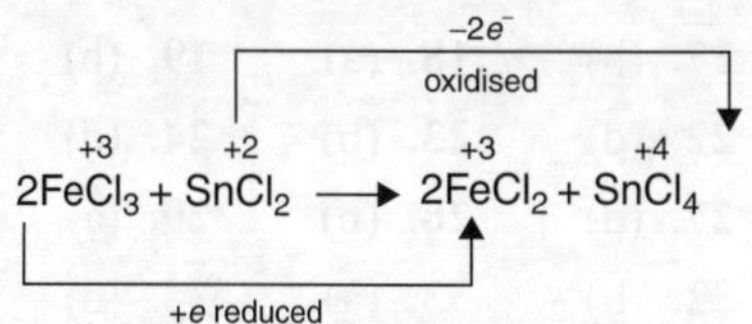

Therefore only (d) is a redox reaction.

Integer Type

18. K, Na, Ca, Mg, Al and Fe are more reactive than hydrogen while Cu, Hg and Ag are less reactive.

19. The balanced equation is $PbS + 4H_2O_2 \rightarrow PbSO_4 + 4H_2O$

20. $FeCl_3 + SnCl_2 \rightarrow FeCl_2 + SnCl_4$

$Sn^{2+} - 2e \rightarrow Sn^{4+}$

$2Fe^{3+} + 2e \rightarrow 2Fe^{2+}$

Hence, the number of electrons involved is 2.

21. Here CO, HNO_2, H_2O_2, SO_2 can act both like an oxidant as well as a reductant.

CO_2, H_2CO_3, SO_3, H_3PO_4 can act as oxidants while H_2S is a reductant.

22. Statement (i) to (v) are features of a chemical reaction. It is not necessary that only gaseous products are formed during a chemical reaction.

23. Nitrates of Na and K on heating give the metal nitrite and O_2. See text part.

Atomic Structure

After studying this unit, you will be able to understand:
- The characteristics, properties and distinguishing features of cathode rays and anode rays
- The characteristics, properties and distinguishing features of the electron, proton and neutron
- The structure of the atom in terms of Thomson's model, Rutherford's model and Bohr's model
- How to differentiate between atomic number and mass number
- How to distinguish and differentiate between isotopes, isobars and isotones
- Electron distribution

Introduction

Atoms are made up of three subatomic particles—electrons, protons and neutrons. An electron has a negative charge, a proton has a positive charge, whereas a neutron has no charge, it is neutral. Protons and neutrons are present in a small nucleus at the centre of the atom. Electrons are outside the nucleus. The atoms of different elements differ in the number of electrons, protons and neutrons.

Cathode Rays

Sir William Crook was first to design a discharge tube, so it is also called a Crook's discharge tube. A discharge tube is a long glass tube fitted with metal electrodes on either end across which high voltage can be applied (Fig. 5.1). The electrode which is connected to the negative terminal of the power source is known as the cathode while the electrode which is connected to the positive terminal is known as the anode.

J J Thomson modified it as follows. The discharge tube is connected to a vacuum pump for controlling the pressure of the gas inside the discharge tube. When the pressure of the gas is reduced to nearly 10^{-2} atm and a potential difference of nearly 10000 volts is applied to the electrodes, an electric current flows

and at the same time, light is emitted by the gas. The fluorescence (green coloured) was caused due to the bombardment of the walls of the tube by rays emanating from the cathode. These rays are known as *cathode rays*. Actually, these rays are a beam or stream of negatively charged small particles.

Features of Cathode Rays

J J Thomson described the cathode rays as follows.

- Cathode rays always travel in a straight line as they create a shadow on the opposite side of the cathode (Fig. 5.2).

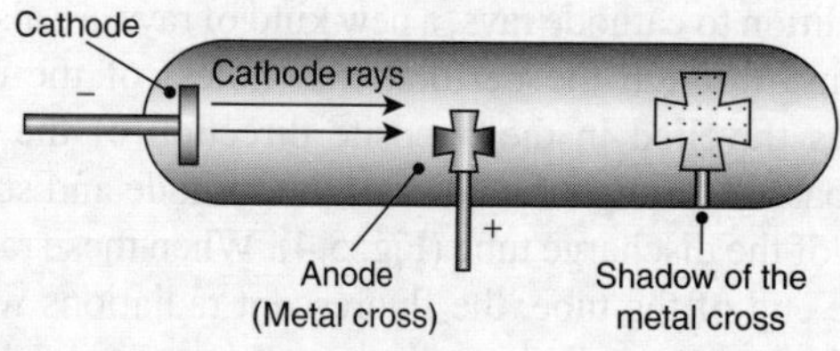

Fig. 5.2 Cathode rays cast shadows of the objects placed in their path

- Cathode rays are composed of negatively charged particles. We know this because they bend towards the positively charged plate when subjected to electric and magnetic fields (Fig. 5.3).
- Cathode rays are composed of material particles and possess energy, so they can produce mechanical effects. For example, they can rotate a light paddle wheel placed in their way (Fig. 5.3).

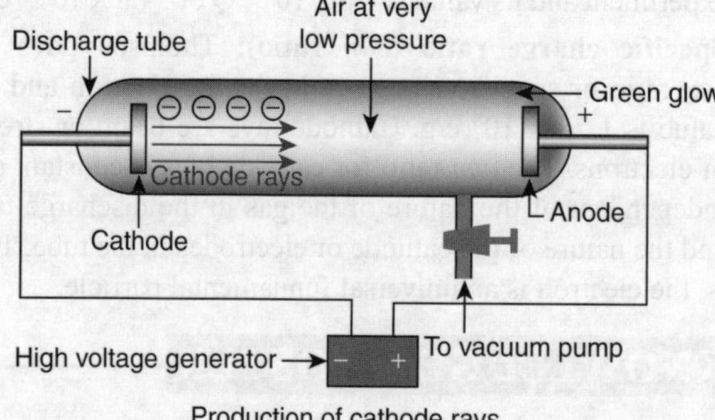

Fig. 5.1 Cathode rays discharge tube

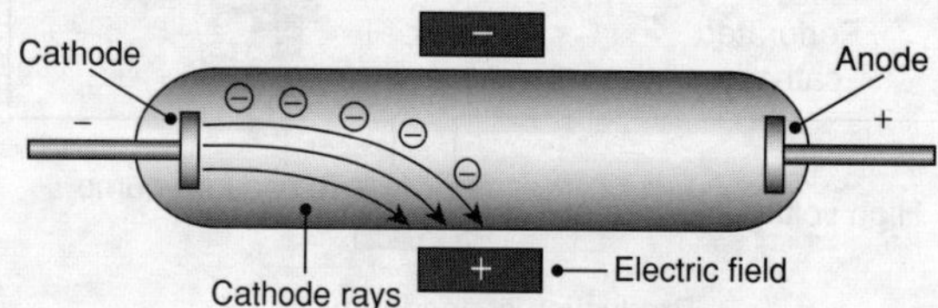

Fig. 5.3 Effects of electric field on cathode rays

- Cathode rays can heat up the object on which they fall. When they strike an object, a part of the kinetic energy is transferred to the object, as a result there is a rise in temperature.
- Cathode rays can penetrate through thin metallic sheets.
- Cathode rays can ionise the gas through which they travel.
- Cathode rays can produce a green fluorescence on a glass surface.
- When cathode rays fall on certain metals like copper, X-rays are formed. X-rays cannot be deflected by any electrical or magnetic fields but they can pass through the opaque material and can be stopped by only solid objects such as bones.

KNOWLEDGE BOOSTER

- The discovery and study of cathode rays gave rise to the discovery of the electron by J J Thomson.
- As the mass of electrons is very small compared to the mass of an atom, the same type of negatively charged particles are formed even if different gases are taken in the discharge tube or different metals are used as the cathode. That is why the e/m ratio (specific charge ratio) remains the same for cathode rays no matter which gas is taken and in what amount.

Anode Rays

Goldstein conducted the cathode ray experiment by using a perforated cathode and discovered anode rays which are composed of positively charged particles (protons).

When the pressure in the tube is decreased, it was observed that in addition to cathode rays, a new kind of rays are also found which came through the perforations (holes) of the cathode. These rays travelled in the opposite direction of the cathode rays and passed through the holes of the cathode and struck the other end of the discharge tube (Fig. 5.4). When these radiations struck the end of the tube, the fluorescent radiations were also seen. Such rays were called *canal rays* as they are passed through the holes or canals in the cathode. These rays were also called anode rays as they move from the anode toward the cathode. It was observed that the anode rays consist of positively charged particles. So, these rays were also called *positive rays*.

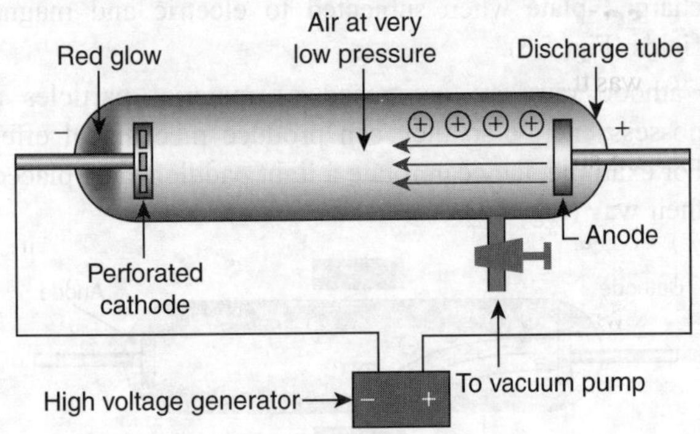

Fig. 5.4 Anode rays discharge tube

Features of Anode Rays

Goldstein described anode rays as follows.

- Anode rays travel in straight lines.
- Anode rays consist of material particles.
- Anode rays are deflected by electric and magnetic fields towards the negatively charged plate which indicates that they are positively charged or composed of positively charged particles.
- The charge to mass ratio of the particles in the anode rays was determined by W Wien using Thomson's technique. Charge to mass ratio of the particles in the anode rays depends upon the nature and mass of the gas taken in the discharge tube.
- The mass of the particles was the same as the atomic mass of the gas taken in the discharge tube.

KNOWLEDGE BOOSTER

- The discovery and study of anode rays gave rise to the discovery of the proton by Goldstein which was further described by Rutherford (the credit of proton discovery was given to Rutherford).
- The e/m ratio or specific charge ratio of anode rays depends upon the nature and mass of the gas taken in the discharge tube.

Electron ($_{-1}^{0}e$)

The electron was discovered by J J Thomson during the study of cathode rays. Cathode rays consist of small, negatively charged particles known as electrons. An electron is a negatively charged particle found in the atoms of all the elements. The electrons are located outside the nucleus in an atom. An electron is usually represented by the symbol e^- or $_{-1}^{0}e$.

Features of an Electron

The following are the features of an electron.

- **Mass of an electron:** The mass of an electron is about 1/1837 of the mass of hydrogen atom. Since the mass of a hydrogen atom is 1 u, the absolute mass of an electron is 0.0005488 amu or 9.1×10^{-28} grams or 9.1×10^{-31} kg.
- **Charge of an electron:** It is confirmed by Milliken's oil drop experiment and its value is 1.6×10^{-19} Q or -4.8×10^{-10} esu.
- **Specific charge ratio (e/m ratio):** Thomson gave the e/m ratio or specific charge ratio for the electron and the value is 1.76×10^8 c/g. Cathode rays are beam or stream of electrons. The e/m ratio for cathode rays is constant and independent of the nature of the gas in the discharge tube and the nature of the cathode or electrodes in the tube. That is, the electron is an universal fundamental particle.

KNOWLEDGE BOOSTER

e/m ratio or specific charge ratio or Thomson ratio order:

$$_{-1}e^- \; > \; _1p^1 \; > \; _2\alpha^4 \; > \; _0n^1$$

- An electron can leave its orbit for a maximum of 10^{-8} s.
- The particle nature of the electron is confirmed by scintillation effect on the ZnS screen.
- Photoelectric effect, emission of β-particles, thermal emission or heating of a metal filament confirm the fundamental particle nature of the electron.
- The density of the electron is 2.17×10^{17} g/cc.

Proton ($_1^1p$)

The proton was discovered by Goldstein and described by Rutherford. A proton is a positively charged particle found in the atoms of all the elements. The protons are located in the nucleus of an atom. The hydrogen atom contains only one proton in its nucleus, the atoms of all other elements contain more than one proton. A proton is usually represented by the symbol p^+ or $_1^1p$.

Features of a Proton

The following are the features of a proton.

- **Mass of proton:** The mass of a proton has been found to be equal to 1.673×10^{-27} kg. This is almost equal to that of an atom of hydrogen. Since the mass of a hydrogen atom is 1 amu, the relative mass of a proton is 1.0072 amu.
- **Charge of proton:** The charge present on a proton is equal and opposite to the charge present on an electron. The value of charge on a proton is 1.602×10^{-19} coulomb or $+4.8 \times 10^{-10}$ esu of positive charge.
- The e/m ratio of a proton is 9.5×10^4 C/g

KNOWLEDGE BOOSTER

1 amu = 1/12 mass of a C atom

1 amu = 1.66×10^{-27} kg = 931.5 MeV

Neutron ($_0^1n$)

In 1932 Chadwick discovered a fundamental particle, the neutron. The neutron is a neutral particle found in the nucleus of an atom. The subatomic particle not present in a hydrogen atom is the neutron. A hydrogen atom contains only one proton and one electron. A neutron is represented by the symbol $_0^1n$. It was discovered by bombarding Be atoms with α-particles.

$$_4Be^9 + {}_2He^4 \rightarrow {}_6C^{12} + {}_0n^1$$

Features of a Neutron

The following are the features of a neutron:

- **Mass of a neutron:** The mass of neutron is slightly higher than the mass of a proton. The relative mass of a neutron is 1 u. The absolute mass of a neutron is 1.674×10^{-27} kg or 1.0086 amu.
- **Charge of a neutron:** A neutron has no charge and it is electrically neutral.

The characteristics of electrons, protons and neutrons are compared in Table 5.1.

Table 5.1 Comparison between proton, neutron and electron

		Electron	Proton	Neutron
(i)	Symbol	e/e^-	p/p^+	n
(ii)	Nature	Negatively charged	Positively charged	Neutral
(iii)	Relative mass	1/1840 of a H atom	Equal to H atom	Equal to H atom
(iv)	Actual mass	9.1×10^{-28} g	1.673×10^{-24} g	1.674×10^{-24} g
(v)	Charge	(−1)	(+1)	No charge
		$(1.602 \times 10^{-19}$ C$)$	$(1.602 \times 10^{-19}$ C$)$	

Structure of Atom and Development of Atomic Model

After the discovery of the electron and proton, scientists started thinking of arranging these particles in an atom. This led to the development of the concept of atomic models. An atomic model depicts the arrangement of fundamental particles in an atom. Mainly, there are three atomic models.

(i) Thomson's atomic model

(ii) Rutherford's atomic model

(iii) Bohr's atomic model

Thomson's Atomic Model

Thomson was the first to propose a detailed model of the atom just after the discovery of electrons. He proposed that an atom consists of a uniform sphere of positive electricity in which the electrons are distributed or embedded more or less uniformly (Fig. 5.5). The negative and the positive charges are equal in magnitude. Thus, the atom as a whole is electrically neutral. This model of atom is known as the *plum pudding model or watermelon model*.

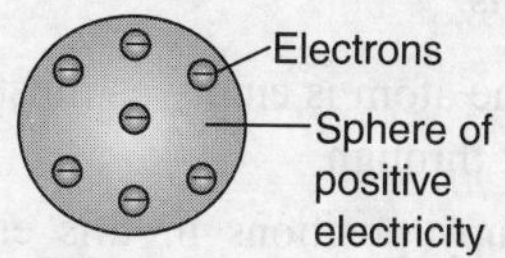

Fig. 5.5 Thomson's model

The salient features of Thomson's atomic model are:

- An atom consists of a positively charged sphere and the electrons are embedded in it.
- The negative and positive charges are equal in magnitude. So, the atom as a whole is electrically neutral.

Although Thomson's model could explain that atoms are electrically neutral, it cannot explain how the positively charged particles are shielded from the negatively charged electrons without getting neutralised. The results of experiments carried out by other scientists (Rutherford and others) could not be explained by this model so it was rejected.

Rutherford's Atomic Model

Rutherford and his students (Hans Geiger and Ernest Marsden) bombarded a very thin gold foil with α-particles (helium nuclei). This famous α-particle scattering experiment is represented in Fig. 5.6.

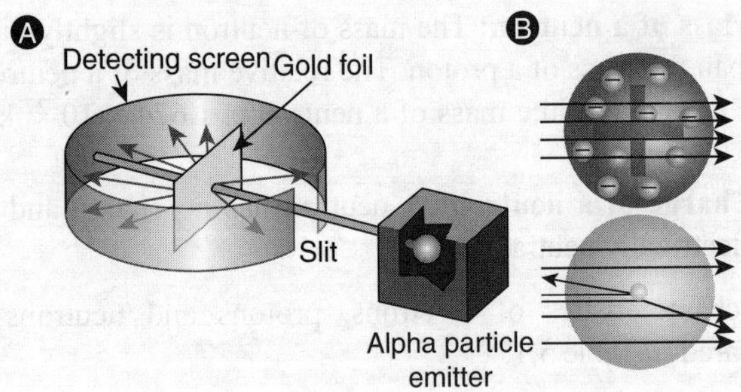

Fig. 5.6 Rutherford's experiment

A stream of high energy α-particles from a radioactive source was directed at a thin foil (thickness < 100 nm) of gold metal. The thin gold foil had a circular fluorescent zinc sulfide screen around it. Whenever α-particles struck the screen, a tiny flash of light was produced at that point.

The results of the scattering experiment were quite unexpected. According to Thomson's model of the atom, the mass of each gold atom in the foil should have been spread evenly over the entire atom, and α-particles had enough energy to pass directly through such a uniform distribution of mass. It was expected that the particles would slow down and change directions only by small angles as they passed through the foil. It was observed that:

- most of the α-particles passed through the gold foil undeflected,
- a small fraction of the α-particles was deflected by small angles, and
- a very few α-particles (~1 in 20,000) bounced back, that is, were deflected by nearly 180°.

Based on these observations during the gold foil experiment, Rutherford introduced his atomic model and arrived at the following conclusions.

(i) Most part of the atom is empty as most of the α-particles passed straight through.

(ii) Electrons occupy positions in this empty space (extra nuclear region).

(iii) A centrally located small, solid, compact part having all positive charge and nearly the whole mass of the atom is called the nucleus (as a few α-particles get defected up to a maximum of 180°).

(iv) The radius of an atom is 10^{-10} m or 10^{-8} cm.

The radius of the nucleus = 10^{-15} m or 10^{-13} cm.

Atomic radius > nucleus radius by 10^{5} times.

Density of nucleus is 10^{17} kg/m^3 or 10^{14} g/cm^3 and the volume of the nucleus is 10^{-39} cm^3.

(v) Centrifugal force develops between electrons and the nucleus. So, the electrons revolve around the nucleus as stars move around the sun (planetary or solar model).

Merits of Rutherford's Model

- Explains the discovery of the nucleus.
- Circulatory rotation of the electrons around the nucleus.

Demerits of Rutherford's Model

There are some drawbacks in Rutherford's atomic model.

(i) **Cannot explain stability of atom:** This model was unable to explain the stability of an atom. According to Rutherford's postulate, electrons revolve at a very high speed around the nucleus of an atom in a fixed orbit. However, Maxwell explained that accelerated charged particles release electromagnetic radiations. Hence, electrons revolving around the nucleus will release electromagnetic radiation. The electromagnetic radiation will possess energy from the electronic motion as a result of which the orbits will gradually shrink. Finally, the orbits will shrink and collapse into the nucleus of the atom (Fig. 5.7). According to the calculations, if Maxwell's explanation is followed, Rutherford's model will collapse within 10^{-8} seconds. Therefore, Rutherford's atomic model was not following Maxwell's theory and it was unable to explain an atom's stability.

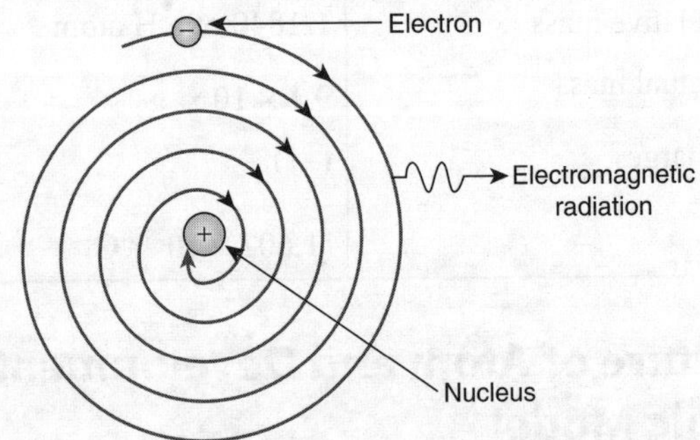

Fig. 5.7 Maxwell's explanation

(ii) Rutherford's theory was incomplete because it did not mention anything about the arrangement of electrons in the orbits. This was one of the major drawbacks of Rutherford's atomic model.

(iii) It cannot explain the number and velocity of electrons.

(iv) According to him, the atomic spectrum is continuous and non-linear. However, it is seen to be linear and discontinuous.

Quantum Theory

It was introduced by Max Planck and then extended by Einstein. It states that a hot vibrating body does not emit or absorb energy continuously but emits or absorbs discontinuously in the form of small energy packets or bundles known as *quanta* (*photon* in the case of light energy).

The energy of radiation (E) is directly proportional to the frequency of radiation (ν).

$$E \propto \upsilon$$
$$E = h\nu$$

Here, h is the Planck's constant and its value is 6.6253×10^{-34} J s or kg m^2 s^{-1}.

Absorption or emission in the form of multiples of quanta is known as quantisation of energy.

$$E = nh\upsilon$$

As
$$v = \frac{c}{\lambda}$$

So
$$E = \frac{nhc}{\lambda}$$

where, n = number of photons absorbed or emitted per second.

$$E \propto v \propto \frac{1}{\lambda}$$

So
$$\frac{E_1}{E_2} = \frac{v_1}{v_2} = \frac{\lambda_2}{\lambda_1}$$

Bohr's Atomic Model

In 1913, Bohr proposed his quantised shell model of the atom to explain how electrons can have stable orbits around the nucleus. The motion of the electrons in the Rutherford model was unstable because, according to classical mechanics and electromagnetic theory, any charged particle moving on a curved path emits electromagnetic radiation; thus, the electrons would lose energy and spiral into the nucleus. To remedy the stability problem, Bohr modified the Rutherford model by requiring the electrons move in orbits of fixed size and energy. The energy of an electron depends on the size of the orbit and is lower for smaller orbits. Radiation can occur only when the electron jumps from one orbit to another. The atom will be completely stable in the state with the smallest orbit, since there is no orbit of lower energy into which the electron can jump.

Features of Bohr's Atomic Model

- Bohr introduced the circular orbit concept based on Planck's quantum theory.
- Around the nucleus there are circular regions—orbits or shells (Fig. 5.8).

$$\begin{array}{ccccc} K & L & M & N & O\ldots \\ n = 1 & 2 & 3 & 4 & 5\ldots \end{array} \longrightarrow$$
Energy and distance from nucleus increase

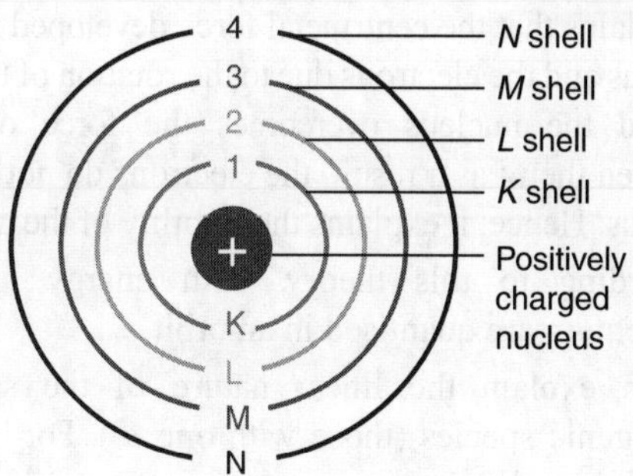

Fig. 5.8 Bohr's model

- Each orbit has a fixed amount (quantised amount) of energy so it is called energy level.
- An electron revolves round the nucleus in a particular orbit having a definite energy without any change of energy or any radiation of energy that is why these orbits are called stationary states.

KNOWLEDGE BOOSTER

Quanta: It is a small bundle of any type of energy.

Photon: It is a small massless bundle of light energy and not a material body.

- Angular momentum (mvr) in each orbit is quantised.

$$Mvr = n\frac{h}{2\pi} = n\hbar$$

where h is the Planck's constant.

- When an electron changes its orbit, energy change occurs in quanta (Fig. 5.9).

$$\Delta E = E_2 - E_1 = h\upsilon \quad \text{or} \quad = \frac{hc}{\lambda} \text{ quanta}$$

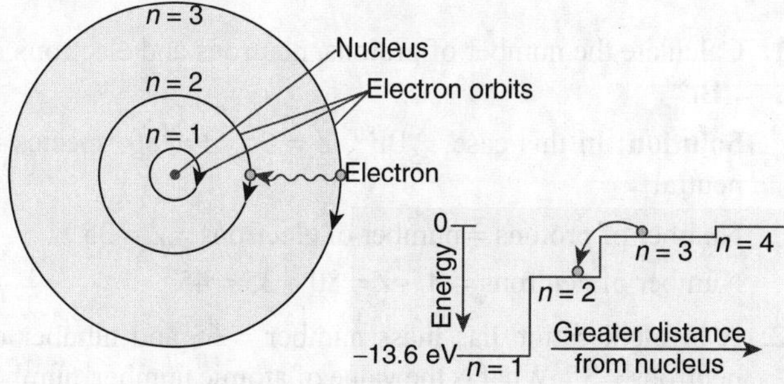

Fig. 5.9 Quantisation of energy

- **Excited state:** Here, an electron jumps from a lower to a higher orbit or energy level, by absorbing energy in quanta.
- **De-excited:** Here, an electron jumps from a higher to a lower energy level by releasing energy in quanta.

Merits of Bohr's Model

Bohr's atomic model not only overcomes all the drawbacks of Rutherford's atomic model, it also has many other advantages over it.

(i) It explains that the centripetal force developed between the nucleus and the electrons due to the rotation of the electrons around the nucleus overcomes the force of attraction between them; as a result, the electrons do not fall into the nucleus. Hence, it explains the stability of the atom.

(ii) According to this theory, both energy and angular momentum are quantised in an orbit.

(iii) It can explain the linear nature of the spectrum of hydrogenic species (those with one e^-). For example, H, He^+, Li^{+2} and so on.

(iv) By using the concept of quantisation of energy, Bohr's model can be used to find the radius of any orbit, velocity and energy of an electron in any orbit.

Drawbacks of Bohr's Theory

- It is not applicable to species having more than one electron, such as Li, He and so on.
- It cannot explain the fine spectrum of H, Li^{++} also.
- It gives no explanation of Zeeman and Stark effects.

Zeeman Effect

It is the splitting of the main spectrum line into several lines in a strong magnetic field.

Stark Effect

It is the splitting of the main spectrum line into several lines in a strong electric field.

- It cannot explain why atoms undergo chemical combination.
- It can only explain the particle nature of the electrons. There is no explanation for the wave nature, that is, it does not follow de Broglie and Heisenberg's theory.

Composition of Nucleus

Atomic Number (Z)

Moseley postulated that the frequency of the X-rays was related to the charge present on the nucleus of the atom of the element used as a cathode and found that

$$\sqrt{v} = a(Z - b)$$

where v is the frequency, Z is the nuclear charge and a and b are constants. a = probability constant, b = a constant having same values for all lines of the X-ray spectrum (Fig. 5.10).

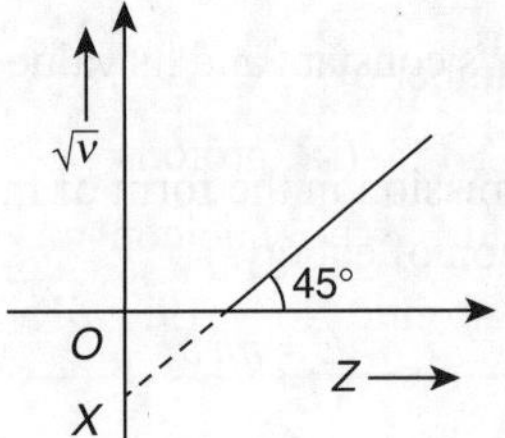

Fig. 5.10 Frequency vs atomic number

The number of unit positive charges carried by the nucleus of an atom is called the atomic number of the element, that is,

$$Z = p = e \quad \text{(from atoms)}$$

$$Z = p \quad \text{(for ions)}$$

Mass Number (A)

It is equal to the sum of protons and neutrons (or atomic number and neutrons) in an atom. Mass number is nearly equal to atomic weight.

$$A = p + n = Z + n$$

Illustrations

1. Calculate the number of protons, neutrons and electrons in $_{35}Br^{80}$.

Solution: In this case, $_{35}Br^{80}$, $Z = 35$, $A = 80$, species is neutral

Number of protons = number of electrons = $Z = 35$

Number of neutrons = $A - Z = 80 - 35 = 45$

2. A bivalent cation has mass number = 65 and number of neutrons = 35. What is the value of atomic number, number of electrons? Identify the bivalent cation.

Solution: As $A = 65$,

$n = 35$

$Z = A - n = 65 - 35 = 30$

Number of electrons = $Z - 2 = 30 - 2 = 28$

It is Zn^{+2}.

For example, (i) helium atom has 2 protons, 2 neutrons and 2 electrons. Its mass number is equal to $2 + 2 = 4$. (ii) Oxygen has 8 protons, 8 neutrons and 8 electrons. Its mass number is 16.

In some cases, particularly heavier elements, the number of neutrons is more than the number of protons. For example, mercury has an atomic number equal to 80. So, it has 80 protons and 80 electrons. But the mass number of mercury is 200. Therefore, the number of neutrons in mercury is $200 - 80 = 120$.

Representation of an Element or Atom

Generally, an atom is represented by its symbol. The atomic number is written on the lower side of the symbol and the mass number is written on the upper side.

$$_Z^A X$$

A = mass number, Z = atomic number, X = symbol of element.

For example, $_3^7Li$ indicates that lithium has an atomic number equal to 3 and a mass number equal to 7.

Relationship Between Mass Number and Atomic Number

Mass number = No. of protons + No. of neutrons

Mass number = Atomic number + No. of neutrons

Illustrations

1. The atomic nucleus of an element has mass number 23 and number of neutrons 12. What is the atomic number of the element?

Solution: We know that

Mass No. = No. of protons + No. of neutrons

23 = No. of protons + 12

$\therefore$ No. of protons = 23 – 12 = 11

Now, Atomic No. = No. of protons = 11

2. Calculate the number of:

   (i) electrons   (ii) protons

(iii) neutrons and   (iv) nucleons in

Mass No. = 39, Atomic No. = 19.

Solution:

(i) We know that

Atomic No. = No. of protons = No. of electrons

No. of electrons = 19

(ii) No. of protons = 19

(iii) Mass No. = No. of neutrons + No. of electrons

No. of neutrons = Mass number – No. of protons

= 39 – 19 = 20

(iv) Nucleons = No. of protons + No. of neutrons

= 19 + 20 = 39

Concept of Isotopes, Isobars, Isosteres and Isotones

Isotopes

They were discovered by Soddy. Atoms of same element with the same atomic number (Z) but different mass number (A) are called isotopes.

Isotopes of Hydrogen

It has three isotopes. These three isotopes are commonly known as hydrogen (H) or protium, deuterium (D) and tritium (T) respectively. As the atomic number is the same for all the three, they all have one electron and one proton but different number of neutrons.

	Protium	Deuterium	Tritium
Atomic number, Z	1	1	1
Mass number, A	1	2	3
Number of protons	1	1	1
Number of electrons	1	1	1
Number of neutrons	0	1	2
Electronic configuration	K	K	K
	1	1	1

Isotopes of Carbon

Carbon has the following two isotopes.

$^{12}_{6}\text{C}$	$^{14}_{6}\text{C}$
$Z = 6$	6
$p = 6$	6
$n = 6$	8
$A = 12$	14

Features of Isotopes

- Isotopes have the same chemical properties.
- Isotopes have the same number of protons and electrons.
- Rate of reaction of lighter isotope > rate of reaction of heavier isotope (isotopic effect).
- Isotopes differ in atomic weight, number of neutrons and physical properties.
- Atomic weight may be fractional due to the presence of isotopes as atomic weight is the average of the mass number of different isotopes.

Isobars

Atoms of different elements with the same mass number (A) but different atomic number (Z) are called isobars. For example,

(i) Argon (atomic number 18) and calcium (atomic number 20) are isobars because they have the same mass number 40.

Argon ($^{20}_{18}\text{Ar}$)	Calcium ($^{40}_{20}\text{Ca}$)
At. no. = 18	At. no. = 20
Mass no. = 40	Mass no. = 40
No. of electrons = 18	No. of electrons = 20
No. of protons = 18	No. of protons = 20
No. of neutrons = 22	No. of neutrons = 20

(ii) $_{18}\text{Ar}^{40}, _{19}\text{K}^{40}, _{20}\text{Ca}^{40}$

(iii) $_{6}\text{C}^{14}, _{7}\text{N}^{14}$

Features of Isobars

- Isobars have different number of p, n but the same ($p + n$) value.
- Isobars have different chemical properties but physical properties related to mass are the same.

Isotones

Species having different atomic number (Z) and mass number (A) but the same number of neutrons are called isotones. The

KNOWLEDGE BOOSTER

Radioactive isotopes: The heaviest isotopes which are unstable due to the presence of extra neutrons in their nuclei and emit various types of radiations, are called radioactive isotopes.

For example, carbon-14, oxygen-18, arsenic-74, sodium-24, iodine-131, cobalt-60 and uranium-235.

Applications of radioactive isotopes: Isotopes are used in various fields such as medicine, agriculture, biology, chemistry, engineering and industry. Radioactive isotopes are used as fuels in nuclear reactors of nuclear power plants for generating electricity.

- Uranium-235 is used as a fuel in the reactors of nuclear power plants for generating electricity.
- Radioactive isotopes are used as 'tracers' in medicine to detect the presence of tumours and blood clots in the human body.
- Radioactive isotopes are used in the treatment of cancer. For example, the cobalt-60 radio isotope is used to cure cancer.
- Radioactive isotopes are used to determine the activity of the thyroid gland which helps in the treatment of diseases like goiter. For example, iodine-131.
- Arsenic-74 tracer is used to detect the presence of tumours and sodium-24 tracer is used to detect the presence of blood clots.
- Radioactive isotopes are used in industry to detect the leakage in underground oil pipelines, gas pipelines and water pipes.

atoms have different number of protons of atomic number. Isotones belong to two or more different elements. They have $(A - Z)$ the same, and A and Z are different. For example,

(i) $_6C^{14}$ $_7N^{15}$ $_8O^{16}$

    n: $14 - 6$ $15 - 7$ $16 - 8$

        $= 8$ $= 8$ $= 8$

(ii) $_{14}Si^{30}$ $_{15}P^{31}$ $_{16}S^{32}$

These have 16 neutrons each.

Isoelectronics

Atoms, ions or molecules having the same number of electrons are called isoelectronic.

For example, CH_4, Ne, Na^+, NH_4^+, Mg^{2+}, F^-, O^- have 10 electrons, so they are isoelectronic.

Isosteres

Compounds or ions having the same number of atoms as well as the same number of electrons are isosteres. All isosteres are isoelectronic but all isoelectronic species cannot be isosteres.

For example, CO, CN^-, N_2 are isosteres as they have two atoms and 14 electrons each.

Isodiaphers

Isodiaphers have same $(n - Z)$ or $(A - 2Z)$ value. Isodiaphers are isotopes having the same isotopic number. For example,

    $_9F^{19}$ $_{19}K^{39}$

    A: 19 A: 39

    Z: 9 Z: 19

    n: 10 n: 20

  $n - Z = 1$ $n - Z = 1$

  $A - 2Z = 1$ $A - 2Z = 1$

Electron Distribution in Orbits

The distribution of electrons in different orbits or shells is governed by a scheme known as the **Bohr Bury scheme**. The arrangement of electrons in various energy levels or orbits of an atom is known as the electronic configuration of the atom.

According to this scheme:

- The electrons are arranged around the nucleus in different energy levels or energy shells. The electrons first occupy the shell with the lowest energy, that is, closest to the nucleus (lower n value).
- The first or the innermost energy shell (K or $n = 1$) can have only two electrons.
- The second shell (L or $n = 2$) can have up to 8 electrons.
- From the third shell (M or $n = 3$) onwards, the shells become larger and the third shell can have a maximum of 18 electrons.

In general, the maximum number of electrons that can be present in any orbit is given by $2n^2$ where n is the number of the energy shell.

Maximum number of electrons in different orbits

Orbit	Value of n	Maximum number of electrons in the orbit
K	1	$2 \times 1^2 = 2$
L	2	$2 \times 2^2 = 8$
M	3	$2 \times 3^2 = 18$
N	4	$2 \times 4^2 = 32$

- The outermost shell or orbit of an atom cannot have more than 8 electrons and the shell next to the outermost shell cannot have more than 18 electrons.

Valence Shell and Valence Electrons

The outermost orbit of an atom is known as its valence shell and the electrons present in the outermost orbit are known as valence electrons. For example,

(i) The electronic configuration of lithium is 2, 1. It may be represented as:

    K L

    2   1

The outermost orbit is the L shell and the number of electrons in its valence shell is one.

KNOWLEDGE BOOSTER

Only the valence electrons of an atom take part in chemical reactions because they have more energy than all the inner electrons of the atom.

(ii) The electronic configuration of beryllium is 2, 2. It may be represented as:

K	L
2	2

The outermost orbit is the L shell and the number of electrons in its valence shell is two.

Valency

The valency of an element is the combining capacity of the atoms of the element with atoms of the same or different elements.

Types of Valency

There are two types of valency—electrovalency and covalency. If an element combines by the loss or gain of electrons to form electrovalent compounds (or ionic compounds), its valency is known as *electrovalency*, and if an element combines by the sharing of electrons to form covalent compounds (or molecular compounds), its valency is known as *covalency*.

Electrovalency

In the formation of an electrovalent compound (or ionic compound), the number of electrons lost or gained by one atom of an element to achieve the nearest inert gas electron configuration is known as its electrovalency. The elements which lose electrons form positive ions, so they have positive electrovalency. The elements which gain electrons form negative ions, so they have negative electrovalency.

Covalency

In the formation of a covalent compound, the number of electrons shared by one atom of an element to achieve the nearest inert gas electron configuration is known as its covalency. If an atom shares 1 electron, its covalency will be 1.

CHAPTER AT A GLANCE

- **Atoms** are made up of three subatomic particles—electrons, protons and neutrons.
- **Cathode** rays are a beam or stream of negatively charged small particles.
- **Anode** rays consist of positively charged particles. So, these rays are also called positive rays.
- The main **sub-atomic particles** are electrons, protons and neutrons. Electrons are present outside the nucleus while protons and neutrons are present inside the nucleus.
- An **electron** is a negatively charged particle found in the atoms of all the elements. The electrons are located outside the nucleus in an atom. An electron is usually represented by the symbol e^- or $_{-1}^{0}e$.
- The absolute **mass of an electron** is 0.0005488 amu or 9.1×10^{-28} g or 9.1×10^{-31} kg.
- **Charge on an electron (e^-)** was confirmed by Milliken's oil drop experiment and its value is 1.6×10^{-19} Q or -4.8×10^{-10} esu.
- e/m ratio or specific charge ratio or Thomson ratio order: $_{-1}e^- > {}_1p^1 > {}_2\alpha^4 > {}_0n^1$
- A **proton ($_1^1p$)** is a positively charged particle found in the atoms of all the elements. The protons are located in the nucleus of an atom.
- The value of the **charge on a proton** is 1.602×10^{-19} coulomb or $+4.8 \times 10^{-10}$ esu of positive charge.
- The **neutron ($_0^1n$)** is a neutral particle found in the nucleus of an atom.
- **Thomson's atomic model:** an atom consists of a uniform sphere of positive electricity in which the electrons are distributed or embedded more or less uniformly. The negative and the positive charge are equal in magnitude. Thus, the atom as a whole is electrically neutral.
- **Rutherford's atomic model:** Rutherford bombarded a very thin gold foil with α-particles (helium nuclei) called the Rutherford's famous α-particle scattering experiment.
- The centrally located small, solid, compact part having all the positive charge and nearly the whole mass is called the **nucleus**.
- **Radius of atom** is $= 10^{-10}$ m or 10^{-8} cm. Radius of nucleus $= 10^{-15}$ m or 10^{-13} cm. Atomic radius > nucleus radius by 10^5 times.
- **Rutherford's model** cannot explain (i) stability of atom, (ii) number and velocity of electrons, (iii) linear nature of spectrum.
- **Quantum theory** states that a hot vibrating body does not emit or absorb energy continuously but emits or absorbs discontinuously in the form of small energy packets or bundles known as quanta (photon in the case of light energy).
- **Bohr** modified and removed all the demerits of Rutherford's model and introduced the circular orbit concept in which both energy and angular momentum are quantised (fixed values).
- The number of unit positive charges carried by the nucleus of an atom is called the **atomic number** of the element. $Z = p = e$ (from atoms), $Z = p$ (for ions).
- **Mass number (A)** is equal to the sum of protons and neutrons, or atomic number and neutrons. Mass number is nearly equal to atomic weight.

  Mass number = Atomic number + No. of neutrons ($A = p + n = Z + n$)

- Generally, an atom is represented by its **chemical symbol**. Atomic number is written on the lower side of the symbol and the mass number is written on the upper side.
- Atoms of same element with same atomic number (Z) but different mass number (A) are called **isotopes**.
- Atoms of different elements with the same mass number but different atomic number are called **isobars**.
- Species having different atomic number and mass number but the same number of neutrons are called **isotones**.
- Atoms, ions or molecules having the same number of electrons are called **isoelectronic**.

- The arrangement of electrons in various energy levels or orbits of an atom is known as the **electronic configuration** of the atom.
- In general, the maximum number of electrons that can be present in any orbit is given by $2n^2$ where n is the number of the energy shell.
- **Valency** of an element is the combining capacity of the atoms of the element with atoms of the same or different elements. Valency is equal to number of valence electrons.

PRACTICE QUESTIONS

Analyse Your Concepts (School Exam Based)

Fill in the Blanks

Instructions: Complete the following statements with an appropriate word/term to be filled in the blank spaces.

1. Cathode rays on passing through an electric field deflect towards the _____________ plate.

2. When cathode rays fall on a metal surface like Cu, _____________ are formed.

3. Thomson's atomic model is also known as _____________ model.

4. The subatomic particle not present in a hydrogen atom is _____________.

5. The number of protons in the nucleus of an atom is equal to its _____________.

6. The total number of protons and neutrons in the nucleus of an atom is equal to its _____________.

7. Rutherford used a thin film of _____________ metal.

8. An atom has atomic mass number 10 and atomic number 20. The atom has _____________ electrons.

9. The concept of a circular orbit was introduced by _______.

10. An atom of an element has 11 protons, 11 electrons and 12 neutrons. The atomic mass of the atom is _____________.

True or False

Instructions: Read the following statements and write your answer as true or false.

1. Electrons are present in all matter and have negative charge.

2. In Na^+, the number of electrons are more than the number of protons.

3. The number of neutrons in the atom of an element is equal to its atomic number.

4. Isotopes of an element differ in the number of neutrons.

5. e/m ratio of a neutron is zero.

6. Element having electronic arrangement 2, 7 is fluorine.

7. Matter is electrical and negative in nature.

8. Mass of an electron is about 1840 times that of a proton.

9. Neutrons were discovered by Henry Becquerel.

10. The first use of quantum theory to explain the structure of the atom was done by Planck.

11. J J Thomson proposed that the nucleus of an atom contains only nucleons.

12. A neutron is formed by an electron and a proton combining together. Therefore, it is neutral.

13. An isotope of iodine is used for making tincture iodine, which is used as a medicine.

Match the Following

Instructions: Each question contains statements given in two columns which have to be matched. Statements (a, b, c, d) in column I have to be matched with statements (p, q, r, s) in column II.

1. Match the following

Column I	Column II
(a) Cathode rays	(p) Milliken
(b) Anode rays	(q) Goldstein
(c) Neutron	(r) Thomson
(d) Charge on electron	(s) Chadwick

Very Short Answer Type Questions

1. What is the nature of charge on cathode and anode rays respectively?

2. By how many times is a proton heavier than an electron?

3. The atomic number of oxygen and sulfur are 8 and 16 respectively. What will be the number of electrons in these?

4. Out of O-16 and O-18 isotopes, which one has a higher number of neutrons?

5. The mass number and atomic number of an element are 24 and 12 respectively. What is the number of nucleons present in it?

6. An element has an atomic number 19 and mass number 39. How many electrons and protons are present in its uni-positive ion?

7. Who determined the charge of the electron for the first time?

8. The nucleus of an element contains 17 protons and 18 neutrons. What are its atomic number and mass number?

9. Write the decreasing order of e/m ratio of e, p, n, α-particles?

10. The radius of an atom is larger than the radius of a nucleus by how many times?

Short Answer Type Questions

1. Give the nuclear compositions of $_8^{18}\text{O}$ and $_7^{15}\text{N}$.

2. An element may be represented as $_{18}^{40}\text{E}$. Find the number of electrons, protons and neutrons in this atom. Give its electronic configuration.

3. What are the characteristic features of Thomson's model of the atom and why was the theory rejected?

4. What are the maximum number of electrons that can be present in the K, L and M shells?

5. Write electronic configuration for the following elements:

$_8^{16}\text{O}, \ _{14}^{28}\text{Si}, \ _{17}^{35}\text{Cl}, \ _{18}^{40}\text{Ar}$

6. What are the main postulates of Bohr's atomic theory?

7. Isotopes have the same chemical properties but different physical properties. Explain.

8. Name the elements which have the following electronic configuration:
   (a) 2, 7 (b) 2, 8, 1
   (c) 2, 8, 7 (d) 2, 8
   Which of these is chemically inert?

9. Why is the atom neutral when it contains charged particles?

Long Answer Type Questions

1. What is Thomson's model of the atom? Why was it rejected?

2. Write the Bohr Bury scheme to fill the electrons in various energy levels of an atom. Using these rules, write the electronic arrangement of the elements having atomic numbers 6, 11, 15 and 18.

3. Explain why atomic masses of most of the elements are fractional. If elemental boron is 20.0% B-10 and 80% B-11, calculate the atomic mass of boron.

4. Hydrogen has three isotopes written as $_1^1\text{H}, _1^2\text{H} : _1^3\text{H}$. Explain why,
   (a) These isotopes have almost identical chemical properties.
   (b) They are electrically neutral.

5. What observations in the scattering experiment led Rutherford to make the following conclusions?
   (a) Most of the space in an atom is empty.
   (b) The whole mass of an atom is present in the centre of the atom.
   (c) The nucleus has a positive charge.

NCERT Corner

In-Text Questions

1. What are canal rays?

2. If an atom contains one electron and one proton, will it carry any charge or not?

3. On the basis of Thomson's model of an atom, explain how the atom is neutral as a whole.

4. On the basis of Rutherford's model of an atom, which subatomic particle is present in the nucleus of an atom?

5. Draw a sketch of Bohr's model of an atom with three shells.

6. What do you think would be the observation if the α-particle scattering experiment is carried out using a foil of a metal other than gold?

7. Name the three sub-atomic particles of an atom.

8. Helium atom has an atomic mass of 4 u and two protons in its nucleus. How many neutrons does it have?

9. Write the distribution of electrons in the carbon and sodium atoms.

10. If the K and L shells of an atom are full, then what would be the total number of electrons in the atom?

11. How will you find the valency of chlorine, sulfur and magnesium?

12. If the number of electrons in an atom is 8 and the number of protons is also 8, then (a) what is the atomic number of the atom and (b) what is the charge on the atom?

13. With the help of Table 5.1, find out the mass number of the oxygen and sulfur atoms.

14. For the symbol H, D and T tabulate three sub-atomic particles found in each of them.

15. Write the electronic configuration of any one pair of isotopes and isobars.

Exercises

1. Compare the properties of electrons, protons and neutrons.

2. What are the limitations of J J Thomson's model of the atom?

3. What are the limitations of Rutherford's model of the atom?

4. Describe Bohr's model of the atom.

5. Compare all the proposed models of the atom given in this chapter.

6. Summarise the rules for writing the distribution of electrons in various shells for the first eighteen elements.

7. Define valency by taking examples of silicon and oxygen.

8. Explain with examples (a) atomic number, (b) mass number, (c) isotopes and (d) isobars. Give any two uses of isotopes.

9. Na^+ has completely filled K and L shells. Explain.

10. If bromine atom is available in the form of, say, two isotopes $^{79}_{35}Br$ (49.7%) and $^{81}_{35}Br$ (50.3%), calculate the average atomic mass of the bromine atom.

11. The average atomic mass of a sample of an element X is 16.2 u. What are the percentages of isotopes $^{16}_{8}X$ and $^{18}_{8}X$ in the sample?

12. If $Z = 3$, what would be the valency of the element? Also, name the element.

13. Composition of the nuclei of two atomic species X and Y are given as under

	X	Y
Protons =	6	6
Neutrons =	6	8

    Give the mass numbers of X and Y. What is the relation between the two species?

14. Complete the following table.

Atomic number	Mass number	Number of neutrons	Number of protons	Number of electrons	Name of the atomic species
9	–	10	–	–	–
16	32	–	–	–	Sulfur
–	24	–	12	–	–
–	2	–	1	–	–
–	1	0	1	0	–

Exemplar Problems

Short Answer Questions

1. Is it possible for the atom of an element to have one electron, one proton and no neutron? If so, name the element.

2. Write any two observations which support the fact that atoms are divisible.

3. Will ^{35}CI and ^{37}CI have different valencies? Justify your answer.

4. Why did Rutherford select a gold foil for his α-ray scattering experiment?

5. Find out the valency of the atoms represented by the Figs. 5.11 (a) and (b).

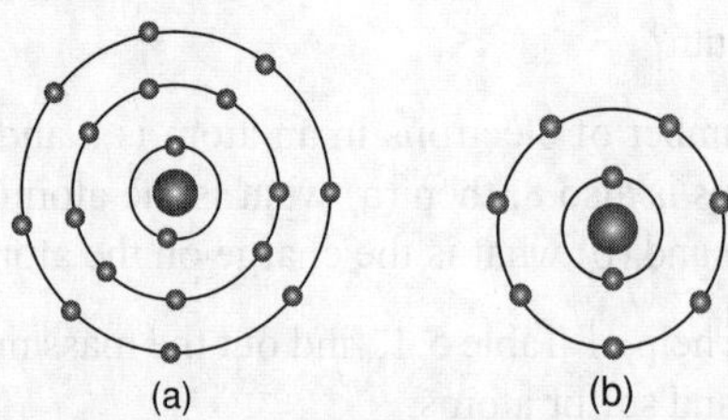

Fig. 5.11

6. One electron is present in the outermost shell of the atom of an element X. What would be the nature and value of charge on the ion formed if this electron is removed from the outermost shell?

7. Write down the electron distribution of the chlorine atom. How many electrons are there in the L shell? (Atomic number of chlorine is 17.)

8. In the atom of an element X, 6 electrons are present in the outermost shell. If it acquires noble gas configuration by accepting the requisite number of electrons, then what would be the charge on the ion so formed?

9. What information do you get from the Fig. 5.12 about the atomic number, mass number and valency of atoms X, Y and Z? Give your answer in a tabular form.

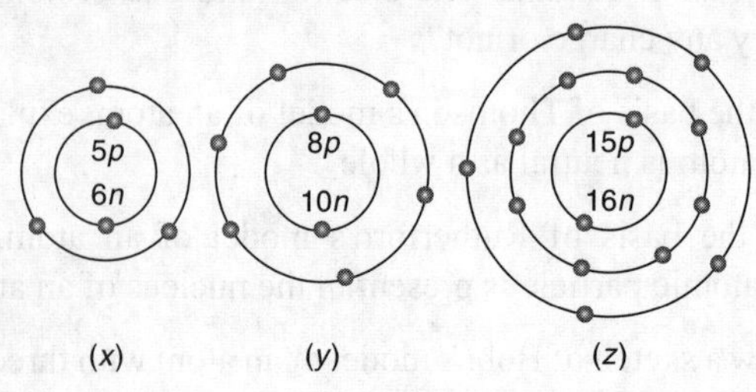

Fig. 5.12

10. In response to a question, a student stated that in an atom, the number of protons is greater than the number of neutrons, which in turn is greater than the number of electrons. Do you agree with the statement? Justify your answer.

11. Calculate the number of neutrons present in the nucleus of an element X which is represented by $_{15}^{31}X$.

12. Match the names of the scientists given in column A with their contributions towards the understanding of the atomic structure as given in column B.

Column A	Column B
(a) Ernest Rutherford	(i) Indivisibility of atoms
(b) J J Thomson	(ii) Stationary orbits
(c) Dalton	(iii) Concept of nucleus
(d) Neils Bohr	(iv) Discovery of electrons
(e) James Chadwick	(v) Atomic number
(f) E Goldstein	(vi) Neutron
(g) Mosley	(vii) Canal rays

13. The atomic number of calcium and argon are 20 and 18 respectively, but the mass number of both these elements is 40. What is the name given to such a pair of elements?

14. Complete the table below on the basis of information available in the symbols given below.

(a) $_{17}^{35}Cl$ (b) $_{6}^{12}C$ (C) $_{35}^{81}Br$

Element	n_p	n_n

15. Helium atom has 2 electrons in its valence shell but its valency is not 2—explain.

16. Fill in the blanks in the following statements.
    (a) Rutherford's α-partide scattering experiment led to the discovery of the ______.
    (b) Isotopes have the same ______ but different ______.
    (c) Neon and chlorine have atomic numbers 10 and 17 respectively. Their valencies will be ______ and ______ respectively.
    (d) The electronic configuration of silicon is ______ and that of sulfur is ______.

17. An element X has a mass number 4 and atomic number 2. Write the valency of this element?

Long Answer Questions

1. Why do helium, neon and argon have a valency of zero?

2. The ratio of the radii of the hydrogen atom and its nucleus is $\approx 10^5$. Assume that the atom and the nucleus are spherical.
   (a) What will be the ratio of their sizes?
   (b) If an atom is represented by planet earth $R_e = 6.4 \times 10^6$ m, estimate the size of the nucleus.

3. Enlist the conclusions drawn by Rutherford from his α-ray scattering experiment.

4. In what way is the Rutherford's atomic model different from Thomson's atomic model?

5. What were the drawbacks of Rutherford's model of an atom?

6. What are the postulates of Bohr's model of an atom?

7. Show diagrammatically the electron distributions in a sodium atom and a sodium ion and also give their atomic number.

8. In the gold foil experiment of Geiger and Marsden, that paved the way for Rutherford's model of an atom, $\approx 1.00\%$ of the α-particles were found to deflect at angles $> 50°$. If the gold foil was bombarded with one mole of α-particles, compute the number of α-particles that would deflect at angles less than 50°.

COMPETITION WINDOW (OBJECTIVE TYPE)

[For NEET, JEE (Main and Advanced), NTSE, KVPY and Olympiads]

Topic-wise MCQs

Sub-atomic particles

1. Which of the following is never true for cathode rays?
   (a) they are electromagnetic waves
   (b) they possess kinetic energy
   (c) they produce mechanical pressure
   (d) they produce heat

2. The canal ray experiment lead to the discovery of ________.
   (a) electrons   (b) nucleus
   (c) protons   (d) neutrons

3. Which of the following is not a basic particle of an element?
   (a) a molecule   (b) an atom
   (c) an ion   (d) none of these

4. What is an electron?
   (a) either a wave or a particle, depending on how it is observed
   (b) a particle
   (c) a wave
   (d) neither of these

5. Proton is
   (a) ionised hydrogen molecules
   (b) ionised hydrogen atom
   (c) nucleus of deuterium
   (d) an α-particle

6. Neutron was discovered by
   (a) Rutherford   (b) Bohr
   (c) Chadwick   (d) Thomson

7. The charge on an electron is
   (a) unit negative
   (b) -1.6×10^{-19} C
   (c) -4.8×10^{-10} esu
   (d) all

8. The mass of an electron is
   (a) 9.1×10^{-10} g
   (b) 9.1×10^{-28} g
   (c) 9.1×10^{-18} g
   (d) 9.1×10^{-25} g

9. The heaviest particle is
   (a) electron
   (b) neutron
   (c) proton
   (d) meson

10. The mass of a neutron is of the order of
    (a) 10^{-27} kg
    (b) 10^{-26} kg
    (c) 10^{-24} kg
    (d) 10^{-23} kg

11. The e/m ratio is highest for
    (a) neutron
    (b) proton
    (c) electron
    (d) He^{2+}

12. Which of these pairs has almost similar masses?
    (a) proton–electron
    (b) neutron–$_1H^1$
    (c) electron–$_1H^1$
    (d) neutron–electron

Atomic models

13. The first model of an atom was given by
    (a) N Bohr
    (b) E Goldstein
    (c) Rutherford
    (d) J J Thomson

14. Rutherford's α-particle scattering experiment was responsible for the discovery of
    (a) atomic nucleus
    (b) electron
    (c) proton
    (d) neutron

15. Select the incorrect statement.
    (a) Thomson's model can explain the neutrality of an atom
    (b) Rutherford used He^{2+} nuclei for bombarding the gold foil
    (c) most of the α-particles got deflected up to an angle of 180°
    (d) the nucleus contains nucleons

16. The electron revolves only in the orbits in which
    (a) $mvr = \dfrac{nh}{2\pi}$
    (b) $mvr < \dfrac{nh}{2\pi}$
    (c) $mvr > \dfrac{nh}{2\pi}$
    (d) $mvr \geq \dfrac{nh}{2\pi}$

17. According to quantum theory of radiation, which is false?
    (a) the magnitude of energy associated with a quanta is dependent on frequency
    (b) photons are the quanta of radiation
    (c) radiations are associated with energy
    (d) radiation is neither emitted nor absorbed discontinuously

18. In an atom, the constituent electrons
    (a) move around the nucleus in fixed energy levels
    (b) move around the nucleus in a random way
    (c) do not move
    (d) are uniformly distributed

19. Which of the following statement is not correct according to Bohr's model of an atom?
    (a) electrons jump from one orbit to another
    (b) an electron neither loses nor gains energy, it jumps from one orbit to another
    (c) the nucleus of an atom is situated at its centre
    (d) the electrons move in circular orbits

Atomic number, mass number

20. Which of the following statement is always correct?
    (a) An atom has an equal number of electrons and protons
    (b) An atom has an equal number of electrons and neutrons
    (c) An atom has an equal number of protons and neutrons
    (d) An atom has an equal number of electrons, protons and neutrons

21. Which of the following are true for an element?
    (i) Atomic number = number of protons + number of electrons
    (ii) Mass number = number of protons + number of neutrons
    (iii) Atomic mass = number of protons = number of neutrons
    (iv) Atomic number = number of protons = number of electrons
    (a) (i) and (ii)
    (b) (i) and (iii)
    (c) (ii) and (iii)
    (d) (ii) and (iv)

22. A neutral atom of an element contains 17 electrons. Its nucleus has 20 neutrons. Its mass number is
    (a) 37
    (b) 3
    (c) 17
    (d) 20

23. The number of electrons in an element X is 15 and the number of neutrons is 16. Which of the following is the correct representation of the element?
    (a) $_{15}^{31}X$
    (b) $_{16}^{31}X$
    (c) $_{15}^{16}X$
    (d) $_{16}^{15}X$

24. The mass number of an atom whose uni-positive ion has 10 electrons and 12 neutrons is
    (a) 21
    (b) 20
    (c) 22
    (d) 23

25. The atomic number of an element is 11 and its mass number is 23. The correct order representing the number of electrons, proton and neutrons respectively in this atom is
    (a) 12, 11, 11
    (b) 23, 11, 23
    (c) 11, 11, 12
    (d) 11, 12, 11

Isotopes, isobars, isotones

26. Isotopes of an element have
    (a) the same physical properties
    (b) different chemical properties
    (c) different number of neutrons
    (d) different atomic numbers

27. Which one of the following pairs represents isobars?

(a) $_{19}K^{40}$ and $_{19}K^{39}$ (b) $_{19}K^{40}$ and $_{18}Ar^{40}$

(c) $_{2}He^{3}$ and $_{2}He^{4}$ (d) $_{12}Mg^{24}$ and $_{12}Mg^{25}$

28. An isotone of $_{32}^{76}Ge$ is

(a) $_{34}^{77}Se$ (b) $_{34}^{78}Se$

(c) $_{32}^{77}Ge$ (d) $_{33}^{87}As$

29. Which one of the following groupings represents a collection of isoelectronic species? (Atomic numbers Cs: 55, Br: 35)

(A) Na^+, Ca^{2+}, Mg^{2+} (B) N^{3-}, F^-, Na^+

(C) Be, Al^{3+}, Cl^- (D) Ca^{2+}, Cs^+, Br

30. Which among the following are isobars?

(a) $_bX^a$, $_{b+1}Y^a$ (b) $_bX^a$, $_cY^b$

(c) $_bX^a$, $_bY^{a+1}$ (d) $_bX^a$, $_{b-1}Y^{a-1}$

31. Which one of the following sets of ions represents the collection of isoelectronic species?

(a) K^+, Ca^{2+}, Sc^{3+}, Cl^- (b) Na^+, Ca^{2+}, Sc^{3+}, F^-

(c) K^+, Cl^-, Mg^{2+}, Sc^{3+} (d) Na^+, Mg^{2+}, Al^{3+}, Cl^-

32. Which of the following are isoelectronic?

(a) Ne and O   (b) K^+ and C

(c) Na^+ and Ne   (d) Na^+ and K^+

33. If two neutrons are added to an element X, then it will get converted to its

(a) isobar   (b) isotone

(c) isotope   (d) none of the above

34. Which of the following pairs are isotopes?

(a) nitric oxide and nitrogen dioxide

(b) hydrogen and deuterium

(c) oxygen and ozone

(d) ice and steam

Electronic configuration and valency

35. In an atom, valence electrons are present in

(a) first orbit

(b) any one of its orbits

(c) outermost orbit

(d) penultimate orbit

36. The maximum number of electrons that can be accommodated in the third shell ($n = 3$) is

(a) 18   (b) 10

(c) 2   (d) 8

37. The number of valence electrons in Cl^- ion is:

(a) 16   (b) 8

(c) 17   (d) 18

38. Which one of the following is the correct electronic configuration of sodium?

(a) 2, 8   (b) 8, 2, 1

(c) 2, 1, 8   (d) 2, 8, 1

39. The maximum number of electrons which a shell (designated by n) can accommodate is

(a) $2n^2$ (b) n^2

(c) $2n + 1$ (d) $2n$

40. The total number of shells, subshells and orbitals that are used in Fe^{2+} is,

(a) 4, 6, 14   (b) 3, 7, 4

(c) 4, 7, 14   (d) 3, 6, 14

41. The ion of an element has 3 positive charges. Mass number of the atom is 27 and the number of neutrons is 14. What is the number of electrons in the ion?

(a) 13   (b) 10

(c) 14   (d) 16

42. The core charge on oxygen is equal to

(a) $-6e$ (b) $+8e$

(c) $+2e$ (d) $-2e$

43. Elements with valency 1 are

(a) always metals

(b) always metalloids

(c) either metals or non-metals

(d) always non-metals

44. An atom with 3 protons and 4 neutrons will have a valency of

(a) 3   (b) 7

(c) 1   (d) 4

45. The electron distribution in an aluminium atom is

(a) 2, 8, 3   (b) 2, 8, 2

(c) 8, 2, 3   (d) 2, 3, 8

46. A diapositive ion has 16 protons. What is the number of electrons in its tetra-positive ion?

(a) 12   (b) 10

(c) 16   (d) 14

Miscellaneous

1. When alpha particles are sent through a thin metal foil, most of them go straight through the foil because

(a) alpha particles are much heavier than electrons

(b) most part of the atom is empty

(c) alpha particles move with high velocity

(d) alpha particles are positively charged

2. In which of the following pairs of shells, is the energy difference between two adjacent orbits minimum?

(a) M, N (b) N, O

(c) K, L (d) L, M

3. Which of the following correctly represents the electronic distribution in the Mg atom?

(a) 3, 8, 1   (b) 2, 8, 2

(c) 1, 8, 3   (d) 8, 2, 2

4. In a sample of ethyl ethanoate ($CH_3COOC_2H_5$) the two oxygen atoms have the same number of electrons but

different number of neutrons. Which of the following is the correct reason for it?
(a) one of the oxygen atoms has gained electrons
(b) one of the oxygen atoms has gained two neutrons
(c) the two oxygen atoms are isotopes
(d) the two oxygen atoms are isobars

5. In the Thomson's model of atom, which of the following statements are correct?
(i) The mass of the atom is assumed to be uniformly distributed over the atom.
(ii) The positive charge is assumed to be uniformly distributed over the atom.
(iii) The electrons are uniformly distributed in the positively charged sphere.
(iv) The electrons attract each other to stabilise the atom.
(a) (i), (ii) and (iii)   (b) (i) and (iii)
(c) (i) and (iv)   (d) (i), (iii) and (iv)

6. When alpha particles are sent through a thin metal foil only one out of ten thousand of them rebounded. This observation led to the conclusion that
(a) only unit positive charge is present in an atom
(b) more number of electrons are revolving around the nucleus of the atom
(c) positively charged particles are concentrated at the centre of the atom
(d) a massive sphere with a negative charge and a small sphere at the centre with a positive charge

7. Which among the following pairs have different number of valence electrons?
(a) Mg^{2+}, Ar   (b) O^{2-}, F^-
(c) Na^+, Al^{3+} (d) P^{3-}, Ar

8. An element has two isotopes with mass numbers 16 and 18. The average atomic weight is 16.5. The percentage abundance of these isotopes is _______ and _______, respectively.
(a) 50, 50   (b) 25, 75
(c) 75, 25   (d) 33.33, 66.67

9. Some of the elements have fractional atomic masses. The reason for this could be
(a) the nuclear reactions
(b) the presence of neutrons in the nucleus
(c) the existence of isobars
(d) the existence of isotopes

10. The number of electrons present in the valence shell of an atom with atomic number 38 is
(a) 1   (b) 8
(c) 2   (d) 10

11. Which of the following arrangements of electrons represents magnesium (Mg)?
(a) 2, 8, 1   (b) 2, 8, 4
(c) 2, 8, 3   (d) 2, 8, 2

12. Which of the following have an equal number of neutrons and protons?
(a) fluorine   (b) chlorine
(c) hydrogen   (d) deuterium

13. The number of electrons in an element with atomic number X and atomic mass Y will be
(a) $(X + Y)$ (b) X
(c) $(X - Y)$ (d) $(Y - X)$

14. The relative atomic masses of many elements are not whole numbers because
(a) of the existence of isotopes
(b) of the presence of impurities
(c) they cannot be determined accurately
(d) the atoms ionise during determination of their masses

15. Which of the following describes an isotope with a mass number of 99 that contains 56 neutrons in its nucleus?

(a) $^{99}_{43}Tc$ (b) $^{56}_{43}Tc$
(c) $^{99}_{56}Ba$ (d) $^{43}_{56}Ba$

16. Which of the following isotopes is used as the standard for atomic mass?
(a) ^{13}C (b) ^{16}O
(c) ^{12}C (d) ^{1}H

17. Members of which of the following have similar chemical properties?
(a) allotropes   (b) isobars
(c) isotope   (d) both isotopes and allotropes

18. While performing cathode ray experiments, it was observed that there was no passage of electric current under normal conditions. Which of the following can account for this observation?
(a) air is a poor conductor of electricity under normal conditions
(b) carbon dioxide is present in air
(c) dust particles are present in air
(d) none of the above

19. The fluorescence on the walls of a discharge tube is due to
(a) canal rays   (b) anode rays
(c) cathode rays   (d) none of the above

20. Which of the following electronic configurations is/are correct?
(a) S (16) = 2, 6, 8   (b) P (15) = 2, 8, 5
(c) Be (3) = 2, 1   (d) O (8) = 2, 6

Advanced and Olympiads

Single Choice

1. Dalton's atomic theory successfully explained the
(i) law of conservation of mass
(ii) law of constant composition

(iii) law of radioactivity
(iv) law of multiple proportion
(a) (i), (ii) and (iii)
(b) (i), (iii) and (iv)
(c) (ii), (iii) and (iv)
(d) (i), (ii) and (iv)

2. Which of the following statements about Rutherford's model of atom are correct?
   (i) considered the nucleus as positively charged
   (ii) established that α-particles are four times as heavy as a hydrogen atom
   (iii) can be compared to the solar system
   (iv) was in agreement with Thomson's model
   (a) (i) and (iii)
   (b) (ii) and (iii)
   (c) (i) and (iv)
   (d) only (i)

3. In the Thomson's model of the atom, which of the following statements are correct?
   (i) the mass of the atom is assumed to be uniformly distributed over the atom
   (ii) the positive charge is assumed to be uniformly distributed over the atom
   (iii) the electrons are uniformly distributed in the positively charged sphere
   (iv) the electrons attract each other to stabilise the atom
   (a) (i) and (iv)
   (b) (i), (iii) and (iv)
   (c) (i), (ii) and (iii)
   (d) (i) and (iii)

4. The atomic masses of two isotopes of O are 15.9936 and 17.0036. Select the incorrect statement
   (a) total number of electrons in each atom is 16
   (b) total mass number in each atom is 33
   (c) total number of neutrons in each atom is 17
   (d) total number of protons in each atom is 17

5. $_7X^{15}$, $_7X^{11}$ are two naturally occurring isotopes of an element X. What is the percentage of each isotope of X if the average atomic mass is 14?
   (a) 95, 5
   (b) 80, 20
   (c) 75, 25
   (d) 65, 35

6. Arrange the following steps which are carried out in the α-ray experiment in the correct sequence:
   (i) passage of α-particles through a slit
   (ii) bombardment of gold foil with α-particles
   (iii) deflection of α-particles
   (iv) production of α-particles
   (a) (iv), (i), (ii), (iii)
   (b) (iv), (i), (iii), (ii)
   (c) (i), (iv), (iii), (ii)
   (d) (i), (iv), (ii), (iii)

7. Rutherford's α-particle scattering experiment showed that
   (i) electrons have negative charge
   (ii) the mass and positive charge of the atom is concentrated in the nucleus
   (iii) neutrons exist in the nucleus
   (iv) most of the space in the atom is empty
   Which of the above statements are correct?
   (a) (i) and (iii)
   (b) (ii) and (iv)
   (c) (i) and (iv)
   (d) (iii) and (iv)

8. Atomic models have been improved over the years. Arrange the following atomic models in their chronological order
   (i) Rutherford's atomic model
   (ii) Thomson's atomic model
   (iii) Bohr's atomic model
   (a) (i), (ii) and (iii)
   (b) (ii), (iii) and (i)
   (c) (ii), (i) and (iii)
   (d) (iii), (ii) and (i)

Multiple Choice

9. Which are not correct amongst the following statements about cathode rays?
   (a) particles forming them possess the same mass and same charge
   (b) e/m ratio of the particles forming them depends upon the gas filled in the discharge tube
   (c) they are deflected by electric and magnetic fields
   (d) e/m ratio is constant, independent of the gas filled and the cathode material

10. Which of the following atoms has two neutrons in its nucleus?
    (a) protium
    (b) tritium
    (c) helium
    (d) lithium

11. Which of the following statements are correct?
    (a) $_6^{12}C$ and $_6^{13}C$ are isotopes
    (b) $_7^{14}N$ and $_6^{14}C$ are isotones
    (c) $_7^{14}N$ and $_6^{14}C$ are isobars
    (d) $_7^{14}N$ and $_6^{13}C$ are isotones

12. Which of the following concepts was considered in Rutherford's atomic model?
    (a) electrons revolve around the nucleus at very high speeds
    (b) existence of nuclear forces of attraction on the electron
    (c) the electrical neutrality of the atom
    (d) the quantisation of energy

13. Which of the following statements are true?
    (a) The total number of electrons and protons in an atom is always equal
    (b) The total number of electrons in any energy level can be calculated by the formula $2n^2$
    (c) Most of the space in an atom is empty
    (d) The total number of neutrons and protons is always equal in a neutral atom

14. In Bohr's model, which of the following are correct?
    (a) energy is quantised in each orbit
    (b) angular momentum is quantised in each orbit
    (c) it is applicable for multi-electronic species also
    (d) by using its radius, the velocity in any orbit can be found out

15. Which pairs are correctly matched here?
   (a) CO_2 and N_2O are isoesters
   (b) $_7^{15}N$ and $_8^{16}O$ are isotones
   (c) P and NO are isosteres
   (d) $_9^{19}F$ and $_{19}^{39}K$ are isodiaphers

Comprehension Type

Comprehension I: Atomic number is equal to number of protons or electrons present in any atom. It is denoted by Z. In the case of ions, the number of electrons differ from those in the atom due to uptake or losing of electrons; thus, Z cannot be equal to electrons for ions. Mass number (A) is the sum of protons + neutrons present in an atom.

16. A diapositive ion has 16 protons. What is the number of electrons in its tetrapositive ion?
   (a) 14   (b) 16
   (c) 12   (d) 20

17. If the atomic weights of C and Si are 12 and 28, respectively, then what is the ratio of the number of neutrons in them?
   (a) $2:4$ (b) $2:3$
   (c) $7:3$ (d) $3:7$

18. Two nuclides X and Y are isoneutronic. Their mass numbers are 76 and 77, respectively. If the atomic number of X is 32, then the atomic number of Y will be
   (a) 32   (b) 30
   (c) 33   (d) 34

Comprehension II: According to Bohr, the extra nuclear region is divided into circular orbits. Each orbit has quantised energy and quantised angular momentum. The angular momentum is given as $mvr = nh/2\pi$. Here, n is 1, 2, 3, ... so on. By using this model, the energy, radius and velocity in any orbit can be calculated for hydrogenic species (1 e species).

19. Which of the following statements is(are) correct in Bohr's model if the mass of an electron becomes 10 times its original mass?
   (a) Energy of the electron increases 10 times.
   (b) Orbit radius decreases 10 times.
   (c) Velocity of the electron increases 10 times.
   (d) Wavelength of the electron will remain the same.

20. If the radius of the Bohr orbit is r then the de Broglie wavelength of the electron in the second orbit will be
   (a) πr (b) $\dfrac{2\pi r}{3}$
   (c) $\dfrac{3\pi r}{2}$ (d) $6\pi r$

21. For which of the following pairs is Bohr's theory applicable?
   (a) Li^+, He   (b) H, He^+
   (c) He^+, Li^{2+} (d) both (b) and (c)

Matrix Matching

22.

Column I	Column II
(a) 25 p, 30 n, 25 e	(p) $_{20}^{40}Ca^{2+}$
(b) 20 p, 20 n, 18 e	(q) $_{25}^{55}Mn$
(c) 34 p, 45 n, 36 e	(r) $_{26}^{56}Fe^{2+}$
(d) 26 p, 30 n, 24 e	(s) $_{34}^{79}Se^{2-}$

23.

Column I	Column II
(a) Isotopes	(p) have the same number of atoms and same number of electrons
(b) Isobars	(q) have the same number of neutrons
(c) Isotones	(r) have the same sum of $p + n$
(d) Isoesters	(s) have the same number of protons
	(t) have the same mass number

Integer Type

24. The e/m ratio of a proton ($_1^1H$) is x times more than that of the α-particle ($_2^4\alpha$).

25. The number of neutrons present in $_6^{14}C$ is x. What is the value of x?

26. The radius of the atom is 10^n times higher than that of the radius of the nucleus. What is the value of n here?

27. How many of the following species are isoelectronic?
Na, K^+, Mg^{2+}, Ca^{2+}, S^{2-}, Ar, P^{3-}, Cl^-, Al

28. In Millikan's experiment, the static electric charge on the oil drops was obtained by shining X-rays. If the static electric charge on the oil drop is -1.282×10^{-18} C, calculate the number of electrons present in it.

29. How many of the following are isotones?
$_6^{12}C$, $_6^{14}C$, $_7^{14}N$, $_7^{15}N$, $_8^{16}O$, $_8^{17}O$

30. How many of the following isotopes are radioactive in nature?
C-14, O-16, O-18, U-235, $_1^3H$, I-127, Fe-56

Answer Keys

Fill in the Blanks

1. positive
2. X-rays
3. plum pudding model
4. neutron
5. atomic number
6. mass number
7. gold metal
8. 10
9. Bohr
10. 23

True or False

1. T 2. F 3. F 4. T 5. T
6. T 7. T 8. F 9. F 10. F
11. F 12. F 13. F

Match the Following

1. (a)—(r); (b)—(q); (c)—(s); (d)—(p)

Very Short Answer Type Questions

1. The nature of the charge present on the cathode and anode rays is negative and positive respectively.

2. A proton is 1840 times heavier than an electron.

3. For a neutral atom, number of electrons = atomic number.
   Number of electrons in oxygen = 8
   So, number of electrons in sulfur = 16

4. O-18 has more neutrons than O-16.
   Number of neutrons = A – Z
   In O-18, number of neutrons = 18 – 8 = 10
   In O-16, number of neutrons = 16 – 8 = 8

5. Number of nucleons = Mass number = $p + n$ = $12 + (24 – 12) = 24$

6. Atomic number (Z) = 19
   Number of protons = Z = 19
   Number of electrons = Z – Positive charge = 19 – 1 = 18

7. Charge on the electron was confirmed for the first time by Millikan through his oil drop experiment.

8. Its atomic number and mass number are 17 and 35 respectively.

9. The decreasing order of e/m ratio is $e > p > \alpha > n$.

10. The radius of the atom is larger than the radius of the nucleus by 100000 times.

Short Answer Type Questions

1. In $^{18}_{8}O$, the number of protons = 8
   Number of neutrons = 18 – 8 – 10
   In $^{15}_{5}N$, the number of protons = 7
   Number of neutrons = 15 – 7 – 8

2. In $^{40}_{18}E$, $Z = 18$, $A = 40$
   As the number of electrons or protons = atomic number (Z),

The number of electrons = 18
Number of protons = 18
Number of neutrons = $A – Z = 40 – 18 = 22$
Electronic configuration of E (18) = 2, 8, 8

3. See text part.

4. The maximum number of electrons that can be present in the K, L and M shells are 2, 8, 18 respectively. It is decided on the basis of $2n^2$ (where n is the orbit number).

5. The electronic configuration can be given as follows:
   $^{16}_{8}O$ = 2, 6
   $^{28}_{14}Si$ = 2, 8, 4
   $^{35}_{17}Cl$ = 2, 8, 7
   $^{40}_{20}Ca$ = 2, 8, 8,2

6. See text part.

7. Isotopes have the same chemical properties as they have the same atomic numbers. Chemical properties are decided by the atomic number and so they are the same. Isotopes have different atomic masses so mass-related physical properties like boiling point and density differ.

8. (i) Element with configuration 2, 7 has atomic number 9—so it is fluorine.
   (ii) Element with configuration 2, 8, 1 has atomic number 11—so it is sodium.
   (iii) Element with configuration 2, 8, 7 has atomic number 17—so it is chlorine.
   (iv) Element with configuration 2, 8 has atomic number 10—so it is neon.
   Here element IV (neon) is an inert gas.

9. An atom is neutral in spite of having charged particles because the number of positively charged particles (protons) are equal to the number of negatively charged species (electrons). For example, in Mg, there are 12 protons and 12 electrons.

Long Answer Type Questions

1. See text part.

2. For theory of the Bohr Bury scheme, see text part.
   The configuration of these elements are as follows:
   (i) Element with atomic number 6 (carbon) has a configuration 2, 4.
   (ii) Element with atomic number 11 (sodium) has a configuration 2, 8, 1.
   (iii) Element with atomic number 15 (phosphorus) has a configuration 2, 8, 5.
   (iv) Element with atomic number 18 (argon) has a configuration 2, 8, 8.

3. Atomic masses of most of the elements are fractional due to the presence of isotopes as atomic mass is taken as the

average atomic mass of all the isotopes of that element. Average atomic mass is given as,

$$(\text{Atomic weight})_{\text{Average}} =$$

$$\frac{\text{Atomic weight} \times \% + \text{Atomic weight} \times \%}{100}$$

$$(\text{Atomic weight})_{\text{Average}} = \frac{10 \times 20 + 11 \times 80}{100} = 10.8$$

4. See text part.

5. See text part in Rutherford's atomic model.

NCERT Corner

In-Text Questions

1. Canal rays are positively charged radiations that can pass through a cathode plate. These rays are made of positively charged particles known as protons.

2. An electron is a negatively charged particle, whereas a proton is a positively charged particle and the magnitude of their charges is equal so, an atom containing one electron and one proton will not carry any charge. Thus, it will be a neutral atom.

3. According to Thomson's model of the atom, an atom consists of both negative and positive charges which are equal in number and magnitude. Hence, they balance each other as a result of which the atom as a whole is electrically neutral.

4. On the basis of Rutherford's model of the atom, protons are present in the nucleus of an atom.

5. See text part in Bohr's model.

6. If the α-particle scattering experiment is carried out using a foil of any metal as thin as the gold foil used by Rutherford, there would be no change in observations. But as other metals are not as malleable as gold, such a thin foil is difficult to obtain. If we use a thick foil, then more α-particles would bounce back and no idea about the location of positive mass in the atom would be available with such accurate certainty.

7. The three sub-atomic particles of an atom are: (i) protons, (ii) electrons, and (iii) neutrons.

8. Number of neutrons = Atomic mass (A) – Number of protons (p)
So, the number of neutrons in the atom $= 4 - 2 = 2$

9. (i) The total number of electrons in a carbon atom is 6. The distribution of electrons in carbon atom is as follows:
First orbit or K-shell = 2 electrons
Second orbit or L-shell = 4 electrons
That is, the distribution of electrons in a carbon atom is 2, 4.

   (ii) The total number of electrons in a sodium atom is 11. The distribution of electrons in a sodium atom is as follows:

First orbit or K-shell = 2 electrons
Second orbit or L-shell = 8 electrons
Third orbit or M-shell = 1 electron
That is, the distribution of electrons in a sodium atom is 2, 8, 1.

10. The maximum capacity of the K shell is 2 electrons and that of the L shell is a maximum of 8 electrons. So, there will be a maximum of ten electrons in the atom.

11. If the number of electrons in the outermost shell (valence electron) of the atom of an element is less than or equal to 4, then the valency of the element is equal to the number of electrons in the outermost shell. On the other hand, if the number of electrons in the outermost shell of the atom of an element is greater than 4, then the valency of that element is determined by subtracting the number of electrons in the outermost shell from 8.
As the distribution of electrons in chlorine, sulfur and magnesium atoms are 2, 8, 7; 2, 8, 6 and 2, 8, 2 respectively, the number of electrons in the outer most shell of chlorine, sulfur and magnesium atoms are 7, 6 and 2 respectively.
Valency of chlorine $= 8 - 7 = 1$
Valency of sulfur $= 8 - 6 = 2$
The valency of magnesium $= 2$

12. (i) As atomic number is equal to the number of protons, the atomic number of the atom is 8.
    (ii) As the number of both electrons and protons is equal, the charge on the atom is 0.

13. Mass number of oxygen = Number of protons + Number of neutrons $= 8 + 8 = 16$
Mass number of sulfur = Number of protons + Number of neutrons $= 16 + 16 = 32$

14.

Symbol	Protons	Neutrons	Electrons
H	1	0	1
D	1	1	1
T	1	2	1

15. $^{12}_{6}C$ and $^{14}_{6}C$ are isotopes of carbon and they have the same electronic configuration 2, 4. $^{22}_{10}Ne$ and $^{22}_{11}Na$ are isobars. They have different electronic configuration as given below:

$^{22}_{10}Ne = 2, 8$ $\qquad$ $^{22}_{11}Na = 2, 8, 1$

Exercises

1. See text part for comparison of these particles.

2. The limitations of J J Thomson's model of the atom are:
   - it could not explain the result of scattering experiment performed by Rutherford,
   - it did not have any experimental support.

3. The limitations of Rutherford's model of the atom are:
   - it failed to explain the stability of an atom,
   - it doesn't explain the spectrum of hydrogen and other atoms.

4. See text part.

5.

Thomson's model	Rutherford's model	Bohr's model
An atom consists of a positively charged sphere and the electrons are embedded in it.	An atom consists of a positively charged centre called the nucleus. The mass of the atom is contributed mainly by the nucleus.	Bohr proved almost all points of Rutherford's theory. Regarding the revolution of electrons, he added that there are only certain orbits known as *discrete orbits* inside the atom in which electrons revolve around the nucleus.
The negative and positive charges are equal in magnitude. As a result, the atom is electrically neutral.	The size of the nucleus is very small (less than 100000 times) compared to the size of the atom.	While revolving in their discrete orbits, the electrons do not radiate any form of energy.

6. See text part.

7. The valency of an element is the combining capacity of that element. The valency of an element is determined by the number of valence electrons present in the atom of that element.
Valency of silicon: It has an electronic configuration of 2, 8, 4. Hence, the valency of silicon is 4 as these electrons can be shared with others to complete the octet.
Valency of oxygen: It has an electronic configuration of 2, 6. Hence, the valency of oxygen is 2 as it will gain 2 electrons to complete its octet.

8. See text part.

9. The atomic number of sodium is 11. So, the neutral sodium atom has 11 electrons and its electronic configuration is 2, 8, 1. But Na^+ is formed by losing 1 electron so it has 10 electrons. Out of 10, the K-shell contains 2 and the L-shell contains 8 electrons. Hence, Na^+ has completely filled K and L shells.

10. It is given that the two isotopes of bromine are $^{79}_{35}Br$ (49.7%) and $^{81}_{35}Br$ (50.3%). Then, the average atomic mass of bromine atom is given by:

$$79 \times \frac{49.7}{100} + 81 \times \frac{50.3}{100}$$

$$= \frac{3926.3}{100} + \frac{4074.3}{100} = \frac{8000.6}{100} = 80.006 \text{ u}$$

11. It is given that the average atomic mass of the sample of element X is 16.2 u. Let the percentage of isotope 18/8 X be $y\%$. Thus, the percentage of isotope 16/8 X will be $(100 - y)\%$. Therefore,

$$18 \times \frac{y}{100} + 16 \times \frac{(100-y)}{100} = 16.2$$

$$\frac{18y}{100} + \frac{16(100-y)}{100} = 16.2$$

$$\frac{18y + 1600 - 16y}{100} = 16.2$$

$$18y + 1600 - 16y = 1620$$

$$2y + 1600 = 1620$$

$$2y = 1620 - 1600$$

$$Y = 10$$

Therefore, the percentage of isotope $^{18}_8 X$ is 10%.

And, the percentage of isotope $^{16}_8 X$ is $(100 - 10)\% = 90\%$.

12. By $Z = 3$, it means that the atomic number of the element is 3. As its electronic configuration is 2, 1, the valency of the element is 1 (as the outermost shell has only one electron). Hence, the element with $Z = 3$ is lithium (Li).

13. Mass number of X = Number of protons + Number of neutrons = 6 + 6 = 12
Mass number of Y = Number of protons + Number of neutrons = 6 + 8 = 14
As these two atomic species X and Y have the same atomic number but different mass numbers, they are isotopes.

14.

Atomic number	Mass number	Number of neutrons	Number of protons	Number of electrons	Name of the element
9	19	10	9	9	Fluorine
16	32	16	16	16	Sulfur
12	24	12	12	12	Magnesium
1	2	1	1	1	Deuterium
1	1	0	1	0	Hydrogen ion

Exemplar Problems

Short Answer Questions

1. Yes, it possible. For example, hydrogen has one electron, one proton and no neutron. It is represented as $^1_5 H$.

2. Atoms are divisible. With the discovery of electrons and protons it was established that the atom is further divisible and is made up of negatively charged electrons and positively charged protons along with some neutral particles called neutrons.

3. No, both ^{35}CI and ^{37}CI will have the same valency, as ^{35}CI and ^{37}CI are isotopes. As the isotopes have the same number of electrons and protons, and differ only in the number of neutrons, their electron distribution (2, 8, 7) will be the same.

4. A light metal cannot be used because on being hit by fast moving α-particles, the atom of the light metal will be simply pushed forward and no scattering can occur. So,

Rutherford selected the gold foil as gold is a heavy metal with high mass number, highly malleable and can be beaten to get very thin foils.

5. It has an electronic configuration = 2,8,8. Its outermost shell has a complete octet so, its valency = 0.
   (b) It has an electronic configuration = 2, 7. It can easily gain one electron to complete its outermost octet so, its valency = –1.

6. $X - 1e^- \rightarrow X^+$
   Here the ion formed by X by the loss of one electron will have a positive nature and one positive (+1) charge exists on the cation formed (X^+).

7. The electron distribution of the chlorine atom is as follows:

Cl	K	L	M
17 $\longrightarrow$	2	8	7

   Hence, the L shell will have 8 electrons.

8. $X - 2e^- \rightarrow X^{2-}$
   When an atom (X) has 6 electrons in its outermost shell and it accepts 2 electrons, it gets two negative charges (X^{2-}).

9.

Element	Atomic number	Mass number	Valency
X	5	11	3
Y	8	18	2
Z	15	31	3, 5

10. The given statement is not correct as number of protons can never be greater than the number of neutrons. The number of neutrons can be equal to or greater than the number of protons. However, the number of protons is equal to the number of electrons for an atom as it is neutral.

11. In $^{31}_{15}X$, number of protons = number of electrons = 15
    Number of neutrons = Mass number (A) – no. of protons (p) = 31 – 15 = 16

12. (a) (iii), (b) (iv), (c) (i), (d) (ii), (e) (vi), (f) (vii) and (g) (v).

13. A pair of elements in which they have the same mass number but different atomic numbers are called isobars. Hence $^{40}_{20}Ca$ and is $^{40}_{18}Ar$ are isobars.

14.

Element	n_p	n_n
$^{35}_{17}Cl$	17	18
$^{12}_{6}C$	6	6
$^{81}_{35}Br$	35	46

15. Helium has only one shell (K-shell) and the maximum number of electrons within the K-shell can be 2. So, it cannot lose or gain electrons. Hence, its valency is zero.

16. (a) nucleus, (b) atomic number, mass number, (c) 0 and 1 and (d) silicon: 2, 8, 4, sulfur: 2, 8, 6.

17. As the element X has an atomic number of 2, an atom of X contains two electrons. These two electrons fill the K-shell completely so the valency of X is zero.

Long Answer Questions

1. Helium has only the K shell and it is completely filled with 2 electrons. Argon and neon have 8 electrons in their outermost shell which is the maximum number of electrons that can be accommodated in the outermost shell, hence their valency is zero as they do not accept or lose any electrons.

2. (i) Volume of the sphere $= \dfrac{4}{3}\pi r^3$

   Let R be the radius of the atom and r be that of the nucleus $= R = 10^3\,r$

   Volume of the atom $= \dfrac{4}{3}\pi R^3 = \dfrac{4}{3}\pi (10^5 r)^3$

   $= \dfrac{4}{3}\pi r^3 \times 10^{15}$

   Volume of the nucleus $= \dfrac{4}{3}\pi r^3$

   Ratio of the size of atom to that of nucleus

   $= \dfrac{\dfrac{2}{3} \times 10^{15} \times \pi r^3}{\dfrac{4}{3}\pi r^3} = 10^{15}$

   (ii) If the atom is represented by the planet earth ($R_e = 6.4 \times 10^6$ m) then the radius of the nucleus would be

   $r_n = \dfrac{R_e}{10^5}$

   $r_n = \dfrac{6.4 \times 10^6 \text{ m}}{10^5} = 6.4 \times 10 \text{ m} = 64 \text{ m}$

3. See text part.

4. See text part.

5. See text part.

6. See text part.

7. $_{11}$Na: 2,8,1 (11 electrons) Na $- 1e^- \rightarrow$ Na$^+$ (10 electrons)
   Electronic configuration: 2, 8.

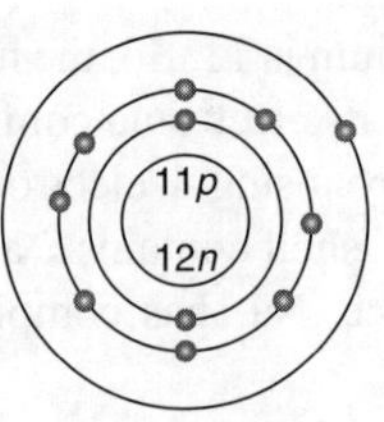

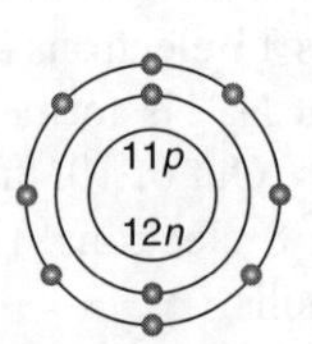

Na atom   Na$^+$ (sodium ion)

The atomic number of an element is equal to the number of protons in its atom. As sodium atom and sodium ion contain the same number of protons, the atomic number of both is 11.

8. % of α-particles deflected more than 50° = 1%
   % of α-particles deflected less than 50° = 100 – 1 = 99%
   Number of α-particles bombarded = 1 mol = 6.022 × 10^{23} particles
   Number of particles that deflected at an angle less than 50°
   = 99/100 × 6.022 × 10^{23} × 100

   $= \dfrac{99}{100} \times 6.022 \times 10^{23} = \dfrac{596.178}{100} \times 10^{23} = 5.96 \times 10^{23}$

Competition Window (Objective Type)

Topic-wise MCQs

1. (b)	**2.** (c)	**3.** (a)	**4.** (a)	**5.** (b)
6. (c)	**7.** (d)	**8.** (b)	**9.** (b)	**10.** (a)
11. (c)	**12.** (b)	**13.** (d)	**14.** (a)	**15.** (c)
16. (a)	**17.** (d)	**18.** (a)	**19.** (b)	**20.** (a)
21. (d)	**22.** (a)	**23.** (a)	**24.** (d)	**25.** (c)
26. (c)	**27.** (b)	**28.** (b)	**29.** (b)	**30.** (a)
31. (a)	**32.** (b)	**33.** (c)	**34.** (b)	**35.** (c)
36. (a)	**37.** (b)	**38.** (d)	**39.** (a)	**40.** (d)
41. (b)	**42.** (b)	**43.** (c)	**44.** (c)	**45.** (a)
46. (a)				

Miscellaneous

1. (b)	**2.** (b)	**3.** (b)	**4.** (c)	**5.** (a)
6. (c)	**7.** (a)	**8.** (c)	**9.** (d)	**10.** (c)
11. (d)	**12.** (d)	**13.** (b)	**14.** (a)	**15.** (a)
16. (c)	**17.** (a)	**18.** (a)	**19.** (c)	**20.** (a)

Advanced and Olympiads

Single Choice

1. (d)	**2.** (a)	**3.** (c)	**4.** (d)	**5.** (c)
6. (a)	**7.** (b)	**8.** (c)		

Multiple Choice

9. (acd)	**10.** (bc)	**11.** (acd)	**12.** (abc)	**13.** (abc)
14. (abd)	**15.** (abd)			

Comprehension Type

16. (c)	**17.** (d)	**18.** (c)	**19.** (d)	**20.** (a)
21. (d)				

Matrix Matching

22. (a)—(q); (b)—(p); (c)—(s); (d)—(r)
23. (a)—(s); (b)—(r, t); (c)—(q); (d)—(p)

Integer Type

24. 2	**25.** 8	**26.** 5	**27.** 6	**28.** 8
29. 3	**30.** 5			

Hints and Solutions

Competition Window (Objective Type)

Topic-wise MCQs

1. As cathode rays are never electromagnetic waves.

6. Neutron was discovered by Chadwick.
$$_4\mathrm{Be}^9 + {}_2\alpha^4 \rightarrow {}_6\mathrm{C}^{12} + {}_0 n^1$$

11. Due to low mass.

13. J J Thomson proposed the first model of the atom.

15. Rutherford's alpha (α) particles scattering experiment resulted in the discovery of the nucleus in the atom. A large number of particles went straight through the atom while a very small number of particles were deflected back showing the presence of a positively charged nucleus.

20. An atom has an equal number of electrons and protons since it is neutral.

21. Mass number = $p + n$, Atomic number = $p = e$.

23. Number of electrons = 15, Number of neutrons = 16. Hence atomic number of element X is 15 and atomic mass is 31. Hence the element is represented as $_{15}^{31}\mathrm{X}$.

24. Mass number = $11 + 12 = 23$.

27. Isobars have the same atomic mass.

28. Isotones have the same number of neutrons.

29. As N^{3-}, F^- and Na^+ has 10 electrons.

31. As these have 18 electrons.

32. Contain same number of electrons (= 10).

33. The number of neutrons is different in the isotopes of the same element.

41. Mass number of the atom = 27
Number of neutrons = 14
Number of protons = 27 – 14 = 13
Number of electrons in the atom = 13
Number of electrons in an ion with 3 positive charges = 13 – 3 = 10.

42. Core charge = $+Ze$ (Z = atomic number) = $+8e$.

43. Element with valency 1 can be both a metal or a non-metal.

44. For the atom $n = 4$, $p = 3$, hence $e = 3$. Distribution of electrons = 2, 1.

45. Aluminium has 13 electrons. Its electron distribution is 2, 8, 3.

46. X^{2+} has 16 protons, then
in X, there are 16 protons and 16 electrons
in X^{2+}, there are 16 protons and 14 electrons
in X^{4+}, there are 16 protons and 12 electrons.

Miscellaneous

3. $_{12}\mathrm{M} - \underset{2}{\mathrm{K}} \quad \underset{8}{\mathrm{L}} \quad \underset{2}{\mathrm{M}}$

4. Two isotopes have the same number of electrons and protons but different number of neutrons.

5. Thomson's model could be compared with a raisin (plum) pudding model according to which the mass of the atom is uniformly distributed over the atom in the form of positive charge and the electrons are uniformly distributed over the atom. Electrons do not attract each other.

10. $Z = 38$, electronic configuration = 2, 8, 18, 8, 2.
So, two valence electrons.

Advanced and Olympiads

Single Choice

1. Dalton's atomic theory successfully explained the law of conservation of mass, law of constant composition and law of multiple proportions.

2. Rutherford's model of the atom explained the presence of the nucleus in the centre and electrons around the nucleus revolving in round orbitals as in the solar system.

4.

	First isotope of O	Second isotope of O
Atomic mass	15.9936	17.0036
(a) Number of neutrons	$16 - 8 = 8$	$17 - 8 = 9$
(b) Number of protons	$16 - 8 = 8$	$17 - 9 = 8$
(Mass number – Number of neutrons)		
(c) Number of electrons	8	8
(d) Mass number	16	17

5. Let the percentage of $_7X^{15}$ be x.

So, the percentage composition of $_7X^{11}$ is $100 - x$

Average atomic weight $= 14 = \dfrac{x(15) + (100 - x)11x}{100}$

$\Rightarrow 1400 = 15x + 1100 - 11x$

$\Rightarrow 1400 = 4x + 1100 \Rightarrow 4x = 300$

$x = \dfrac{300}{4} = 75$

So, the percentage of $_7X^{11} = 100 - x$

$= 100 - 75 = 25$ and that of $_7X^{15}$ is 75.

6. (i) production of α-particles
(ii) production of a narrow beam of α-particles
(iii) bombardment of gold foil with α-particles
(iv) deflection of α-particles

7. Two important observations of Rutherford's α-particles scattering experiment were, (i) the mass and positive charge of the atom is concentrated in the nucleus, and (ii) most of the space in the atom is empty.

8. Thomson's atomic model, followed by Rutherford's model, followed by Bohr's model.

Multiple Choice

12. According to Rutherford's theory, an atom is electrically neutral and electrons revolve around the nucleus. Nuclear forces of attraction exist between the nucleus and the electrons. The only assumption that Rutherford did not consider is quantisation of energy.

15. P and NO are not isosters as they have different numbers of atoms. However, they are isoelectronic (15 e).

$_9^{19}F$ and $_{19}^{39}K$ are isodiaphers as they have the same $n - p$ value (1).

Comprehension Type

16. Let the element be X and the ion X^{2+}. It has 16 protons, then

in X, there are 16 protons and 16 electrons

in X^{2+}, there are 16 protons and 14 electrons

X^{4+}, there are 16 protons and 12 electrons

17. Number of neutrons in $_6^{12}C = 12 - 6 = 6$

Number of neutrons in $_{14}^{28}Si = 28 - 14 = 14$

The ratio of the number of neutrons in C and Si is 6 : 14 or 3 : 7.

18.

$_{32}^{76}X$	$_P^{77}Y$
Protons = 32	Protons + Neutrons = 77
Protons + Neutrons = 76	Neutrons = 44
Neutrons = 44	Protons = 33

20. Bohr radius $= r$

Bohr velocity, $V = \dfrac{nh}{2\pi m r}$

de-Broglie wavelength, $\lambda = \dfrac{nh}{mV}$

So, for the third orbit, $n = 2$

So $\lambda = \dfrac{h}{\dfrac{m \times 2h}{2\pi m r}} = \dfrac{2\pi r}{2} = \pi r$

21. This theory is applicable for 1 electron systems such as H, He^+, Li^{2+}.

Integer Type

24. As the charge on an α-particle is twice that of a proton, but the mass is 4 times that of a proton, it will have half the e/m ratio of a protons. Or, a proton has twice the e/m ratio of an α-particle.
So x = 2.

25. Number of neutrons $= A - Z = 14 - 6 = 8$
So x = 8.

26. The radius of an atom is 10^5 times higher than the radius of the nucleus, so the value of n is 5.

27. Here K^+, Ca^{2+}, S^{2-}, Ar, P^{3-}, Cl^- are all isoelectronic as they all have 18 electrons.

28. Total electrons present $= \dfrac{\text{Total Charge}}{\text{Charge on 1 electron}}$

$= \dfrac{-1.282 \times 10^{-18}\ C}{-1.6022 \times 10^{-19}\ C} = 8$

29. $_6^{14}C$, $_7^{15}N$, $_8^{16}O$ are isotones as they have 8 neutrons.

30. These are radioactive isotopes in nature.

C-14, O-18, U-235, $_1^3H$, I-127

The Periodic Table

After studying this unit, you will be able to understand:

- All periodic laws, need for periodic table and classification of elements
- The different periodic tables
- Periodic properties and their trends
- Comparison of periodic properties for various elements

	1	2	3+	4	5	6	7	8	9	10	11	12‡	13	14	15	16	17	18
1	1 H																	2 He
2	3 Li	4 Be											5 B	6 C	7 N	8 O	9 F	10 Ne
3	11 Na	12 Mg											13 Al	14 Si	15 P	16 S	17 Cl	18 Ar
4	19 K	20 Ca	21 Sc	22 V	23 Ti	24 Cr	25 Mn	26 Fe	27 Co	28 Ni	29 Cu	30 Zn	31 Ga	32 Ge	33 As	34 Se	35 Br	33 Kr
5	37 Rb	38 Sr	39 Y	40 Nb	41 Zr	42 Mo	43 Tc	44 Ru	45 Rh	46 Pd	46 Pd	47 Ag	49 In	50 Sn	51 Sb	52 Te	53 I	54 Xe
6	55 Cs	56 Ba	57 La (58-71)	72 Hf	73 Ta	74 W	75 Re	76 Os	77 Ir	78 Pt	78 Pt	79 Pt	81 Tl	82 Pb	83 Bi	84 Po	85 At	86 Rn
7	87 Fr	88 Ra	89 Ac (90-103)	104 Rf	105 Db	106 Sg	107 Bh	108 Hs	109 Mt	110 Ds	111 Rg	112 Cn	113 Nh	114 Fl	115 Mc	116 Lv	117 Ts	118 Og

58 Ce	59 Pr	60 Nd	61 Pm	62 Sm	63 Eu	64 Gd	65 Tb	66 Dy	67 Ho	68 Er	69 Tm	70 Yb	71 Lu
90 Th	91 Pa	92 U	93 Np	94 Pu	95 Am	96 Cm	97 Bk	98 Cf	99 Es	100 Fm	101 Md	102 No	103 Lr

Introduction

Matter is present in the form of elements, compounds and mixtures. Elements like silver, gold, copper, tin, mercury and lead have been known since ancient times. From the discovery of the first element phosphorus, in 1649, we now know 118 elements. All these elements have seemingly different properties. Scientists started recognising similarities and dissimilarities among these elements and started various types of arrangements to make their study easy and systematic. This helped in the development of the periodic table. The periodic table is an arrangement of elements with similar properties placed together. It helps in the easier study of elements.

Genesis of Periodic Classification

Dobereiner–The Law of Triads

According to this law, elements could be arranged in groups of three, in which the atomic weight of the middle element was the arithmetic mean of the first and the last element and also the difference in the atomic weight of consecutive elements was a constant. Example,

$$_3\text{Li}^9, \;_{11}\text{Na}^{23}, \;_{19}\text{K}^{39}$$

$$_4\text{Be}^9, \;_{12}\text{Mg}^{24}, \;_{20}\text{Ca}^{40}$$

KNOWLEDGE BOOSTER

Hennig Brand discovered the first element 'phosphorus' in 1649.

The mean of the atomic masses of the first (Li) and the third (K) elements is $(7 + 39)/2$ u. The atomic mass of the middle element, sodium, Na is equal to 23 u. Two more examples of Dobereiner's triads:

$$_{20}Ca^{40}, _{38}Sr^{88}, _{56}Ba^{137}$$

Mean of the atomic masses of the first and third elements = $(40 + 137)/2 = 88.5$ u

$$_{17}Cl^{35.5}, _{35}Br^{80}, _{53}I^{127}$$

Mean of the atomic masses of the first and third elements = $(35.5 + 127)/2 = 81.25$ u

Actual atomic mass of the second element (Br) = 80 u

Dobereiner's idea of classification of elements into triads did not receive wide acceptance since he could arrange only a few elements in this manner.

Telluric Helix

The idea of periodic arrangement of elements was introduced for the first time by Chancourtois in 1862. By using atomic weights, he drew the elements as a continuous spiral around a cylinder divided into 16 parts. A list of elements was wrapped around a cylinder so that several sets of similar elements lined up, creating the first geometric representation of the periodic law.

Newlands (1864)—Law of Octaves

According to this law, 'if the elements are arranged in the order of increasing atomic weight, the eighth element starting from a given one, is a kind of repetition of the first, like the eighth node in an octave of music'. For example, starting from lithium, the eighth element is sodium so the properties of lithium and sodium should be similar.

$$Li^{6.9} \quad Be^{9} \quad B^{10.8} \quad C^{12} \quad N^{14} \quad O^{16} \quad F^{19} \quad Na^{23} \quad Mg^{24.3}$$

$$Al^{27} \quad Si^{28.1} \quad P^{31} \quad S^{32.1} \quad Cl^{35.5} \quad K^{39.1} \quad Ca^{40.1}$$

Newlands could arrange elements up to calcium by using this law. At that time more than 60 elements were known, so this law was not appreciated.

Prout (1915)

His law was known as Prout's unitary hypothesis. According to it, all elements are made up of hydrogen atoms.

Lothar Meyer (1869)

He plotted various physical properties of elements against their atomic weights and pointed out that the similar elements occupy similar positions on the curve (Fig. 6.1). For example, alkali metals are present at peaks of the curves.

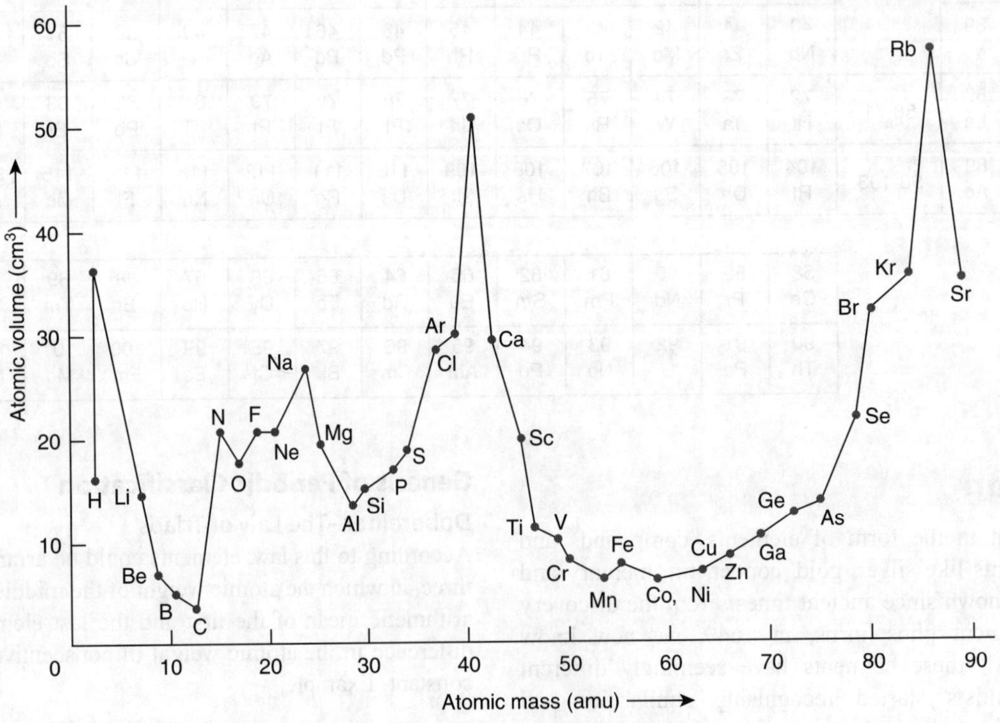

Fig. 6.1 Atomic volume vs atomic mass

Illustrations

1. (a) What is the position of the halogens (F, Cl) on Lothar Meyer's curves?

(b) Which elements occupy the minima in Lothar Meyer's curves?

Solution:

(a) The halogens occupy positions on the ascending portions of the curve.

(b) Transition elements occupy the minima of the curves in Lothar Meyer's curves.

2. If the atomic weight of phosphorus (P) is 31 and that of antimony (Sb) is 120, find the atomic weight of arsenic using the triad system.

Solution: $^{31}P \; ^?As \; ^{126}Sb$

$$\text{Atomic weight of As} = \frac{31+120}{2} = 75.5$$

Mendeleev's Periodic Law

According to this law, 'the physical and chemical properties of elements are the periodic function of their atomic weights'.

Mendeleev's Periodic Table

Mendeleev considered atomic weight as the fundamental property of an element and he arranged different elements in the order of their increasing atomic weights and formed the periodic table.

Features of Mendeleev's Periodic Table

Mendeleev arranged elements in horizontal rows and vertical columns of a table in order of their increasing atomic weights in such a way that the elements with similar properties occupied the same vertical column or group. Mendeleev's system of classifying elements was more elaborate than that of Lothar Meyer's. He fully recognised the significance of periodicity and used a broader range of physical and chemical properties to classify the elements. In particular, Mendeleev relied on the similarities in the empirical formulae and properties of the compounds formed by the elements. He realised that some of the elements did not fit in with his scheme of classification if the order of atomic weight was strictly followed. He ignored the order of atomic weights, thinking that the atomic measurements might be incorrect, and placed the elements with similar properties together. For example, iodine with a lower atomic weight than tellurium (Group VI) was placed in Group VII along with fluorine, chlorine, bromine because of similarities in properties (Table 6.1). At the same time, keeping his primary aim of arranging the elements of similar properties in the same group, he proposed that some of the elements were still undiscovered and, therefore, left several gaps in the table. For example, both gallium and germanium were unknown at the time Mendeleev published his Periodic Table. He left a gap under aluminium and a gap under silicon, and called these elements *eka-aluminium* and *eka-silicon*. Mendeleev predicted not only the existence of gallium and germanium, but also described some of their general physical properties. These elements were discovered later. Some of the properties predicted by Mendeleev for these elements and those found experimentally are listed in Table 6.1.

Table 6.1 Mendeleev's periodic table

Group:	I	II	III	IV	V	VI	VII	VIII	Zero
Oxide:	R_2O	RO	R_2O_3	R_2H_5	R_2H_5	RO_3	R_2O_3	RO_4	RO_4
Hydride:	A RH B	A RH_2 B	A RH_4 B	A RO_4 B	A RO_4 B	A RH_2 B	A RH B	Transition traids	Noble gases
1	H 1(At. No.) 1.008(AL.Wt.)								He 2 4.026
2	Li 3 6.939	Be 4 9.012	B 5 10.811	C 6 12.011	N 7 14.007	O 8 15.999	F 9 18.998		Ne 10 20.183
3	Na 11 22.99	Mg 12 24.312	Al 13 26.981	Si 14 28.086	P 15 30.974	S 16 32.06	Cl 17 35.453		Ar 18 39.948
4 First series / second series	K 19 39.102 Cu 29 63.54	Ca 20 40.08 Zn 30 65.37	Sc 21 44.96 Ga 31 69.72	Ti 22 47.90 Ge 32 72.59	V 23 50.94 As 33 74.92	Cr 24 51.99 Se 34 78.96	Mn 25 54.939 Br 35 79.909	Fe 26 C 27 Ni 28 55.85 58.93 58.71	Kr 36 83.80
5 First series / second series	Rb 37 85.47 Ag 47 107.87	Sr 38 87.62 Cd 48 112.40	Y 39 88.905 In 49 114.82	Zr 40 91.22 Sn 50 118.69	Nb 41 92906 Sb 51 121.75	Mg 42 95.94 Te 52 127.60	Te 43 (99) I 53 124.9014	Ru 44 Rh 45 Pd 46 101.07 102.91 106.4	Xe 54 (13130)
6 First series / second series	Ca 55 132.90 Au 79 196.97	Ba 53 137.34 Hg 80 200.59	*Rare Earths 57-71 Tl 81 204.37	Hf 72 178.49	Ta 73 180.948 Bi 83 208.98	W 74 183.85 Po 84 (210)	Re 75 186.2 At 85 (210)	Os 76 Ir 77 Pt 78 190.2 192.2 195.09	Rn 86 (222)
7	Fr 87 (223)	Ra 88 (226)	Actinided elements 89-103	Pb 82 207.19	Ha 105				

Merits and Uses of Mendeleev's Periodic Table

1. Mendeleev classified elements in the form of a table so that their study became easier.
2. Mendeleev's periodic table helped in the discovery of new elements. For example, eka-aluminium (Ga).
3. Deciding atomic weights of new elements and correction of atomic weights of existing elements. For example, the atomic weight of indium was corrected from 76 to 113. It was also corrected for other elements such as Be and U.

Demerits and Limitations of Mendeleev's Periodic Table

1. The position of hydrogen is not justified as it has been placed in group I and also in group VII as it resembles both the alkali metals and the halogens in properties.
2. There is no position or place for isotopes.
3. There are many anomalous pairs as elements with higher atomic weight are placed before elements having lower atomic weight. For example,

Ar (39.94)	K (39.10)
Te (127.6)	I (126.6)
Co (58.9)	Ni (58.6)

4. Lanthanides and actinides have been put along with lanthanum and actinium respectively and there are no suitable separate positions or places for these elements.
5. Many elements with different properties are placed together (anomalous placement). For example, less reactive coinage metals are placed with highly reactive alkali metals.

Transition elements and typical elements are placed within the same box.

KNOWLEDGE BOOSTER

Today Mendeleev's periodic table has no practical significance. However, its significance lies in the fact that the concept of periodicity in properties and the resultant arrangement of elements in the form of a table was pioneered by Mendeleev.

Illustrations

1. Which group was not present in Mendeleev's periodic table?

   Solution: The zero group was not present in Mendeleev's periodic table.

2. Give any two examples of anomalous pairs in Mendeleev's periodic table.

   Solution: Anomalous pairs are: K, Ar and Co, Ni.

3. In Mendeleev's periodic table, the elements were arranged in the increasing order of their atomic masses. However, cobalt with atomic mass of 58.93 amu was placed before nickel having an atomic mass of 58.71 amu. Give a reason for this.

   Solution: Elements with similar properties can be grouped together.

4. Write the formulae of chlorides of eka-silicon and eka-aluminium, the elements predicted by Mendeleev.

   Solution: $GeCl_4$, $GaCl_3$.

Modern Periodic Law

H G J Moseley in 1913 introduced the modern periodic law based on his work on X-ray spectra of elements. He introduced the atomic number (Z) and said that it is equal to the total nuclear charge or number of protons ($Z = p$). He established atomic number as a fundamental property and introduced the modern periodic law. According to it, 'the physical and chemical properties of elements are the periodic functions of their atomic numbers'.

Modern Periodic Table

It is based on arranging different elements in the increasing order of their atomic numbers. Its main features are as follows:

- Elements have been arranged in the order of increasing atomic numbers.
- It has seven horizontal rows known as periods.
- There are sixteen vertical columns which are called groups or families.
- The transition elements are placed in the middle of the long periods.
- Isotopes of an element are assigned the same place, as they have the same atomic number.
- Dissimilar metals are placed in different groups. For example, noble metals and alkali metals are separated by placing them in group I and group II respectively.

Illustrations

1. (a) The number of horizontal and vertical columns in the modern periodic table are respectively _____.

   (b) Write the number of elements present in shortest and longest periods respectively.

Solution:

   (a) The number of horizontal and vertical columns in the modern periodic table are 7 and 16 respectively.

   (b) The shortest period is 1 with 2 elements while the longest period is 6 with 32 elements.

2. Three elements A, B and C have 3, 4 and 2 electrons respectively in their outermost shell. Give the group number to which they belong in the Modern periodic table. Also, give their valencies.

Solution:

Element	Group No.	Valency
A	Group 13	3
B	Group 14	4
C	Group 2	2

Long Form of Periodic Table

It is also known as the extended form of the periodic table or Mosley's periodic table. It was developed by Range, Werner and Burey on the basis of electronic configuration of elements.

Features

- It is divided into four blocks—s, p, d and f on the basis of electronic configuration.
- It has seven horizontal rows known as periods.
- The first period is the shortest period and has only two elements (H, He).
- The second and third periods are short periods and have eight elements each.
- The fourth and fifth periods are long periods and have eighteen elements each.
- The sixth period is the longest period and has 32 elements.
- The seventh period is incomplete and has 26 elements.
- The first elements of each period is an alkali metal while last element is an inert gas.

Period	Name	Total number of elements
1	Shortest	2
2, 3	Short	8, 8
4, 5	Long	18, 18
6	Longest	32
7	Incomplete	26*

- There are 18 vertical columns and the vertical columns are known as **groups**. Elements having similar outer electronic configurations in their atoms are arranged in vertical columns, referred to as **groups** or **families**.
- According to the recommendation of the International Union of Pure and Applied Chemistry (IUPAC), the groups are numbered from 1 to 18 replacing the older notation of groups IA … VIIA, VIII, IB … VIIB and 0.
- Group 3 (IIIrd vertical row) is the largest group having 32 elements including lanthanides and actinides.
- Groups 1, 2, 13, 14, 15, 16, 17 have metals, non-metals and metalloids. These elements are known as representative elements because their valence electrons represent their group numbers.
- Groups 3, 4, 5 (only Sc, Y, La and Ac), 6, 7, 8, 9 and 10 have only transition metals.
- In group 3, except Sc, Y, La and Ac, the other elements are known as inner transition metals and are placed separately in two rows under the periodic table. They belong to the 6th period ($_{58}$Ce, to $_{71}$Lu) and the 7th period ($_{90}$Th to $_{103}$Lr).
- The 0 group (Group 18) has the noble gases or inert gases.

Different Portions of the Long Form of Periodic Table

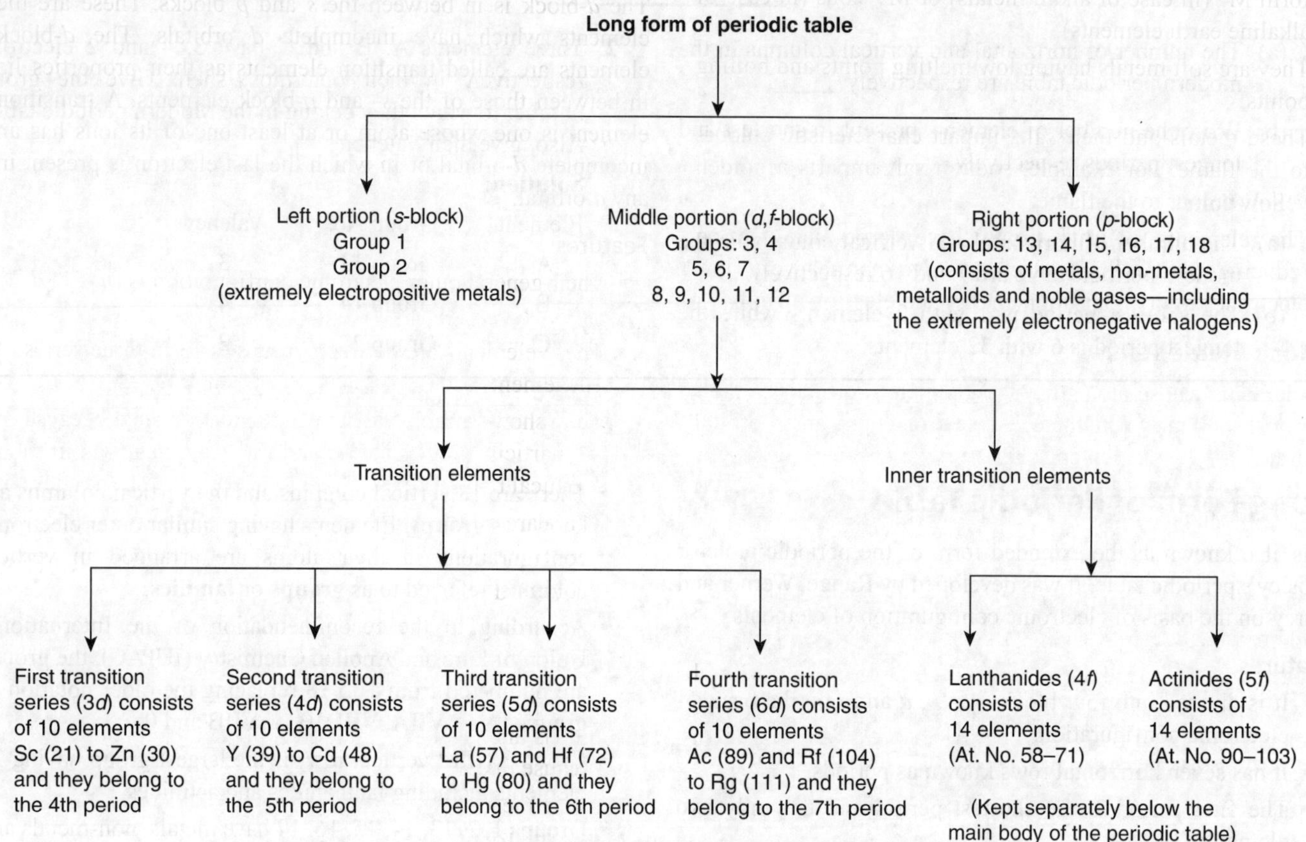

Illustrations

1. (a) What is the basis of arranging elements in the long form of the periodic table?

   (b) What is the position of lanthanides and actinides in the long form of the periodic table?

   Solution:

   (a) In the long form of the periodic table, the elements are arranged on the basis of their electronic configuration.

   (b) Both lanthanides and actinides are placed separately below the main periodic table.

2. What are the defects in the long form of the periodic table?

   Solution: The long form of periodic table has the following defects:

   (i) **Position of hydrogen:** The position of hydrogen is not fixed in this table also. It is placed either in Group 1 or in Group 17.

   (ii) **Position of lanthanides and actinides:** No individual places have been assigned to these 28 elements in the periodic table.

   (iii) This table does not reflect the exact distribution of electrons among all the orbitals of the atoms of all the elements.

Types of Elements

There are four types of known elements (s-, p-, d-, f-).

s-Block Elements

The elements having ns^1 and ns^2 electronic configurations in their outermost shell are called s-block elements. Elements with ns^1 configuration are called **group 1** (alkali elements). Elements with ns^2 configuration are called **group 2** (alkaline earth elements).

Features

- They are highly reactive and readily form univalent or bivalent positive ions by losing the valence electrons.
- Elements of this block are soft, malleable and good conductors of heat and electricity.
- Elements of this block have the largest atomic and ionic radii but the lowest ionisation energies.
- They show fixed valency and oxidation states.

- The loss of the outermost electrons (s) occurs readily to form M^+ (in case of alkali metals) or M^{+2} ions (in case of alkaline earth elements).
- They are soft metals having low melting points and boiling points.
- These metals and their salts impart characteristic colours to the flame. For example, sodium salt imparts a golden yellow colour to the flame.
- The elements of this group have large size, strong reducing nature, high electropositive nature, very low electronegativity, ionisation energy and electron affinity.

p-Block Elements

The elements whose atoms have incomplete p-orbitals in their outermost shell or in which the last electron enters any p-orbital are known as p-block elements. For example, elements of groups 13 (IIIA), 14 (IVA), 15 (VA), 16 (VIA), 17 (VIIA), 18 (VIIIA) are the p-block elements.

Features

- The general outer electronic configuration for these elements is $ns^2\, np^{1-6}$.
- Group 13 (IIIA) elements have one electron in the p-orbital whereas group 18 (VIIIA) (inert gas) elements have 6 electrons in their outer p-orbitals. The outer p-orbitals in an inert gas are fully filled with electrons.

Boron	$1s^2\, 2s^2\, 2p^1$
Oxygen	$1s^2\, 2s^2\, 2p^4$
Neon	$1s^2\, 2s^2\, 2p^6$

- They include both metals and non-metals but there is a regular gradation from metallic to non-metallic character as we move from left to right across the period.
- These elements do not impart colour in the flame test.
- Except F and the inert gases, all other elements of this block show variable oxidation states.

Group	13 (IIIA)	14 (IVA)	15 (VA)	16 (VIA)	17 (VIIA)
Ox. state	+3	+4	+5	+6	+7
		to –4	to –3	to –2	to –1

- They have quite high ionisation energies and the values tend to increase as we move from left to right across the period.
- They form mostly covalent compounds like oxides, halides, sulfides, carbonates and so on.

d-Block Elements

The d-block is in between the s and p blocks. These are the elements which have incomplete d orbitals. The d-block elements are called transition elements as their properties lie in between those of the s- and p-block elements. A transition element is one whose atom or at least one of its ions has an incomplete d-orbital or in which the last electron is present in any d-orbital.

Features

- Their general outer electronic configuration is $(n-1)\, d^{1-10}\, ns^{1-2}$.
- These elements are between groups 2–13 in three series of 10 elements each.
- They show variable valency and oxidation state because of the participation of the ns and $(n-1)d$ electrons in their chemical bond formation due to nearly similar energies.
- These are metals with high values of melting points, boiling points, densities, thermal stabilities and hardness.
- They are ductile and malleable.
- They good conductors of heat and electricity due to the presence of mobile or free electrons.
- They form coloured ions and complexes.
- Metals and their ions are generally paramagnetic in nature because of the presence of unpaired electrons.
- These metals form a number of alloys as they have almost similar sizes.
- These metals and their compounds are widely used as catalysts.

f-Block Elements

The elements placed in two separate rows at the bottom of the periodic table are the f-block elements. They have incomplete f-orbitals in their electronic configurations. The elements from cerium to lutetium that have incomplete $4f$ orbitals are the **lanthanones** or **lanthanoids**. The elements from thorium to lawrencium having incomplete $5f$ orbitals are the **actinoids.**

Features

- Their general outer configuration is $(n-2)\, f^{1-14}\, (n-1)\, d^{1-2}\, ns^2$.
- Many of them are synthetic elements.
- They are metals having high melting and boiling points.
- They show variable oxidation states (variable valency). However, their most common and stable oxidation state is +3.
- They form coloured ions and complexes.

Illustrations

1. Why are d-block elements called transition elements?

   Solution: d-block elements are called transition elements because their properties lie or transit in between those of the s and p-block elements.

2. Which block elements can show variable oxidation states?

   Solution: Elements of the p, d and f-block can show variable oxidation states.

   For example, N (–3 to +5), Fe (0, +2, +3) and Th (+3, +4).

To Find Position in Periodic Table

Period number = Largest value of n (last orbit number)

Group number = Sum of valence electrons

Block = Sub-orbit in which last electron is filled

Example,

s^1	s^2	s^2p^1	s^2p^6
I	II	III	VIII
1	2	13	18

Example, Na = 2, 8, 1   or $1s^2, 2s^2, 2p^6, 3s^1$

Orbit = 3,   Group number = 1,   Block = s

Illustrations

1. If the atomic number of Br is 35 predict its period, group number and block.

Solution:

Br $\rightarrow$ 2, 8, 18, 7

35        $1s^2, 2s^2, 2p^6, 3s^2, 3p^6, 3d^{10}, 4s^2, 4p^5$

Period = Largest n value = 4 = 4 (IV)

Group number = Number of valence $e^- = 2 + 5 = 7 = 17$ (VII)

Block = Sub-orbit in which last e^- is filled = p

2. Find the block, period and group number of iron.

Solution: Iron is present in the d-block, 4th period and in Group 8 (VIII).

Fe $\rightarrow$ 2, 8, 14, 2

26        $1s^2, 2s^2, 2p^6, 3s^2, 3p^6, 3d^6, 4s^2$

Period = Largest n value = 4 = 4 (IV)

Group number = Number of valence $e^- = 2 + 6 = 8 = 8$ (VIII)

Block = Sub-orbit in which last e^- is filled = d

Trends in Periodic Properties

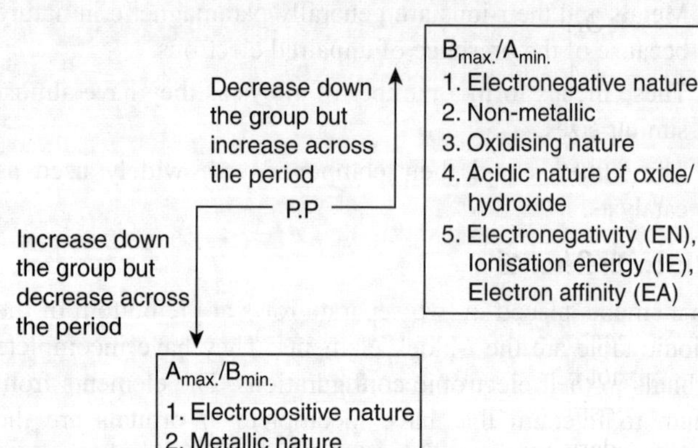

These properties of the elements vary periodically with their atomic numbers when we move from left to right across the period or top to bottom in any group.

Number of Orbits and Valence Electrons

In a period: On moving left to right, the number of orbits remains the same while the number of valence electrons increase. For example,

Element of group 17	Number of shells equals the period number	Electronic configuration
F	2	2, 7
Cl	3	2, 8, 7
Br	4	2, 8, 18, 7
I	5	2, 8, 18, 18, 7
At	6	2, 8, 18, 32, 18, 7

In a group: On moving down the group the number of orbits increases while the number of valence electrons remains same (that is equal to group number). For example,

Elements of the III period	Na	Mg	Al	Si	P	S	Cl	Ar
Atomic number	11	12	13	14	15	16	17	18
Electronic configuration	2, 8, 1	2, 8, 2	2, 8, 3	2, 8, 4	2, 8, 5	2, 8, 6	2, 8, 7	2, 8, 8

Valency

The combining capacity of an element is known as *valency*. It is measured in terms of number of valence electrons (present in the outer orbit) or number of H-atoms or O-atoms or Cl-atoms linked with one atom of an element.

Valency = number of valence electrons = number of H atoms or O atoms or Cl atoms linked with one atom of an element.

For example, in CH_4

Valency of carbon = 4

> ## KNOWLEDGE BOOSTER
>
> ### Word of Advice
>
> **If the number of valence electrons is more than 4, valency = valence electrons – 8**
>
> For example, in H_2S
>
> Valency of sulfur = $6 - 8 = -2 = 2$
>
> Valency is simply a number with no significance of + and – signs.

This is because the carbon atom has 4 valence electrons, or is linked with 4 H atoms.

For example, in NH_3

Valency of nitrogen = 3

Variation in Valency in a Period

In the case of representative elements (s, p-block) the valency increases as the number of valence electrons increase from 1 to 8 across the period.

Group	1	2	13	14	15	16	17	18
Number of electrons	1	2	3	4	5	6	7	8
Valence on hydrogen scale	1	2	3	4	3	2	1	0
Valence on oxygen scale	1	2	3	4	5	6	7	0

On the hydrogen scale, the valency increases from 1 to 4 and then decreases from 4 to 0, while on the oxygen scale, it increases from 1 to 7.

	1	2	13	14	15	16	17
3rd Period	NaH	MgH_2	AlH_3	SiH_4	PH_3	H_2S	HCl
	1	2	3	4	3	2	1
	Na_2O	MgO	Al_2O_3	SiO_2	P_2O_5	SO_3	Cl_2O_7
	1	2	3	4	5	6	7

Properties of an Element

The properties of an element depend upon the number of valence electrons present in it.

In a period: On moving across a period, the number of valence electrons changes so the properties of elements vary significantly. For example, the reactivity first decreases up to group 14 and then increases up to group 17. In a period, the group 1 element (alkali metal) and the group 17 element (halogen) are more reactive.

In a group: In a group, the number of valence electrons is the same. So, most of the elements have similar physiochemical properties and on moving down the group they change uniformly. For example, in group 1 (alkali metals), the reactivity increases down the group while in halogens it decreases.

For example, $$\frac{Li < Na < K < Rb < Cs}{\text{Reactivity increases}}$$

$$\frac{F > Cl > Br > I}{\text{Reactivity decreases}}$$

Metallic Nature

In a period: On moving from left to right, the metallic nature decreases and across the right side non-metals dominate.

For example, elements of the 2nd period

Li, Be, B, C, N, O, F, Ne

(Li, Be = Metal; B = Metalloid; C, N, O, F = Non-metal)

In a group: On moving down a group, the metallic nature increases. That is, non-metals followed by metalloids and then metals. For example, the metallic nature increases from flourine to iodine: $F < Cl < Br < I$.

Nature of Oxides and Hydroxides

In a period: On moving left to right in a period, the basic nature decreases while the acidic nature increases.

Example, Na_2O, MgO, Al_2O_3, SiO_2, P_2O_5, SO_2, Cl_2O_7

Most basic (left)   Most acidic (right)

$$NaOH > Mg(OH)_2 > Al(OH)_3 > Si(OH)_4$$

Most basic (left)   Most acidic (right)

In a group: On moving down the group, the basic nature increases while the acidic nature decreases.

Example, in group 1, Cs_2O and CsOH will be the most basic.

Atomic Size (Radius)

It is the distance from the centre of the nucleus of an atom to its outer most shell of electrons. The absolute value of atomic radius cannot be determined because it is not possible to locate the exact position of electrons in an atom as an orbital has no sharp boundaries. It is not possible to isolate an individual atom for its size determination due to its small size. As the absolute value of atomic size cannot be determined, it is expressed in terms of the operational definitions such as ionic radius, covalent radius, van der Waals radius and metallic radius. We will discuss these in Class X.

Variation in the Value of Radii

In a period: On moving from left to right, the atomic size decreases as the number of atomic orbitals is the same, while the number of electrons increase and as a result, the effective nuclear charge increases.

$$1 > 2 > 3 > 14 > 15 \approx 16 > 17 < 18 \text{ (zero group)}$$

Example, $Li > Be > B > C > N \approx O > F$

$Na > Mg > Al > Si > P > S > Cl$

In a group: The atomic radius increases from top to bottom as the number of shells or orbitals increase. The screening effect increases so Z_{eff} decreases.

Example, $Li < Na < K < Rb < Cs$

$F < Cl < Br < I$

Cation vs Atom

The size of the cation is always smaller than its atom. During cation formation, the outermost orbit is destroyed and the number of valence electrons decrease. So Z_{eff} increases and the size decreases.

Example,   Na   > Na^+
            2, 8, 1   2, 8

Orbits   3   2

The sodium atom is larger than the sodium ion due to a higher number of orbits and less Z_{eff}.

Anion vs Atom

The size of the anion is always greater than the size of its atom. During anion formation, electrons are taken up, so Z_{eff} decreases and the size increases.

Example, Cl^- > Cl
            2, 8, 8   2, 8, 7

Orbits   3   3

The chloride ion is larger than the chlorine atom as electron repulsion increases the size.

Size of cation α 1/magnitude of positive charge	Size of anion α magnitude of negative charge	In general anion > atom > cation
$M^{+3} < M^{+2} < M^+ < M$	$M^{-4} > M^{-3} > M^{-2} > M^- > M$	Example, $X^- > X > X^+$
Example, $Fe^{+3} < Fe^{+2} < Fe$	Example, $O^{-2} > O^- > O$	

Illustrations

1. Explain why the sodium atom is larger than the sodium cation.

   Solution: The sodium atom (Na) is larger than the sodium cation (Na^+) since it has a higher number of orbits and less effective nuclear charge.

   $$Na \xrightarrow{-e^-} Na^+$$
   $$_{2,8,1} _{2,8}$$

   Orbits   3   2

   Z_{eff} less   more

2. Write in the decreasing order of size.

   (a) F^-, N^{3-}, O^{2-} (b) Na^+, Al^{3+}, Mg^{2+}

   Solution:

   (a) $N^{3-} > O^{2-} > F^-$

   (More the –ve charge, larger is the size)

   (b) $Na^+ > Mg^{2+} > Al^{3+}$

   (More the +ve charge, smaller is the size)

Ionisation Enthalpy

It is also called *ionisation energy* or *ionisation potential*. It is defined as, the energy required to remove the most loosely bound electron from an isolated atom in the gaseous state resulting in the formation of a positive ion.

$$M - 1e^- \rightarrow M^+ - I_1$$
$$M^+ - 1e^- \rightarrow M^{+2} - I_2$$

Here I_1 and I_2 are first and second ionisation energies respectively.

Unit: eV/atom or kcal/mol or kJ/mol.

Variation in Value of Ionisation Energy

In a period: On moving from left to right in a period, the ionisation energy increases as Z_{eff} increases and size decreases. So, removal of the electron becomes more and more difficult.

   For example, Li < Be > B < C < N > O < F < Ne

In a group: On moving from top to bottom in a group, ionisation energy decreases as Z_{eff} decreases and size increases. So, removal of electron becomes more and more easy.

   Example, F > Cl > Br > I

Electronegativity

According to Pauling (1931), it is the power or tendency of an atom in a molecule to attract the shared pair of electrons towards itself.

Variation in Electronegativity Values

In a period: On moving left to right in a period, the electronegativity increases as Z_{eff} increases and size decreases.

   For example, Li < Be < B < C < N < O < F

In a group: On moving from the top to bottom in a group, the electronegativity decreases as Z_{eff} decreases and size increases.

   Example, F > Cl > Br > I.

Electron Gain Enthalpy or Electron Affinity (EA)

It is the amount of energy released when a neutral isolated gaseous atom accepts an extra electron to form a gaseous anion.

$$M + 1e^- \rightarrow M^- + E_1$$
$$M^- + 1e^- \rightarrow M^{-2} + E_2$$

Here E_1 and E_2 are the first and second electron gain enthalpy respectively.

Unit: eV/atom or kcal/mol or kJ/mol.

KNOWLEDGE BOOSTER

The value of electron affinity decreases since the addition of an extra electron becomes more and more difficult and is possible only by absorbing some part of energy. E_2 becomes endothermic in comparison to E_1. (E_1 is exoergonic and E_2 is endoergonic)

Example, $O \xrightarrow[E_1]{+e^-} O^- + \xrightarrow[E_2]{e^-} O^{2-}$

Variation in Value of Electron Affinity

In a period: On moving from left to right in a period, the electron affinity increases as Z_{eff} increases and size decreases.

In general, electron affinity follows the following trend:

$$Be < N < Li < B < C < O < F$$

In a group: On moving down a group, the electron affinity decreases as Z_{eff} decreases and the size increases.

Example, $Cl > Br > I$

Diagonal Relationship

Certain 2nd period elements show some similarities with the 3rd period elements that are diagonal to them. This is called *diagonal relationship*. This relationship exists due to their similar ionic sizes, electronegativities and polarising power.

$$\text{Polarising power} = \frac{\text{Ionic charge}}{(\text{Ionic radius})^2}$$

I	II	III	IV
Li	Be	B	C
Na	Mg	Al	Si

Illustrations

1. Write in the decreasing order of basic nature.

   Na_2O, MgO, K_2O, CaO

   Solution: Basic nature decreases from left to right across the period and increases down the group. So,

   $CaO > MgO > K_2O > Na_2O$

2. Write in the increasing order of acidic nature.

   CH_4, H_2O, NH_3, HF

   Solution: Acidic nature increases from left to right across the period. So,

   $CH_4 < NH_3 < H_2O < HF$

Study of Specific Groups (Group 1, 2, 17, 18)

Alkali Metals [1 (ns^1)]

This group includes six elements—Li, Na, K, Rb, Cs, Fr.

These metals are known as alkali metals as their hydroxides are soluble in water and their aqueous solutions are strongly alkaline in nature. These metals are always found in the combined state and not in the free state due to their high reactivity. Out of these metals Na and K are quite abundant in nature and occupy the 7th and 8th positions in abundance by weight in the earth crust. Out of these metals, francium (Fr) is least abundant.

General Features

The general features of these elements are as follows.

(i) **Electronic configuration:** All these elements have one valence electron and the general electronic configuration ns^1. That is, the valence electron is in the s orbital and the inner orbits are complete. Due to the same electronic configurations, they have a great similarity in their properties.

For example,

$_3$Li: 2, 1 $_{11}$Na: 2, 8, 1 $_{19}$K: 2, 8, 8, 1

$_{37}$Rb: 2, 8, 18, 8, 1 $_{55}$Cs: 2, 8, 18, 18, 8, 1

$_{87}$Fr: 2, 8, 18, 32, 18, 8, 1

(ii) **Physical state:** These are soft, silvery white solids with a metallic lustre and possessing high malleability and ductility. They are cuttable with a knife (except lithium), and on cutting, their metallic lustre fades rapidly due to surface oxidation by atmospheric air. These metals have a silvery lustre due to the presence of highly mobile electrons in their metallic lattices. As they have only one valence electron, metallic bonding is not so strong which makes them soft. This softness increases with the increase in atomic number the metallic bond strength decreases with an increase in atomic size.

$$\xrightarrow{\text{Li, Na, K, Rb, Cs}}$$
Decreasing order of hardness

(iii) Atomic and ionic radii: The atoms of these elements are largest in size in their periods and the ionic radii of these metals are smaller than the atomic radii of the metals atoms, that is, $M > M^+$.

The value of atomic and ionic radii increase on moving down the group from Li to Cs due to the increase in the number of orbits and a decrease in the value of the effective nuclear charge.

	Li	Na	K	Rb	Cs
Atomic radii (Å)	1.23	1.57	2.03	2.16	2.35

(iv) Electropositive and metallic nature: Due to their low ionisation energies, these are highly electropositive metals with high reactivity and these properties increase from Li to Cs as ionisation energy decreases.

(v) Valency and oxidation states: These metals can easily lose their only valence electron to form the M^+ ion; so, they are univalent in nature and exhibit only the +1 oxidation state.

$$M - e^- \rightarrow M^+$$

(vi) Reducing nature: Alkali metals are powerful reducing agents as they have large size and low ionisation energies. Therfefore, they can easily lose their valence electron and get oxidised. In the free state, this nature increases from Li to Cs as the removal of the electron becomes easier.

$$M - e^- \rightarrow M^+$$

(vii) Effect of air: Alkali metals (except lithium) in air get tarnished at once and give oxides, hydroxides and carbonates. So, they are kept in inert solvents like kerosene oil or paraffin oil to prevent their reaction with air.

$$M \xrightarrow{O_2} M_2O \xrightarrow{H_2O} MOH \xrightarrow{CO_2} M_2CO_3$$

(viii) Reaction with water: Alkali metals react with water to form their metal hydroxides which are strongly alkaline in nature. The reactivity towards water increases down the group.

$$2M + 2H_2O \rightarrow 2MOH + H_2 \uparrow$$

(ix) With halogens: Alkali metals react with halogens to form metal halides which are ionic in nature. The reactivity towards halogens increases down the group.

$$2M + X_2 \rightarrow 2MX + heat$$

(x) With acids: Alkali metals react with dilute acids like HCl, H_2SO_4 to form the metal salt and hydrogen.

$$2M + H_2SO_4 \rightarrow M_2SO_4 + H_2 \uparrow$$

(xi) Flame test: These metals and their salts impart a characteristic colour during the flame test. Due to this effect, alkali metals are used in fireworks.

Li	Na	K	Rb	Cs
crimson red	golden yellow	pale violet	violet	violet

Alkaline Earth Metals [2 (*ns²*)]

This group includes six elements—Be, Mg, Ca, Sr, Ba, Ra.

These are called alkaline earth metals as their oxides are present in the earth's crust and are alkaline in nature.

General Features

The general features of these elements are as follows.

(i) Electronic configuration: These elements have two electrons in their valence orbit and their general electronic configuration is ns^2. Except beryllium, the rest of these metals have eight electrons in the penultimate shell.

For example,

$_4$Be: 2, 2 $_{12}$Mg: 2, 8, 2 $_{20}$Ca: 2, 8, 8, 2

$_{38}$Sr: 2, 8, 18, 8, 2 $_{56}$Ba: 2, 8, 18, 18, 8, 2

$_{88}$Ra: 2, 8, 18, 32, 18, 8, 2

(ii) Physical state: These are silvery greyish white metals which are soft in nature but harder than alkali metals due to stronger metallic bonds. These metals are less malleable and ductile than alkali metals. Their hardness decreases down the group.

(iii) Atomic and ionic radii: The atomic and ionic radii of these elements are quite large but smaller than alkali metals due to higher nuclear charge than alkali metals. Values of atomic and ionic radii increase down the group from Be to Ra due to an increase in the number of orbits and a decrease in the effective nuclear charge successively.

	Be	Mg	Ca	Sr	Ba	Ra
Atomic radii (Å)	1.12	1.60	1.97	2.15	2.22	–

(iv) Electropositive and metallic nature: These elements are highly electropositive in nature as they have low ionisation energy. They can easily lose their two valence electrons to form M^{2+} cations and their electropositive nature increases down the group.

(v) Valency and oxidation states: All these elements show a valency of 2, and exist in the +2 oxidation states in their compounds. It is their most common and most stable oxidation state as M^{2+} ions have noble gas configuration. The removal of further electrons is extremely difficult.

$$M - 2e^- \rightarrow M^{2+}$$

(vi) Reducing nature: These metals are strong reducing agents as they can easily lose their two valence electrons. Their reducing nature increases down the group as the value of oxidation potential increases from Be to Ba.

$$M \rightarrow M^{2+} + 2e^-$$

(vii) Effect of air: In air, these metals get tarnished. However, Be and Mg are less affected.

$$M \xrightarrow{O_2} MO \xrightarrow{H_2O} M(OH)_2 \xrightarrow{CO_2} MCO_3$$

(viii) Reaction with water: Except Be all these metals react with water to give hydroxides and the order of reactivity towards water decreases down the group.

$$M + 2H_2O \rightarrow M(OH)_2 + H_2 \uparrow$$

(ix) With halogens: All these metals combine directly with halogens at higher temperatures to form MX_2 type of halides.

$$M + X_2 \xrightarrow{\Delta} MX_2$$

(x) With acids: Alkaline metals react with dilute acids like HCl, H_2SO_4 to form the metal salt and hydrogen.

$$M + H_2SO_4 \rightarrow MSO_4 + H_2 \uparrow$$

(xi) Flame test: Except Be and Mg, the other elements of this group and their salts impart characteristic colours during the flame test as the electrons get excited to higher energy levels and return back to their original state after some time.

Be,	Mg,	Ca,
Colourless	Colourless	Brick red
Sr,	Ba,	Ra
crimson red	green	crimson red

Halogen Family [17 (ns^2, np^5)]

This group includes five elements—F, Cl, Br, I, At.

The elements of this group are known as halogens (in Greek, *halo* = sea or salt, *gens* = producing).

General Features

The general features of these elements are as follows.

(i) Electronic configuration: These elements have seven electrons in their valence orbit and their general electronic configuration is ns^2np^5.

For example,

$_9$F: 2, 7 $_{17}$Cl: 2, 8, 7 $_{35}$Br: 2, 8, 18, 7

$_{53}$I: 2, 8, 18, 18, 7 $_{85}$At: 2, 8, 18, 32, 18, 7

(ii) Physical state: All the halogens exists as diatomic covalent molecules with their characteristic colours. For example, F_2 is light yellow, Cl_2 is greenish yellow, Br_2 is reddish brown and I_2 is deep violet.

(iii) Melting and boiling points: On moving down the group, the melting and boiling points increase.

Melting point (in K)	F_2	Cl_2	Br_2	I_2
	54	172	266	386
Boiling point (in K)	F_2	Cl_2	Br_2	I_2
	85	239	332	458

(iv) Atomic and ionic radii: Halogen atoms have the smallest radius when compared to any other element in their respective periods. On moving down the group, the atomic and ionic radii increase from F to I.

Atomic radii			
F	Cl	Br	I
64	99	114	133

(v) Non-metallic nature: Halogens are non-metallic in nature because of their high ionisation enthalpies. On moving down the group, the non-metallic character decreases. They are highly reactive non-metals and their reactivity also decreases down the group.

(vi) Electronegative nature: These elements are electronegative by nature as they have high electronegativity and by taking up one electron they form the halide ion (X^-).

$$X + e^- \rightarrow X^-$$

(vii) Valency and oxidation states: All these elements show a valency of 1 and exist in the −1 oxidation state in their compounds. It is their most common and most stable oxidation state as the X^- ion has the noble gas configuration. For example, HF, HCl.

- Except fluorine, other halogens can show +1 to +7 oxidation states.

(viii) Oxidising nature: Halogens act as oxidising agents and their oxidising nature decreases down the group.

$$F_2 > Cl_2 > Br_2 > I_2$$

(ix) Nature of compounds: Halogens form both ionic and covalent compounds. The halides of highly electropositive metals are ionic while those of weakly electropositive metals and non-metals are covalent. For example, NaCl, KBr, $CaCl_2$ are ionic while PCl_5, $SiCl_4$ are covalent.

(x) Reaction with hydrogen: All the halogens react with hydrogen to form the hydrogen halide.

$$H_2 + X_2 \rightarrow 2HX$$

Noble Gases (Zero Group, Group 18) [18 (ns^8)]

This group includes six elements—He, Ne, Ar, Kr, Xe, Rn. They are monoatomic and chemically almost inert so they are known as inert gases. Now a few compounds of Xe and Kr are known, so they are called noble gases.

Helium was discovered by Janssen and Lockyer (1868) in the sun's atmosphere. On earth it was discovered by Ramsay. Argon was discovered by Lord Rayleigh and Ramsay (1894). Neon, krypton and xenon were discovered by Ramsay and Travers (1998). Ramsay and Rayleigh discovered Ar (first to be discovered and most abundant among them 93%). Radon was discovered by Dorn. They are present in the atmosphere (1% by volume), except Rn which is obtained by radioactive decay of Ra, so they are also called *aerogens*.

General Features

The general features of these elements are as follows.

(i) Electronic configuration: All the noble gases except helium have eight valence electrons so their general electronic configuration is $ns^2 np^6$.

For example,

$_{10}$Ne: 2, 8 $_{18}$Ar: 2, 8, 8 $_{36}$Kr: 2, 8, 18, 8

$_{54}$Xe: 2, 8, 18, 18, 8 $_{86}$Rn: 2, 8, 18, 32, 18, 8

(ii) Physical state: They are colourless, tasteless, odourless monoatomic gases which are neither inflammable nor support combustion and cannot be easily liquified.

(iii) Atomic radii or van der Waals radii: Atomic radii of inert gases are expressed as van der Waals radii and on moving down the group, the atomic radii increase.

He	Ne	Ar	Kr	Xe	Rn
120	160	190	200	220	- (in pm)

(iv) Melting and boiling points: They have minimum BP, MP in their periods due to weak van der Waals forces. On moving down the group, the melting and boiling point increase as follows.

Melting point (in K)

He	Ne	Ar	Kr	Xe	Rn
1.1	24.6	83.8	115.9	161.3	202

Boiling point (in K)

He	Ne	Ar	Kr	Xe	Rn
4.2	27.1	87.2	119.7	165.0	211

(v) Solubility: These gases are partially soluble in water and the solubility (or polarisability) in water increases from He to Xe.

(vi) Diffusion: These gases undergo diffusion and on moving down, the diffusion of noble gases decreases as follows:

$$He > Ne > Ar > Kr > Xe > Rn$$

KNOWLEDGE BOOSTER

Uses of Inert Gases

Helium: Helium (80%) with O_2 (20%) is used for respiration by sea divers as it is not soluble in blood even at high pressure. It is filled in balloons and tyres of aeroplanes.

Neon: To make discharge lamps and sign boards due to more penetration power. Neon gives different colours when mixed with mercury in a discharge tube. Ne lamps are used in botanical gardens to stimulate growth and formation of chlorophyll.

Argon: To create an inert atmosphere in the metallurgy of Ti, Zr. Ar + 15 % N_2 increases the life of filaments in bulbs.

Krypton: It is used in filling incandescent metal filament electric bulbs. It is also used in electrical appliances, cosmic ray measurement.

Xenon: It used in research labs to detect radiations. It is also used in flash tubes.

Radon: It is used in the preparation of ointments, used in radiotherapy of cancer.

Illustrations

1. Name:

   (i) Three elements that have a single electron in their outermost shells.

   (ii) Two elements that have two electrons in their outermost shells.

   (iii) Three elements with filled outermost shells.

   Solution:

   (i) Li, Na and K
      (2,1) (2,8,1) (2,8,8,1)

   (ii) Be and Mg
      (2,2) (2,8,2)

   (iii) He, Ne and Ar
      (2) (2,8) (2,8,8)

2. (i) Lithium, sodium and potassium are all metals that react with water to liberate hydrogen gas. Is there any similarity in the atoms of these elements?

   (ii) Helium is an unreactive gas and neon is a gas of extremely low reactivity. What, if anything, do their atoms have in common?

   Solution:

   (i) All the given metals are highly reactive. So, these metals can react with water to liberate hydrogen gas. They have the same number of valence electrons (1) and can readily lose this to become uni-positive ions.

   (ii) Helium and neon have completely filled outermost shells which makes them exceptionally unreactive and thus, stable. They belong to the noble gases group.

3. Properties of the elements are given below. Where would you locate the following elements in the periodic table?

   (i) A soft metal stored under kerosene.

   (ii) An element with variable (more than one) valency stored under water.

   (iii) An element which is tetravalent and forms the basis of organic chemistry.

   (iv) An element which is an inert gas with atomic number 2.

   (v) An element whose thin oxide layer is used to make other elements corrosion resistant by the process of anodising.

   Solution:

   (i) Sodium (Na) Group 1 and Period 3 or potassium (K) Group 1 and Period 4.

   (ii) Phosphorus (P) Group 15 and Period 3.

   (iii) Carbon (C) Group 14 and Period 2.

   (iv) Helium (He) Group 18 and Period 1.

   (v) Aluminium (Al) Group 13 and Period 3.

CHAPTER AT A GLANCE

- **The periodic table** is an arrangement of elements with similar properties placed together. It helps in easier study of elements.
- **The law of triads:** Elements could be arranged in groups of three, in which the atomic weight of the middle element was the arithmetic mean of the first and last element. Also, the difference in the atomic weight of consecutive elements was atom constant.
- **Law of octaves:** If the elements are arranged in the order of increasing atomic weight, the eighth element starting from a given one, is a kind of repetition of the first like the eighth node in an octave of music.
- **Prout's law (1915):** According to it, all elements are made up of hydrogen atoms.
- **Lothar Meyer (1869):** He plotted various physical properties of elements against their atomic weights and pointed out that the similar elements occupy similar positions on the curve.
- **Mendeleev's periodic law:** The physical and chemical properties of elements are the periodic functions of their atomic weights.
- Mendeleev considered **atomic weight** as the fundamental property of an element and he arranged different elements in the order of their **increasing** atomic weights and formed periodic table.
- **Modern periodic law:** The physical and chemical properties of elements are the periodic functions of their atomic numbers.
- **Modern periodic table:** It is based on arranging different elements in the increasing order of their atomic numbers.
- **Long form of periodic table:** It is also known as extended form of periodic table or Mosley's periodic table. It was developed by Range, Werner and Burey on the basis of electronic configuration of elements.
- The **periodic table** has seven horizontal rows known as **periods**. Here, the first period is the shortest period having only two elements (H, He). The sixth period is the longest period having 32 elements.
- There are 18 vertical columns and the vertical columns are known as **groups**. Elements having similar outer electronic configurations in their atoms are arranged in vertical columns, referred to as **groups** or **families**.
- The elements having ns^1 and ns^2 electronic configurations in their outermost shell are called **s-block elements**.

- The elements whose atoms have incomplete p-orbitals in their outermost shell or in which the last electron enters any p orbital are known as **p-block elements**.
- The **d-block** is between the s and p blocks. These are the elements which have **incomplete d orbitals**. The d-block elements are called transition elements as their properties lie between those of the s- and p-block elements.
- The elements placed in two separate rows at the bottom of the periodic table are the **f-block elements**. They have incomplete f orbitals in their electronic configurations.
- The elements from cerium to lutetium having incomplete $4f$ orbitals are **lanthanones** or **lanthanoids**.
- The elements from thorium to lawrencium having incomplete $5f$ orbitals are **actinoids.**
- **Valency** = number of valence electrons = number of H atoms or O atoms or Cl atoms linked with one atom of an element.
- **Atomic size (radius):** It is the distance between the centre of the nucleus of an atom and its outer most shell of electrons.
- **Size of cation is always smaller than its atom:** During cation formation, the outermost orbit is destroyed and the number of valence electron decreases. So Z_{eff} increases and size decreases.
- **Size of anion is greater than its atom:** During anion formation, electrons are taken up, so, Z_{eff} decreases and size increases.
- **Ionisation energy** or ionisation potential is defined as, the energy required to remove the most loosely bound electron from an isolated atom in the gaseous state resulting in the formation of a positive ion.
- **Electronegativity** is the power or tendency of an atom in a molecule to attract the shared pair of electrons towards itself.
- **Electron gain enthalpy** is the amount of energy released when a neutral isolated gaseous atom accepts an extra electron to form a gaseous anion.
- **Diagonal relationship:** Certain 2nd period elements show some similarities with the 3rd period elements that are diagonal to them. This is called diagonal relationship. This is due to the similar ionic sizes, electronegativities and polarising power.

PRACTICE QUESTIONS

Analyse Your Concepts (School Exam Based)

Fill in the Blanks

Instructions: Complete the following statements with an appropriate word/term to be filled in the blank spaces.

1. The law of octaves was introduced by ______________.

2. In the modern periodic table, elements are arranged in the ______________ order of atomic number.

3. Halogens are present in column/group ______________.

4. Elements with atomic number 58 to 71 are called ______________.

5. For inert gases, atomic radius is given mainly in terms of ______________.

6. The ______________ group elements have the least ionisation energy.

7. The element with highest electronegativity is ______________.

8. The element with highest electron affinity is ______________.

9. The strongest basic oxide is ______________.

10. The largest anion is ______________.

11. The element showing diagonal relationship with beryllium is ______________.

12. Elements with atomic numbers more than 92 are called ______________.

True or False

Instructions: Read the following statements and write your answer as true or false.

1. The long form of periodic table is based on the electronic configuration.

2. On moving the down the group, the metallic nature decreases.

3. Nitrogen has less first ionisation energy than oxygen.

4. Ionisation energy is directly proportional to effective nuclear charge.

5. Helium has the lowest ionisation energy in the periodic table.

6. Aluminium is smaller than gallium.

7. K^+ is larger than Ca^{2+}.

8. Fluorine is the strongest oxidising agent.

9. Cs is the most electropositive element in the periodic table.

10. Due to stable electronic configuration, inert gases have high ionisation energy and electron affinity.

Match the Following

Instructions: Each question contains statements given in two columns which have to be matched. Statements (a, b, c, d) in column I have to be matched with statements (p, q, r, s) in column II.

1.

Column I	Column II
(a) Law of triads	(p) Moseley
(b) Law of octaves	(q) Mendeleev
(c) Periodic law	(r) Newlands
(d) Modified periodic law	(s) Dobereiner

2.

Column I	Column II
(a) Alkali metals	(p) Group 8
(b) Halogens	(q) Group 7
(c) Chalcogens	(r) Group 6
(d) Inert gases	(s) Group 1

Very Short Answer Type Questions

1. If the atomic weights of sulfur (S) and tellurium (Te) are 32 and 127, find the atomic weight of selenium (Se) using the law of triads?

2. Why did Newlands' law of octaves fail?

3. What is the basic difference between Mendeleev's periodic law and the modern periodic law?

4. How many horizontal rows and vertical columns are present in the modern periodic table?

5. In the long form of periodic table, which period is incomplete?

6. What is the group number and period for calcium atom?

7. Which block elements are called the representative elements?

8. Which group of elements are known as inert gases?

9. What is the relation between atomic size and effective nuclear charge?

10. Why did Mendeleev leave some places empty in his periodic table?

11. Phosphorus cannot conduct electricity while potassium can. Why?

12. Why are lanthanides called rare earth elements?

13. Why is lead a metal and silicon a metalloid?

14. Which periods are called long periods?

15. Write the element with the smallest size and highest electronegativity.

16. Out of Li and K, which has stronger metallic character and why?

Short Answer Type Questions

1. What is the basic theme of organisation in the periodic table?

2. Why do elements in the same group have similar physical and chemical properties?

3. What is the basis for the difference in approach between Mendeleev's periodic law and the modern periodic law?

4. Write the general electronic configuration of s-, p-, d- and f-block elements.

5. Consider the following species. N^{3-}, O^{2-}, F^-, Na^+, Mg^{2+} and Al^{3+}.
   (a) What is common between them?
   (b) Arrange them in the order of increasing ionic radii.

6. Explain why cations are smaller and anions are larger in radii than their parent atoms.

7. On moving down a group, a sharp increase in atomic radius is found from Li to K, but later on a gradual change in atomic radius value is observed. Why?

8. Name any two elements which can show similar chemical properties to Mg; also give the reason of these similarities.

9. Na, Mg and Al are elements of the same period in the modern periodic table having one, two and three valence electrons respectively. Which of these elements (a) has the largest atomic radius, (b) is the most reactive? Justify your answer stating the reason for each case.

10. The elements $_4Be$, $_{12}Mg$ and $_{20}Ca$, each having two valence electrons in their valence shells, are in the periods 2, 3 and 4 respectively, of the modern periodic table. Answer the following questions associated with these elements giving reasons, in each case.
    (a) In which group should they be?
    (b) Which one of them is the least reactive?
    (c) Which one of them has the largest atomic size?

11. How many groups and periods are there in the modern periodic table? How do the atomic size and metallic character of elements vary as we move:
    (a) down a group?
    (b) from left to right in a period?

Or

Write the number of periods and groups in the modern periodic table. How does the metallic character of elements vary on moving:
(a) from left to right in a period?
(b) down a group? Give reasons to justify your answer.

12. (a) Name the metals among the first five elements of the modern periodic table.
    (b) Write their chemical symbols.
    (c) Write the formula of their oxides.

13. An element X (atomic number = 20) burns in the presence of oxygen to form a basic oxide.
    (a) Identify the element and write its electronic configuration.
    (b) State its group number and period number in the modern periodic table.
    (c) Write a balanced chemical equation for the reaction when this oxide is dissolved in water.

14. Calcium is an element with atomic number 20. Stating the reason, answer each of the following questions.
    (a) Is calcium a metal or a non-metal?
    (b) Will its atomic radius be larger or smaller than that of potassium with atomic number 19?
    (c) Write the formula of its oxide.

15. On the basis of the trends in the periodic table, answer the following about the elements with atomic numbers 3 to 9.
    (a) Name the most electropositive element among them.
    (b) Name the most electronegative element.
    (c) Name the element with smallest atomic size.
    (d) Name the element which is a metalloid.
    (e) Name the element which shows maximum valency.

Long Answer Type Questions

1. Which important property did Mendeleev use to classify the elements in his periodic table and did he stick to that?

2. (a) The modern periodic table has been evolved through the early attempts of Dobereiner, Newlands and Mendeleev. List one advantage and one limitation of all three attempts.
   (b) Name the scientist who first showed that the atomic number of an element is a more fundamental property than its atomic mass.
   (c) State the modern periodic law.

3. Answer the following questions based on the elements with atomic number 3 to 9.
   (a) Name the element with the smallest atomic radius.
   (b) Name the element which shows maximum valency.
   (c) Name the element which is a metalloid.
   (d) Name the element which is the most electropositive.
   (e) Write the chemical formula of the compound formed when the elements of atomic number 6 and 8 react together.

4. The position of three elements A, B and C in the periodic table are shown below:

Group 16	Group 17
–	–
–	A
–	–
B	C

   (a) State whether A is a metal or non-metal.
   (b) State whether C is more reactive or less reactive than A.
   (c) Will C be larger or smaller in size than B?
   (d) Which type of ion, cation or anion, will be formed by element A?

5. The position of eight elements in the modern periodic table is given below where atomic numbers of elements are given in the parentheses.

Period number	Elements	
2	Li (3)	Be (4)
3	Na (11)	Mg (12)
4	K (19)	Ca (20)
5	Rb (37)	Sr (38)

   (a) Write the electronic configuration of Ca.
   (b) Predict the number of valence electrons in Rb.
   (c) What is the number of shells in Sr?
   (d) Predict whether K is a metal or a non-metal.
   (e) Which one of these elements has the largest atoms in size?
   (f) Arrange Be, Ca, Mg and B in the increasing order of the size of their respective atoms.

COMPETITION WINDOW (OBJECTIVE TYPE)

[For NEET, JEE (Main and Advanced), NTSE, KVPY and Olympiads]

Topic-wise MCQs

Periodic table

1. The properties of elements of Mendeleev's periodic law are the periodic functions of their
   (a) atomic number
   (b) atomic weight
   (c) mass of the atom
   (d) shape of the atom

2. Which of the following elements was not present in Mendeleev's periodic table?
   (a) francium
   (b) cesium
   (c) rubidium
   (d) uranium

3. The long form of periodic table was developed by
   (a) Bohr
   (b) Mendeleev
   (c) Range and Werner
   (d) Rutherford

4. The long form of periodic table is based on
   (a) electronegativity
   (b) mass of the atom
   (c) shape of the atom
   (d) atomic number

5. The *d*-block elements in the long form of periodic table are placed
   (a) on the extreme right
   (b) in the middle
   (c) on the extreme left
   (d) at the bottom

6. Modern periodic table is based on atomic number of the elements. The significance of atomic number was proved by the experiment of
   (a) Moseley's work on X-ray spectra
   (b) Bragg's work on X-ray diffraction
   (c) Millikan's oil drop experiment
   (d) discovery of X-rays by Roentgen

7. In a period, the elements are arranged in the order of
   (a) constant charges in the nucleus
   (b) equal charges in the nucleus
   (c) decreasing charges in the nucleus
   (d) increasing charges in the nucleus

8. The name of element with atomic number 100 was adopted in the honour of
   (a) Enrico Fermi
   (b) Albert Einstein
   (c) Alfred Nobel
   (d) Dimitri Mendeleev

9. All the elements in a group in the periodic table have the same
   (a) atomic weight
   (b) same number of valence electrons
   (c) atomic number
   (d) none of these

10. Fluorine, chlorine, bromine and iodine are placed in the same group (17) of the periodic table, because
    (a) they are electronegative elements
    (b) they are non-metals
    (c) they have seven valence electrons
    (d) their atoms are generally univalent

11. Which of the following pairs of atomic numbers represent elements belonging to the same group?
    (a) 15 and 13
    (b) 12 and 21
    (c) 14 and 31
    (d) 13 and 31

12. Elements placed in the same group have
    (a) have nearly the same chemical properties
    (b) the same number of valence electrons
    (c) same physical properties
    (d) both (a) and (b)

13. Which of the following has the highest melting point?
    (a) Cl
    (b) Br
    (c) I
    (d) F

Atomic and ionic radii

14. Which of the following has the smallest size?
    (a) Mg^{2+}
    (b) Na^+
    (c) F^-
    (d) Al^{3+}

15. The smallest radius among the given species is
    (a) lithium
    (b) lithium ion
    (c) helium
    (d) hydrogen

16. The atomic radii of elements in moving down a group from the first transition series to the second transition series
    (a) increases
    (b) remains the same
    (c) decreases
    (d) first increase then decreases

17. The arrangement of Na, Rb, K, Cs in the increasing order of atomic radius, is
    (a) $Rb < K < Cs < Na$
    (b) $K < Na < Cs < Rb$
    (c) $Na < Cs < K < Rb$
    (d) $Na < K < Rb < Cs$

18. Which of the following is the correct order of atomic radii?
    (a) $Na > Mg > K > Li$
    (b) $Li > Mg > Na > K$
    (c) $K > Na > Li \approx Mg$
    (d) $K > Li > Mg > Na$

19. Cationic radii are
    (a) inversely proportional to effective nuclear charge
    (b) directly proportional to effective nuclear charge
    (c) inversely proportional to square of effective nuclear charge
    (d) directly proportional to square of effective nuclear charge

20. Which of the following represent the correct order of increasing ionic size?
    (a) $Al^{3+} < Mg^{2+} < Na^+ < F^- < O^{2-}$
    (b) $Mg^{2+} < Al^{3+} < Na^+ < F^- < O^{2-}$
    (c) $Na^+ < Mg^{2+} < F^- < O^{2-} < Al^{3+}$
    (d) $O^{2-} < F^- < Al^{3+} < Mg^{2+} < Na^+$

21. Which of the following is correct order of ionic size?
    (a) $Ca^{2+} > Cl^- > K^+ > S^{2-}$
    (b) $Ca^{2+} > K^+ > Cl^- > S^{2-}$
    (c) $S^{2-} > Cl^- > K^+ > Ca^{2+}$
    (d) $Cl^- > K^+ > Ca^{2+} > S^{2-}$

Ionisation potential and electron affinity

22. The energy with which an atom binds the electrons is experimentally determined by measuring its
    (a) electronegativity
    (b) electron affinity
    (c) atomic size
    (d) ionisation potential

23. Ionisation potential of hydrogen is
    (a) equal to that of chlorine
    (b) slightly higher than that of chlorine
    (c) less than that of chlorine
    (d) much higher than that of chlorine

24. Electron affinity of noble gases is
    (a) almost zero
    (b) low
    (c) high
    (d) very high

25. Which of the following has maximum electron affinity?
    (a) Li
    (b) H
    (c) K
    (d) Na

26. Which of the following element has the minimum electron affinity?
    (a) I
    (b) Cl
    (c) Br
    (d) F

Electonegativity, valency and electropositive nature

27. Which of the following elements has the strongest tendency to form anions?
    (a) F
    (b) P
    (c) Cr
    (d) S

28. Which of the following sets has the strongest tendency to form anions?
    (a) Na, Mg and Al
    (b) N, O and F
    (c) Ga, In and Te
    (d) V, Cr and Mn

29. Which of the following has the maximum electronegativity?
    (a) C
    (b) Ba
    (c) Cr
    (d) Cl

30. The electronegativity follows the order
    (a) $F > I > Cl > Br$
    (b) $I > F > Cl > Br$
    (c) $F > Cl > Br > I$
    (d) $Cl > F > I > Br$

31. The most electronegative element is
    (a) sodium
    (b) lithium
    (c) potassium
    (d) rubidium

32. Which metal has maximum valency state?
    (a) Na
    (b) Mg
    (c) Al
    (d) Si

33. The most electropositive amongst the alkaline earth metals is
    (a) beryllium
    (b) magnesium
    (c) barium
    (d) calcium

34. The maximum covalency of an element of the atomic number 7 is
    (a) 5
    (b) 4
    (c) 3
    (d) 2

Acidic and basic nature of oxides

35. The correct order of increasing basicity is
    (a) $MgO < BeO < CaO < BaO$
    (b) $BeO < MgO < CaO < BaO$
    (c) $BaO < CaO < MgO < BeO$
    (d) $CaO < BaO < BeO < MgO$

36. Which of the following is the most basic oxide?
    (a) $CsOH$ (b) $Be(OH)_2$
    (c) $Al(OH)_3$ (d) $Si(OH)_4$

37. Which is the following is the most acidic oxide?
    (a) CO_2 (b) SO_3
    (c) Al_2O_3 (d) Cl_2O_7

38. The basicities of the oxides As_2O_5, MgO and BeO decrease in the order
    (a) $BeO > As_2O_5 > MgO$
    (b) $MgO > BeO > As_2O_5$
    (c) $BeO > MgO > As_2O_5$
    (d) $As_2O_5 > BeO > MgO$

39. Which of the following is most acidic?
    (a) HF (b) NH_3
    (c) CH_4 (d) PH_3

40. Which of the following is most acidic?
    (a) H_3BO_3 (b) HNO_3
    (c) H_2SO_4 (d) $HClO_4$

Miscellaneous

1. Up to which element was the law of octaves found to be applicable?
   (a) oxygen   (b) calcium
   (c) cobalt   (d) potassium

2. In Mendeleev's periodic table, gaps were left for the elements to be discovered later. Which of the following elements later found a place in the periodic table?
   (a) germanium   (b) chlorine
   (c) oxygen   (d) silicon

3. The period number in the long form of the periodic table is equal to the
   (a) magnetic quantum number of any element of the period
   (b) atomic number of any element of the period
   (c) maximum principal quantum number of any element of the period
   (d) maximum azimuthal quantum number of any element of the period

4. The elements in which electrons are progressively filled in the $4f$ orbitals are called
   (a) actinoids   (b) transition elements
   (c) lanthanoids   (d) halogens

5. Where would you locate the element with electronic configuration 2,8 in the modern periodic table?

   (a) Group 8   (b) Group 2
   (c) Group 18   (d) Group 10

6. Which of the following is the outermost shell for elements of period 2?
   (a) K shell   (b) L shell
   (c) M shell   (d) N shell

7. Which one of the following elements exhibit the minimum number of valence electrons?
   (a) Na   (b) Al
   (c) Si   (d) P

8. Which of the following gives the correct increasing order of the atomic radii of O, F and N?
   (a) O, F, N   (b) N, F, O
   (c) O, N, F   (d) F, O, N

9. Which of the following is the correct order of size of the given species?
   (a) $I > I^- > I^+$ (b) $I^+ > I^- > I$
   (c) $I > I^+ > I^-$ (d) $I^- > I > I^+$

10. Which of the following elements would lose an electron easily?
    (a) Mg   (b) Na
    (c) K   (d) Ca

11. Which of the following elements does not lose an electron easily?
    (a) Na   (b) F
    (c) Mg   (d) Al

12. Three elements B, Si and Ge are
    (a) metals
    (b) non-metals
    (c) metalloids
    (d) metal, non-metal and metalloid respectively

13. Which of the following elements will form an acidic oxide?
    (a) an element with atomic number 7
    (b) an element with atomic number 3
    (c) an element with atomic number 12
    (d) an element with atomic number 19

14. The element with atomic number 14 is hard and forms acidic oxide and a covalent halide. To which of the following categories does the element belong?
    (a) metal   (b) metalloid
    (c) non-metal   (d) left-hand side element

15. Which one of the following does not increase while moving down the group of the periodic table?
    (a) atomic radius
    (b) metallic character
    (c) valence
    (d) number of shells in an element

16. Which of the following set of elements is written in order of their increasing metallic character?

(a) Be, Mg, Ca   (b) Na, Li, K
(c) Mg, Al, Si   (d) C, O, N

17. The statement that is not correct for periodic classification of elements is:
(a) The properties of elements are periodic functions of their atomic numbers.
(b) Non-metallic elements are less in number than metallic elements.
(c) For transition elements, the $3d$ orbitals are filled with electrons after the $3p$ orbitals and before the $4s$ orbitals.
(d) The first ionisation enthalpies of elements generally increase with increase in atomic number as we go along a period.

18. What type of oxide would eka-aluminium form?
(a) EO_3 (b) E_3O_2
(c) E_2O_3 (d) EO

19. Which of the following statements about the modern periodic table is correct:
(a) It has 18 horizontal rows known as Periods.
(b) It has 7 vertical columns known as Periods.
(c) It has 18 vertical columns known as Groups.
(d) It has 7 horizontal rows known as Groups.

20. Which of the given elements A, B, C, D and E with atomic number 2, 3, 7, 10 and 30 respectively belong to the same period?
(a) A, B, C   (b) B, C, D
(c) A, D, E   (d) B, D, E

21. Which of the following is the strongest reducing agent?
(a) Na   (b) Ca
(c) Cs   (d) Li

22. Which of the following is not an actinoid?
(a) Curium $(Z = 96)$ (b) Californium $(Z = 98)$
(c) Uranium $(Z = 92)$ (d) Terbium $(Z = 65)$

23. The elements A, B, C, D and E have atomic numbers 9, 11, 17, 12 and 13 respectively. Which pair of elements belong to the same group?
(a) A and B   (b) B and D
(c) A and C   (d) D and E

24. Which of the following is the most basic hydroxide?
(a) $NaOH$ (b) $CsOH$
(c) $Mg(OH)_2$ (d) $Ca(OH)_2$

Advanced and Olympiads

Single Choice

1. Which of the following statement(s) about the modern periodic table are incorrect?
(i) The elements in the modern periodic table are arranged on the basis of their decreasing atomic number.
(ii) The elements in the modern periodic table are arranged on the basis of their increasing atomic masses.
(iii) Isotopes are placed in adjoining group(s) in the periodic table.
(iv) The elements in the modern periodic table are arranged on the basis of their increasing atomic number.
(a) (i) only   (b) (i), (ii) and (iii)
(c) (i), (ii) and (iv)   (d) (iv) only

2. The correct order of the decreasing ionic radii among the following isoelectronic species is
(a) $K^+ > Ca^{2+} > Cl^- > S^{2-}$
(b) $Ca^{2+} > K^+ > S^{2-} > Cl^-$
(c) $Cl^- > S^{2-} > Ca^{2+} > K^+$
(d) $S^{2-} > Cl^- > K^+ > Ca^{2+}$

3. Among the following, the third ionisation energy is highest for
(a) aluminium   (b) beryllium
(c) boron   (d) magnesium

4. Arrange the following elements in the order of their increasing non-metallic character Li, O, C, Be, F.
(a) $F < O < C < Be < Li$ (b) $Li < Be < C < O < F$
(c) $F < O < C < Be < Li$ (d) $F < O < Be < C < Li$

5. The increasing order of the atomic radii of the following elements is
(i) C   (ii) O   (iii) F   (iv) Cl   (v) Br
(a) (ii) < (iii) < (iv) < (i) < (v)
(b) (iv) < (iii) < (ii) < (i) < (v)
(c) (iii) < (ii) < (i) < (iv) < (v)
(d) (i) < (ii) < (iii) < (iv) < (v)

6. Select the set showing correct matching

Column I	Column II
(i) Alkali metals	(p) Group 8
(ii) Halogens	(q) Group 7
(iii) Chalcogens	(r) Group 6
(iv) Inert gases	(s) Group 1

(a) (i)—(s), (ii)—(q), (iii)—(r), (iv)—(p)
(b) (i)—(p), (ii)—(r), (iii)—(q), (iv)—(s)
(c) (i)—(q), (ii)—(p), (iii)—(s), (iv)—(r)
(d) (i)—(r), (ii)—(p), (iii)—(s), (iv)—(q)

7. The successive ionisation values for an element E are given below

$\Delta_i H_1$ 410 kJ mol^{-1}

$\Delta_i H_2$ 820 kJ mol^{-1}

$\Delta_i H_3$ 1100 kJ mol^{-1}

$\Delta_i H_4$ 1500 kJ mol^{-1}

$\Delta_i H_5$ 3200 kJ mol^{-1}

Find out the number of valence electrons for the element E.
(a) 4   (b) 3
(c) 5   (d) 2

Multiple Choice

8. Which of the following sequences contain atomic numbers of only representative elements?
   (a) 3, 33, 53, 87
   (b) 2, 10, 22, 36
   (c) 7, 17, 25, 37, 48
   (d) 9, 35, 51, 88

9. Which of the following are correctly matched?
   (a) 11, 19, 37 (alkali metals)
   (b) 26, 27, 28 (d-block elements)
   (c) 16, 34, 52 (chalcogens)
   (d) 10, 18, 38 (aerogens)

10. Which of the following properties decreases down the group?
    (a) ionisation energy
    (b) size or radius
    (c) electronegativity
    (d) oxidising nature

11. In which of the following options does the order of arrangement not agree with the variation of property indicted against it?
    (a) $Al^{3+} < Mg^{2+} < Na^+ < F^-$ (increasing ionic size)
    (b) $B < C < N < O$ (increasing first ionisation enthalpy)
    (c) $I < Br < Cl < F$ (increasing electron gain enthalpy)
    (d) $Li < Na < K < Rb$ (increasing metallic radius)

12. How many of these elements have less electronegativity than oxygen?
    (a) fluorine   (b) chlorine
    (c) nitrogen   (d) sulfur

13. How many of these elements have less electron affinity than fluorine?
    (a) O   (b) Cl
    (c) N   (d) S

14. How many of these isoelectric ions are larger in size than F^-?
    (a) Na^+ (b) O^{2-}
    (c) N^{3-} (d) Mg^{2+}

15. Which of the following orders are correct?
    (a) $Li > Na > K$ (reducing nature)
    (b) $F_2 > Cl_2 > Br_2$ (oxidising nature)
    (c) $Be > B > Li$ (ionisation energy)
    (d) $F > O > N$ (electron affinity)

Comprehension Type

Comprehension I: Ionisation potential is one of the fundamental properties of the elements that can be measured directly. It is defined as the minimum energy required to remove an electron from an atom, ion or molecule. For a given species, the energy required to remove the first electron is called the first ionisation potential and so on. The magnitude of the ionisation potential is mainly governed by three factors— atomic radius, nuclear charge and number of inner electrons. In general, it is possible to attach an additional electron to any atom, ion or molecule. The energy that is released in this process is called the electron affinity of the species. The magnitude of the electron affinity depends upon the size of the atom, stability of the resulting structure and the strength of binding of the outermost electron.

16. In moving across any period from left to right, the ionisation potential
    (a) gradually increases
    (b) gradually decreases
    (c) remains the same
    (d) first increases and then decreases

17. The magnitude of the electron affinities for halogens in comparison to those of alkali metals are
    (a) very high   (b) very low
    (c) nearly the same   (d) exactly the same

Comprehension II: In the modern periodic table, elements are arranged in the order of increasing atomic numbers which is related to the electronic configuration. Depending upon the type of orbital receiving the last electron, the elements in the periodic table have been divided into four blocks—s, p, d and f. The modern periodic table consists of 7 periods and 18 groups. Each period begins with the filling of a new energy shell. In accordance with the Aufbau principle, the seven periods (1 to 7) have 2, 8, 8, 18, 18, 32 and 32 elements respectively. The seventh period is still incomplete. To avoid the periodic table being too long, the two series of f-block elements, called lanthanoids and actinoids, are placed at the bottom of the main body of the periodic table.

18. The last element of the p-block in the 6th period is represented by the outermost electronic configuration.
    (a) $7s^2\, 7p^6$
    (b) $5f^{14}\, 6d^{10}\, 7s^2\, 7p^0$
    (c) $4f^{14}\, 5d^{10}\, 6s^2\, 6p^6$
    (d) $4f^{14}\, 5d^{10}\, 6s^2\, 6p^4$

19. The electronic configuration of the element which is just above the element with atomic number 43 in the same group is
    (a) $1s^2\, 2s^2\, 2p^6\, 3s^2\, 3p^6\, 3d^5\, 4s^2$
    (b) $1s^2\, 2s^2\, 2p^6\, 3s^2\, 3p^6\, 3d^5\, 4s^3\, 4p^6$
    (c) $1s^2\, 2s^2\, 2p^6\, 3s^2\, 3p^6\, 3d^6\, 4s^2$
    (d) $1s^2\, 2s^2\, 2p^6\, 3s^2\, 3p^6\, 3s^7\, 4s^2$

20. Electronic configurations of four elements A, B, C and D are given below:
    (A) $1s^2\, 2s^2\, 2p^6$ (B) $1s^2\, 2s^2\, 2p^4$
    (C) $1s^2\, 2s^2\, 2p^6\, 3s^1$ (D) $1s^2\, 2s^2\, 2p^5$

    Which of the following is the correct order of increasing tendency to gain electrons:
    (a) $A < C < B < D$
    (b) $A < B < C < D$
    (c) $D < B < C < A$
    (d) $D < A < B < C$

Matrix Matching

21. Match the following

Column I	Column II
(a) Representative element	(p) Fe
(b) Transition element	(q) Na
(c) Pseudo transition metal	(r) Ar
(d) Aerogen	(s) Zn

22. Match the following

Column I (Properties)	Column II (Elements)
(a) Maximum electronegativity	(p) Chlorine
(b) Maximum ionisation energy	(q) Helium
(c) Maximum electron gain enthalpy	(r) Fluorine
(d) Most electropositive	(s) Cesium

Integer Type

23. The atomic numbers of some elements are given below. How many of these belong to the *s*-block?
3, 19, 21, 26, 31, 38, 55, 56, 58

24. Out of the given elements, how many are metals?
Li, K, Ca, B, Si, Cu, Br, Pd, Ne

25. How many of these elements are larger in size than fluorine?
Li, Cl, Ar, B, C, N, K, O, P

26. How many of these elements have less ionisation energy than nitrogen?
Be, C, O, F, Ne, Li, Na, P, Si

27. How many of these elements have positive values of electron gain enthalpies?
F, Cl, O, N, Ne, Ar, Be, C, S

Answer Keys

Fill in the Blanks

1. Newlands
2. increasing
3. 7
4. lanthanides
5. van der Waals radius
6. 1 (one)
7. fluorine
8. chlorine
9. Cs_2O
10. I^-
11. aluminium
12. Trans-uranium elements

True or False

1. T   2. F   3. F   4. T   5. F
6. F   7. T   8. T   9. T   10. F

Match the Following

1. (a)—(s), (b)—(r), (c)—(q), (d)—(p)
2. (a)—(s), (b)—(q), (c)—(r), (d)—(p)

Very Short Answer Type Questions

1. ^{32}S $^?Se$ ^{127}Te
   Using the law of traids
   Atomic weight of Se $= \dfrac{32+127}{2} = 79.5$

2. It failed as during classification of elements, only atomic masses are considered and not their properties.

3. Mendeleev's periodic law is based on atomic weights while the modern periodic law is based on atomic numbers.

4. In the modern periodic table there are seven horizontal rows and sixteen vertical columns.

5. In the long form of periodic table, the 7th period is incomplete.

6. It is present in group 2 and the 4th period.

7. *s* and *p*-block elements are called representative elements.

8. Zero group or group 18 elements are called inert or noble gases.

9. Atomic size is inversely proportional to effective nuclear charge.

10. Mendeleev left some places empty in his periodic table so that newly discovered elements could be accommodated there.

11. Phosphorus is a non-metal so it cannot conduct electricity while potassium being a metal can conduct electricity.

12. As lanthanides are present in nature in low abundance, they are called rare earth elements.

13. Lead is a metal and silicon is a metalloid. On moving down the group, the metallic nature increases.

14. The 4th and 5th periods are called long periods.

15. Fluorine has the smallest size and highest electronegativity.

16. K has strong metallic bonds because it is larger in size and can lose electrons very easily.

Short Answer Type Questions

1. The basic theme of organisation of elements in the periodic table is to simplify and systematise the study of the properties of all the elements and millions of their compounds. On the basis of similarities in chemical properties, the various elements have now been divided into different groups. This has made the study simple because the properties of elements are now studied in the form of groups rather than individually.

2. Elements in the same group have similar electronic configuration and hence have similar physical and chemical properties.

3. Mendeleev's periodic law states that the physical and chemical properties of the elements are a periodic function of their atomic weights. The modern periodic law states that physical and chemical properties of elements are a periodic function of their atomic numbers. Thus, the basic difference in approach between Mendeleev's periodic law and the modern periodic law is the change in the basis of classification of elements from atomic weight to atomic number.

4. (i) s-block elements: ns^{1-2} where $n = 2\text{–}7$

   (ii) p-block elements: $ns^2\, np^{1-6}$ where $n = 2\text{–}6$

   (iii) d-block elements: $(n-1)\, d^{1-10}\, ^{0-2}$ where $n = 4\text{–}7$

   (iv) f-block elements: $(n-2)\, f^{0-14}\, (n-1)\, d^{0-2}\, ns^2$ where $n = 6\text{–}7$

5. (a) Each one of these ions contains 10 electrons and hence all are isoelectronic.

   (b) The ionic radii of isoelectronic ions decrease with an increase in the magnitude of the nuclear charge.
   For example, consider the isoelectronic ions N^{3-}, O^{2-}, F^-, Na^+, Mg^{2+} and Al^{3+}. All these ions have 10 electrons but their nuclear charges increase in the order:
   N^{3-} (+7), O^{2-} (+8), F^- (+9), Na^+ (+11), Mg^{2+} (+12) and Al^{3+} (+13). Therefore, their ionic radii decrease in the order $N^{3-} > O^{2-} > F^- > Na^+ > Mg^{2+} > Al^{3+}$.

6. The ionic radius of a cation is always smaller than radius of the parent atom because the loss of one or more electrons increases the effective nuclear charge. As a result, the force of attraction of the nucleus for the electrons increases and hence the ionic radii decrease. In contrast, the ionic radius of an anion is always larger than its parent atom because the addition of one or more electrons decreases the effective nuclear charge. As a result, the force of attraction of the nucleus towards the electrons decreases and hence the ionic radii increase.

7. On moving from Li to K, a sharp increase in atomic radius is observed as the number of the orbit increases. While going from K to Cs, this change is gradual as screening by the inner orbit electron is not effective, that is, the Z_{eff} is more.

8. Be and Ca can show similar chemical properties with Mg as all these elements have the same number of valence electrons (2) and they are present in same group (2).

9. (a) Na has the largest atomic radius because it has 11 protons and 11 electrons, and the least effective nuclear charge among these elements.

   (b) Al is the least reactive element because it is smallest in size, therefore, has the most effective nuclear charge and least tendency to lose electrons.

10. (a) They should be in group 2, since they have 2 valence electrons.

    (b) Beryllium is the least reactive because it has more effective nuclear charge due to smaller size and least tendency to lose electrons.

    (c) Calcium has the largest atomic size due to the presence of 4 shells.

11. There are 18 groups and 7 periods in the modern periodic table.

    (a) Atomic size and metallic character both go on increase on moving down the group.

    (b) Atomic size and metallic character both decrease along a period while going from left to right.

    Note: Metallic character is the level of reactivity of a metal.

12. (a) Lithium and beryllium.

    (b) Li and Be.

    (c) Li_2O and BeO.

13. (a) The element is calcium. Its electronic configuration is 2, 8, 8, 2.

    (b) Group number is 2 and period number is 4.

    (c) $CaO + H_2O \rightarrow Ca(OH)_2$

14. (a) Ca(20): 2, 8, 8, 2
    It is a metal because it can lose 2 electrons to become stable.

    (b) It will be smaller due to greater effective nuclear charge.

    (c) The formula of its oxide is CaO.

15. (a) Lithium   (b) Fluorine

    (c) Fluorine   (d) Boron

    (e) Carbon

Long Answer Type Questions

1. Mendeleev used the atomic weight as the basis of classification of elements in the periodic table. He arranged the known elements in the order of increasing atomic weight, grouping together elements with similar properties. He sincerely stuck to this basis leaving blank spaces where elements were not known at that time. For example, the element gallium after aluminium and germanium after silicon were not known at the time Mendeleev prepared his periodic table. He called these elements as eka-aluminium

and eka-silicon. Later, these elements were discovered and their properties were found to be similar to those predicted by Mendeleev.

2. (a) **Dobereiner's triads**
   Advantage: The three elements of a triad were found to possess similar properties.
   Limitation: Some elements which are dissimilar were being grouped into a triad.
   Newlands' octaves
   Advantage: Newlands was the first one to come up with the idea of periodicity while classifying the elements.
   Limitation: It was applicable only for lighter elements up to calcium.
   Mendeleev's attempt
   Advantage: Although noble gases were discovered very late, they can be placed easily in Mendeleev's periodic table without disturbing the order of the rest of the elements.
   Limitation: Position of hydrogen was not explained.

   (b) Henry Moseley showed that the atomic number of an element is a more fundamental property than its atomic mass.

   (c) It states that the properties of the elements are periodic functions of their atomic number.

3. (a) Element with smallest atomic radius ______ Fluorine/F
   (b) Element with maximum valency ______ Carbon/C
   (c) Element which is metalloid ______ Boron/B
   (d) Element which is most electropositive ______ Lithium/Li
   (e) CO and CO_2

4. (a) Group 16 and 17 elements are non-metals.

Group 16	Group 17
O	F
S	Cl (A)
Se	Br
Te (B)	I (C)

   hence, A is a non-metal.
   (b) C is less reactive than A, as reactivity decreases down the group for halogens.
   (c) C will be smaller in size than B as moving across a period, the nuclear charge increases so, electrons are pulled closer to the nucleus.
   (d) A will form an anion as it has 7 electrons in its valence shell and thus, needs 1 more electron to complete its octet.

$$A + e^- \rightarrow \underset{\text{Anion}}{A^-}$$

5. (a) 2, 8, 8, 2.
   (b) It has one valence electron.
   (c) 2, 8, 18, 8, 2. It has five shells (orbits).
   (d) K is a metal.
   (e) Rubidium (Rb) has the largest atom in size.
   (f) Be > Mg > Ca > Rb is the increasing order of atomic size.

Competition Window (Objective Type)

Topic-wise MCQs

1. (b)	**2.** (a)	**3.** (c)	**4.** (d)	**5.** (b)
6. (a)	**7.** (d)	**8.** (a)	**9.** (b)	**10.** (c)
11. (d)	**12.** (d)	**13.** (c)	**14.** (d)	**15.** (b)
16. (a)	**17.** (d)	**18.** (c)	**19.** (a)	**20.** (a)
21. (c)	**22.** (d)	**23.** (b)	**24.** (a)	**25.** (b)
26. (a)	**27.** (a)	**28.** (b)	**29.** (d)	**30.** (c)
31. (b)	**32.** (d)	**33.** (c)	**34.** (c)	**35.** (b)
36. (a)	**37.** (d)	**38.** (b)	**39.** (a)	**40.** (d)

Miscellaneous

1. (b)	**2.** (a)	**3.** (c)	**4.** (c)	**5.** (c)
6. (b)	**7.** (a)	**8.** (d)	**9.** (d)	**10.** (c)
11. (b)	**12.** (c)	**13.** (a)	**14.** (b)	**15.** (c)
16. (a)	**17.** (c)	**18.** (c)	**19.** (c)	**20.** (b)
21. (d)	**22.** (d)	**23.** (c)	**24.** (b)	

Advanced and Olympiads

Single Choice

1. (b)	**2.** (d)	**3.** (b)	**4.** (b)	**5.** (c)
6. (a)	**7.** (a)			

Multiple Choice

8. (ad)	**9.** (abc)	**10.** (acd)	**11.** (bc)	**12.** (bcd)
13. (acd)	**14.** (bc)	**15.** (bcd)		

Comprehension Type

16. (a)	**17.** (a)	**18.** (c)	**19.** (a)	**20.** (a)

Matrix Matching

21. (a)—(q), (b)—(p), (c)—(s), (d)—(r)
22. (a)—(r), (b)—(q), (c)—(p), (d)—(s)

Integer Type

23. 5	**24.** 5	**25.** 9	**26.** 7	**27.** 4

Hints and Solutions

Competition Window (Objective Type)

Topic-wise MCQs

1. It is Mendeleev's periodic law.

3. Range, Werner and Burry developed the long form of the periodic table on the basis of electronic configuration.

6. Moseley equation is $\sqrt{f} = a(Z - b)$.

Here a and b are constants, f is the frequency of the X-rays and Z is the atomic number.

9. All the elements in a group have same number of valence electrons. For example, each halogen has seven valence electrons.

11. $_{13}Al$ and $_{31}Ga$ are elements of Group 13.

13. Melting point increases down the group due to increase of molecular weight.

14. Here Al^{+3} is the smallest in size as the size of the cation is inversely proportional to the magnitude of the positive charge present on the cation.

15. Li^+ has the smallest size.

17. Atomic radius increases down the group. So, Na < K < Rb < Cs is the correct order.

20. The correct order of size is
$Al^{3+} < Mg^{2+} < Na^+ < F^- < O^{2-}$.

21. The correct order of size is
$S^{2-} > Cl^- > K^+ > Ca^{2+}$ (for details see text part).

27. Fluorine has the strongest tendency to form anions due to its highest electronegativity value.

28. F, O, N are three most electronegative elements in the periodic table.

35. The basic nature of oxides increases down the group; so, the basicity increases BeO < MgO < CaO < BaO.

36. CsOH is the most basic since the basic nature of hydroxides decreases from left to right in a period.

37. Cl_2O_7 is the most acidic since the acidic nature increases from left to right in a period.

39. HF is the most acidic since the electronegativity of the fluorine atom is the highest.

40. Acidic nature of the oxyacids increases across the period so $HClO_4$ is the strongest acid.

Miscellaneous

5. Element with configuration 2, 8 is the inert gas Ne which is in Group 18.

7. The number of valence electrons in these elements Na, Al, Si and P are 1, 3, 4 and 5 respectively.

9. As size of anion > atom > cation so $I^- > I > I^+$

10. Alkali metals can lose electrons easily and this tendency increases down the group (K > Na).

11. Being the most electronegative element, fluorine has the tendency to accept an electron and not lose an electron.

13. Elements with atomic numbers 3, 12, 19 belong to the s-block and form basic oxides while element with atomic number 7 is nitrogen and its oxide is acidic (NO_2).

14. Element with atomic number 14 is Si which is a metalloid.

16. On moving down the group, the metallic nature increases.

17. In the case of transition elements, electrons in the $3d$ subshell is filled after $4s$.

18. For eka-aluminium (Ga) the oxide formed is E_2O_3 (Ga_2O_3).

20. Elements B, C, D with atomic number 3, 7, 10 belong to the IInd period.

21. Li is the strongest reducing agent.

22. Terbium ($Z = 65$) belongs to the lanthanides while the rest are actinides.

23. Elements with atomic number 9 (F) and 17 (Cl) belong to the same group (Group 17).

24. CsOH is the strongest base and the order is CsOH > NaOH > $Ca(OH)_2$ > $Mg(OH)_2$.

Advanced and Olympiads

Single Choice

2. In the case of isoelectronic species the value of Ionic radius follows the order $M^{-4} > M^{-3} > M^{-2} > M^{-1} > M^+ > M^{+2} > M^{+3}$.
Here $S^{-2} > Cl^- > K^+ > Ca^{+2}$ is the decreasing order of size.

3. Beryllium has the highest third ionisation energy as Be^{2+} has $1s^2$ (He gas configuration) so removal of electron is very tough.

$$Be \xrightarrow{I_1} Be^+ \xrightarrow{I_2} Be^{2+} \xrightarrow{I_3}$$

$$I_1 < I_2 <<< I_3$$

4. On moving left to right across the period, the non-metallic nature increases; so, Li < Be < C < O < F.

5. Generally in a period, on moving from left to right, the atomic radius decreases. In a group, on moving from top to bottom, atomic radius increases.

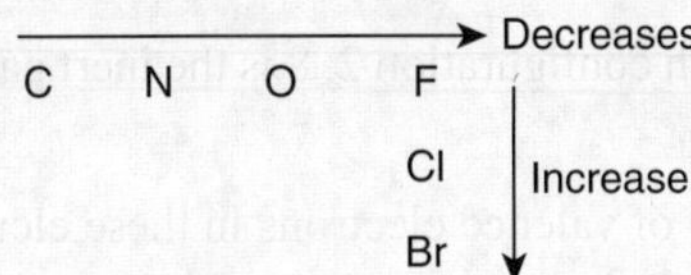

Br > Cl > C > O > F

$e > d > a > b > c$

7. A sudden jump in the ionisation enthalpy occurs from the 4th to 5th value so it is of Group 14 and they have 4 valence e^-.

Multiple Choice

8. Elements of the *s* and *p*-block are representative elements. (3, 33, 53, 87) and (9, 35, 51, 88)

9. Elements with atomic number 10, 18 are aerogens while with atomic number 38 is an alkaline earth metal.

10. On moving down the group, the size increases while ionisation energy, electronegativity and oxidising nature decrease.

11. The correct order of ionisation energy must be B < C < O < N and for electron affinity the correct order is I < Br < F < Cl.

12. Electronegativity of these elements decreases as follows: F > O > N > Cl > S

13. Electron affinity order is as follows: Cl > F > S > O > N

14. The order of size is $N^{3-} > O^{2-} > F^- > Na^+ > Mg^{2+}$

15. On moving down the group, the reducing nature increases so Li < Na < K. The rest of the other orders are correct.

Comprehension Type

18. In each period, the last element is an inert gas so in the VIth period it will be $4f^{14}\,5d^{10}\,6s^2\,6p^6$.

Integer Type

23. Elements with atomic number 3, 19, 55 (1), 38, 56 (2) are *s*-block elements while elements with 21, 26 atomic number are *d*-block, element with 31 is *p*-block and element with atomic number 58 is *f*-block (4*f*).

24. Metals: Li, K, Ca, Cu, Pd
Non-metals: B, Si, Br, Ne

25. All these elements are larger in size than fluorine.

26. Be, C, O, Li, Na, P, Si have lower ionisation energy than nitrogen while F, Ne have more ionisation energies than nitrogen.

27. N, Be, Ne, Ar have positive values of electron gain enthalpies due to their stable electronic configurations while rest other have negative values.

Chemical Bonding

Introduction

We know that different elements have different atomic numbers and electronic configurations. The properties of atoms depend upon their electronic configurations. Some atoms are more reactive than others. Noble gas (He, Ne, Ar, Kr, Xe and Rn) atoms are not reactive at all; they are inert and stable. Then the question arises—why don't noble gases react to form compounds, while other elements do so? This can be answered by comparing the electronic configurations of noble gases with those of other elements. Also, it is important to understand how and why atoms react to form molecules and compounds. Atoms gain electrons in their outermost shells or lose them from their outermost shells, or share electrons with other atoms in such a way that their outermost shells become filled to capacity. They can do this by reacting with other atoms. As long as the outermost shell can accommodate more electrons (it is not full), an atom tends to combine with other atoms in order to fill its outermost shell. When the outermost shell is filled to capacity, the atom becomes stable.

The atoms of all other elements (elements other than the noble gases) have in their outermost shells, less than 8 electrons (their outermost shells are not filled to capacity). Therefore, the atoms of these elements combine with other atoms to achieve stable configurations like those of the noble gases. It is the tendency on the part of an atom to achieve a stable configuration (like that of the noble gases) which is responsible for its chemical reactivity.

Concept of Valency and Valence Electron

Valency

The combining capacity of an element is known as *valency*. It is measured in terms of number of valence electrons (present in the outer orbit) or number of H-atoms or O-atoms or Cl-atoms linked with one atom of an element. For example, Cl or Br atom can share one valence electron, their valency is 1; oxygen or sulfur can share two valence electrons, their valency is 2; nitrogen or phosphorus can share 3 valence electrons, their valency is 3; carbon can share 4 valence electrons, therefore its valency is 4 and so on.

Valency = number of valence electrons = number of H atoms or O atoms or Cl atoms linked with one atom of an element.

For example, in CH_4
Valency of carbon = 4
This is because the carbon atom has 4 valence electrons, or is linked with 4 H atoms.
For example, in NH_3
Valency of nitrogen = 3

Variation in Valency in a Period

In the case of representative elements (*s, p*-block) the valency increases as the number of valence electrons increase from 1 to 8 across the period.

Group	1	2	13	14	15	16	17	18
Number of electrons	1	2	3	4	5	6	7	8
Valence on hydrogen scale	1	2	3	4	3	2	1	0
Valence on oxygen scale	1	2	3	4	5	6	7	0

On the hydrogen scale, the valency increases from 1 to 4 and then decreases from 4 to 0, while on the oxygen scale, it increases from 1 to 7.

	1	2	13	14	15	16	17
3rd Period	NaH	MgH$_2$	AlH$_3$	SiH$_4$	PH$_3$	H$_2$S	HCl
	1	2	3	4	3	2	1
	Na$_2$O	MgO	Al$_2$O$_3$	SiO$_2$	P$_2$O$_5$	SO$_3$	Cl$_2$O$_7$
	1	2	3	4	5	6	7

Octet Rule

It was introduced by Lewis and Kossel. According to this, any atom or ion having an octet state (8 valence electrons) are quite stable. Every atom tries to attain the octet state by chemical combination (chemical bond formation) to gain stability—the nearest inert gas configuration.

KNOWLEDGE BOOSTER

Helium has 2 electrons (duplet) and all other inert gases have 8 valence electrons (octet). All elements having this state of configuration all quite stable and chemically unreactive.

Exceptions to the Octet Rule

(i) Contraction of octet state: Here, the central atom is electron deficient, or does not have an octet state, or has 4 or 6 electrons. For example,

BeX$_2$	BX$_3$	AlX$_3$	SnCl$_2$
4	6	6	6 e^-

(ii) Expansion of octet state: Here, the central atom has more than 8 electrons due to the presence of vacant or empty d-orbitals. For example,

P Cl$_5$,	S F$_6$,	Os F$_8$
10	12	16 e^-

(iii) Odd electron species: Odd electron species like NO, NO$_2$ ClO$_2$ do not follow the octet rule.

(iv) Inter-halogen compounds: There is expansion of octet state and such bonds are formed between halogens atoms only. For example,

I F$_7$,	BrF$_3$
14	10

(v) Compounds of xenon: Examples are XeF$_2$, XeF$_4$ XeF$_6$.

Illustrations

1. Define octet rule. Write its significance and limitations.

   Solution:

   Octet rule: See text part.

   Significance of octet rule: It helps to explain why different atoms combine with each other to form ionic compounds or covalent compounds.

   Limitations of octet rule: See text part.

2. Among the following compounds, which follow the octet rule?

   CO$_2$, AlCl$_3$, SiCl$_4$, PbCl$_4$, NH$_3$, H$_2$S, HF, BeCl$_2$, CCl$_4$

   Solution: CO$_2$, SiCl$_4$, PbCl$_4$, NH$_3$, H$_2$S, HF, CCl$_4$ follow the octet rule. We can see that the central atom has the octet state (8 electrons).

Chemical Bond

During chemical combination between two atoms, to get the nearest inert noble gas configuration for stability, a chemical bond is formed between them by redistribution of electrons. A chemical bond is defined as *the force of attraction that binds two atoms together*. A chemical bond implies that the force of attraction and force of repulsion are balanced at a particular distance.

A chemical bond is formed to:

- attain the octet state,
- minimise energy,
- gain stability, and
- decrease reactivity.

When two atoms come closer to each other, attraction and repulsion forces operate between them (Fig. 7.1). The distance at which the attraction forces overcome the repulsion forces is called *bond distance*. Here, the potential energy for the system is the lowest, so the bond is formed.

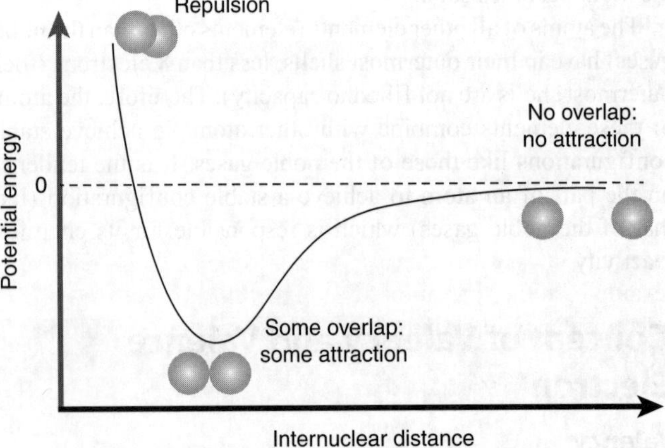

Fig. 7.1 Potential energy curve showing the formation of a covalent bond

Type of Bonds

A chemical bond is formed by complete transfer of electrons (ionic bond), or by equal or mutual sharing of electrons (covalent), or by donating and accepting electron pairs (coordinate bond). Thus, chemical bonds are of these types—ionic bond, covalent bond, coordinate bond.

KNOWLEDGE BOOSTER

Bond strength order:

ionic > covalent > coordinate > metallic bond > hydrogen bond > van der Waals bond or interactions. Metallic bond, hydrogen bond and van der Waals bonds are interactions.

Electrovalent Bond or Ionic Bond

It was introduced by Kossel. It is formed by the complete transfer of valence electrons from a metal to a non-metal or from one atom to another. One atom (metallic or electropositive having 1, 2, 3 valence electrons) donates its excess electrons to another atom (more electronegative non-metal having 5, 6, 7 valence electrons) so that both the atoms may acquire a stable noble gas configuration. The atom which loses electrons becomes positively charged and is called the *cation*. The atom which takes up the electrons lost by the first atom becomes negatively charged and is called the *anion*. These two oppositely charged ions are now held together by an electrostatic force of attraction or coulombic force. This force of attraction binding the two ions together is known as an electrovalent or ionic bond. It is also called a polar bond. It is a non-directional bond. An electrovalent bond is polar, that is, the positive and negative charges are separated. Compounds containing such bonds are called electrovalent, or ionic, or polar compounds. For example, $NaCl$, Na_2O, MgO, CaO, $CaCl_2$, $MgCl_2$ (Table 7.1).

KNOWLEDGE BOOSTER

Electrovalency

When an element forms an electrovalent bond, its valency is known as *electrovalency*. The number of electrovalent bonds an atom can form is equal to its electrovalency. That is, the electrovalency of an element is equal to the number of electrons lost or gained by the atom to form an ion. Elements which lose electrons show positive electrovalency and those which gain electrons show negative electrovalency. For example, in the formation of sodium chloride (Na^+Cl^-), the electrovalency of sodium (Na) is +1, while that of chlorine (Cl) is –1.

Elements that lose or gain one, two, three, … , electrons, are said to be monovalent (univalent), divalent (bivalent), trivalent, … , respectively.

Monovalent elements: Na, K, Cl, F
Divalent elements: Mg, Ca, Ba, O, S
Trivalent elements: Al

Conditions for the Formation of an Ionic Bond

The process of ionic bond formation must be exothermic ($\Delta H = -\text{ve}$). The other essential conditions are as follows:

- the metal must have low ionisation energy (for easy removal of electron/s to form the cation),
- the non-metal must have high electron affinity (to accept the electron easily to form an anion),
- the ions must have high lattice energy,
- the cation should be large with low electronegativity,
- the anion must be small with high electronegativity.

For example,

Na	+	Cl	→	Na^+ Cl^-
(2, 8, 1)		(2, 8, 7)		(2, 8) (2, 8)
Mg	+	O	→	Mg^{+2} O^{-2}
(2, 8, 2)		(2, 6)		(2, 8) (2, 8)
Al	+	N	→	Al^{+3} N^{-3}
(2, 8, 3)		(2, 5)		(2, 8) (2, 8)

KNOWLEDGE BOOSTER

In the formula of an ionic compound, the positive ion is written first. Charges on the ions of an ionic compound are usually not shown with the formula. So, sodium chloride is usually expressed as $NaCl$, not as Na^+Cl^-.

Formation of Ionic Bonds in NaCl, $MgCl_2$, CaO

Sodium Chloride (NaCl)

The combination of sodium (Na) and chlorine (Cl) atoms gives rise to sodium chloride (NaCl).

The atomic number of sodium is 11, so its electronic configuration is 2, 8, 1. As it has only one electron in its outermost shell, sodium atom transfers this electron and becomes a positively charged sodium ion (Na^+).

$$\underset{2,8,1}{Na} \xrightarrow{-1e} \underset{2,8}{[Na]^+}$$

Thus, the electronic configuration of the Na^+ ion is the same as that of neon which is the noble gas nearest to sodium in the periodic table.

Now let us consider the chlorine atom (Cl). The atomic number of chlorine is 17 so its electronic configuration is 2, 8, 7. As it has 7 electrons in its outermost shell, it requires 1 electron to acquire a stable noble gas configuration so the chlorine atom takes the 1 electron transferred by the sodium atom and becomes a negatively charged chloride ion (Cl^-).

$$\underset{(2, 8, 7)}{\ddot{\underset{..}{\text{Cl}}}\cdot} + 1e \longrightarrow \underset{(2, 8, 8)}{[:\ddot{\underset{..}{\text{Cl}}}:]^-}$$

Thus, the chloride ion (Cl^-) attains the configuration of the nearest noble gas, argon. The two ions (Na^+ and Cl^-) being oppositely charged, are now held together by electrostatic force of attraction as Na^+Cl^-.

$$\underset{(2, 8, 1)}{\text{Na}} + \underset{(2, 8, 7)}{\text{Cl}} \longrightarrow \underset{(2, 8)}{\text{Na}^+} \underset{(2, 8, 8)}{\text{Cl}^-}$$

$$\dot{\text{Na}} + :\overset{..}{\underset{..}{\text{Cl}}}\cdot \longrightarrow \underset{(2, 8, 8)}{\text{Na}^+[:\overset{..}{\underset{..}{\text{Cl}}}:]^-} \text{ or } \text{Na}^+\text{Cl}^-$$

The formation of sodium chloride can be shown diagrammatically (Fig. 7.2).

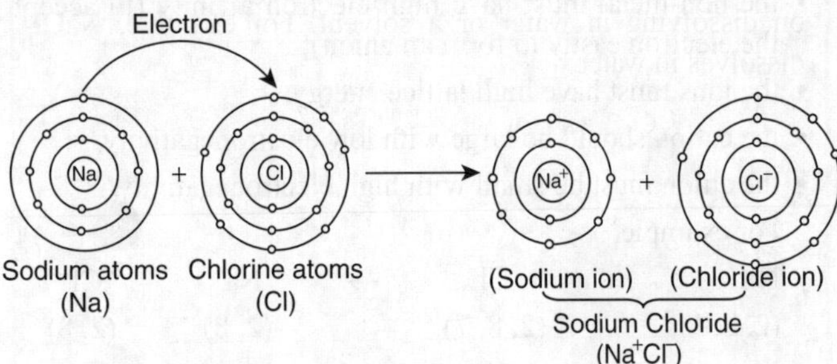

Fig. 7.2 Formation of sodium chloride

The force that holds Na^+ and Cl^- ions together is called an electrovalent bond. As this bond exists between ions, it is also called an ionic bond.

Magnesium Chloride (MgCl$_2$)

The number of valence electrons in magnesium (atomic number 12) is 2 and that in chlorine (atomic number 17) is 7. The magnesium atom acquires a stable configuration of 8 electrons by losing two electrons from its outermost shell (one each to each atom of chlorine) and so becomes a positive magnesium ion, Mg^{2+}.

$$\underset{\substack{(2, 8, 2) \\ \text{atom}}}{\text{Mg}} - 2e^- \rightarrow \underset{\substack{(2, 8) \\ \text{cation}}}{\text{Mg}^{2+}}$$

However, each chlorine atom, which contains 7 valence electrons in its outermost shell, can accept only 1 of the 2 electrons donated by a magnesium atom to attain the nearest inert gas configuration. So, for each magnesium atom forming a magnesium ion (Mg^{2+}), there must be two chlorine atoms to form two chloride ions.

$$\underset{\substack{(2, 8, 7) \\ \text{2 Cl atoms}}}{2\text{Cl}} + 2e^- \rightarrow \underset{\substack{(2, 8, 8) \\ \text{2 Chloride anions}}}{2\text{Cl}^-}$$

The ratio of magnesium to chloride ions in magnesium chloride must be 1 : 2, so the molecular formula of the compound magnesium chloride is MgCl$_2$.

Electron Dot Structure of Magnesium Chloride

Calcium Oxide (CaO)

The number of valence electrons of a calcium atom (atomic number 20) is 2, and that of an oxygen atom is 6. Oxygen requires 2 electrons to attain the octet to get to the nearest inert gas configuration. In the presence of oxygen, each calcium atom loses its 2 valence electrons to one oxygen atom (Fig. 7.3). As a result, the calcium atom forms a calcium ion with a charge +2 (Ca^{2+}), and the oxygen atom forms an oxide ion with a charge -2 (O^{2-}). Since only one oxygen atom is needed to accept the 2 valence electrons donated by a calcium atom, the formula of calcium oxide is CaO.

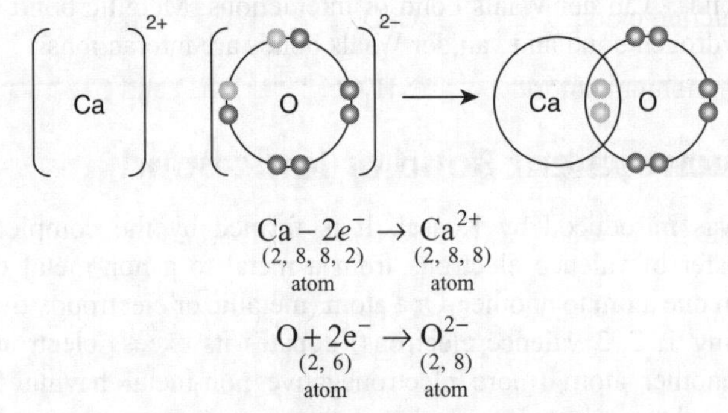

$$\underset{\substack{(2, 8, 8, 2) \\ \text{atom}}}{\text{Ca} - 2e^-} \rightarrow \underset{\substack{(2, 8, 8) \\ \text{atom}}}{\text{Ca}^{2+}}$$

$$\underset{\substack{(2, 6) \\ \text{atom}}}{\text{O} + 2e^-} \rightarrow \underset{\substack{(2, 8) \\ \text{atom}}}{\text{O}^{2-}}$$

Fig. 7.3 Structure of CaO

Electron Dot Structure of Calcium Oxide

KNOWLEDGE BOOSTER

Redox Concept of Ionic Bond Formation

During the formation of an electrovalent or ionic bond, the transfer of electron/s is involved. The electropositive atom undergoes oxidation, while the electronegative atom undergoes reduction. This means, it is a redox process.

For example, sodium chloride is formed by the combination of sodium and chlorine.

$$2\text{Na} + \text{Cl}_2 \rightarrow 2\text{Na}^+ + 2\text{Cl}^- \quad \text{(or 2 NaCl)}$$

This reaction can be written as two half reactions:

$$2\text{Na} \rightarrow 2\text{Na}^+ + 2e^- \quad \text{(Oxidation half)}$$

$$\text{Cl}_2 + 2e^- \rightarrow 2\text{Cl}^- \quad \text{(Reduction half)}$$

$$\underset{\text{Reduction}}{\overset{\text{Oxidation}}{2\text{Na} + \text{Cl}_2 \rightarrow 2\text{Na}^+ + 2\text{Cl}^-}} \quad \text{(Redox reaction)}$$

Features of Electrovalent or Ionic Compounds

The important features of ionic compounds are discussed below.

(i) **Physical state:** Ionic compounds have solid crystalline structures (flat surfaces) with definite geometry due to

Table 7.1 List of some important compounds having ionic bonds

Compound	Formula	Ions involved
Sodium chloride	$NaCl$	Na^+ and Cl^-
Magnesium chloride	$MgCl_2$	Mg^{2+} and Cl^-
Magnesium oxide	MgO	Mg^{2+} and O^{2-}
Calcium chloride	$CaCl_2$	Ca^{2+} and Cl^-
Calcium oxide	CaO	Ca^{2+} and O^{2-}
Ammonium chloride	NH_4Cl	NH_4^+ and Cl^-
Barium chloride	$BaCl_2$	Ba^{2+} and Cr
Potassium nitrate	KNO_3	K^+ and NO_3^-
Ammonium sulfate	$(NH_4)_2SO_4$	NH_4^+ and SO_4^{2-}
Cupric sulfate	$CuSO_4$	Cu^{2+} and SO_4^{2+}
Cupric chloride	$CuCl_2$	Cu^{2+} and Cl^-

strong electrostatic forces of attraction as the constituents are arranged in a definite pattern of positively and negatively charged ions. For example, sodium chloride (NaCl) is made up of Na^+ and Cl^- ions arranged in a definite order in three dimensions to form crystals.

(ii) **High melting and high boiling points:** Electrovalent compounds have high melting and boiling points. This is due to the presence of strong electrostatic forces of attraction between the positive and negative ions. A large amount of heat energy is required to break this force of attraction. Hence, the melting and boiling points of these compounds are very high.

BP, MP ∝ Electrostatic force of attraction

Volatile nature ∝ 1/ Electrostatic force of attraction

(iii) **Solubility:** Electrovalent compounds are usually soluble in water but insoluble in organic solvents such as benzene, acetone, carbon disulfide and carbon tetrachloride. Ionic compounds are soluble in polar solvents like water due to the high dielectric constant of these solvents as the force of attraction between ions gets destroyed and ionic compounds get dissolved in them.

(iv) **Electrical conductance:** Electrovalent compounds can conduct electricity in the molten or fused state and in their aqueous solutions but not in the solid state. In the solid state of electrovalent compounds, the ions are held together in fixed positions and cannot move freely so they do not conduct electricity. When an electrovalent compound is dissolved in water or is melted, the crystal

structure breaks down and as a result the ions now become free to move and can, therefore, conduct electricity easily.

In fact, the ions of ionic compounds get hydrated or solvated on dissolving in water or a solvent. For example, NaCl dissolves in water.

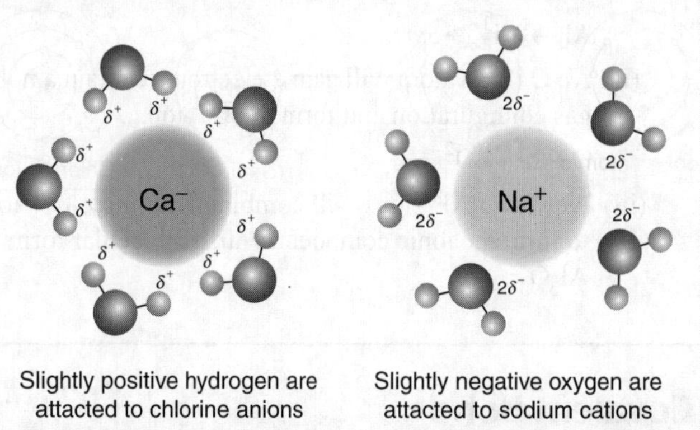

(v) **Isomorphism:** Ionic compounds can show isomorphism, that is, they can have the same type of crystalline structures. For example,
- all alums have the same crystalline structure,
- NaF and MgO also have the same crystalline structure.

(vi) **Nature of reactions:** These compounds can show fast ionic reactions as activation energy is zero for ions.

(vii) **Spatial or stereo isomerism:** These compounds cannot show spatial isomerism (geometrical and optical isomerisms) due to the non-directional nature of the ionic bond.

Lattice energy

It is the energy released during the formation of an ionic solid molecule from its constituent ions, or, it is the energy needed to break an ionic solid molecule into its constituent ions. It is denoted by U.

$$U \propto \text{Charge on ion}$$

$$\propto \frac{1}{\text{Size of ion}}$$

For example, $\dfrac{NaCl < MgCl_2 < AlCl_3 < SiCl_4}{\text{As charge on metal atom increases and its size decreases}} \longrightarrow$

In the case of ionic compounds lattice energy decreases as follows:

$$\text{Bi–bi} > \text{Uni–bi} \quad \text{or} \quad \text{Bi–uni} > \text{Uni–uni}$$

For example, $MgO > MgCl_2 > NaCl$

Illustrations

1. What is the molecular formula of aluminium oxide?

   Solution:

   (i) An Al (2, 8, 3) atom has 3 electrons in the valence shell which it can lose to form the Al^{3+} ion and attain a noble gas configuration.

   $$Al \rightarrow Al^{3+} + 3e$$

   (ii) An O (2, 6) atom will gain 2 electrons to attain a noble gas configuration and form the O^{2-} ion.

   $$O + 2e \rightarrow O^{2-}$$

   (iii) Now, two Al^{3+} ions will combine with three O^{2-} ions to form the ionic compound with a molecular formula Al_2O_3.

2. Arrange the following molecules in order of increasing ionic character of their bonds

   $LiF, K_2O, N_2, SO_2, ClF_3$.

   Solution: $N_2 < SO_2 < ClF_3 < K_2O < LiF$

3. Write the favourable factors for the formation of an ionic bond.

   Solution: The favourable factors for the formation of ionic bond are the following.

   - Low ionisation enthalpy of the metal atom.
   - High electron gain enthalpy of the non-metal atom.
   - High lattice enthalpy of the compound formed.
   - High heat of formation (exothermic).

Covalent Bond

The chemical bond formed when two atoms share electrons between them is known as a covalent bond. It is formed by an equal sharing of electrons between two similar or different atoms. The sharing of electrons between the two atoms takes place in such a way that both the atoms acquire the stable electronic configurations of their nearest noble gases. The molecule or ion formed due to mutual sharing of electrons is known as a covalent molecule. Here, the number of electrons shared (covalent bonds) represents the covalency. One atom can share a maximum of four electrons with another atom, that is, the maximum covalency can be four. For example,

- in ammonia (NH_3), the covalency of nitrogen atom is three,
- in methane (CH_4) the covalency of the carbon atom is four.

KNOWLEDGE BOOSTER

Atoms of non-metals having 4, 5, 6, 7 valence electrons prefer to form covalent bonds and not ionic bonds because the atoms of such elements do not favour the loss of valence electrons due to energy constraints which prevents transfer of electrons.

Conditions for the Formation of Covalent Bond

The process of covalent bond formation is possible if these conditions are fulfilled.

- Atoms involved in bonding must have 4 or more valence electrons.
- Atoms must have high electronegativities, high electron gain enthalpies and high ionisation energies.

Types of Covalent Bonds

Covalent bonds are of these three types:

- single covalent bond,
- double covalent bond and
- triple covalent bond.

Single Covalent Bond (—)

A single covalent bond is formed when one pair of electrons is shared between two atoms. That is, each atom contributes one electron during sharing. It is shown by putting a short line (—) between two atoms.

Examples: H_2, X_2, HX, CH_4, CCl_4, NH_3, BF_3, BCl_3, H_2O, H_2S.

Formation of a Hydrogen Molecule (H_2)

A molecule of hydrogen has two hydrogen atoms. Each hydrogen atom has one electron. When two atoms of hydrogen combine, one electron of each hydrogen atom takes part in sharing. Hence, two electrons (one pair of electrons) are shared between the two hydrogen atoms.

$$H^{\bullet} + {}^{\bullet}H \rightarrow H : H$$

The shared electron pair always exists between these two atoms. The two dots between the two H atoms represent the pair of shared electrons. One pair of shared electrons gives a single bond. Such a bond is represented by a short line between the two atoms. Hence, a hydrogen molecule may be shown as follows.

$$H : H \text{ or } H-H \quad (H \odot H)$$

Once the covalent bond is formed, both H-atoms have a stable configuration of the noble gas helium ($1s^2$).

Formation of a Methane Molecule (CH_4)

A carbon atom has four electrons in its outermost shell or valence shell. It shares all its valence electrons with those of four H-atoms. Hence, an atom of carbon forms four single covalent bonds with four H atoms.

$$\overset{\bullet}{\underset{\bullet}{C}}{}^{\bullet} + 4\ \dot{H} \rightarrow H : \overset{\bullet\bullet}{\underset{\bullet\bullet}{C}} : H \text{ or } H - \overset{\displaystyle H}{\underset{\displaystyle H}{C}} - H$$

Pictorially, a methane molecule can be shown as follows.

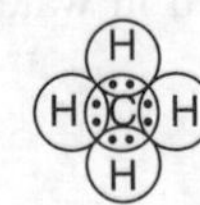

Double Covalent Bond (=)

A double covalent bond is formed when two pairs of electrons are shared between the two combining atoms. A sharing of two pairs of electrons is shown by marking two short lines (=) between the symbols of the two atoms.

Examples: O_2, CO_2, CS_2, C_2H_4, C_6H_6.

Formation of an Oxygen Molecule (O_2)

An atom of oxygen contains six electrons in its valence shell. It needs two more electrons to attain a stable eight-electron inert gas configuration (octet, neon). This is achieved when each of the two oxygen atoms shares its two electrons with the other, resulting in the formation of a stable oxygen molecule.

$$:\overset{..}{O} + \overset{..}{O}: \longrightarrow \overset{..}{O}::\overset{..}{O} \text{ or } O=O$$

Double bond

Pictorially, the oxygen molecule can be shown as follows.

Formation of an Ethylene Molecule (C_2H_4)

In the formation of an ethylene or ethene molecule (C_2H_4), each of the two C atoms combines with two H atoms to form two single covalent bonds. The remaining two electrons of each C atom form a double bond between the two C atoms.

$$2\ \overset{.}{\underset{.}{C}}\cdot + 4\overset{.}{H} \rightarrow \ \overset{H}{\underset{H}{:C}}::\overset{H}{\underset{H}{C:}} \text{ or } \overset{H}{\underset{H}{}}{>}C=C{<}\overset{H}{\underset{H}{}}$$

(Ethylene molecule)

Pictorially, a molecule of ethylene (C_2H_4) can be shown as follows.

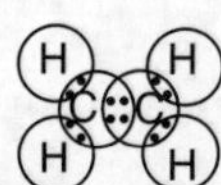

Triple Covalent Bond (≡)

A triple covalent bond is formed when three pairs of electrons (six electrons) are shared between the two combining atoms. A triple bond is shown by marking three short lines (≡) between the two symbols of the atoms.

Examples: N_2, C_2H_2, HCN.

Formation of a Nitrogen Molecule (N_2)

An atom of nitrogen has five electrons in its valence shell. It needs three more electrons to attain the stable octet state configuration of an inert gas (neon). This is achieved when two nitrogen atoms combine together by sharing three valence electrons each to form a nitrogen molecule.

$$:\overset{}{N}: + :\overset{}{N}: \longrightarrow :N: :N: \text{ or } N\equiv N$$

Triple bond

Pictorially, a nitrogen molecule can be shown as follows.

Formation of an Acetylene Molecule (C_2H_2)

In an acetylene (ethyne) molecule, two C atoms combine with two H atoms. Each C atom shares three of its valence electrons with the other C atom. One electron of each C atom is shared with one electron of a H atom.

Hence, in a molecule of acetylene, there is a triple covalent bond between the two C atoms and each C atom is joined to one H atom by a single covalent bond. Pictorially, a molecule of acetylene can be shown as follows.

Polar and Non-polar Covalent Bonds or Compounds

Non-polar	Polar
• In a non-polar covalent bond, sharing of electron/electrons is between atoms of the same element or atoms with nearly same electronegativity. • The shared electrons are exactly located in between two atoms undergoing covalent bond formation, that is, no polarity is possible. • These compounds do not ionise in water due to lack of charge separation. • It is possible when the electronegativity difference between the two atoms is less than 0.5. **Example:** $X-X$, $O=O$, $N\equiv N$	• In a polar covalent bond, sharing of electron/electrons is between atoms of different elements or atoms with different electronegativities. • The shared electrons are shifted towards the more electronegative atom, that is, polarity develops. • These compounds ionise in water due to charge separation. • It is possible when the electronegativity difference between the two atoms is more than 0.5. **Example:** $\overset{\delta+}{H}-\overset{\delta-}{O}-\overset{\delta+}{H}$, $\overset{\delta+}{H}-\overset{\delta-}{X}$, $H^{\delta+}-\overset{\delta-}{N}-H^{\delta+}$ with $\underset{\delta+}{H}$ below

Features of Covalent Compounds

The important features of covalent compounds are discussed below.

(i) **Physical state:** Covalent compounds are made up of neutral atoms. So, the forces of attraction (van der Waals force) between the molecules are weaker than those (coloumbic force) found in ionic compounds. Therefore, covalent compounds are usually volatile liquids or gases. Covalent compounds are mostly liquids or gases but if molecular weight is high, they may exists as solids. Solid nature is favoured by increase of molecular weight. For example,

F_2	Cl_2	Br_2	I_2
Gas	Gas	Liquid	Solid

(ii) High melting and high boiling points: The melting and the boiling points of covalent compounds are generally low. As covalent compounds are made up of neutral atoms or molecules, the forces of attraction (van der Waals force) between the molecules or atoms are very weak. So, a comparatively small amount of heat energy is needed to break these weak intermolecular forces of attraction. Hence, they have low melting and boiling points. Covalent compounds have lower boiling and melting point values than ionic compounds as the covalent bond is a weak van der Waals force while the ionic bond is a strong coulombic force. For example,

$$KOH \quad > \quad HX$$

Strong force of attraction   weak van der Waals forces

$\Rightarrow$ BP, MP $\propto$ Hydrogen bonding $\propto$ Molecular weight

$$HF > HI > HBr > HCl$$

Due to H-bonding

(iii) Solubility: Solubility in covalent compounds follows the *like dissolves like* concept. That is, a non-polar molecule dissolves in a non-polar solvent while a polar molecule dissolves in a polar solvent. For example,

- CCl_4 (non-polar) dissolves in an organic solvent like benzene.
- Alcohol or NH_3 (polar) dissolves in H_2O.

(iv) Electrical conductance: Covalent compounds do not conduct electricity. This is because they are made up of neutral molecules or atoms and not of ions and do not produce ions in the molten state or in aqueous solutions. An exception is graphite which is a conductor of electricity. Here, free electrons are available in its hexagonal sheet like slippery structure. In the case of diamond, the structure is tetrahedral, so free electrons are not available. Hence, it is a non-conductor.

(v) Isomorphism: Usually covalent compounds do not show isomorphism.

(vi) Nature of reactions: They show slow molecular reactions as they need activation energy.

(vii) Spatial or stereo isomerism: Covalent bonds are directional, so these compounds can show spatial isomerism.

Some Important Covalent Compounds and Their Structures

1. Hydrogen (H_2)

2. Chlorine (Cl_2)

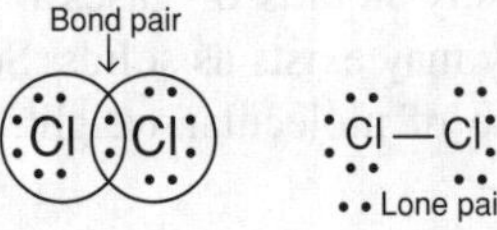

3. Hydrogen chloride (HCl)

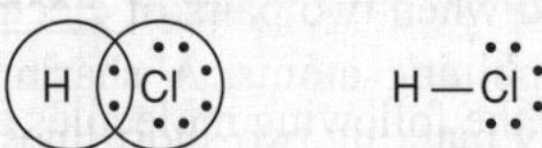

4. Oxygen (O_2)

5. Nitrogen (N_2)

6. Ammonia (NH_3)

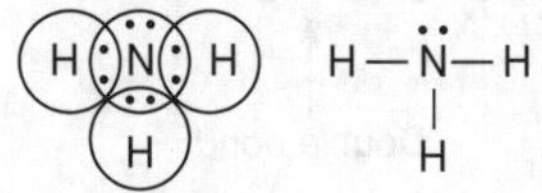

7. Water (H_2O)

8. Methane (CH_4)

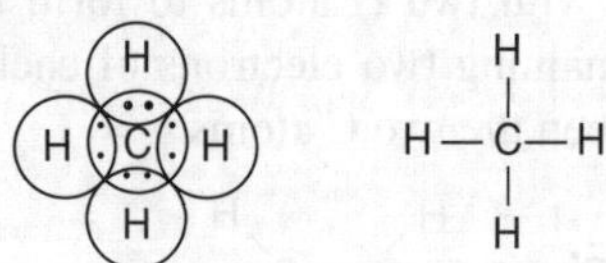

9. Carbon Tetrachloride (CCl_4)

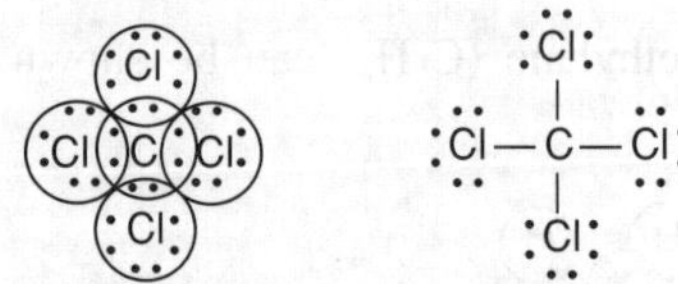

10. Carbon dioxide (CO_2)

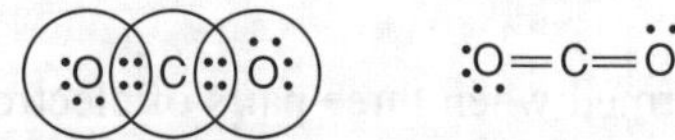

11. Ethylene (C_2H_4)

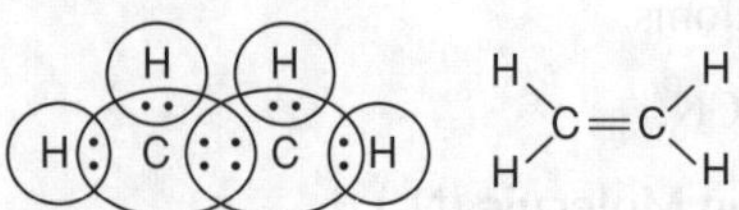

12. Ethyne (C_2H_2)

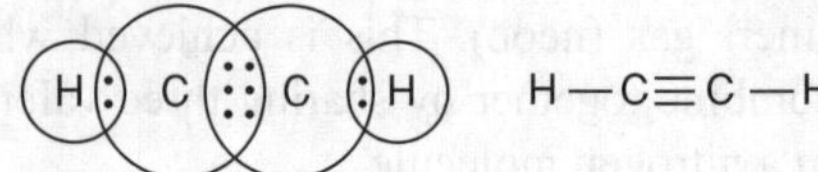

Illustrations

1. Draw the Lewis structures for the following molecules and ions:

H_2S, $SiCl_4$, BeF_2, CO_3^{2-}, HCOOH

Solution:

$$H_2S = H \overset{\displaystyle :\!\ddot{S}\!:}{} H, \quad SiCl_4 = :\ddot{C}l\!-\!\underset{:\ddot{C}l:}{\overset{:\ddot{C}l:}{Si}}\!-\!\ddot{C}l:, \quad BeF_2 = F\!-\!Be\!-\!F$$

$$CO_3^{2-} = \left[\ \underset{:\ddot{O}:\ \ :\ddot{O}:}{\overset{:O:}{\underset{C}{\|}}}\ \right]^{2-} \qquad HCOOH = H\!-\!\overset{:O:}{\underset{\|}{C}}\!-\!\ddot{O}\!-\!H$$

2. The skeletal structure of CH_3COOH as shown below is correct, but some of the bonds are shown incorrectly. Write the correct Lewis structure for acetic acid.

$$H\!=\!\overset{H}{\underset{H}{\overset{|}{C}}}\!-\!\overset{:O:}{\underset{\|}{C}}\!-\!O\!-\!H$$

Solution: The structure given in the question is incorrect as the carbon atom cannot be pentavalent. The correct structure is:

$$H\!-\!\overset{H}{\underset{H}{\overset{|}{C}}}\!-\!\overset{:O:}{\underset{\|}{C}}\!-\!\ddot{O}\!-\!H$$

3. What do you understand by bond pairs and lone pairs of electrons? Illustrate by giving one example of each type.

Solution: Covalent bonds are formed by mutual sharing of electrons between the two atoms. The shared pairs of electrons thus present between the bonded atoms are called bond pairs.

All the electrons of an atom may not participate in the bonding. Those that do not participate in the bonding are called lone pairs of electrons. For example, in CH_4 $\left(\overset{H}{\underset{H}{\overset{|}{H:\ddot{C}:H}}} \right)$, there are only 4 bond pairs but in H_2O $\left(\underset{H\ H}{:\ddot{O}:} \right)$, there are two bond pairs and two lone pairs.

Coordinate or Dative or Semi-polar Bonds

It was introduced by Lowry and Sidgwick. It is a special type of bond which is formed by the donation of electron pairs from the donor to the receiver. It involves partial transfer or unequal sharing of electrons between two molecules to form a bond. It is shown by an arrow ($\longrightarrow$) from the donor to the receiver. It is just like a covalent bond. However, both the electrons involved in sharing belong to one atom or group (donor).

$$X:\ \ \ +\ \ \ Y \to X \longrightarrow Y$$

$$\text{Donor} \qquad \text{Receiver}$$

or Lewis base Lewis acid

It is directional in nature and the arrow is shown from the donor to the receiver side. A coordinate bond is an intermediate bond between the ionic and covalent bond but closer to the covalent bond. The features of coordinate compounds are closer to those of covalent compounds. Compounds having covalent and coordinate bonds are SO_2, SO_3, CO, NH_3 and BF_3. Compounds having ionic, covalent and coordinate bonds are NH_4Cl, NH_4NO_3, $CuSO_4.5H_2O$ and all complexes like $K_4Fe(CN)_6$. For example,

- **Formation of ammonium ion:** Ammonium ion has all three types of bonds—ionic, covalent and coordinate. When ammonia reacts with H^+ (having no electrons) formation of an ammonium ion takes place. Here, the lone pair present on the nitrogen atom is donated for sharing to the H^+. As a result, a coordinate bond is formed between nitrogen atom and H^+. Here ammonia acts like a donor or Lewis base, while H^+ acts like a Lewis acid or acceptor. Here, after coordinate bond formation, all the four N—H bonds become identical and the positive charge belongs on the whole ammonium ion.

$$H\!-\!\overset{H}{\underset{H}{\overset{|}{N}}}\!\overset{\times}{\underset{\times}{}}\ +\ H^+ \longrightarrow \left[\ H\!-\!\overset{H}{\underset{H}{\overset{|}{N}}}\!\rightarrow\!H\ \right]^+ \text{ or } NH_4^+$$

Ammonium ion

- **Formation of hydronium ion:** When an acid like HCl is dissolved in water, it releases H^+ quickly. This released H^+ forms a coordinate bond with a water molecule to give H_3O^+ (hydrated proton). Here, water acts as a donor or Lewis base using the lone pair present on the oxygen atom which is partially transferred towards the formation of the coordinate bond.

$$\underset{\underset{\text{Water}}{H}}{H:\ddot{O}:}\ +\ \underset{\text{Proton}}{H^+} \longrightarrow \left[\underset{\underset{\text{Hydronium ion}}{H}}{H:\ddot{O}:H}\right]^+ \equiv \left[\underset{\underset{\text{Hydronium ion}}{H}}{H\!-\!\overset{\ddot{O}}{\underset{\downarrow}{}}\!-\!H}\right]^+$$

Illustrations

1. Which type of bond is present in hydrated copper sulfate $(CuSO_4.5H_2O)$?

Solution: Hydrated copper sulfate can have ionic, covalent, coordinate and hydrogen bonding.

2. Explain why the formation of $NH_3.BF_3$ adduct takes place when ammonia reacts with boron trifluoride.

Solution: When ammonia (NH_3) reacts with boron trifluoride (BF_3), formation of the adduct $NH_3.BF_3$ takes place due to coordinate bond formation between NH_3 and BF_3.

KNOWLEDGE BOOSTER

Metallic bonding

As we know, metals are hard solids and they are made up of atoms. It has been established that the atoms in a metal are very closely packed. The force or strong interaction that holds the atoms closely together in a metal is known as the metallic bond. Metal atoms lose one, two or three electrons to form positively charged ions known as cations. The electrons thus lost move freely in the metal, that is, these electrons become mobile, but the cations do not leave their positions. In a metal lattice, it is assumed that the metal ions are immersed in a sea of electrons. Due to the presence of mobile electrons, metals are good conductors of heat and electricity. Hardness, density and melting point depend upon the strength of the metallic bond.

Hydrogen bonding

It was introduced by Latimer and Rodebush. It is a weak interaction shown by dotted (.....) lines between hydrogen and highly electronegative and small atoms such as F, O, N. It develops due to the presence of partial opposite charges on the H-atom and such highly electronegative atoms due to polarity in the molecule. Its nature is dipole, ion– or dipole–induced dipole interactions. HCl has no H-bonding as Cl is large in size. Hydrogen bonding influences solubility in water, viscosity, boiling point and many other physical properties. One interesting property is, when two ice cubes are pressed, they form one block.

(i) Intermolecular H-bonding: It is formed between two or more different molecules of the same or different types. For example, HF, H_2O, NH_3, R–OH, R–COOH.

(ii) Intramolecular H-bonding: This type of hydrogen bond is formed within the same molecule.

For example,

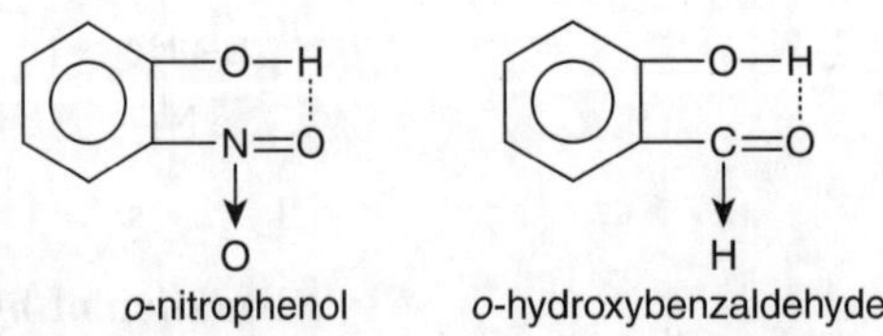

CHAPTER AT A GLANCE

- **Valency:** The combining capacity of an element is known as its valency. It is measured in terms of the number of valence electrons (present in the outer orbit) or number of H-atoms or O-atoms or Cl-atoms linked with one atom of an element. The number of electrons shared by an atom is known as its valency.

- **Octet rule:** Any atom or ion having the octet state (8 valence electrons) is quite stable. Every atom tries to attain the octet state by chemical combination (chemical bond formation) to gain stability (nearest inert gas configuration).

- **Chemical bond:** It is the force of attraction that binds two atoms together. A chemical bond implies a balance between the force of attraction and the force of repulsion at a particular distance. It is formed to attain the octet state, minimise energy, gain stability and decrease reactivity.

- **Ionic bond:** It is formed by the complete transfer of valence electrons from a metal to a non-metal, or from one atom to another. The electrovalency of an element is equal to the number of electrons lost or gained by the atom to form an ion.

- Ionic compounds have solid, crystalline structures, high BP, MP, are soluble in water, show electrical conductance in molten or aqueous solution state and show ionic reactions.
- **Lattice energy:** It is the energy released during the formation of an ionic solid molecule from its constituent ions. It is also the energy needed to break an ionic solid molecule into its constituent ions.
- The chemical bond formed when two atoms share electrons between them is known as a **covalent bond**. It is formed by equal sharing of electrons between two similar or different atoms.
- Covalent bonds are of these **three types**—single covalent bond, double covalent bond and triple covalent bond.
- **Non-polar covalent bond:** In a non-polar covalent bond, sharing of electron/electrons takes place between atoms of the same element or atoms with nearly same electronegativity.
- **Polar covalent bond:** In a polar covalent bond, sharing of electron/electrons takes place between atoms of different elements or atoms with different electronegativities.
- **Covalent compounds** may be gas, liquid or solid with low bp, mp values. Polar compounds dissolve in polar solvents and non-polar dissolve in non-polar solvents. They are non-conductors and show slow molecular reactions.
- **Coordinate bond:** It is a special type of bond which is formed by donation of an electron pair from a donor to a receiver. It involves partial transfer or unequal sharing of electrons between two molecules to form a bond. It is shown by ($\longrightarrow$) from the donor to the receiver.

PRACTICE QUESTIONS

Analyse Your Concepts (School Exam Based)

Fill in the Blanks

Instructions: Complete the following statements with an appropriate word/term to be filled in the blank spaces.

1. The number of electrons in the valence shell of the noble gases is ______________ except helium which has ______________.

2. F^- and Ne contain the same number of ______________.

3. Potassium chloride, being an ionic solid, has a ______________ melting point.

4. Polar covalent compounds are generally soluble in ______________ solvents.

5. In the solid state, NaCl is a ______________ conductor of electricity.

6. In an aqueous solution, NaCl dissociates as follows:
   $NaCl \rightarrow$ ......$+$.......

7. The bond in HCl is ______________.

8. An electronegativity difference of 1.9 or above between two atoms ensures ______________ bond between them.

9. Sharing of two electron pairs between two atoms results in the formation of a ______________ bond.

10. Noble gases exist as individual ______________.

11. The valency of Cl is 1 because it contains ______________ electron less than the stable noble gas configuration.

12. The electrostatic forces of attraction between the metallic ions and free electrons is called ______________.

13. Ionic compounds are generally ______________ in polar solvents like water.

14. The forces of attraction between the elemental gaseous molecules is known as ______________.

15. In ammonia–boron trifluoride complex, the electron donor molecule is ______________.

True or False

Instructions: Read the following statements and write your answer as true or false.

1. Being a noble gas, He has eight electrons in its valence shell.

2. An O atom will gain two electrons to attain the octet in its valence shell.

3. CCl_4 is a covalent compound.

4. The N—H bonds in NH_3 are polar covalent bonds.

5. $BaCl_2$ is an ionic compound.

6. $FeCl_2$ is reduced by Cl_2 to $FeCl_3$.

7. CCl_4 is a good conductor of electricity.

8. BCl_3 obeys the octet rule.

9. The number of valence electrons in all noble gases except helium is 8.

10. A solution of calcium chloride conducts electricity.

11. Hydrogen tends to achieve stable duplet arrangement.

12. Ca^{2+} and O^{2-} have achieved stable octet arrangement.

13. Covalent bonds are directional bonds.

14. Water contains one single covalent bond.

15. Magnesium oxide is a covalent compound.

Match the Following

Instructions: Each question contains statements given in two columns which have to be matched. Statements (a, b, c, d) in column I have to be matched with statements (p, q, r, s) in column II.

1.

Column I (Type of Solid)	Column II (Examples)
(a) Covalent	(p) SiO_2
(b) Molecular	(q) MgO
(c) Ionic	(r) CCl_4
(d) Metallic	(s) Brass

2.

Column I	Column II
(a) Ionic bond	(p) van der Waals force
(b) Covalent bond	(q) Electrostatic force
(c) Metallic bond	(r) Weak interaction
(d) Hydrogen bond	(s) Strong interaction

Very Short Answer Type Questions

1. What is the nature of forces existing between ions in ionic bonds?

2. Between ZnO and ZnS which is more covalent and why?

3. How many covalent bonds are there in a molecule of nitrogen?

4. Explain why ionic bond formation takes place between potassium and chlorine while covalent bond formation takes place between hydrogen and chlorine.

5. The atomic number of the sodium atom is 11. What is its electrovalency?

6. An atom has a configuration 2, 6. What will be its covalency?

7. Element X has 10 protons and 10 electrons. Will it be reactive?

8. Which of the elements would be most stable?
   $_9X$, $_{10}Y$, $_{11}Z$

9. Three elements A, B and C have the following configurations:
   (i) A = 2, 8, 1   (ii) B = 2, 8, 7   (iii) C = 2, 8, 2
   What type of molecule will form between the following?
   (a) A and B   (b) C and B   (c) B and B

10. What is the difference in covalent bonding in CCl_4 and CH_4?

11. Why does solid NaCl not conduct electricity but aqueous or molten NaCl solution does?

12. What is the main condition for the formation of an ionic bond with respect to heat of formation?

Short Answer Type Questions

1. Distinguish between ionic and covalent compounds.

2. Two neon atoms do not form a covalent bond to give a neon molecule Ne_2. Why?

3. The elements A and B have the following configurations:
   A   2, 6
   B   2, 8, 2
   What is the nature of the bond between A and B?

4. Describe the nature of bond between the following
   (a) sodium and chlorine
   (b) carbon and chlorine
   (c) hydrogen and chlorine

5. Draw the electron dot structures of MgF_2, CaO, H_2O and CO_2.

6. From the list of compounds: Cl_2, N_2, CO_2, C_2H_2 and O_2, choose the molecule that contains
   (a) only a single bond   (b) only a triple bond
   (c) only a double bond   (d) two double bonds
   (e) single and triple bonds

7. HF can dissolve in water but benzene does not dissolve in water. Why?

8. Why do ionic compounds have higher boiling and melting points than covalent compounds?

Long Answer Type Questions

1. Explain the formation of calcium sulfide starting from calcium and sulfur. Draw the diagrammatic representation of the atom showing the electronic arrangement in various shells.

2. Show the formation of covalent bonds in HCl, CCl_4, CH_4, H_2, O_2 and Cl_2.

3. Explain the nature of the covalent bond using the bond formation in CH_3Cl.

4. What is an electrovalent bond and how is it formed? Explain with an example.

5. Draw the electron dot structures for
   (a) ethanoic acid
   (b) H_2S
   (c) F_2

COMPETITION WINDOW (OBJECTIVE TYPE)

[For NEET, JEE (Main and Advanced), NTSE, KVPY and Olympiads]

Topic-wise MCQs

Introduction, octet rule

1. Linus Pauling received the Nobel Prize for his work on
   (a) chemical bonds
   (b) thermodynamics
   (c) atomic structure
   (d) photosynthesis

2. The combination of atoms take place so that they
   (a) can gain two electrons in the outermost shell
   (b) get eight electrons in the outermost shell
   (c) acquire stability by lowering of energy
   (d) get eighteen electrons in the outermost shell

3. A chemical bond is formed to get
   (a) octet state configuration
   (b) minimum energy state
   (c) maximum stability
   (d) all of these

4. Among the following, the electron deficient compound is
   (a) BCl_3
   (b) CCl_4
   (c) PCl_5
   (d) CH_4

5. In which of the following molecules does the central atom not follow the octet rule?
   (a) CO_2
   (b) H_2S
   (c) BF_3
   (d) PCl_3

6. Which of the following bonds/forces is/are the weakest?
   (a) covalent bond
   (b) van der Waals force
   (c) hydrogen bond
   (d) ionic bond

7. Which of the following are examples of expansion of octet state?
   (a) PCl_5
   (b) NO_2
   (c) SF_6
   (d) all of these

8. Select the correct statement.
   (a) a chemical bond is the force that holds two species together
   (b) helium and argon have duplet state
   (c) ionic bond and covalent bond are bonds while metallic bond is an interaction
   (d) both (a) and (c)

Ionic bond, ionic compounds

9. Cation and anion combine in a crystal to form the following type of compound.
   (a) ionic
   (b) metallic
   (c) covalent
   (d) dipole–dipole

10. Which of the following pairs of elements form a compound with maximum ionic character?
    (a) Na and F
    (b) Cs and F
    (c) Na and C
    (d) Cs and I

11. An ionic bond X^+Y^- is most likely to be formed when
    (a) the ionisation energy of X is high and the electron affinity of Y is low
    (b) the ionisation energy of X is low and the electron affinity of Y is high
    (c) the ionisation energy of X and the electron affinity of Y are high
    (d) the ionisation energy of X and the electron affinity of Y are low

12. Which of the following will conduct electricity?
    (a) crystalline NaCl
    (b) fused NaCl
    (c) molten sulfur
    (d) diamond

13. The compound which contains ionic as well as covalent bonds is
    (a) $C_2H_4Cl_2$
    (b) CH_3I
    (c) KCN
    (d) H_2O_2

14. Four elements P, Q, R, S have atomic numbers $Z - 1$, Z, $Z + 1$ and $Z + 2$ respectively. If Z is 9, then bond between which pair of elements will be ionic?
    (a) P and R
    (b) S and Q
    (c) S and R
    (d) Q and R

15. The hydration of ionic compounds involves
    (a) evolution of heat
    (b) weakening of attractive forces
    (c) dissociation into ions
    (d) all of these

16. Which of the following pairs will form the most stable ionic bond?
    (a) Mg and F
    (b) Na and F
    (c) Na and Cl
    (d) Li and F

17. In which of the following species are the bonds non-directional?
    (a) PCl_3
    (b) RbCl
    (c) $BeCl_2$
    (d) BCl_3

18. An electrovalent compound does not exhibit space isomerism because of
    (a) the presence of oppositely charged ions
    (b) its crystalline nature
    (c) the non-directional nature of the bond
    (d) its high melting point

19. Ionic compound can show conductance in
    (a) aqueous solution state
    (b) molten state
    (c) solid state
    (d) both (a) and (b)

Covalent bond, covalent compounds

20. A covalent bond is formed by
  (a) complete transfer of electrons
  (b) partial sharing of electrons
  (c) mutual and equal sharing of electrons
  (d) all of these

21. Select the correct statement regarding the covalent bond.
  (a) it is a directional bond
  (b) a maximum of 3 covalent bonds can be formed between two atoms
  (c) it is van der Waals force
  (d) all of these

22. The most polar covalent bond is
  (a) C–S
  (b) C–O
  (c) C–F
  (d) C–Br

23. Which of the following represents the Lewis structure of the N_2 molecule?
  (a) $\overset{\times}{\underset{\times}{N}} \equiv \overset{\times}{\underset{\times}{N}}$
  (b) $\overset{\times\times}{\underset{\times}{N}} \equiv \overset{\times\times}{\underset{\times}{N}}$
  (c) $\overset{\times\times}{\underset{\times\times}{N}} = \overset{\times\times}{\underset{\times\times}{N}}$
  (d) $\overset{\times\times}{\underset{\times\times}{N}} - \overset{\times\times}{\underset{\times\times}{N}}$

24. Which contains both polar and non-polar bonds?
  (a) NH_4Cl
  (b) HCN
  (c) H_2O_2
  (d) CH_4

25. The valency of sulfur in sulfuric acid is
  (a) 2
  (b) 8
  (c) 4
  (d) 6

26. How many of these molecules have both polar and non-polar covalent bonds?
  (a) CH_3OH
  (b) HF
  (c) H_2O_2
  (d) both (a) and (c)

27. Polar covalent compounds are soluble in
  (a) non-polar solvents like CS_2
  (b) polar solvents like water
  (c) both of these
  (d) none of these

28. Which of the following are soluble in water?
  (a) HF
  (b) CCl_4
  (c) NH_3
  (d) both (a) and (c)

29. Which of the following molecule contains only non-polar covalent bonds?
  (a) CCl_4
  (b) CH_3CHO
  (c) C_2H_6
  (d) HF

Coordinate bond, metallic bond and hydrogen bond

30. NH_3 and BF_3 form an adduct readily through
  (a) ionic bond between BF_3 and NH_3
  (b) coordinate bond between B and N
  (c) covalent bond between B and N
  (d) H-bonds between F atoms of BF_3 and H-atoms of NH_3

31. Hydrogen bonding is absent in
  (a) H_2O
  (b) NH_3
  (c) C_2H_5OH
  (d) $C_2H_5OC_2H_5$

32. Malleability and ductility of metals can be accounted for by the
  (a) capacity of layers of metal ions to slide over one another
  (b) crystalline structure in metal
  (c) presence of electrostatic forces
  (d) interaction of electrons with metal ions

33. Which of the following species contain covalent coordinate bonds?
  (a) $AlCl_3$
  (b) CO
  (c) $[Fe(CN)_6]^{3-}$
  (d) N_3

34. H-bond is strongest in
  (a) C_2H_5OH
  (b) $H–F$
  (c) H_2O
  (d) CH_3COCH_3

35. Which of the following statement is correct regarding the metallic bond?
  (a) strength of metallic bond $\propto$ number of valence electrons
  (b) hardness density $\propto$ metallic bond strength
  (c) it is stronger than ionic bond
  (d) both (a) and (b)

Miscellaneous

1. A bond formed between two like atoms cannot be
  (a) coordinate
  (b) metallic
  (c) ionic
  (d) covalent

2. In X–H—Y, X and Y both are electronegative elements:
  (a) electron density of X will increase and on H decrease
  (b) electron density will increase in both
  (c) electron density will decrease in both
  (d) electron density on X will decrease and on H increase

3. The weakest among the following types of bonds is
  (a) metallic bond
  (b) hydrogen bond
  (c) ionic bond
  (d) covalent bond

4. Which of the following contains (electrovalent) and non-polar (covalent) bonds?
  (a) CH_4
  (b) HCN
  (c) NH_4Cl
  (d) H_2O_2

5. Which of the following does not have a coordinate bond?
  (a) NH_4^+
  (b) H_2O
  (c) BH_4^-
  (d) both (b) and (c)

6. The two atoms X and Y lie at the top of group 2 and group 16 respectively. On combination, they form a compound of the type
  (a) X_2Y_2
  (b) XY
  (c) X_2Y
  (d) XY_2

7. Which one of the following molecules contains both ionic and covalent bonds?

(a) CH_2Cl_2 (b) K_2SO_4
(c) $BeCl_2$ (d) SO_2

8. In which of the following are ionic, covalent and coordinate bonds present?
(a) water (b) ammonia
(c) sodium isocyanide (d) potassium bromide

9. Amongst H_2O, H_2S, H_2Se and H_2Te, the one with the highest boiling point is
(a) H_2O because of hydrogen bonding
(b) H_2Te because of higher molecular weight
(c) H_2S because of hydrogen bonding
(d) H_2Se because of lower molecular weight

10. Which of the following is the electron deficient molecule?
(a) C_2H_6 (b) B_2H_6
(c) SiH_4 (d) PH_3

11. The types of bonds present in N_2O_5 are
(a) only covalent (b) only ionic
(c) ionic and covalent (d) covalent and coordinate

12. Which of the following molecules does not have coordinate bonds?
(a) CH_3–NC (b) CO
(c) O_3 (d) CO_3^{2-}

13. Two ice cubes are pressed over each other and unite to form one cube. Which force is responsible for holding them together?
(a) van der Waals forces
(b) covalent attraction
(c) hydrogen bond formation
(d) dipole–dipole attraction

14. Which of the following conditions is needed during ionic bond formation?
(a) low ionisation energy of metal atom
(b) high electron gain enthalpy of non-metal atom
(c) high lattice energy for ions
(d) all of these

15. Which of the following has maximum solubility in CCl_4 solvent?
(a) F_2 (b) Cl_2
(c) Br_2 (d) I_2

16. Which of the following is insoluble in water?
(a) HF (b) NH_3
(c) C_6H_6 (d) NaOH

17. Which of the following properties is not related to metallic bond strength?
(a) hardness (b) density
(c) oxidation state (d) melting point

18. Which of the following is the correct decreasing order of strength bonds?
(a) covalent > ionic > metallic > hydrogen bond
(b) ionic > covalent > metallic > hydrogen bond
(c) covalent > ionic > hydrogen bond > metallic

(d) ionic > hydrogen bond > metallic > covalent

19. Which of the following properties is affected by hydrogen bonding?
(a) solubility in water (b) boiling point
(c) viscosity (d) all of these

20. Which of the following can show electrical conductance?
(a) solid NaCl (b) molten NaCl
(c) graphite (d) both (b) and (c)

Advanced and Olympiads

Single Choice

1. Which of the following pairs contain only covalent bonds?
(a) NaCl, MgO (b) NaCN, CO_2
(c) CO_2, SO_2 (d) K_2CO_3, CO_2

2. Which of the following pairs contain both covalent and coordinate bonds?
(a) PCl_5, CCl_4 (b) HNO_3, H_2SO_4
(c) SF_6, $SiCl_4$ (d) Na_2SO_4, NaCl

3. The electric configuration of four elements are as follows:
A: $1s^2\, 2s^2\, 2p^4$ B: $1s^2\, 2s^2\, 2p^5$
C: $1s^2\, 2s^2\, 2p^6\, 3s^1$ D: $1s^2\, 2s^2\, 2p^6\, 3s^2$
Now decide the possible formulae of ionic compounds that could be formed between them
(a) A_2C, DA, CB, D_2B (b) AC, DA, CB, CB
(c) C_2A, DA, CB, DB_2 (d) AC, D_2A, C_2B, DB

4. Which of the following types of bonds are present in $CuSO_4 \cdot 5H_2O$?
(1) Electrovalent (2) Covalent (3) Coordinate
Select the correct answer using the code given below.
(a) 1 and 2 only (b) 1 and 3 only
(c) 1, 2 and 3 (d) 2 and 3 only

5. What is the dominant intermolecular force or bond that must be overcome in converting liquid CH_3OH to a gas?
(a) dipole–dipole interaction
(b) hydrogen bonding
(c) London dispersion force
(d) covalent bonds

6. Which of the following set represents correct statements?
 (i) Ionic bond is electrostatic force of attraction by nature
 (ii) Covalent bond is an interaction
 (iii) Hydrogen bond is an interaction only
 (iv) Coordinate bond is directional
 (v) Metallic bond is a strong interaction
(a) (i), (ii), (iv) and (v) (b) (i), (ii), (iii) and (iv)
(c) (ii), (iii), (iv) and (v) (d) (i), (iii), (iv) and (v)

7. In which of the following is the correct valency not given?
(a) chlorine (valency = 1 and 7)
(b) oxygen (valency = 2 and 6)
(c) phosphorus (valency = 3 and 5)
(d) sulphur (valency = 2 and 6)

8. Which of the following is incorrectly matched here?
  (a) HCN   — Contains both polar and non-polar bonds
  (b) CCl_4 — All bonds are polar
  (c) NH_4Cl — Contains ionic, covalent and coordinate bonds
  (d) H_3O^+ — Has covalent bonds only

Multiple Choice

9. The octet rule is not obeyed in
  (a) CO_2 (b) BCl_3
  (c) PCl_5 (d) SiF_4

10. Which of the following are characteristics of covalent compounds?
  (a) they have low melting and boiling points
  (b) they are formed between two atoms having no or very small electronegativity difference
  (c) they may or may not be insoluble in water
  (d) their molecules have indefinite geometry

11. Which type of chemical bonds are present in N_2O_5?
  (a) H-bond   (b) coordinate bond
  (c) covalent bond   (d) ionic bond

12. Which compound(s) among the following contain an ionic bond?
  (a) NaOH   (b) HCl
  (c) K_2S (d) LiH

13. Which of the following species is/are capable of forming a coordination bond with BF_3?
  (a) F^- (b) NH_3
  (c) NH_4^+ (d) Ca^{2+}

14. Which of the following statement(s) is/are correct regarding ionic compounds?
  (a) They are good conductors at room temperature in aqueous solution.
  (b) They are generally soluble in polar solvents.
  (c) They consist of ions.
  (d) They generally have high melting and boiling points.

15. Which of the following compounds contain/s both ionic and covalent bonds?
  (a) NH_4Cl (b) NaCN
  (c) $CuSO_4 \cdot 5H_2O$ (d) NaOH

16. Most ionic compounds have
  (a) high melting points and low boiling points
  (b) high melting points and non-directional bonds
  (c) high solubilities in polar solvents and low solubilities in non-polar solvents
  (d) three dimensional network structures, and are good conductors of electricity in the molten state

17. Which of the following statements is/are true?
  (a) covalent bonds are directional
  (b) ionic bonds are non-directional

  (c) a polar bond is formed between two atoms which have the same electronegativity values
  (d) the presence of polar bonds in a polyatomic molecule suggests that it has zero dipole moment

18. Which one of the following can have intermolecular H-bonding?
  (a) H_2O (b) *o*-nitro phenol
  (c) HF   (d) CH_3COOH

Comprehension Type

Comprehension I: The nature of chemical bonding is mainly of three types—electrovalent or polar, covalent or non-polar and coordinate covalent or semi-polar. A large number of inorganic compounds are polar in character whereas the compounds of carbon are non-polar. When inorganic substances such as copper sulfate and silver nitrate are dissolved in water, the individual ions which hold them in place in the solid thus become free to move about independently in the solution. If now two metallic electrodes – which are connected to the positive and negative terminals of a source of current – are placed in the solution, the positive ions are attracted to the negative pole and negative ions to the positive pole of this simple electrolytic cell.

19. The compound which ionises almost completely in water is
  (a) NH_3 (b) KCl
  (c) CH_3COOH (d) PCl_5

20. An electrovalent compound is made up of
  (a) neutral atoms
  (b) neutral molecules
  (c) electrically charged atoms or groups of atoms
  (d) electrically charged molecules

21. Multiple covalent bonds exist in a molecule of
  (a) Cl_2 (b) O_2
  (c) N_2 (d) both (b) and (c)

Comprehension II: van der Waals forces are electrical in nature and are due to the interaction between the electrons of one atom and the nucleus of a second atom but from a relatively long distance because the valence orbitals are completely filled. Other types of interactions between atoms include the hydrogen, ionic and covalent bonds.

22. Of the following, the weakest interaction involves the
  (a) covalent bond
  (b) ionic bond
  (c) van der Waals interaction
  (d) hydrogen bond

23. In which substance below are the bonds predominantly ionic?
  (a) carbon tetrachloride   (b) sodium chloride
  (c) carbon dioxide   (d) diamond

24. In which of the following molecules does hydrogen bonding occur?
  (a) ammonia   (b) methane
  (c) chlorobenzene   (d) hexane

Matrix Matching

25. Match the following

Column I	Column II
(a) $NaOH$	(p) Non-polar covalent bond
(b) $MgCl_2$	(q) Ionic bond
(c) H_2O_2	(r) Polar covalent bond
(d) H—F	(s) Hydrogen bond

26. Match the following

Column I	Column II
(a) $CaCl_2$	(p) Ionic
(b) CaC_2	(q) Covalent
(c) $CuSO_4.5H_2O$	(r) Coordinate
(d) KHF_2	(s) Hydrogen bond

Integer Type

27. Which of the following don't follow the octet rule—BI_3, CCl_4, $ZnBr_2$, NO, NO_2, PCl_5, PH_3, SF_6, CH_4?

28. How many covalent bonds are present in a molecule of PCl_5?

29. Number of double bonds present in CO_2 molecules is __________.

30. How many single bonds are present in C_2H_6?

31. How many of these compounds are soluble in water? CO_2, NH_3, SO_3, $SiCl_4$, C_2H_5OH, C_6H_6, HF, NaOH, KCl

32. How many of these compounds can have covalent bonds? NaOH, $CaCl_2$, CaO, KCN, KNO_3, $MgSO_4$, NaF, NaHg, sucrose

33. How many of the following are odd e^- molecules: NH_3, NO, NO_2, ClO_2, Cl_2O, SO_3, CO_2, H_2O?

34. How many following statements are correct?
(a) A chemical bond is formed to attain the octet state and stability.
(b) Ionic bond is directional.
(c) Covalent bond is directional.
(d) Ionic bonds involve sharing of electrons.
(e) Two atoms can have a maximum of 3 covalent bonds between them.
(f) Coordinate bond is intermediate bond between ionic and covalent bond.

Answer Keys

Fill in the Blanks

1. 8, 2		**2.** 10	
3. high		**4.** polar	
5. bad		**6.** Na^+, Cl^-	
7. polar covalent		**8.** ionic	
9. double		**10.** atoms	
11. 1		**12.** metallic bond	
13. soluble		**14.** van der Waals	
15. NH_3			

True or False

1. F	**2.** T	**3.** T	**4.** T	**5.** T
6. F	**7.** F	**8.** F	**9.** T	**10.** T
11. T	**12.** T	**13.** T	**14.** F	**15.** F

Match the Following

1. (a)—(p), (b)—(r), (c)—(q), (d)—(s)
2. (a)—(q), (b)—(p), (c)—(s), (d)—(r)

Very Short Answer Type Questions

1. The nature of forces existing between ions in ionic bonds is electrostatic or coulombic force of attraction.

2. Between ZnO and ZnS, ZnS is the more covalent. As the cation is the same in both, the anion determines the extent of polarisation. Larger anions have greater polarisability. Between O^{2-} and S^{2-} ions, S^{2-} ion is more polarisable. Greater polarisability imparts greater covalent character to the compound.

3. There are 3 covalent bonds in a molecule of nitrogen.

4. Potassium belongs to group 1 and is highly electropositive. Chlorine belongs to group 17 and is highly electronegative. Potassium has very low IP value and can form the K^+ ion easily. Chlorine has very high IP value and high EA value and can form Cl^- ion easily. Thus, an ionic bond results between potassium and chlorine. IP value of hydrogen is not as low as that of K, so its tendency to form H^+ ion is very less. Sharing of electrons takes place between hydrogen and chlorine resulting in the formation of a polar covalent bond.

5. Its electrovalency is 1 as electrovalency is equal to number of electrons lost or gained.

6. Its covalency is 2 as it needs 2 electrons to attain the octet state.

7. Element X having 10 protons and 10 electrons is neon which is an inert gas. So, it is inactive.

8. Y is the most stable as it has the octet configuration (2, 8).

9. (a) A and B combine to give AB type of molecule (it is actually NaCl).

(b) C and B combine to give CB_2 type of molecule (it is actually $CaCl_2$)

(c) B and B combine to give B_2 type of molecule (it is actually Cl_2)

10. In CCl_4 all C—Cl bonds are polar covalent bonds while in CH_4 all C—H bonds are non-polar covalent bonds.

11. In molten or aqueous NaCl solution, ions are free to move so electric conduction is possible. It is not possible in solid NaCl as ions are not free.

12. For formation of an ionic bond, it is necessary that heat of formation must be negative or exothermic.

Short Answer Type Questions

1. See text part.

2. Two neon atoms do not form a covalent bond to give a neon molecule Ne_2. This is because neon is an inert gas with an octet configuration (2, 8). So, neon atoms do not undergo any type of chemical combination.

3. An ionic bond is formed between A and B as A needs 2 electrons to attain the octet state and B can lose 2 electrons to attain the octet state. The B atom will lose 2 electrons and A will get these 2 electrons (see example of MgO formation in text part).

4. (a) sodium and chlorine combine to form NaCl so an ionic bond is formed between them.

   (b) carbon and chlorine combine to form CCl_4 so four covalent bonds are formed between them.

   (c) hydrogen and chlorine combine to form HCl so a covalent bond is formed between them.

   (For structure of these compounds see text part.)

5. See text part.

6. (a) only a single bond: Cl_2

   (b) only a triple bond: N_2

   (c) only a double bond: O_2

   (d) two double bonds: CO_2

   (e) single and triple bonds: C_2H_2

7. Polar compounds dissolve in polar solvents. HF being polar dissolves in a polar solvent (water) but benzene being non-polar does not dissolve in water but dissolves only in non-polar solvents.

8. Ionic compounds have higher boiling and melting points than covalent compounds. In ionic compounds, the force of attraction is strong electrostatic or coloumbic force but in covalent compounds weak van dar Waals forces are present. Therefore, more energy is needed in ionic compounds to overcome this force than in covalent compounds.

Long Answer Type Questions

1. The number of valence electrons in a calcium atom (atomic number 20) is 2, and that of sulfur atom is 6, that is, sulfur requires 2 electrons to attain its octet. In the presence of sulfur, each calcium atom loses its 2 valence electrons to

one sulfur atom (figure). As a result, the calcium atom forms a calcium ion with charge +2 (Ca^{2+}), and the sulfur atom forms an oxide ion with a charge of –2 (S^{2-}). Since only one sulfur atom is needed to accept the 2 valence electrons donated by a calcium atom, the formula of calcium sulfide is CaS.

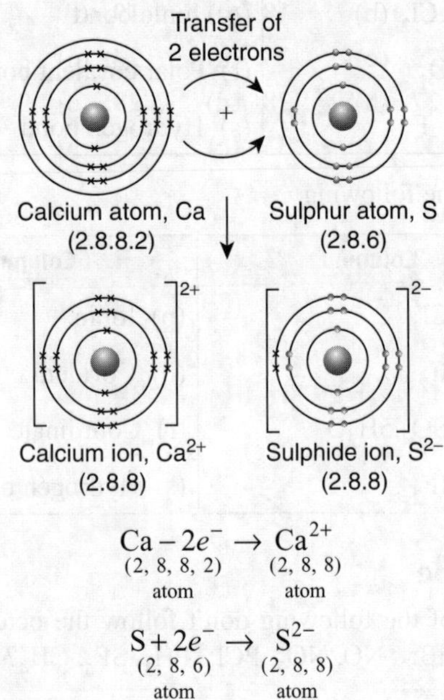

$$Ca - 2e^- \rightarrow Ca^{2+}$$
$$\underset{\text{atom}}{(2,\,8,\,8,\,2)} \qquad \underset{\text{atom}}{(2,\,8,\,8)}$$

$$S + 2e^- \rightarrow S^{2-}$$
$$\underset{\text{atom}}{(2,\,8,\,6)} \qquad \underset{\text{atom}}{(2,\,8,\,8)}$$

2. See text part.

3. Carbon has four electrons in its valence shell. To complete its octet, it shares these four electrons with other atoms to create four bonds, thus becoming stable. The bonds that are formed by sharing electrons are known as covalent bonds.

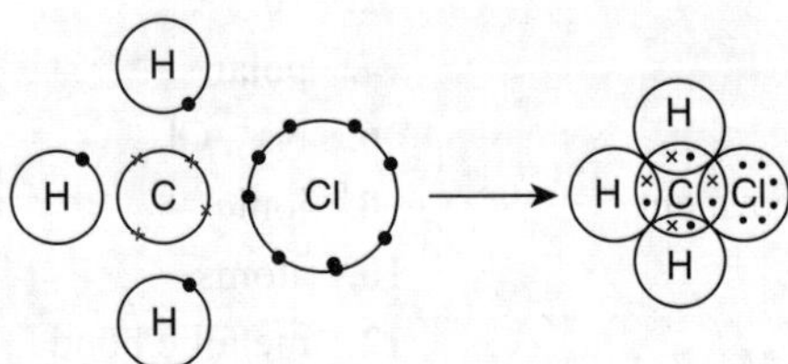

In CH_3Cl, carbon requires 4 electrons to complete its octet, while each hydrogen atom requires one electron to complete its duplet. Also, chlorine requires an electron to complete the octet.

Therefore, all of these share the electrons and as a result, carbon forms 3 bonds with hydrogen and one with chlorine.

4. See text part.

5. (a) Ethanoic acid

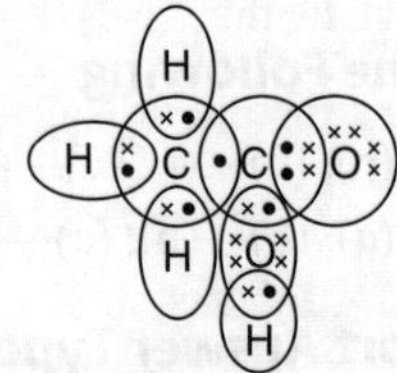

   (b) H_2S

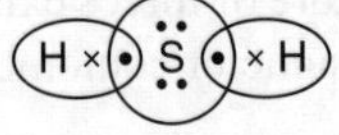

   (c) F_2

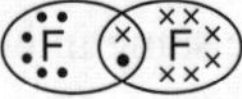

Competition Window (Objective Type)

Topic-wise MCQs

1. (a)	**2.** (c)	**3.** (d)	**4.** (a)	**5.** (c)
6. (b)	**7.** (d)	**8.** (d)	**9.** (a)	**10.** (b)
11. (b)	**12.** (b)	**13.** (c)	**14.** (b)	**15.** (d)
16. (a)	**17.** (b)	**18.** (c)	**19.** (d)	**20.** (c)
21. (d)	**22.** (c)	**23.** (a)	**24.** (c)	**25.** (d)
26. (d)	**27.** (b)	**28.** (d)	**29.** (c)	**30.** (b)
31. (d)	**32.** (a)	**33.** (c)	**34.** (b)	**35.** (d)

Miscellaneous

1. (c)	**2.** (b)	**3.** (b)	**4.** (c)	**5.** (d)
6. (b)	**7.** (b)	**8.** (c)	**9.** (a)	**10.** (b)
11. (d)	**12.** (d)	**13.** (c)	**14.** (d)	**15.** (d)
16. (c)	**17.** (c)	**18.** (b)	**19.** (d)	**20.** (d)

Advanced and Olympiads

Single Choice

1. (c)	**2.** (b)	**3.** (c)	**4.** (c)	**5.** (b)
6. (d)	**7.** (b)	**8.** (d)		

Multiple Choice

9. (bc)	**10.** (abc)	**11.** (bc)	**12.** (acd)	**13.** (ab)
14. (abcd)	**15.** (abcd)	**16.** (bcd)	**17.** (ab)	**18.** (acd)

Comprehension Type

19. (b)	**20.** (c)	**21.** (d)	**22.** (c)	**23.** (b)
24. (a)				

Matrix Matching

25. (a)—(q, r), (b)—(q), (c)—(r, s), (d)—(r, s)

26. (a)—(p), (b)—(p, q), (c)—(p, q, r, s), (d)—(p, q, s)

Integer Type

27. 6	**28.** 5	**29.** 2	**30.** 7	**31.** 7
32. 5	**33.** 3	**34.** 4		

Hints and Solutions

Competition Window (Objective Type)

Topic-wise MCQs

4. Boron in BCl_3 has 6 electrons in its outermost shell. Hence BCl_3 is an electron pair deficient compound.

7. Both PCl_5 and SF_6 are examples of expansion of octet state while NO_2 is an odd electron molecule.

8. Only helium can have a duplet configuration and not argon.

10. They have maximum electronegativity difference.

14. Q is fluorine and S is sodium so they form NaF which is an ionic compound.

16. Stability of the ionic bond depends upon the lattice energy which is more between Mg and F due to the +2 charge on the Mg atom.

19. Ionic compounds can show conductance in aqueous solution and molten state as ions are free to move. In the solid state, there is no conductance as ions cannot move.

23. Lewis structure of N_2 molecule is $\overset{\times}{\underset{\times}{N}} \equiv \overset{\times}{\underset{\times}{N}}$.

24. The O–O bond is non-polar while the O–H bond is polar.

26. In CH_3OH, the C—H bond is non-polar while the O—H bond is polar. In H_2O_2, the H—O bond is polar while the O—O bond is non-polar.

27. Polar covalent compounds are soluble in polar solvents and non-polar covalent compounds dissolve in non-polar solvents.

28. Being polar, HF and NH_3 dissolve in water which is a polar solvent while CCl_4 being non-polar, does not dissolve in water.

29. C_2H_6 contains only non-polar bonds C—C and C—H, while CCl_4 and HF have only polar bonds. CH_3CHO contains both polar (C=O) and non-polar (C—H) covalent bonds.

32. When a metal is beaten, it does not break but is converted into a sheet. It is said to possess the property of malleability. Due to their ductile nature, metals can be drawn into wires. These two properties of metals can be accounted for by the capacity of the layers of metal ions to slide over one another.

34. F has more electronegativity than oxygen.

35. Hardness density $\propto$ metallic bond strength $\propto$ number of valence electrons.

It is weaker than an ionic bond so statement (c) is not correct.

Miscellaneous

3. Hydrogen bond is the weakest bond.

5. Both H_2O and BH_4^- do not have any coordinate bonds; they have only covalent bonds.

9. H_2O has highest boiling point due to H-bonding. The order is $H_2S < H_2Se < H_2Te < H_2O$.

10. The compound in which the central atom does not have an octet is known as an electron deficient compound. So, B_2H_6 is an electron deficient compound.

15. In the case of the halogens, solubility in non-polar solvents (CCl_4) increases with increase in magnitude of van der Waals force or molecular weight.

$$I_2 > Br_2 > F_2 > Cl_2$$

16. C_6H_6 being non-polar, does not dissolve in benzene.

20. Molten NaCl can show electrical conductance due to free ions and graphite can show conductance due to the presence of free electrons.

Advanced and Olympiads

Single Choice

1. Both CO_2 and SO_2 are covalent molecules. See their structures in the text part.

2. Both HNO_3 and H_2SO_4 have covalent bonds as well as coordinate bonds.

3. A, B, C and D are O, F, Na and Mg respectively. So, the compounds formed by them are C_2A (Na_2O), DA (MgO), CB (NaF) and DB_2 (MgF_2) respectively.

4. $CuSO_4 \cdot 5H_2O$:

$$Cu^{++} + O = \overset{\overset{\displaystyle O}{\|}}{\underset{\underset{\displaystyle O^-}{|}}{S}} - O^- . 5H_2O$$

Ionic and covalent bonds are present in $CuSO_4$ while H_2O molecules are attached by coordinate bonds.

5. Methanol can undergo intermolecular association through H-bonding as the –OH group in alcohols is highly polarised.

$$\underset{\text{...O—H...O—H...O—H...}}{\overset{\displaystyle CH_3 \quad CH_3 \quad CH_3}{|\qquad |\qquad |}}$$

As a result, in order to convert liquid CH_3OH to the gaseous state, the strong hydrogen bonds must be broken.

6. Statements (a), (c), (d) and (e) are correct. Statement (b) is incorrect as a covalent bond is a bond with van der Waals force of attraction.

7. Oxygen can show a valency of 2 but not of 6.

8. H_3O^+ can have both covalent and coordinate bonds.

Multiple Choice

9. BCl_3 and PCl_5 do not follow the octet rule. For details see the text part.

Integer Type

27. BCl_3 ($6\ e^-$), $ZnCl_2$ ($4\ e^-$), NO (odd e^-), NO_2 (odd e^-), PCl_5 ($>8e^-$) and SF_6 (>8) don't follow the octet rule.

28. PCl_5 has 5 covalent bonds.

29. It has 2 double bonds as shown.

$$\underset{\pi \quad\ \ \pi}{O = C = O}$$

30.

$$\underset{\underset{H\ \ H}{|\ \ |}}{\overset{\overset{H\qquad\quad H}{\diagdown\qquad\diagup}}{H - C - C - H}}$$

$1C - C \rightarrow 1$ single bond

$6C - H \rightarrow 6$ single bonds

31. CO_2, NH_3, SO_3, C_2H_5OH, HF, NaOH and KCl are soluble in water.

32. NaOH, KCN, KNO_3, $MgSO_4$ and sucrose can have covalent bonds.

33. NO ($15\ e^-$), NO_2 ($23\ e^-$) and ClO_2 ($33\ e^-$) have an odd number of electrons.

34. Statements (a), (c), (e) and (f) are correct. For details see the text part.

CHAPTER 8

Gaseous State

Introduction

Chemistry is the branch of science dealing with the study of matter from various perspectives. On the basis of physical composition, matter can be divided into three states—gaseous state, liquid state and solid state. The state of matter in which inter-particle attraction is the weakest and inter-particle distance is large enough so that the particles can be completely free to move randomly in the entire available space is known as the gaseous state. A gas can occupy all the space of the vessel in which it is kept. All gases behave uniformly under similar conditions of temperature and pressure irrespective of their chemical composition or colour.

Kinetic Theory of Gases

Kinetic theory of gases was put forward by Bernoulli and developed by Clausius Kelvin. It was explained by Maxwell and Boltzmann. It is also known as the *dynamic particle model* and *microscopic model*.

Main Assumptions

(i) Each gas is composed of a large number of tiny particles called molecules and these molecules are identical in mass, size and shape.

(ii) The volume of gaseous molecules is negligible compared to the volume of the gas.

(iii) Molecules show random motion in straight lines in all possible directions but at a constant speed (Fig. 8.1).

(iv) Molecules undergo collisions which are perfectly elastic—there is no change in energy.

(v) The force of attraction between molecules is negligible. Thus, gas molecules can move freely and independently of each other.

(vi) The effect of gravity on molecular motion is negligible.

(vii) Pressure of the gas is due to collisions of molecules with the walls of the container.

(viii) At any particular instance, molecules have different speeds and so have different kinetic energies. However, the average kinetic energy of these molecules is directly proportional to the absolute temperature.

Kinetic energy $\propto$ absolute temperature (T)

$\propto$ mass, nature of gas

Kinetic Energy and Average Kinetic Energy

$$KE = \frac{3}{2}RT$$

$$(KE)_{\text{ave.}} = \frac{3}{2}\frac{RT}{N} = \frac{3}{2}kT$$

Here, $k = R/N = $ Boltzmann constant

$$= 1.38 \times 10^{-23} \text{ J K}^{-1} \text{ molecule}^{-1}$$

$N = $ number of gas molecules.

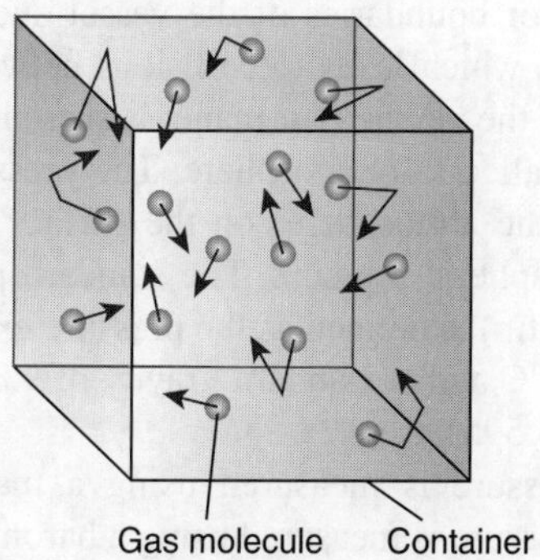

Fig. 8.1 Random motion of molecules

Illustration

1. Find the kinetic energy of 10 mol of gas at 200°C.

   Solution:

   $$KE = \frac{3}{2} RT$$

   where $R = 8314 \times 10^7$ erg/K/mol, $T = 200 + 273 = 473$ K and $n = 10$ mol of gas.

   $$KE = \frac{3}{2} \times 10 \times 8314 \times 10^7 \times 473$$

   $$= 58987.83 \times 10^7 \text{ erg}$$

Characteristics of a Gas

Some important characteristics of gases are discussed below.

 (i) **Composition:** Gases are always non-metals. For example, O_2, N_2, He, Cl_2.

 (ii) **Homogeneous nature:** Gases have similar composition throughout, so they are homogeneous in nature regardless of the identities and proportions of the constituents. A gaseous mixture is always homogenous as it has only the gas phase.

(iii) **Shape and volume:** The molecules of gases have maximum energy and minimum force of attraction so don't have a definite volume or shape. As gases have neither a fixed shape nor a fixed volume, they acquire the shape and volume of the vessel in which they are kept. Since gaseous molecules have only weak interactions between each other, they can move freely and occupy any space.

(iv) **Compressibility:** Gases can be compressed easily by applying external pressure as the space between gas molecules is large due to weak intermolecular forces between them. That is, after compression, gas molecules come closer to each other and occupy less space.

 (v) **Density:** Gases have very low densities, they are extremely light and they have large intermolecular spaces. the density of gases with respect to hydrogen is called relative density.

(vi) **Diffusion:** Gaseous can diffuse easily as the gaseous molecules are weakly held and have enough kinetic energy to diffuse. The gases can fill their container completely. For example, if LPG (Liquified Petroleum Gas) leaks, its molecules can spread all around the kitchen and it can be discovered by the smell.

   When someone opens a bottle of perfume in one corner of a room, its smell spreads thoughout the room quickly. The smell of perfume spreads due to the diffusion of perfume vapours into the air (Fig. 8.2).

(vii) **Expansion:** Gases are highly expansible. When the temperature is increased and pressure is decreased, the molecules will move further apart. So, the volume of the intermolecular space will increase.

(viii) **Liquefaction:** A gas can be liquefied by cooling and by applying pressure.

Standard Variables for Gases

The physical behaviour of a gas can be explained by certain standard variables—(i) volume (V), (ii) pressure (P) and (iii) temperature (T).

All gases are similar in their physical behaviour and they can expand and contract equally under similar conditions of temperature and pressure. For a given mass of a gas, a change in one or more than one variable (pressure, volume and temperature), results in a change in the remaining variables as well.

Volume

The volume (V) of a gas is equal to the volume of the container in which gas is taken as a gas tries to occupy the entire available space. It is expressed in litres, m^3, mL or cm^3.

Units of Volume

In the SI system, the volume of a gas is measured in cubic metres (m^3). Other units are cubic centimetre (cm^3), millilitre (mL), cubic decimetre (dm^3) and litre (L).

Relationship between units:

$$1 \text{ m}^3 = 1000 \text{ dm}^3 = 1000 \text{ litres}$$
$$1 \text{ Litre} = 1000 \text{ mL} = 1000 \text{ cm}^3$$

Pressure

The pressure (P) of a gas is the force that the gas exerts per unit area on the walls of its container. A gas can exert pressure on all the interior boundaries of the vessel due to the motion of its molecules which leads to collisions between themselves and the walls of the vessel. Our planet is surrounded by a thick blanket of air called the atmosphere. The pressure exerted by air (present in the atmosphere) on the surface of the earth is known as atmospheric pressure. The standard pressure of one atmosphere (1 atm) is defined as the pressure exerted by 76 cm of mercury at 0°C and at standard gravity of 9.8 m s^{-2} (density of mercury = 13.5951 g cm^{-3}).

Gaseous pressure is measured using a manometer while atmospheric pressure is measured using a barometer.

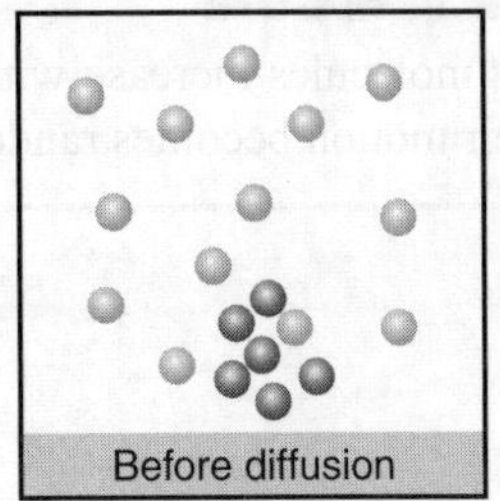
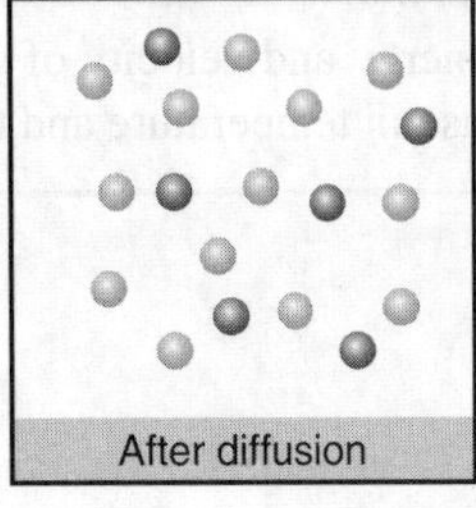

Fig. 8.2 Diffusion of gases

Units of Pressure

Gaseous pressure is measured in newton per m^2 or nm Hg or atmosphere or torr. The SI unit of pressure is pascal (Pa). Pascal (Pa) is defined as the pressure exerted when a force of 1 newton acts on an area of 1 m^2. Pascal (Pa), atmosphere (atm), in centimetres of mercury (cm Hg), in millimetres of mercury (mm Hg), torr (named after Torricelli) are the units.

Relationship between these units:

$$1 \text{ atm} = 76.0 \text{ cm Hg} = 760 \text{ mm Hg} = 760 \text{ torr}$$
$$= 1.013 \times 10^5 \text{ N/m}^2 \text{ or pascal (Pa)}$$
$$= 1.013 \times 10^6 \text{ dyne/cm}^2$$
$$= 1.013 \text{ bar}$$

Temperature

The temperature (T) of a gas is defined as the degree of heat of that gas. Temperature of a gas tells the energetic level of the gaseous molecules.

Scale of Temperature

The most commonly used scale of temperature is the celsius scale, formerly known as the centigrade scale. The zero on the celsius scale is purely arbitrary. In other words, the temperature of a substance may go below zero, that is, a substance can have negative temperature. The behaviour of gases show that it is not possible to have a temperature below −273°C. This point has led to the formulation of another scale called the kelvin scale. The zero on this scale corresponds to −273°C.

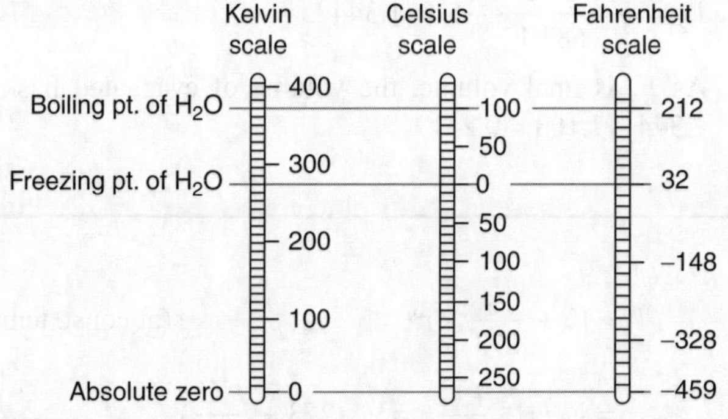

Relationship between kelvin (K) and degree celsius (°C)

Any celsius scale value can be converted into the kelvin scale value by adding 273 to it.

$$\text{kelvin (K)} = °C + 273$$

For example, 43°C = 43 + 273 = 316 K

Temperature in Fahrenheit

A temperature interval of 1°F is equal to an interval of 5/9 degrees celsius. The fahrenheit and celsius scales intersect at −40° (−40°F = −40°C).

	From Fahrenheit	To Fahrenheit
Celsius	$[°C] = ([°F] - 32) \times \dfrac{5}{9}$	$[°F] = [°C] \times \dfrac{9}{5} + 32$

Standard Temperature and Pressure (STP)

As the volume of a gas changes remarkably with the change in temperature and pressure, it becomes essential to choose standard values of temperature and pressure to which a gas volume can be referred.

The standard values preferred are 0°C or 273 K for temperature and 1 atmospheric unit (atm) or 760 mm Hg for pressure. These standard values are called standard temperature and pressure (STP).

Standard temperature = 0°C = 273 K

Standard pressure = 760 mm Hg = 76 cm Hg = 1 atm

Mass

The mass of a gas can be determined by the usual weight determination methods using weighing difference methods. It is related to number of moles of the gas.

$$n \text{ (mol)} = \frac{W}{m}$$

Here W = weight in g, m = molar mass of the gas

Gas Laws

As gases show their dependency on pressure, volume and temperature, the relationship of these factors can be explained through the gas laws.

Boyle's Law

In 1662, Robert Boyle systematically studied the relationship between the pressure and volume of gases and observed that at constant temperature, the volume of a fixed mass of a dry gas decreased by half when the pressure on the gas was doubled and it became four times its original volume when its pressure was decreased to one-fourth. On the basis of this observation, he introduced a law known as Boyle's law according to which, at constant temperature, the volume of a given mass of a gas is inversely proportional to its pressure,

$$V \propto \frac{1}{P}$$

Or

$$V = \frac{K}{P}$$

Here, K = constant.

PV = constant (K) at constant temperature

At constant temperature,

$$P_1V_1 = P_2V_2$$

Initial conditions   Final conditions

Relation Between Density and Pressure

According to Boyle's law at constant temperature and constant mass

$$V \propto \frac{1}{P} \qquad \text{(as } T \text{ and mass are constant)}$$

$$V \propto \frac{1}{d} \qquad \text{(}d\text{ is the density)}$$

As
$$V = \frac{\text{Mass}}{\text{Density}}$$

So
$$\frac{1}{d} \propto \frac{1}{P}$$

That is, $d \propto P$ or $d = \dfrac{K}{P}$

$$\boxed{\frac{d_1}{P_1} = \frac{d_2}{P_2}}$$

KNOWLEDGE BOOSTER

Significance of Boyle's law

- Air at the sea level is dense as it is compressed by the mass of air present above it but the density and pressure decrease with an increase in altitude.
- The size of weather balloons becomes larger and larger as it rises or goes to higher altitude as the external pressures become lower and lower.

Various Plots of Pressure vs Volume

Boyle's law can be graphically verified by plotting graphs of P versus V and these plots are called *isotherms*.

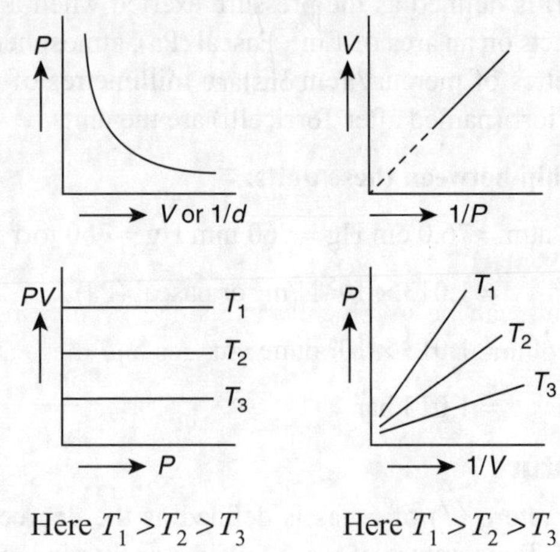

Here $T_1 > T_2 > T_3$ Here $T_1 > T_2 > T_3$

Illustration

1. A 1.103 L flask containing nitrogen at a pressure of 710.6 torr is connected to an evacuated flask of unknown volume. The nitrogen which acts ideally is allowed to expand into the combined system of both flasks isothermally. If the final pressure of nitrogen is 583.1 torr, determine the volume of the evacuated flask.

   Solution: As the flasks are isothermally combined, the temperature is constant. According to Boyle's law

   $$P_1V_1 = P_2V_2$$
   $$(710.6 \text{ torr})(1.103 \text{ L}) = (583.1 \text{ torr}) \times V_2$$
   $$V_2 = \frac{710.6 \times 1.103}{583.1} \text{ L} = 1.344 \text{ L}$$

   As V_2 is final volume, the volume of evacuated flask is $1.344 - 1.103 = 0.241$ L

Charles' Law

In 1787 J Charles studied the relationship between the volumes of gases and temperatures. In order to explain this law, let us consider the following plot between volume and temperature.

It shows that the volume of a definite amount of a gas varies linearly with temperature on the celsius scale.

It can be given as $V_t = a + bt$

Here, a and b are constants.

$$a = V_0 \text{ at } 0°C$$

$$b = \text{slope} \left(\frac{\partial V}{\partial T} \right)_P$$

So according to Charles' law, at constant pressure, the volume of a given mass of a gas increases or decreases by 1/273 of its volume of 0°C for every one degree centigrade rise or fall in temperature.

$$V_t = V_0 + \frac{V_0}{273} \times t°C \qquad \text{(at const. temp.)}$$

$$= V_0 \left(1 + \frac{t°C}{273}\right) = V_0 \left(273 + \frac{t°C}{273}\right)$$

$$\frac{V_0 T}{273}$$

Therefore,

$$V_t \propto T \quad \text{or} \quad V \propto T \text{ at constant pressure}$$

Or
$$\frac{V}{T} = \text{constant}$$

Hence, at constant pressure, the volume of a given mass of a gas is directly proportional to its temperature in kelvin.

At constant pressure

$$\boxed{\frac{V_1}{T_1} = \frac{V_2}{T_2}}$$

Relation Between Density and Temperature

As

$$V \propto \frac{1}{d} \quad (d = \text{density}),$$

$$V \propto T$$

So

$$dT = \text{constant}$$

That is,

$$\boxed{d_1 T_1 = d_2 T_2}$$

Plots of *V* and *T*

Charles' law can be graphically explained by plotting graphs between volume and temperature of a gas and these are called *isobars*.

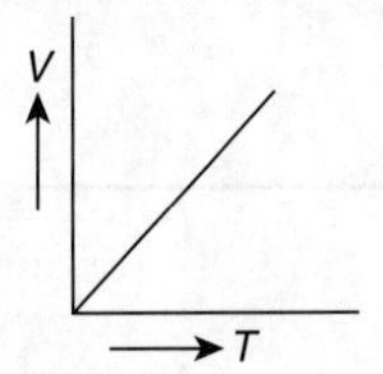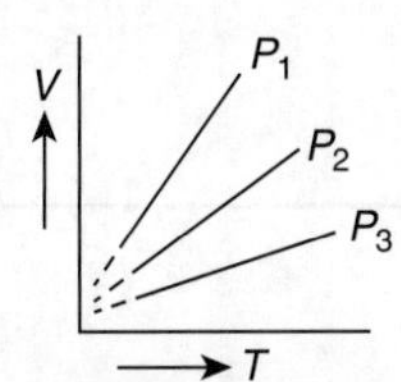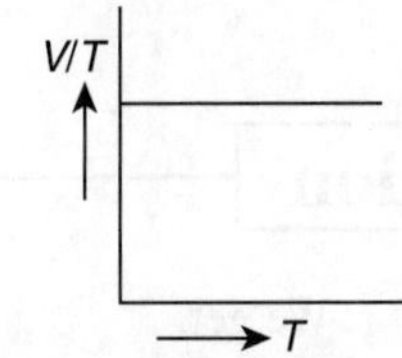

$$\text{Here } P_1 < P_2 < P_3$$

As $P \propto T$

Illustrations

1. To what temperature must an ideal gas be heated under isobaric conditions to increase the volume from 100 cm³ at 25°C to 1.00 dm³?

   Solution: According to Charles' law

   $$\frac{V_1}{T_1} = \frac{V_2}{T_2}$$

   $$\frac{100 \text{ cm}^3}{300 \text{ K}} = \frac{1.00 \times 10^3 \text{ cm}^3}{T_2}$$

   $$T_2 = \frac{1000 \times 298}{100} = 2980 \text{ K}$$

   Temperature = 2980 – 273 = 2707°C

2. A flask has a capacity of one litre. What volume of air will escape from the flask if it is heated from 27°C to 37°C? Assume that the pressure is constant.

   Solution:

   $T = 27°C = 300 \text{ K}$

   $T_1 = 37°C = 310 \text{ K}$

   $V = 1 \text{ litre}$

   $V_1 = ?$

   $$\frac{V}{T} = \frac{V_1}{T_1} \quad \text{(At constant pressure)}$$

   $$\frac{1}{300} = \frac{V_1}{310}$$

   $$V_1 = \frac{310}{300} = 1.0333 \text{ litre}$$

   As capacity of the flask is 1 litre.

   So, volume of air that has escaped = 1.033 – 1 = 33.3 mL

Pressure–Temperature Law (Gay-Lussac's Law)

In order to explain this law, let us consider the graph. It shows that the pressure of a definite amount of a gas at constant volume varies linearly with temperature on the celsius scale.

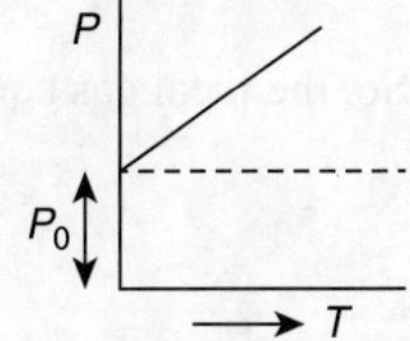

It can be given as $P_t = a + bt$

Here, *a* and *b* are constants.

$$a = P_0 \text{ at } 0°C$$

$$b = \text{slope} \left(\frac{\partial P}{\partial T}\right)_V$$

According to the pressure–temperature law, at constant volume, the pressure of a given mass of a gas increases or decreases by $\dfrac{1}{273}$ of its pressure at 0°C for every 1°C rise or fall in temperature.

$$P_t = P_0 + \frac{P_0}{273} \times t°C$$

$$P_0 \left(\frac{273 + t°C}{273} \right) = \frac{P_0 T}{273}$$

$$P \propto T \text{ that is, } \frac{P}{T} = \text{constant}$$

At constant volume,

$$\boxed{\frac{P_1}{T_1} = \frac{P_2}{T_2}}$$

Plots of *P* and *T*

These curves are called *isochores* and the slope will be greater for lower volumes.

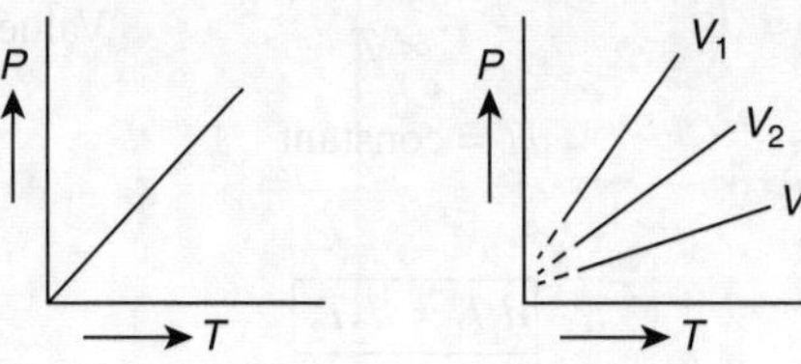

Here $V_3\, V_2 > V_1$

As $\qquad\qquad V \propto T \qquad\qquad$ (that is, $T_1 < T_2 < T_3$)

Illustration

1. A 10.0 L vessel is filled with a gas to a pressure of 2.00 atm at 0°C. At what temperature will the pressure inside the container be 2.50 atm?

   Solution: As the volume of the container remains constant, applying the pressure–temperature law,

   $$\frac{P_1}{T_2} = \frac{P_2}{T_2}$$

   We get,

   $$\frac{2 \text{ atm}}{273 \text{ K}} = \frac{2.50 \text{ atm}}{T_2}$$

   Or $T_2 = 341$

Avogadro's Law

According to this law, at constant temperature and pressure, equal volumes of gases contain equal number of molecules or moles.

$$V \propto N \text{ or } n$$

$$\frac{V}{n} = \text{constant}$$

$$\boxed{\frac{V_1}{n_1} = \frac{V_2}{n_2}} \quad \boxed{\frac{V_1}{N_1} = \frac{V_2}{N_2}}$$

Here, n = number of moles of gas used, N = number of molecules of gas used.

KNOWLEDGE BOOSTER

Don't forget: Avogadro's law is applicable under the conditions of high temperature and low pressure (for an ideal gas).

Combined Gas Laws and Ideal Gas Equation

When both pressure and temperature of a fixed mass of a gas are varied, on combining Boyle's, Charles' and other gas laws we get the following.

As $\qquad\qquad V \propto \dfrac{1}{P} \qquad\qquad$ [Boyle's law]

$\qquad\qquad\qquad V \propto T \qquad\qquad$ [Charles' law]

$\qquad\qquad\qquad V \propto n \qquad\qquad$ [Avogadro's law]

So $\qquad\qquad V \propto \dfrac{nT}{P}$

Or $\qquad\qquad PV \propto nT$

Or $\qquad\qquad PV = nRT = nST$

$$PV = \frac{w}{M}\, RT$$

$$P = \frac{w}{MV}\, RT$$

$$P = CRT \qquad \left[C = \frac{w}{MV} \right]$$

$$P = \frac{d}{M}\, RT \qquad \left[d = \frac{w}{V} \right]$$

As $\qquad\qquad \dfrac{PV}{T} = \text{constant}$

So, the ideal gas equation can be written as

$$\frac{P_1 V_1}{T_1} = \frac{P_2 V_2}{T_2}$$

KNOWLEDGE BOOSTER

R or *S*: molar gas constant or universal gas constant

Values of $R = 0.0821$ L atm K^{-1} mol^{-1}

$= 8.314$ joule K^{-1} mol^{-1}

$= 8.314 \times 10^7$ erg K^{-1} mol^{-1}

$= 2$ cal K^{-1} mol^{-1}

Illustrations

1. Calculate the density of CO_2 at 100°C and 800 mm Hg pressure.

Solution:

$$P = \frac{800}{760} \text{ atm}$$

$$T = 373 \text{ K}$$

$$PV = \frac{w}{M} RT$$

$$P = \frac{d}{m} RT \qquad \left(\text{As } \frac{w}{v} = \text{density}\right)$$

$$\frac{800}{760} = \frac{d}{44} \times 0.0821 \times 300$$

$$d = 1.5124 \text{ g L}^{-1}$$

2. How large a balloon could you fill with 4.0 g of He gas at 22°C and 720 mm of Hg?

Solution:

$$P = \frac{720}{760} \text{ atm}$$

$$T = 295 \text{ K}$$

$$w = 4 \text{ g}$$

$$m = 4 \text{ (He)}$$

$$PV = \frac{w}{m} RT$$

$$V = \frac{wRT}{m \times P} = \frac{4 \times 0.0821 \times 295 \times 760}{4 \times 720}$$

$$= 25.565 \text{ litre}$$

3. A certain amount of an ideal gas occupies a volume of 10 m^3 at a given temperature and pressure. What would be its volume after reducing its pressure to half the initial value and raising the temperature to twice the initial value?

Solution:

$$P_2 = 0.5P_1$$

$$T_2 = 2T_1$$

$$\frac{P_1V_1}{T_1} = \frac{P_2V_2}{T_2}$$

$$\frac{P_1V_1}{T_1} = \frac{0.5P_1V_2}{2T_1}$$

$$V_2 = 4 V_1$$

$$V_1 = 1 \text{ m}^3$$

$$V_2 = 4.0 \text{ m}^3$$

Dalton's Law of Partial Pressure

When a gas like N_2 or H_2 which is insoluble or less soluble in water is collected over water by the downward displacement method (Fig. 8.3), it mixes with water vapour; so, the correct pressure of moist gas is given as

$$P_{\text{moist gas}} = P_{\text{dry gas}} + P_{\text{VP of water}}$$

$$P_{\text{dry gas}} = P_{\text{moist gas}} - P_{\text{VP of water}}$$

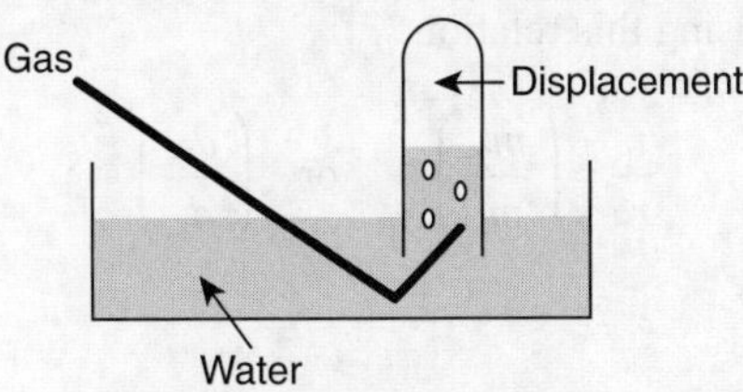

Fig. 8.3 Collection of a gas over water

Aqueous tension is partial pressure of water vapour in moist gas. Vapour pressure of water varies with temperature. For example, At 0°C it is 4.6 torr while at 25°C it is 23.8 torr.

On the basis of this observation, Dalton introduced his law according to which, the total pressure of a mixture of non-reacting gases is equal to the sum of partial pressure of these gases at constant temperature and constant volume.

$$P_{mix} = P_1 + P_2 + P_3 \ldots\ldots$$

Here P_{mix} = pressure of the gaseous mixture

P_1, P_2, P_3 = partial pressure of gases

$$\text{Partial pressure of any gas} = \frac{\% \text{ of that gas}}{100} \times P_{\text{mix}}$$

% of a gas in a mixture =

$$\frac{\text{Partial pressure of the gas}}{\text{Total pressure of gaseous mixture}} \times 100$$

- Partial pressure of any component A is given as

$$P_A = \frac{\text{moles of } A}{\text{Total moles}} \times P_{\text{Total}}$$

- Total pressure of a mixture having different components is given as

$$P_{\text{mix}} = (n_1 + n_2 + n_3 \ldots)\,\frac{RT}{V}$$

$$P_{\text{mix}} = \left(\frac{w_1}{m_1} + \frac{w_2}{m_2} + \frac{w_3}{m_3} \ldots \frac{RT}{V}\right)$$

Here w_1, w_2, w_3 = weight of each component or non-reacting gas, and m_1, m_2, m_3 are their molar masses, T = temperature in kelvin and V = volume in litre.

KNOWLEDGE BOOSTER

Don't forget: Dalton's law is not applicable for a mixture of reacting gases like N_2 and O_2, SO_2 and O_2.

Illustration

1. Calculate the total pressure in a 10 litre cylinder which contains 0.4 g of helium, 1.6 g of oxygen and 1.4 g of nitrogen at 27°C. Also calculate the partial pressure of helium gas in the cylinder. Assume ideal behaviour of gases. Given $R = 0.082$ litre atm K^{-1} mol^{-1}.

Solution:

Given $V = 10$ litre

$$\text{Moles of He} = \frac{0.4}{4} = 0.1$$

$$\text{Moles of O}_2 = \frac{1.6}{32} = 0.05$$

$$\text{Moles of N}_2 = \frac{1.4}{28} = 0.05$$

Total moles = $0.1 + 0.05 + 0.05 = 0.2$

As $PV = nRT$

$$P_{\text{total}} = \frac{nRT}{V} = \frac{0.2 \times 0.082 \times 300}{10}$$

$$= 0.492 \text{ atm}$$

P_{He} = Mole fraction of He × Total pressure

$$= \frac{0.1}{0.2} \times 0.492 = 0.246 \text{ atm}$$

Graham's Law of Diffusion

According to this law, at constant pressure and temperature, the rate of diffusion of a gas is inversely proportional to the square root of its density or molecular weight. It is applicable only at low pressure.

$$r \propto \frac{1}{\sqrt{M}} \quad \text{or} \quad \frac{1}{\sqrt{d}}$$

Here r = rate of diffusion or effusion of a gas or liquid, M and d are the molecular weight and density respectively.

KNOWLEDGE BOOSTER

Diffusion: It is the movement of gaseous or liquid molecules without any porous bars, that is, spreading of molecules in all directions.

Effusion: It is movement of gas molecules or liquid molecules through a porous bar, that is, a small hole or orifice.

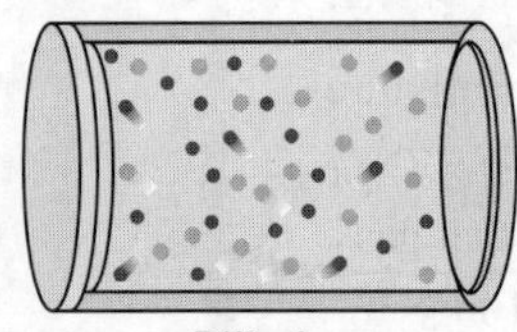

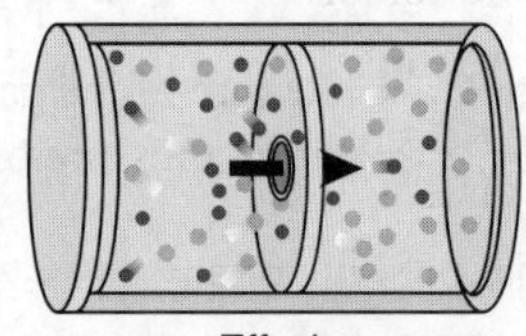

Hence the diffusion or effusion of a gas or gaseous mixture is directly proportional to the pressure difference of the two sides and is inversely proportional to the square root of the gas or mixture effusing or diffusing out.

For any two gases, the ratio of rate of diffusion at constant pressure and temperature can be shown as

$$\frac{r_1}{r_2} = \sqrt{\frac{M_2}{M_1}} \quad \text{or} \quad \sqrt{\frac{d_2}{d_1}}$$

If pressures are different and we take two gases, then,

$$\frac{r_1}{r_2} = \frac{P_1}{P_2} \cdot \left(\sqrt{\frac{M_2}{M_1}} \quad \text{or} \quad \sqrt{\frac{d_2}{d_1}}\right)$$

Uses of Graham's Law

(i) Detecting the presence of marsh gas in mines.

(ii) Separation of isotopes by different diffusion rates. For example, U-235, U-238.

(iii) Detection of molecular weight and vapour density of gases using this relation.

$$\frac{r_1}{r_2} = \left(\frac{m_2}{m_1}\right)^{1/2} \quad \text{or} \quad \left(\frac{d_2}{d_1}\right)^{1/2}$$

Illustrations

1. What is the molecular mass of a gas that diffuses through a porous membrane 1.86 times faster than Xe? What might the gas be?

Solution:

$$\frac{r_{Xe}}{r_Z} = \sqrt{\frac{M_Z}{M_{Xe}}}$$

$$\frac{1}{1.86} = \sqrt{\frac{M_Z}{131.29}}$$

$$\sqrt{M_Z} = \sqrt{\frac{131.29}{1.86}}$$

$$M_Z = \frac{131.29}{(1.86)^2} = 37.9 \text{ g/mol}$$

Molecular mass = 37.9 amu

So the gas may be F_2.

2. 20 dm^3 of SO_2 diffuses through a porous partition in 60 s. What volume of O_2 will diffuse under similar conditions in 30 s?

Solution:

$$r_{SO_2} = \frac{V_{SO_2}}{t_{SO_2}} = \frac{20}{60} = 0.333 \text{ dm}^3/s$$

$$r_{O_2} = \frac{V_{O_2}}{t_{O_2}} = \frac{V}{30} \text{ dm}^3/s$$

$$\frac{r_{SO_2}}{r_{O_2}} = \sqrt{\left(\frac{M_{O_2}}{M_{SO_2}}\right)}$$

$$\frac{0.333}{\dfrac{V}{30}} = \sqrt{\left(\frac{32}{64}\right)}$$

$$V = 14.14 \text{ dm}^3$$

CHAPTER AT A GLANCE

- A **gas** can occupy all the space of the vessel in which it is kept. All gases can behave uniformly under similar conditions of temperature and pressure irrespective of their chemical composition or colour.

- Each gas is composed of a large number of **tiny particles** called **molecules** and these molecules are identical in mass, size and shape.

- At any particular instant, molecules have different speeds and so have different **kinetic energies**. However, the average kinetic energy of these molecules is directly proportional to the **absolute temperature**.

- Gases form homogenous mixtures, have no definite shape or volume, have high compressibility and diffusion rate.

- The **temperature (T)** of a gas is defined as the degree of 'hotness' of that gas. Temperature of a gas is an indicator of the energetic level of the gaseous molecules.

- **Relationship between kelvin (K) and degree celsius (°C).** Any celsius scale values can be converted into kelvin scale values by adding 273 to the degree celsius value. kelvin (K) = °C + 273.

- **Boyle's law:** At constant temperature, the volume of a given mass of a gas is inversely proportional to its pressure.

  At constant temp.

  P_1V_1 (Initial conditions) $= P_2V_2$ (Final conditions)

- Boyle's law can be graphically verified by plotting graphs between P versus V and these plots are called **isotherms**.

- **Charles' law:** At constant pressure, the volume of a given mass of a gas increases or decreases by 1/273 of its volume ot 0°C for every one-degree centigrade rise or fall in temperature, or, at constant pressure, the volume of a given

mass of a gas is directly proportional to its temperature in kelvin.

At constant pressure $= \dfrac{V_1}{T_1} = \dfrac{V_2}{T_2}$

- Charles' law can be graphically explained by plotting graphs between volume and temperature of a gas and these are called **isobars**.

- **Gay-Lussac's law:** At constant volume, the pressure of a given mass of a gas increases or decreases by 1/273 of its pressure at 0°C for every 1°C rise or fall in temperature.

At constant volume $= \dfrac{P_1}{T_1} = \dfrac{P_2}{T_2}$

- Plots between pressure and temperature at constant volume are called **isochores** and the slope will be greater for lower volumes.

- **Avogadro's law:** At constant temperature and pressure, equal volumes of gases contain the same number of molecules or moles.

- **Combined gas laws and ideal gas equation:** When both pressure and temperature of a fixed mass of a gas are varied, on combining Boyle's, Charles' and other gas laws, we get the combined gas law ($PV \propto nT$).

- **Dalton's law:** Total pressure of a mixture of non-reacting gases is equal to the sum of the partial pressures of these gases at constant temperature and constant volume.

  $P_{mix} = P_1 + P_2 + P_3 \ldots\ldots$

- **Graham's law:** At constant pressure and temperature, the rate of diffusion of a gas is inversely proportional to the square root of its density or molecular weight.

- **Diffusion:** It is the movement of gaseous or liquid molecules without any porous bars—spreading of molecules in all directions.
- **Effusion:** It is movement of gas molecules or liquid molecules through a porous bar—a small hole or orifice.
- **Various plots:**

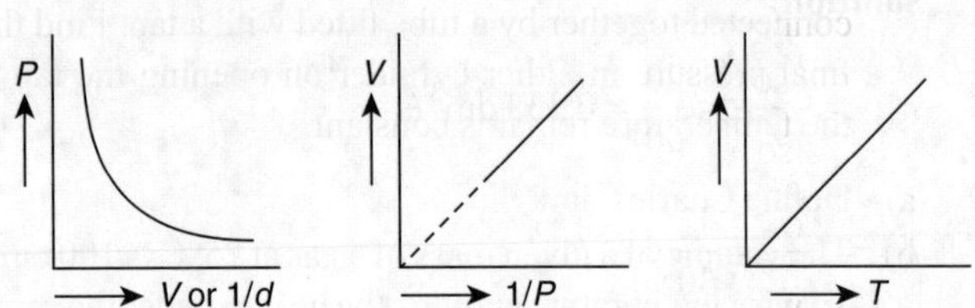

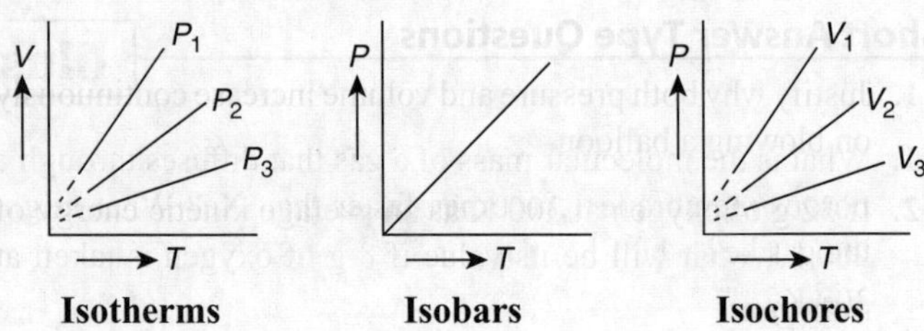

PRACTICE QUESTIONS

Analyse Your Concepts (School Exam Based)

Fill in the Blanks

Instructions: Complete the following statements with an appropriate word/term to be filled in the blank spaces.

1. The temperature of a gas is defined as the degree of ______________ of the gas.

2. Gases are ______________ compressible.

3. If temperature of a gas is 37°C, its value in kelvin is ______________.

4. At constant temperature if the pressure of a gas increases, then its volume ______________.

5. The collisions between gas molecules are perfectly ______________.

6. The average kinetic energy of gas molecules is directly proportional to the ______________.

7. The plots of P versus V at constant temperature are called ______________.

8. The plots of V versus T at constant pressure are known as ______________.

9. The ratio of number of moles of an individual gas to the total number of moles gaseous mixture represents ______________.

10. ______________ formula represents the exact number of atoms of different elements present in 1 molecule of the compound.

True or False

Instructions: Read the following statements and write your answer as true or false.

1. According to Charle's law, 0°C is the lowest possible temperature.

2. The plot of V versus T is a rectangular hyperbola.

3. Gases can be liquefied by cooling.

4. The plots of P versus V at constant temperature are known as isochores.

5. The value of standard temperature is 298 K.

6. The temperature –273° is known as absolute 0.

7. According to Boyle's law, PV is constant at constant temperature.

8. Gases do not have a tendency to mix with one another.

9. At higher altitudes, the air is less dense.

10. Hot air is filled into balloons used for meteorological purposes.

Very Answer Short Type Questions

1. Define absolute zero.

2. Why can gaseous molecules move freely?

3. Find the volume occupied by 6.4 g of SO_2 at STP.

4. What is the effect of an increase in temperature on average kinetic energy?

5. Why can gases exert pressure?

6. What are the plots of pressure (P) versus volume (V) known as?

7. If the pressure of a gas is 3800 mm of mercury, what is its pressure in atmospheres?

8. If a gas has a volume of 3000 L, what will be its value in m^3?

9. Define aqueous tension.

10. Why it is difficult to achieve absolute zero temperature for gases practically?

Short Answer Type Questions

1. Justify why both pressure and volume increase continuously on blowing a balloon.

2. If 32 g of oxygen at 300 K as an average kinetic energy of 200 kJ, what will be its value if 8 g of oxygen is taken at 300 K?

3. Explain why temperature in kelvin is always taken as positive.

4. Find the ratio of rate of diffusion of helium gas with methane at the same temperature and pressure.

5. When N_2 and O_2 react under certain conditions of temperature and pressure to form N_2O_3, what is the volume ratio of the reactant and the product?

6. A monovalent, diatomic gaseous element X reacts with ozone to form its oxide (gaseous) and oxygen. What will be the volume ratio of the reactants and the products, if all volumes are measured at the same temperature and pressure?

7. Why do mountaineers have to carry oxygen cylinders with them?

8. A given mass of a gas occupies 300 mL at 800 mm of Hg. What volume will the gas occupy if the pressure is increased to 1200 mm of Hg, the temperature remaining constant?

9. If we take CO_2 and N_2O at 300 K and 1 atm pressure, what is the relation in rate of diffusion for these gases?

10. 6 g ethane is confined in a bulb of one litre capacity. The bulb is so weak that it will burst if the pressure exceeds 10 atm. At what temperature will the pressure of the gas reach the bursting value?

Long Answer Type Questions

1. (a) Define Boyle's law.
   (b) The capacity of one cylinder is 4 dm^3 and that of the other is 1 dm^3; the pressure in the first cylinder is 560 mm Hg and that in the second one is 1000 mm Hg, both of these cylinders having carbon dioxide are connected together by a tube fitted with a tap. Find the final pressure in either cylinder on opening the tap if the temperature remains constant.

2. (a) Define Charles' law.
   (b) The volume of a given mass of a gas at 25°C is 100 cm^3. To what temperature should it be heated under the same pressure, so that it occupies a volume of 125 cm^3?

3. (a) Write the value of standard temperature and standard pressure at STP.
   (b) Why it is necessary to specify both pressure and temperature while expressing the specific volume of a gas?
   (c) A gas is enclosed in a cylinder under STP conditions. At what temperature does the volume of the enclosed gas become 1/8th of its initial volume, the pressure remaining constant?

4. Give the reasons for the following.
   (a) Gases have lower density compared to that of solids or liquids.
   (b) Gases exert pressure in all directions.
   (c) Inflating a balloon seems to violate Boyle's law.
   (d) A gas completely fills the vessel in which it is kept.

5. 20 g of oxygen gas is enclosed in a 1 dm^3 flask at 298 K. Calculate the pressure exerted by the gas, if the molecular mass (molar mass) of any gas occupies 22.4 litre at STP.

COMPETITION WINDOW (OBJECTIVE TYPE)

[For NEET, JEE (Main and Advanced), NTSE, KVPY and Olympiads]

Topic-wise MCQs

Features of gases, Boyle's and Charles' laws, ideal gas equation

1. Which of the following statements is not correct about the states of matter?
   (a) molecules of solids are least energetic
   (b) molecules of gases have least energy
   (c) molecules of liquids have slow molecular motion
   (d) interparticle forces are maximum in the case of solids

2. Which is not a true statement about the gaseous state?
   (a) they are highly compressible
   (b) they are a little compressible
   (c) thermal energy is equal to molecular attraction
   (d) they exert pressure on the walls of the vessel due to collisions

3. Which is the process of separation of individual gases from the gaseous mixture?
   (a) atmolysis (b) effusion
   (c) pyrolysis (d) cataphoresis

4. The plot of which pair of species will be a rectangular parabola:
   (a) PV vs P (b) PV vs P^{-1}
   (c) P vs V^{-1} (d) P vs V

5. Among the plots of P vs V as given below, which one corresponds to Boyle's law?

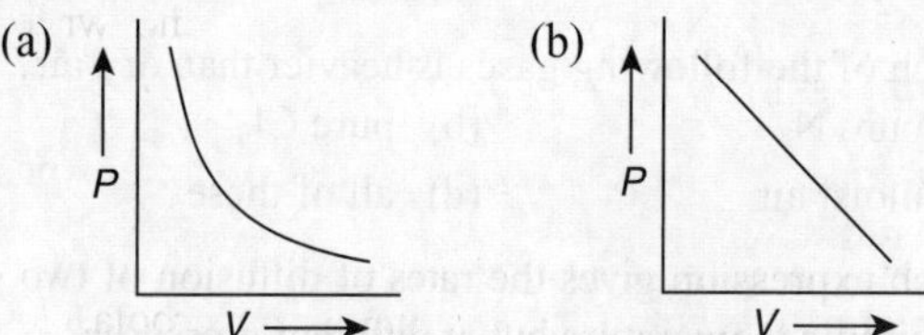

(c)

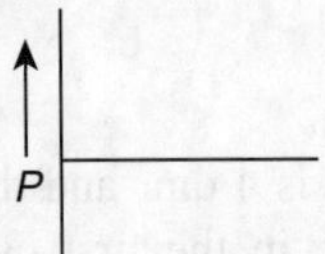

(d) all of these

6. Which of the following plots will be linear with a slope equal to zero in the case of an ideal gas?
   (a) V vs T at constant P
   (b) P vs T at constant V
   (c) V/T vs T at constant P
   (d) log P vs log V at constant T
7. Which of the following is the Avogadro number?
   (a) molecules present in 22.4 L of a gas at 298 K and 1 atm pressure
   (b) molecules present in one g-molecule of gas at any temperature and pressure
   (c) molecules present in 2.24 m^3 of gas at STP
   (d) molecules present in 1 cm^3 of a gas at STP
8. Measured at common temperature and pressure, the volume of same number of gram-molecules of different gases would be same. The above statement refers to
   (a) Gay Lussac's law
   (b) Avogadro's law
   (c) Charles'–Gay Lussac's law
   (d) Amagat's law
9. For ideal gases, isotherm refers to the
   (a) gases at same temperature
   (b) gases at same pressure
   (c) plot of P vs V, at constant T
   (d) gases having the same heat capacities
10. Which expression gives the combined effect of the Boyle's law and Charles' law on a fixed mass of a gas?
    (a) $P_1T_2 = P_2T_1$ (b) $V_2/V_1 = P_1/P_2 \times T_2/T_1$
    (c) $P_2/P_1 = V_1/V_2$ (d) all of these

Dalton's law, Graham's law of diffusion

11. Three gases A, B, C are enclosed in a container and have partial pressures p_1, p_2, p_3 respectively at T kelvin. The total pressure at T kelvin will be
    (a) always equal to $(p_1 + p_2 + p_3)$
    (b) equal to $(p_1 + p_2 + p_3)$ only if the gases do not react with one another
    (c) greater than $(p_1 + p_2 + p_3)$
    (d) less than $(p_1 + p_2 + p_3)$
12. Under similar conditions, which of the following gases will diffuse four times as quickly as oxygen?
    (a) He (b) H_2
    (c) N_2 (d) CH_4
13. Which of the following gases is heavier than dry air?
    (a) pure N_2 (b) pure Cl_2
    (c) moist air (d) all of these
14. Which expression gives the rates of diffusion of two gases at the same temperature but at different pressures?
    (a) $\frac{r_1}{r_2} = \frac{\sqrt{P_1 d_1}}{\sqrt{P_1 d_2}}$ (b) $\frac{r_1}{r_2} = \frac{P_1}{P_2}\frac{\sqrt{d_2}}{\sqrt{d_1}}$
    (c) $\frac{r_1}{r_2} = \frac{P_2}{P_1}\frac{\sqrt{d_2}}{\sqrt{d_1}}$ (d) $\frac{r_1}{r_2} = \frac{P_1^2}{P_2^2}\frac{\sqrt{d_1}}{\sqrt{d_2}}$
15. A gas diffuses at twice the rate of oxygen under the same conditions of temperature and pressure. The molecular mass of the gas will be
    (a) 64 (b) 8
    (c) 16 (d) 2

Kinetic theory of gases, velocities

16. Which of the following is not a postulate of the kinetic theory of gases?
    (a) gas molecules are in a permanent state of random motion
    (b) pressure of the gas is due to molecular impacts on the walls of the container
    (c) the gaseous molecules are perfectly elastic
    (d) the molecular collisions are perfectly elastic
17. Which of the following will not change the rms speed of gas molecules?
    (a) change in temperature
    (b) change in pressure at constant T
    (c) change in volume at constant T
    (d) all of these
18. Which of the following expressions does not give root mean square velocity of a gas at a particular temperature?
    (a) $(3RT/M)^{1/2}$ (b) $\sqrt{(3P/dM)}$
    (c) $(3PV/M)^{1/2}$ (d) $\sqrt{(3P/d)}$
19. Which of the following is true?
    (a) $u_{rms} > \bar{\upsilon} > \alpha$ (b) $u_{rms} < \bar{\upsilon} < \alpha$
    (c) $\bar{\upsilon} > u_{rms} > \alpha$ (d) $\alpha > u_{rms} > \bar{\upsilon}$
20. In the kinetic gas equation, $PV = 1/3\ mNX^2$, X is
    (a) average speed
    (b) most probable speed
    (c) root mean square speed
    (d) any of these

Gas laws

21. At what temperature in the celsius scale, will the volume (V) of a certain mass of gas at 27°C will be doubled, keeping the pressure constant?
    (a) 54°C (b) 327°C
    (c) 427°C (d) 527°C
22. 120 g of an ideal gas of molecular weight 40 mol^{-1} is confined to a volume of 20 L at 400 K. Using $R = 0.0821$ L atm K^{-1} mol^{-1}, the pressure of the gas is
    (a) 4.90 atm (b) 5.02 atm
    (c) 4.92 atm (d) 4.96 atm
23. How many moles of He gas occupy 22.4 litres at 30°C and 1 atmospheric pressure?
    (a) 0.90 (b) 1.11
    (c) 0.11 (d) 1.0

24. 16 g of oxygen and 3 g of hydrogen are mixed and kept at 760 mm pressure and 0°C. The total volume occupied by the mixture will be nearly
(a) 22.4 L (b) 33.6 L
(c) 448 L (d) 44800 mL

25. What will be the partial pressure of H_2 in a flask containing 2 g of H_2, 14 g of N_2 and 16 g of O_2?
(a) 1/2 the total pressure
(b) 1/3 the total pressure
(c) 1/4 the total pressure
(d) 1/16 the total pressure

26. 50 volumes of H_2 take 25 minutes to diffuse out of the vessel. How long will 40 volumes of oxygen take to diffuse out from the same vessel under similar conditions?
(a) 20 min (b) 80 min
(c) 200 min (d) 100 min

27. The total kinetic energy of 2 moles of an ideal gas at 127°C is (use $R = 8.3$ J K^{-1} mol^{-1})
(a) 9.96 kJ (b) 19.92 kJ
(c) 3.32 kJ (d) 39.84 kJ

28. The density of air is 0.00130 g/mL. The vapour density of air will be
(a) 0.00065 (b) 0.65
(c) 14.4816 (d) 14.56

29. Volume of 4.4 g of CO_2 at NTP is
(a) 22.4 L (b) 44.8 L
(c) 2.24 L (d) 4.48 L

Miscellaneous

1. What will be total pressure exerted by 16 g O_2, 22 g CO_2 in a 2 litre flask at 27°C?
(a) 12.3 atm (b) 13.8 atm
(c) 15.6 atm (d) 16.4 atm

2. 5.75 g of a gas has a volume of 3.4 litre at 50°C and 0.94 atm. What is its molecular weight?
(a) 44 (b) 47.6
(c) 64 (d) 32

3. Two similar balloons are filled with H_2 and O_2 at same temperature and pressure. If O_2 escapes at the rate of 65 mL / hour, what will be the rate of escape of hydrogen?
(a) 130 mL /h (b) 230 mL/h
(c) 260 mL/h (d) 32.5 mL/h

4. If 1.5 g of a gas has 195 mL volume at 12°C and 49 mm pressure, its molecular weight is
(a) 278 (b) 270
(c) 178 (d) 275

5. If at a certain temperature, a student wants to decrease the volume of a gas by 5%, how much pressure should he increase?
(a) 15% (b) 8.34%
(c) 5.26% (d) 56%

6. A student has a flask containing air. He heated it from 300 K to 500 K. What % of air has escaped to the atmosphere?
(a) 40% (b) 50%
(c) 30% (d) 100%

7. 14 g of monovalent element X combines with 16 g of oxygen. On the basis of this information, which of the following is a correct statement?
(a) the element X could have the atomic weight of 7 and its oxide is X_2O
(b) the element X could have an atomic weight of 14 and its oxide is XO_2
(c) the element X could have an atomic weight of 7 and its oxide is XO
(d) the element X could have an atomic weight of 14 and its oxide is X_2O

8. At 273 K, it is found that density of N_2 at 5 bar is equal to a gaseous oxide at 2 bar. What will be the molecular weight of the gaseous oxide?
(a) 60 (b) 68.5
(c) 82 (d) 70

9. The amount of a given product calculated to be obtained in a chemical reaction that goes to completion is
(a) the theoretical yield of the reaction
(b) the percent efficiency of the reaction
(c) the yield the reaction
(d) none of these

10. A mixture of gas in a cylinder has 20% CO_2, 15% O_2 and 65% N_2 at 760 mm and 300 K. The partial pressure of each gas will be, respectively,
(a) 152, 114 and 494 mm
(b) 494, 114 and 152 mm
(c) 114, 494 and 152 mm
(d) 210, 214 and 336 mm

11. The volume of O_2 required for the complete combustion of 4 L of CO at NTP is
(a) 1 litre (b) 2 litre
(c) 4 litre (d) 8 litre

12. A 1-litre flask having vapours of CH_3OH at a pressure of 1 atm and 298 K was evacuated till the pressure was 0.0001 mm. Now how many moles of CH_3OH will be left in the flask?
(a) 6.02×10^{21} (b) 3.24×10^{15}
(c) 3.24×10^{18} (d) 6.02×10^{15}

13. V vs T curves at constant pressure P_1 and P_2 for an ideal gas are shown on the right. Which is correct?
(a) $P_1 > P_2$
(b) $P_1 < P_2$
(c) $P_1 = P_2$
(d) all of these

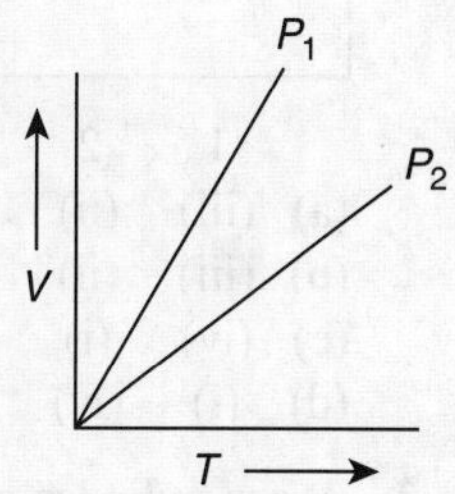

14. A and B are ideal gases. The molecular weights of A and B are in the ratio of $1:4$. The pressure of a gas mixture containing equal weights of A and B is P atm. What is the partial pressure (in atm) of B in the mixture?

(a) $P/5$
(b) $P/2$
(c) $P/2.5$
(d) $3P/4$

15. A mixture of gases contains H_2 and O_2 gases in the ratio of $1:4$ (w/w). What is the molar ratio of the two gases in the mixture?

(a) $4:1$
(b) $16:1$
(c) $2:1$
(d) $1:4$

16. Which of the following is the correct plot of, volume of fixed amount of ideal gas as a function of temperature (at constant pressure)?

(a) 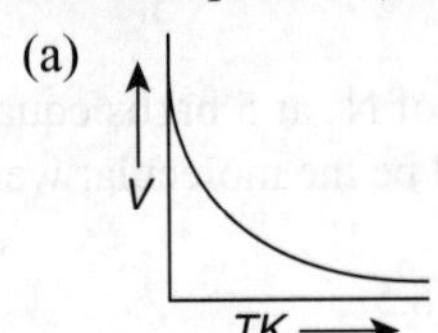
(b) 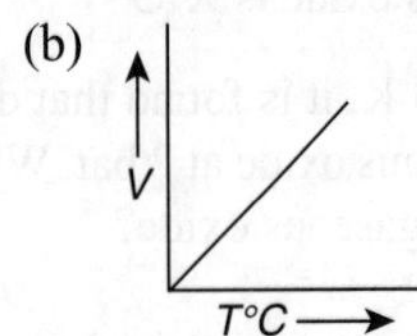
(c) 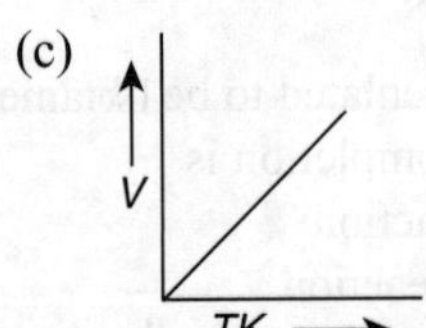
(d) 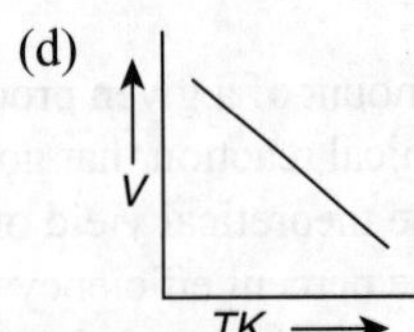

Advanced and Olympiads

Single Choice

1. The pressure of a gas at STP is doubled and the temperature is raised to 546 K. What is the final volume of the gas?

(a) $V_2 = 2V_1$
(b) $V_2 = V_1$
(c) $V_2 = V_1/2$
(d) $V_2 = 4V_1$

2. Match the following

List I	List II
1. RMS speed	(i) $3/2RT$
2. Average speed	(ii) $\sqrt{(8KT/nm)}$
3. Mole fraction	(iii) $\sqrt{(3KT/m)}$
4. Kinetic energy of more atoms of gas	(iv) $n/\sum n$
	(v) $pv = nRT$

$$
\begin{array}{ccccc}
 & 1 & 2 & 3 & 4 \\
\text{(a)} & \text{(iii)} & \text{(ii)} & \text{(i)} & \text{(v)} \\
\text{(b)} & \text{(iii)} & \text{(ii)} & \text{(iv)} & \text{(i)} \\
\text{(c)} & \text{(iv)} & \text{(i)} & \text{(iii)} & \text{(ii)} \\
\text{(d)} & \text{(i)} & \text{(iv)} & \text{(ii)} & \text{(iii)} \\
\end{array}
$$

3. A gas cylinder containing a cooling gas can withstand a pressure of 14.9 atm. The pressure gauge of the cylinder indicates 12 atm at 27°C due to a sudden fire in the building,

its temperature starts rising. At what temperature will the cylinder explode?

(a) 362.5 K
(b) 390 K
(c) 372.5 K
(d) 426 K

4. A bubble of air released by a diver at the bottom of a pool of water becomes larger as it approaches the surface of the water. Assume that the temperature is constant and select the true statement. The pressure inside the bubble is

(a) greater near the bottom of the water
(b) greater near the top of the water
(c) same at all depths
(d) cannot be determined from the given data

5. Consider the graph and select the correct order.

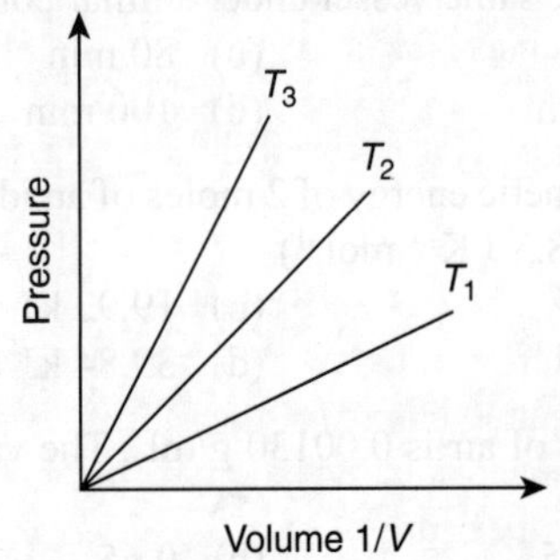

(a) $T_3 > T_2 > T_1$
(b) $T_1 > T_2 > T_3$
(c) $T_3 = T_2 = T_1$
(d) $T_2 > T_3 > T_1$

6. One litre of CO_2 is passed over hot coke. The volume becomes 1.4 litre. The per cent composition of the products is

(a) 0.6 litre CO_2 and 0.8 litre CO
(b) 0.8 litre CO_2
(c) 0.6 litre CO
(d) none of these

Multiple Choice

7. Which of the following statement/s is/are correct?

(a) Gases cannot be directly condensed into solids without passing through the liquid state
(b) Gases and liquids have viscosity as a common property
(c) Gases and liquids have pressure as a common property
(d) Particles in all the three states have random translational motion

8. Which of the following graphs represents Boyle's law?

(a) 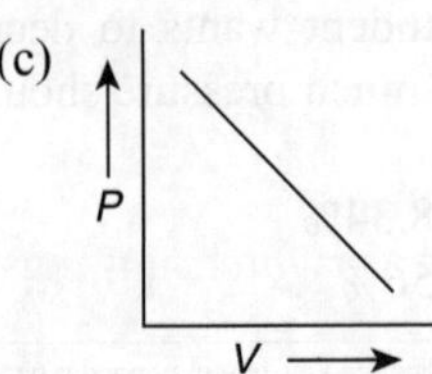
(b) 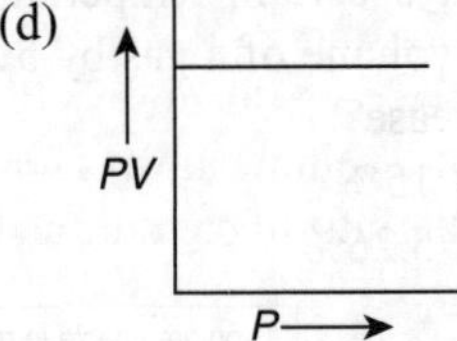
(c) 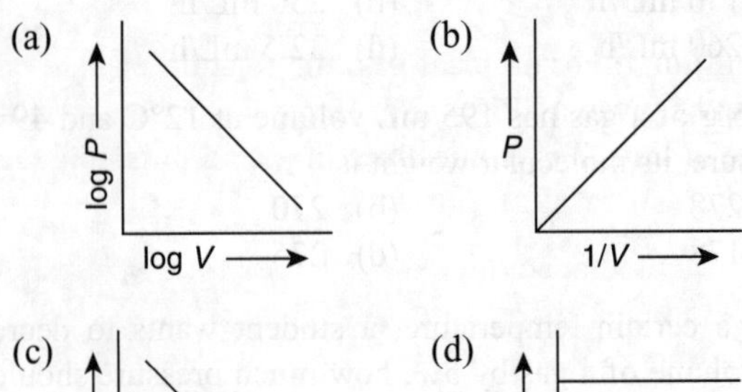
(d)

9. Which of the following statement(s) is/are correct?
   (a) Helium escapes at a rate 2 times as fast as O_2 does.
   (b) Helium escapes at a rate 4 times as fast as SO_2 does.
   (c) Helium escapes at a rate 2.65 times as fast as CO does.
   (d) Helium diffuses at a rate 8.65 times as much as CO does.

10. Select the correct statements
   (a) If the temperature of a gas is 47°C, in the kelvin scale, it is 320 K
   (b) 1 atm = 755 mm of Hg
   (c) Rate of diffusion $\propto \sqrt{\dfrac{1}{\text{density}}}$
   (d) Dalton's law is not applicable for a mixture reacting gases such as NH_3 and HCl

11. Which is/are not correct according to the kinetic theory of gases?
   (a) the pressure exerted by a gas is proportional to the mean square velocity of the molecules
   (b) the pressure exerted by a gas is proportional to the root mean square velocity of the molecules
   (c) the root mean square velocity is inversely proportional to the temperature
   (d) the mean translational KE of the molecule is directly proportional to the absolute temperature

12. Which of the following statements are correct?
   (a) gases are highly expansible
   (b) gases exert pressure in all direction
   (c) gases have low compressibility
   (d) the kelvin scale is also known as the absolute scale of temperature

13. Which of these quantities are the same for all ideal gases at any particular temperature?
   (a) The number of molecules in 1 g
   (b) The kinetic energy of 1 mol
   (c) The kinetic energy of 1 g
   (d) The number of molecules in 1 mol

14. Which of the following masses of gas would occupy about 3 dm^3 at 25°C and 1 atm?
   (a) 2.25 g of Ne
   (b) 8.0 g of SO_2
   (c) 5.5 g of CO_2
   (d) 4.0 g of O_2

Comprehension Type

Comprehension I: For an ideal gas, the relation of combined gas laws can be given as $PV = nRT$. Here, P is pressure in atm, V is volume in litre, T is temperature in kelvin, n is number of moles and R is universal gas constant.

15. At a constant temperature, in a given mass of an ideal gas
   (a) the product of pressure and volume always remains constant
   (b) the pressure always remains constant
   (c) the volume always remains constant
   (d) the ratio of pressure and volume always remains constant

16. Which of the following does not represent the ideal gas equation?
   (a) $PV = RT$
   (b) $PV = nRT$
   (c) $PV = \dfrac{1}{3}mNv$
   (d) $P = \rho\dfrac{RT}{M}$

17. Which of the following is incorrect according to the ideal gas equation?
   (a) $P \propto V$
   (b) $V \propto T$
   (c) $P \propto \dfrac{1}{T}$
   (d) $V \propto n$

Matrix Matching

18. Match the following

Column I	Column II
(a) Plot of P versus V at constant temperature	(p) Isobar
(b) Plot of V versus T at constant pressure	(q) Isochore
(c) Plot of P versus T at constant volume	(r) −273°
(d) Absolute zero	(s) Isotherm

19. Match the following

Column I	Column II
(a) $P \propto \dfrac{1}{V}$	(p) Avogadro's law
(b) $V \propto T$	(q) Gay-Lussac's law
(c) $P \propto T$	(r) Boyle's law
(d) $V \propto n$	(s) Charles' law

Integer Type

20. If the value of pressure of a gas is 4560 mm of mercury and its pressure is X atm., find the value of X.

21. If a gas has a volume of 5000 L and its value is n m^3, what is the value of n?

22. A 2 L flask contains 22 g of carbon dioxide and 1 g of helium at 293 K. Find the partial pressure exerted by CO_2 in atm if the total pressure is 3 atm.

23. Calculate the value of carbon monoxide gas required to react with oxygen to give 6 L of CO_2 gas.

24. Find the volume of O_2 in litre, that can be prepared at 333 K and 1 atm pressure by the decomposition of 20 g of H_2O_2.

25. A sample of gas occupies 8 litre under a pressure of 1 atm. What will be its volume if the pressure is increased to 2 atm.? Assume that the temperature of the gas sample does not change.

Answer Keys

Fill in the Blanks

1. hotness
2. highly
3. 310
4. decreases
5. elastic
6. absolute temperature
7. isotherm
8. isobars
9. mole fraction
10. molecular

True or False

1. F
2. F
3. T
4. F
5. F
6. T
7. T
8. F
9. T
10. T

Very Short Answer Type Questions

1. Absolute zero is $-273°C$ or 0 K. It is the minimum possible temperature for any gas.

2. Gaseous molecules can move freely as there is large inter-particle space due to weak forces of attraction.

3. At STP 64 g of SO_2 has a volume = 22400 mL

   So, 6.4 g of SO_2 has a volume $= \dfrac{6.4 \times 22400}{64} = 2240$ mL
   $= 2.24$ L

4. On increasing temperature, the average kinetic energy increases as

   Average kinetic energy $\propto$ Absolute temperature

5. A gas can exert pressure as gaseous molecules undergo collisions with the walls of the vessel.

6. The plots of Pressure (P) versus Volume (V) are known as isotherms.

7. As 760 mm of Hg = 1 atm
   3800 mm of Hg = 3800/760 = 5 atm.

8. As 1000 L = 1 m^3
   3000 L = 3000/1000 = 3 m^3

9. Aqueous tension is equal to the partial pressure of water vapour.

10. It is difficult to achieve absolute zero temperature for gases practically as the gases liquefy or solidify before reaching this temperature.

Short Answer Type Questions

1. When a balloon is blown continuously, more and more amount of air enters the balloon that increases the volume and the number of collisions of air molecules with the walls of the balloon. This also increases the pressure. Here Boyle's law is not applicable as the amount of gas or air is not fixed.

2. At 300 K, 32 g O_2 has a kinetic energy = 200 kJ
   As kinetic energy $\propto$ Absolute temperature $\not\propto$ mass of gas
   So 8 g and 32 g of O_2 will have same kinetic energy at 300 K.

3. Temperature in kelvin is given as $t°C + 273$. As the lowest possible temperature for any gas is $-273°C$ which means any degree centigrade temperature more than $-273°C$ will be positive in kelvin).

4. $\dfrac{r_{He}}{r_{CH_4}} = \sqrt{\dfrac{M_{CH_4}}{M_{He}}}$

   $= \sqrt{\dfrac{16}{4}} = \dfrac{2}{1}$

   $r_{He} : r_{CH_4} = 2 : 1$

5. The balanced chemical equation for this reaction is:
   $$2N_2(g) \ + \ 3O_2(g) \ \rightarrow \ 2N_2O_3(g)$$

   Gay-Lussac's law   2 volumes   3 volumes $\rightarrow$ 2 volumes
   (At the same P and T)

   Hence volume ratio is $N_2 : O_2 : N_2O_3 = 2 : 3 : 2$.

6. As the element X is monovalent, its oxide will have the formula X_2O
   The balanced chemical equation for the reaction is
   $$X_2(g) \ + \ O_3(g) \ \rightarrow \ X_2O(g) \ + \ O_2(g)$$

   Gay-Lussac's law   1 volume   1 volume   1 volume   1 volume
   (at the same P and T)

   So, under the same conditions of temperature and pressure, the volume ratio is
   $$X_2 : O_3 : X_2O : O_2 = 1 : 1 : 1 : 1.$$

7. Mountaineers have to carry oxygen cylinders with them for breathing since on mountains, the air is less dense as the atmospheric pressure is low. It means less oxygen is available for breathing.

8. $P_1 = 800$ mm, $P_2 = 1200$ mm, $V_1 = 300$ ml, $V_2 = ?$
   Applying Boyle's law $P_1V_1 = P_2V_2$
   Substituting the value
   $800 \times 300 = 1200 \times V_2$

   $V_2 = \dfrac{800 \times 300}{1200} = 200$

   Hence, the new volume = 200 mL

9. Under the same conditions of 300 K and 1 atm pressure, both these gases will have the same rate of diffusion as their molar masses are the same (44). As per Graham's law, the rate of diffusion is inversely proportional to the square root of molar mass.

10. $PV = nRT$

    $10 \times 1 = \dfrac{6}{30} \times 0.082 \times T$

    $T = \dfrac{50}{0.082} = 609.75$ K $= 336.75°C$

Long Answer Type Questions

1. (a) See text part

(b) Total volume of carbon dioxide (after opening the tap)
$$= 4 + 1 = 5 \ dm^3$$
For the first cylinder, $P_1 V_1 = P_2 V_2$
$$560 \times 4 = P_2 \times 5$$

So $P_2 = \dfrac{560 \times 4}{5} = 448 \ mm \ Hg$

For second cylinder, $P_1 V_1 = P_2 V_2$
$$1000 \times 1 = P_2 \times 5$$

So $P_2 = \dfrac{1000 \times 1}{5} = 200 \ mm \ Hg$

Hence, final pressure $= 448 + 200 = 648 \ mm \ Hg$

2. (a) See text part

(b) Let the required temperature be $t°C$.
$$V = 100 \ cm^3, \ T = (273 + 25) \ K = 298 \ K$$
$$V' = 125 \ cm^3, \ T' = (273 + t) \ K$$

By Charles' law, $\dfrac{V}{T} = \dfrac{V'}{T'}$ or $T' = \dfrac{V'}{V} \times T$

Substituting the values, $T' = \dfrac{125 \times 298}{100} = 372.5 \ K$

temperature in $°C$ = temperature in kelvin $- 273$
$$= 372.5 - 273 = 99.5°C$$

3. (a) At STP standard temperature $= 0°C = 273 \ K$

Standard pressure $= 1 \ atm = 760 \ mm \ of \ Hg$

(b) Any change in pressure or temperature can also change the volume of the gas. So, while expressing the specific volume of the gas, it is necessary to specify both its pressure and its temperature.

(c) Let the initial volume be V_1 final volume $= \dfrac{V_1}{8}$

Initial temperature $T_1 = 273 \ K$, Final temperature $T_2 = ?$

By Charles' law, $\dfrac{V_1}{T_1} = \dfrac{V_2}{T_2}$

So $\dfrac{V_1}{273} = \dfrac{V_1/8}{T_2}$ or $T_2 = \dfrac{V_1 \times 273}{8 \times V_1} = 34.3$
$$= 34.3 - 273 = -238.7°C$$

4. (a) Gases have a lower density compared to that of solids or liquids because the mass of a gas per unit volume is small due to large intermolecular spaces.

(b) At a given temperature, the number of molecules of the gas striking against the walls of the container per unit time per unit area is the same. Thus, gases exert the same pressure in all directions.

(c) According to Boyle's law $P \propto \dfrac{1}{V}$

When a balloon is inflated, the pressure inside the balloon decreases, and according to Boyle's law, the volume of the gas should increase, but it will decrease violating Boyle's law.

(d) A gas completely fills the vessel in which it is kept because inter-particle attraction is weak and inter-particle space is large in gases which makes the particles completely free to move randomly in the entire available space.

5. Oxygen (O_2) is a diatomic gas.

Its molar mass $= 16 \times 2 = 32 \ g$

32 g of oxygen occupies 22.4 dm^3 at STP

So, 20 g of oxygen will occupy $= \dfrac{22.4}{32} \times 20 = 14 \ dm^3$

$P_1 = 1 \ atm, \ P_2 = ?$
$V_1 = 14 \ dm^3, \ V_2 = 1 \ dm^3$
$T_1 = 273 \ K, \ T_2 = 298 \ K$

By using the gas equation, $\dfrac{P_1 V_1}{T_1} = \dfrac{P_2 V_2}{T_2}$

$$P_2 = \dfrac{P_1 V_1 T_2}{T_1 V_2} = \dfrac{1 \times 14 \times 298}{273 \times 1} = 15.28 \ atm$$

Competition Window (Objective Type)

Topic-wise MCQs

1. (b)	**2.** (b)	**3.** (a)	**4.** (d)	**5.** (d)
6. (c)	**7.** (b)	**8.** (b)	**9.** (c)	**10.** (b)
11. (b)	**12.** (b)	**13.** (b)	**14.** (b)	**15.** (b)
16. (c)	**17.** (b)	**18.** (d)	**19** (a)	**20.** (c)
21. (b)	**22.** (c)	**23.** (a)	**24.** (d)	**25.** (a)
26. (b)	**27.** (a)	**28.** (d)	**29.** (c)	

Miscellaneous

1. (a)	**2.** (b)	**3.** (c)	**4.** (a)	**5.** (c)
6. (a)	**7.** (a)	**8.** (d)	**9.** (a)	**10.** (a)
11. (b)	**12.** (b)	**13.** (b)	**14.** (a)	**15.** (a)
16. (c)				

Advanced and Olympiads

Single Choice

1. (b)	**2.** (b)	**3.** (c)	**4.** (a)	**5.** (a)
6. (a)				

Multiple Choice

7. (b, c)	**8.** (a, b, d)	**9.** (a, b, c)	**10.** (a, c, d)
11. (a, b, c)	**12.** (a, b, d)	**13.** (b, d)	**14.** (b, c, d)

Comprehension Type

15. (a)	**16.** (c)	**17.** (a)

Matrix Matching

18. (a)—(s); (b)—(p); (c)—(q), (d)—(r)

19. (a)—(r), (b)—(s), (c)—(q), (d)—(p)

Integer Type

20. 6	**21.** 5	**22.** 2	**23.** 6	**24.** 8
25. 4				

Hints and Solutions

Competition Window (Objective Type)

Topic-wise MCQs

1. Gas molecules are highly energetic.

2. Gases are highly compressible.

6. V/T is constant at constant P and n.

9. Plots at constant T are termed as isotherms.

12. $r \propto 1/\sqrt{M}$

13. The vapour density of chlorine is 35.5 while dry air has a density of 14.4, that is, the molar mass of chlorine is nearly twice that of air.

15. $r_1/r_2 = \sqrt{(M_2/M_1)}$

$2r/r = \sqrt{(32/M_1)}$

On squaring both sides
$4/1 = 32/M_1$
$M_1 = 8$

17. As at constant temperature, PV remains constant.

21. As $\dfrac{V_1}{T_1} = \dfrac{V_2}{T_2}$

So, $V/300 = 2V/T_2$

$T_2 = 600 \text{ K} = 327°\text{C}$

22. $P = wRT/m$

$P = \dfrac{3 \times 0.082 \times 400}{20} = 4.92 \text{ atm}$

23. $n = PV/RT$

$= \dfrac{1 \times 22.4}{0.082 \times 303}$

$= 0.9 \text{ mol}$

24. $V = \dfrac{nRT}{P}$

$V = (n_1 + n_2)\dfrac{RT}{P}$

$= \left(\dfrac{16}{32} + \dfrac{3}{2}\right) \times \dfrac{0.082 \times 273}{1}$

$= 2 \times 0.082 \times 273 = 447.72 \approx 448$

25. $nH_2 = 2/2 = 1$

$nO_2 = 16/32 = 0.5$

$nN_2 = 14/28 = 0.5$

now $P\,H_2 = \dfrac{nH_2}{nO_2 + nN_2 + nH_2} \times P \text{ total}$

$= \frac{1}{2} P$

26. $rH_2/rO_2 = \sqrt{(MO_2/MH_2)}$

on

$\dfrac{50/25}{40/t} = \sqrt{(32/2)}$

on solving

$\dfrac{50t}{40 \times 25} = 4,\ t = 80 \text{ min}$

27. $KE = 3/2\ nRT$

$= \dfrac{3}{2} \times 2 \times 8.3 \times 400$

$= 99605 = 9.96 \text{ kJ}$

28. VD of air = Mass of 11200 cc of air
$= 11200 \times$ density of air
$= 11200 \times 0.00130 = 14.56$

29. 44 g CO_2 = 22.4 L
4.4 g CO_2 = 2.24 L

Miscellaneous

5. If initial volume is V, then final volume is 0.95 V

$P_1 V_1 = P_2 V_2$

$P_2 = \dfrac{P_1 \times V}{0.95 V} = 1.0526\ P_1$

The increase in pressure is 0.0526
so % = 0.0526×100
= 5.26% of initial pressure

6. Here P, V, R are the same
So $n_1 T_1 = n_2 T_2$
That is, $n_1 \times 300 = n_2 \times 500$
$n_2 = 3/5\ n_1$
Number of moles escaped

$= n_1 - \dfrac{3}{5} n_1 = \dfrac{2}{5} n_1$

$= 0.4\ n_1 = 40\%$

7. 16 g of O_2 combines with 14 g of the monovalent element X.
8 g of O_2 can combine with 7 g of X
Equivalent weight of X = 7
Since it is monovalent, its atomic weight is also 7
So, the element X could have an atomic weight of 7 and its oxide is X_2O.

8. As $d = \dfrac{PM}{RT}$

So $\dfrac{PN_2 \times 28}{RT} = \dfrac{P_{oxide} \times M_{oxide}}{RT}$

$M_{oxide} = \dfrac{5 \times 28}{2} = 70$

9. The amount of product to be obtained according to a chemical reaction is called the theoretical yield of the reaction.

10. Apply $P_A = \dfrac{\% \text{ of } A \times P_{mix}}{100}$

On solving

$PCO_2 = 152$ mm

$PO_2 = 114$ mm

$PN_2 = 494$ mm

11. $2CO + O_2 \rightarrow 2CO_2$

At STP 2×22.4 litre of CO requires 22.4 litre of O_2

So 4 litre of CO requires $\dfrac{4 \times 22.4}{2 \times 22.4} = 2$ litre

12. First find moles

$(n) = \dfrac{PV}{RT}$

$= \dfrac{10^{-4}}{760} \times \dfrac{1 \times 1}{0.082 \times 298}$

Number of molecules $= n \times N_0$

On solving we get

$= 3.24 \times 10^{15}$ molecules

13. As $P \propto T$, $P_2 > P_1$

15. $\dfrac{w_{H_2}}{w_{O_2}} = \dfrac{1}{4}$

$\dfrac{n_{H_2}}{n_{O_2}} = \dfrac{1/2}{4/32} = \dfrac{4}{1}$

16. According to Charles' law, $V \propto T$ at constant pressure and definite mass. The graph of V versus T will be a straight line passing through the origin, that is, $y = mx$ type line.

Advanced and Olympiads
Single Choice

1. $P_1 = 760$ mm Hg, $P_2 = 2 \times 760$ mm Hg

$T_1 = 273$ K, $T_2 = 546$ K

$V_1 = V_1$, $V_2 = ?$

$\dfrac{P_1 V_1}{T_1} = \dfrac{P_2 V_2}{T_2}$

$\dfrac{760 \times V_1}{273} = \dfrac{2 \times 760 \times V_2}{546}$

Or $V_2 = \dfrac{760 \times V_1 \times 546}{2 \times 760 \times 273}$

So $V_2 = V_1$

3. As $P \propto T$

So $\dfrac{P_1}{T_1} = \dfrac{P_2}{T_2}$

$\dfrac{12}{300} = \dfrac{14.9}{T_2}$

On solving, we get

$T_2 = 372.5$ K

4. The pressure inside the bubble is greater near the bottom of the water because more pressure is exerted on the bubble by the water.

5. As $V \propto T$

Here, $V_3 > V_2 > V_1$

So $T_3 > T_2 > T_1$

6. $C + CO_2 \rightarrow 2CO$

Initial volume of $CO_2 = 1$ litre

Let x be the volume of CO_2 reacted

Then volume of CO formed is $2x$ since 1 volume of CO_2 produces 2 volumes of CO

The CO_2 left behind is $1 - x$

The total volume $= (1 - x) + 2x = 1.4$ litre

Or $1 + x = 1.4$ lit

$x = 0.4$ lit

Multiple Choice

10. $47°C = 47 + 273 = 320$ K. So (a) is correct.

As 1 atm $= 760$ mm of Hg. So (b) is incorrect.

Statements (c) and (d) are also correct.

12. Here statement (a), (b) and (d) are correct. See text part for features of a gas.

Gases have high compressibility so (b) is not correct.

Comprehension Type

15. According to Boyle's law, the product of P and V always remains constant.

16. It represents the kinetic gas equation.

17. $PV = \dfrac{1}{3} mNv$

Integer Type

20. 760 mm of Hg $= 1$ atm.

4560 mm of Hg $= 4560/760 = 6$ atm.

21. 1000 L $= 1$ m^3

So, 5000 L $= 5000/1000 = 5$ m^3

22. Number of moles of

$CO_2 = \dfrac{\text{Weight of } CO_2}{\text{Molecular weight of } CO_2} = \dfrac{22}{44} = 0.5$

Number of moles of He $= \dfrac{1}{4} = 0.25$

Partial pressure $=$ Mole fraction $\times$ Total pressure

$P_{CO_2} = \dfrac{0.5}{0.5 + 0.25} \times 3 = 2$

23. $\underset{\text{2 moles}}{2CO} + O_2 \rightarrow \underset{\text{2 moles}}{2CO_2}$

2×22.4 L CO_2 gas is produced from 2×22.4 L of CO gas

6 L CO_2 gas is produced from 6 L of CO gas

24. $2H_2O_2 \rightarrow 2H_2O + O_2$

As $V = \dfrac{nRT}{P}$

$V = \dfrac{20}{34} \times \dfrac{0.082 \times 333}{1}$

On solving, $n = 8$ litre

25. Mass and temperature is constant, hence we can use Boyle's law

$P_1 V_1 = P_2 V_2$

Given $P_1 = 1$ atm, $V_1 = 8$ litre, $P_2 = 2$ atm, $V_2 = ?$

$1 \times 8 = 2 \times V_2 \Rightarrow V_2 = 4$ litre

Water

LEARNING OBJECTIVES

After studying this unit, you will be able to understand:
- Water, its structure, physiochemical properties
- Solutions, types and concentration units
- Solubility and the factors affecting it
- Crystals and crystallisation
- Hydrated and anhydrous substances
- Efflorescence and deliquescence
- Drying and dehydrating agents
- Soft and hard water

Introduction

Water is the most abundant and most important natural resource in the biosphere. More than 70% of the area of the earth is occupied with water however, only 2.5% of this water is fresh water including the frozen water in the polar ice caps and glaciers. Water is an oxide of hydrogen with a molecular formula of H_2O having one part of hydrogen and eight parts of oxygen by weight.

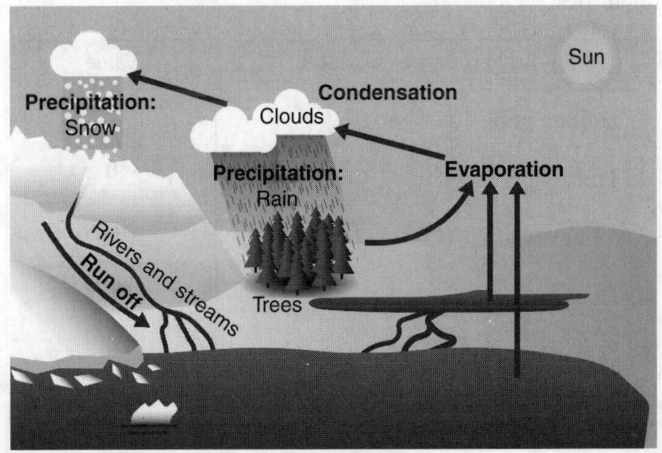

Water (H_2O)

Its chemical name is dihydrogen oxide and its molecular weight is 18 amu. It is a major constituent of all living beings and of the environment in which we live. Almost 70% of our body weight is due to water. It can exist in all the three physical states—solid (ice), liquid (water) and gas (steam). It is present in both the free and the combined states.

Sources of Water

Water in its liquid state is found in oceans and seas. Apart from this, it is also found in rivers, lakes, streams and other water bodies. Water available in the above sources is known as *surface water*. Water present in oceans and seas has appreciable proportions of dissolved salts, predominantly sodium chloride. The presence of these salts imparts a salty taste to water and this water is hence known as saline water. In addition to surface water, water is also available under the earth's surface and is known as *underground water*. This water accumulates due to the seepage of rain water through the soil. This underground water may come out in the form of springs. This can be drawn out artificially by digging wells or with the help of tube wells.

Structure of Water

A water molecule consists of two hydrogen atoms joined to an oxygen atom by covalent bonds. It has an angular or bent shape. In water, the central oxygen atom has two bond pairs of electrons (bp) and two lone pairs of electrons. Due to the presence of two lone pairs of electrons (lp), the shape of the water molecule becomes angular or bent with a bond angle of 104.5° due to lp–lp electron repulsion.

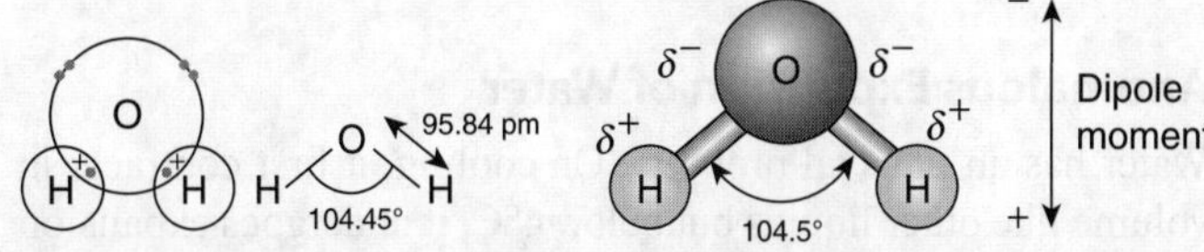

In the water molecule, the O–H bond length is 95.84 pm. The bond moments of the two O–H bonds makes the molecule a permanent dipole and the dipole moment value of the H_2O molecule is 1.85 D.

Hydrogen Bonding in Water

In the liquid state, water molecules are held together by intermolecular hydrogen bonding and each water molecule can have hydrogen bonding with four other water molecules.

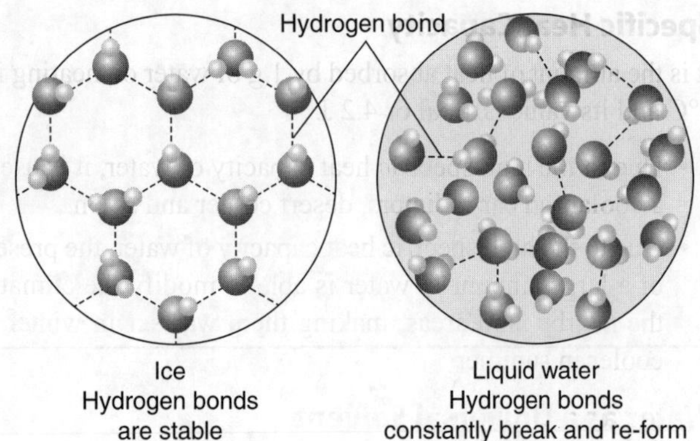

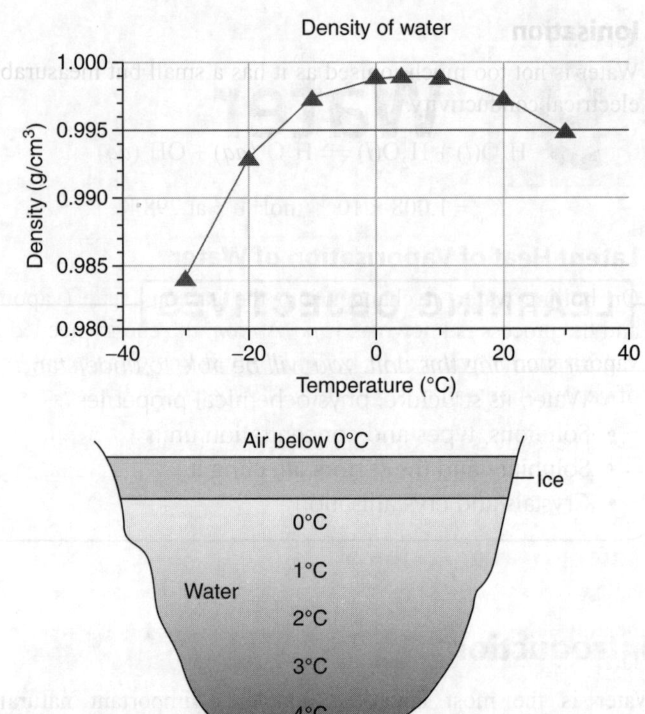

In the solid state (ice), the water molecules come closer and get arranged in a tetrahedral manner. The structure of ice is normally hexagonal in which each oxygen atom of water is tetrahedrally surrounded by four other oxygen atoms through a hydrogen atom. Each hydrogen atom is covalently bonded to one oxygen atom and is linked to another oxygen atom by a hydrogen bond. This type of packing gives ice an open cage structure with large open spaces. This makes the density of ice less than that of water; so, it floats on water. The floating ice prevents or delays the freezing of the underlying water. This property helps in saving the lives of aquatic animals even in winter.

Physical Properties of Water

Nature

Water is a clear, transparent, colourless, odourless and tasteless liquid. The taste in water is due to the presence of gases and solid impurities dissolved in it.

Boiling Point

It has a boiling point of 100°C or 373 K under normal pressure of 1 atm or 760 mm of mercury. The boiling point of water depends upon pressure and is directly proportional to it. On hills, where the atmospheric pressure is low, water boils below 100°C.

Density

The maximum density of H_2O is at 4°C and its value is 1 g/cm³ or 1000 kg/m³.

$$\text{Density of } H_2O > \text{Density of ice}$$

Anomalous Expansion of Water

Water has an unusual property. On cooling, it first contracts in volume like other liquids but below 4°C, it undergoes expansion till 0°C (up to the point when it freezes). This property of water enables marine life to exist even in the coldest regions in the world. Even when the water freezes on the surface, it can remain a liquid below the ice layer as the density of water is more than the density of ice and ice is also a bad conductor of heat.

Freezing Point of Water or Melting Point of Ice

The freezing point of water or melting point of ice at 1 atmosphere pressure is 0°C or 273 K. The freezing point of water decreases with the increase in pressure and in the presence of dissolved impurities.

Physical properties of water	Value
1. Boiling point °C	100
2. Freezing point °C	0
3. Viscosity at 20°C	10.09
4. Density at 20°C (g/mL)	0.997
5. Maximum density at °C	4
6. Ionic product kw at 25°C	1×10^{-14}
7. Dielectric constant at 20°C	82
8. Solubility of NaCl per 100 g of water at 25°C	35.9
9. Molecular weight	18

Stability

The water molecule has a very high thermal stability due to high negative heat of formation ($\Delta H^o_f = -285.9$ kJ mol^{-1}). For example, at 1500 K, it dissociates only to an extent of less than 0.02% and at 2270 K and 1 atmospheric pressure, it dissociates only 0.6%.

Ionisation

Water is not too much ionised as it has a small but measurable electrical conductivity.

$$H_2O(l) + H_2O(l) \rightleftharpoons H_3O^+(aq) + OH^-(aq)$$

$$K_w = 1.008 \times 10^{-14} \text{ mol}^2 \text{ L}^{-2} \text{ at 298 K}$$

Latent Heat of Vaporisation of Water

On boiling water, it changes into the gaseous state (vapour) and the process is known as *vaporisation*. the energy needed in vaporisation of water at its boiling point is known as latent heat of vaporisation and it is 540 cal/g or 2268 J/g.

- The body sweats during hot weather and evaporation of sweat takes heat from the body thus giving a cooling effect to the body.

Latent Heat of Fusion of Ice

Fusion is the process of converting ice into the liquid form by heating and the heat required in this process is known as the latent heat of fusion of ice. Its value is 80 cal/g or 336 J/g.

- Due to high specific latent heat of fusion, lakes and rivers do not freeze normally.

Specific Heat Capacity

It is the amount of heat absorbed by 1 g of water on heating it by 1°C and its value is 1 cal or 4.2 J.

- Due to the high specific heat capacity of water, it is used as a coolant in car radiators, desert cooler and so on.
- Due to the high specific heat capacity of water, the presence of a large amount of water is able to modify the climate of the nearby land areas, making them warmer in winter and cooler in summer.

Water as a Universal Solvent

It is a universal solvent due to its high dielectric constant, high liquid range and ability to dissolve most of the compounds. As water has a unique ability to dissolve a wide range of substances, it is called a universal solvent. The ability of water to form solutions is responsible for all the life activities occurring in nature and also for the sustenance of life on earth.

- Water is a good solvent for ionic compounds but bad for covalent compounds as the force of attraction between ions is destroyed or greatly reduced and ions get dissolved in water.

KNOWLEDGE BOOSTER

Air dissolved in water	Salt dissolved in water
Air is present in the dissolved state in all the natural sources of water. Oxygen being more soluble, is present in a larger concentration in water than nitrogen. The composition of air dissolved in water is 66% N_2, 33% O_2 and 1% CO_2. **Importance of dissolved air in water:** This air dissolved in water has some important biological applications. • The dissolved oxygen is essential for marine life. • Aquatic plants make use of dissolved CO_2 for photosynthesis to get their food. $$6CO_2 + 12H_2O \xrightarrow[\text{Sunlight}]{\text{Chlorophyll}} \underset{\text{Glucose}}{C_6H_{12}O_6} + 6O_2 + 6H_2O$$ • The dissolved CO_2 reacts with limestone to form calcium bicarbonate. $$\underset{\text{Limestone}}{CaCO_3} + CO_2 + H_2O \rightarrow \underset{\substack{\text{Calcium} \\ \text{bicarbonate}}}{Ca(HCO_3)_2}$$	Substances which are apparently insoluble in water actually dissolve in water in traces. Even when we put water in a glass vessel, an extremely small amount of glass dissolves in it. due to this reason when distilled water is kept in a sealed bottle for a long time, it leaves etchings on the inside surface of glass. Tap water, river water and well water contain dissolved solids but rainwater and distilled water do not contain dissolved solids and so concentric rings are not formed in their case. **Importance of dissolved salts in water:** • Salts and minerals are essential for the growth and the development of plants. • They provide taste to water. • They supply the essential minerals needed by our bodies.

Chemical Properties of Water

Water can behave as an acid, a base, an oxidising agent, a reducing agent and as a ligand to metal ions. Some of the important chemical properties of water are given below.

Amphoteric Nature

Although water is neutral towards litmus (pH = 7 at 298 K), it acts both as an acid and a base and hence shows amphoteric nature. It can act as a base towards acids stronger than itself and as an acid towards bases stronger than itself.

For example,

$$\underset{\text{Base}}{H_2O(l)} + \underset{\text{Acid}}{HX(aq)} \rightleftharpoons \underset{\text{Acid}}{H_3O^+(aq)} + \underset{\text{Base}}{X^-(aq)}$$

$$\underset{\text{Acid}}{H_2O(l)} + \underset{\text{Base}}{NH_3(aq)} \rightleftharpoons \underset{\text{Acid}}{NH_4^+(aq)} + \underset{\text{Base}}{OH^-(aq)}$$

Hydrolysing Nature

Water can hydrolyse many halides, nitrides, phosphides and carbides.

For example,

$$SiCl_4 + 2H_2O \rightarrow H_4SiO_4 + 4HCl$$

Silicon
tetrachloride

$$Al_4C_3 + 12H_2O \rightarrow 4Al(OH)_3 + 3CH_4$$

$$AlN + 3H_2O \rightarrow NH_3 + Al(OH)_3$$

$$Ca_3P_2 + 6H_2O \rightarrow 2PH_3 + 3Ca(OH)_2$$

Oxidising and Reducing Nature

Water acts as an oxidising as well as a reducing agent. It acts as an oxidising agent when it reacts with active metals.

$$2Na + 2H_2O \rightarrow 2NaOH + H_2$$

$$Ca + H_2O \rightarrow CaO + H_2$$

Water acts as a reducing agent when it reacts with highly electronegative elements.

$$2F_2 + 2H_2O \rightarrow 4HF + O_2$$

$$2Cl_2 + 2H_2O \xrightarrow{sunlight} 4HCl + O_2$$

Formation of Hydrates

Water reacts with many metal salts to form their hydrates. For example,

- In some hydrates, water molecules get attached to certain oxygen containing anions through hydrogen bonds. For example, $CuSO_4 \cdot 5H_2O$. In this hydrate, four water molecules are coordinated to the central Cu^{2+} ion while the fifth water molecule is attached to the sulfate group by hydrogen bonds. Some other examples are $FeSO_4 \cdot 7H_2O$, $MgSO_4 \cdot 7H_2O$.
- In some hydrates, water molecules occupy the interstitial sites (voids) in the crystal lattice. For example, $BaCl_2 \cdot 2H_2O$.

Illustrations

1. Give the properties of water which are responsible for controlling the temperature of our body.

   Solution: Our body is almost 70% water. Water has a high specific heat. Due to its high specific heat capacity, the presence of a large amount of water is able to modify and control the temperature of our body and keeps it warm in winter and cooler in summer.

2. How is air dissolved in water different from ordinary air?

   Solution: Rivers and lakes have a large amount of water which has a high specific heat capacity, hence, they do not freeze easily. Even if they freeze, only the top layer freezes. There is liquid water below the frozen layer due to anomalous expansion of water.

3. Explain why:
   (a) Boiled or distilled water tastes flat.
   (b) Ice at zero degrees centigrade has greater cooling effect than water at 0°C.
   (c) Air dissolved in water contains a higher proportion of oxygen.

   Solution:
   (a) Boiled water tastes flat as it does not contain matter like air, carbon dioxide and other minerals.
   (b) Ice at 0°C gives more cooling effect than water at 0°C. Ice at 0°C absorbs 336 J per gram of energy to melt to 0°C water and so gives more cooling effect.
   (c) Air dissolved in water contains a higher percentage of oxygen because, solubility of oxygen in water is more than in air.

4. Why does temperature in Mumbai and Chennai not fall as low as it does in Delhi?

   Solution: The temperature in Mumbai and Chennai does not fall as low as in Delhi as Mumbai and Chennai are situated at the sea shore. Due to high specific heat capacity, the presence of a large amount of water is able to modify the climate of the nearby land areas. They are warmer in the winter and cooler in the summer. However, Delhi is not situated next to an ocean, therefore, the temperatures are more extreme.

Solutions and Their Types

Solutions

A solution is a homogeneous mixture of two or more pure substances. For example, lemonade and soda water are examples of solutions. In a solution there is homogeneity at the particle level. For example, lemonade tastes the same throughout. This shows that particles of sugar or salt are evenly distributed in the solution. A solution has a solvent and a solute as its two components (binary solution). The component of the solution that dissolves the other component (usually the component present in larger amount) is known as the *solvent* while the component of the solution that is dissolved in the solvent (usually present in lesser quantity) is known as the *solute*. For example, a solution of sugar in water is a solid in liquid solution. Sugar is the solute and water is the solvent. A solution of iodine in alcohol known as 'tincture of iodine', has iodine (solid) as the solute and alcohol (liquid) as the solvent.

Different Types of Solutions

We can divide solutions into different categories as follows.

(i) **Based on amount of solute in the solution:** Depending upon the dissolution of the solute in the solvent, solutions can be categorised into unsaturated, saturated and supersaturated solutions.

Unsaturated solution: An unsaturated solution is a solution in which a solvent is capable of dissolving some more solute at a given temperature.

Saturated solution: A saturated solution can be defined as a solution in which a solvent is not capable of dissolving any more solute at a given temperature.

Supersaturated solution: A supersaturated solution contains more than the maximum amount of solute that is capable of being dissolved at a particular temperature. When the solvent is reduced, the extra solute will crystallise quickly.

(ii) **Based on nature of solvent:** Solutions are of two types, depending on whether the solvent is water or not.

Aqueous solution: When a solute is dissolved in water, the solution is known as an aqueous solution. For example, salt in water, sugar in water and copper sulfate in water.

Non-aqueous solution: When a solute is dissolved in a solvent other than water, it is known as a non-aqueous solution. For example, iodine in carbon tetrachloride, sulfur in carbon disulfide, phosphorus in ethyl alcohol.

(iii) **Dilute and concentrated solution:** Solutions are spoken of as having two components, the solvent, and the solute. Another classification of the solution depends on the amount of solute added to the solvent. A dilute solution contains a small amount of solute in a large amount of solvent. A concentrated solution contains a large amount of solute dissolved in a small amount of solvent.

(iv) **Based upon solute particle size:** Solutions are of three types—true solutions, suspensions and colloids.

(v) **True solution:** A solution in which the particles of the solute can be broken down to such a fine state that they cannot be seen even under a powerful microscope is known as a true solution.

Properties of True Solutions

- A solution is always homogeneous in nature.
- The properties of the solute are retained in the true solution. For example, a sugar solution is sweet in taste and a solution of salt in water is saline in taste.
- The particles of a solution are smaller than 1 nm (10^{-9} m) in diameter. So, they cannot be seen by the naked eye.
- A true solution does not show the Tyndall effect because of very small particle size. A true solution does not scatter a beam of light passing through the solution so the path of light is not visible. That is, solutions are transparent to light.
- The solute particles in a solution can easily pass through a filter paper; so, a true solution passes through a filter paper. Hence the solute particles cannot be separated from the mixture by the process of filtration.
- The solute particles do not settle down when left undisturbed, that is, a solution is stable.

Types of True Solutions

Usually we think of a solution as a liquid that contains either a solid, liquid or a gas dissolved in it. However, we can also have solid solutions (alloys) and gaseous solutions (air).

- **Solution of solid in a solid:** Metal alloys are the solutions of a solid in a solid. For example, brass is a solution of zinc in copper. It is prepared by mixing molten zinc with molten copper and cooling the mixture.
- **Solution of liquid in a liquid:** Vinegar is a solution of acetic acid in water.
- **Solution of gas in a liquid:** Aerated drinks like soda water are gas in liquid solutions. These contain carbon dioxide (gas) as the solute and water (liquid) as the solvent.
- **Solution of gas in a gas:** Air is a mixture of a gas in gas and it is a homogeneous mixture. Its two main constituents are oxygen (21%) and nitrogen (78%) and the other gases (CO_2, Ar) are present in very small quantities.

Concentration of a Solution

Concentration

The concentration of a solution is the amount of solute present in a given amount (mass or volume) of solution, or the amount of solute dissolved in a given mass or volume of solvent. The solution whose concentration is exactly known is referred to as a standard solution and such solutions are prepared in a volumetric flask (standard flask).

Concentration of solution = Amount of solute/Amount of solution **or** Amount of solute/Amount of solvent

$$C = \frac{w\,(g)}{V\,(L)} = \frac{w\,(g)}{V\,(ml)} \times 1000$$

There are various ways of expressing the concentration of a solution, but here we will learn only two methods. The most common way of expressing the concentration of a solution is the 'percentage method'.

Mass Percentage (w/W%)

$w\%$ by mass% indicates that w grams of solute is taken in 100 grams of solution. For example, a 5% solution of glucose means that 5 grams of glucose is present in 100 grams of the solution. The concentration of a solution in terms of mass percentage of solute can be calculated by using the following formula.

$$\text{Concentration of solution} = \frac{\text{Mass of solute}}{\text{Mass of solution}} \times 100$$

$$\text{Mass of solution} = \text{Mass of solute} + \text{Mass of solvent}$$

So, we can obtain the mass of the solution by adding the mass of solute and the mass of solvent.

In the above example,

Mass of solute (salt) = 5 g and mass of solvent (water) = 95 g

So, Mass of solution = Mass of solute + Mass of solvent = 5 + 95 = 100 g

$$\text{Concentration of solution} = \frac{5}{100} \times 100 = 5\% \text{ (by mass)}$$

Percentage Strength (w/V%)

It is known as percentage by strength. $w/V\%$ by strength indicates that w grams of solute is taken in 100 mL of solution.

For example, a 10% solution of glucose by strength means that 10 grams of glucose is present in 100 mL of the solution. the concentration of a solution in terms of percentage by strength of solute can be calculated by using the following formula.

$$\text{Concentration of solution} = \frac{\text{Mass of solute}}{\text{Volume of solution}} \times 100$$

Volume Strength (v/V%)

In the case of a liquid solute dissolved in a liquid solvent, the concentration of a solution is defined as volume strength. $v/V\%$ by strength means v mL of solute is present in 100 millilitres of the solution. For example, a 42% v/V solution of ethyl alcohol means that 42 mL of ethyl alcohol is present in 100 mL of the solution.

$$\text{Concentration of solution} = \frac{\text{Volume of solute}}{\text{Volume of solution}} \times 100$$

Illustrations

1. A syrup is prepared by dissolving 125 g of sucrose in 175 g of water. Find the mass % of sucrose in this syrup.

Solution: Mass of solute (sucrose) = 125 g

Mass of solvent = 175 g

So, Mass of solution = Mass of solute + Mass of solvent = 125 + 175 = 300 g

$$\text{Concentration of solution} = \frac{\text{Mass of solute}}{\text{Mass of solution}} \times 100$$

$$\text{Concentration of solution} = \frac{125}{300} \times 100 = 41.66\% \text{ by mass}$$

2. A solution is prepared by dissolving 90 g of glucose in 200 mL of water. Find the percentage strength of the glucose solution.

Solution: Mass of solute (glucose) = 90 g

Volume of solvent = 200 mL

$$\text{Concentration of solution} = \frac{\text{Mass of solute}}{\text{Volume of solution}} \times 100$$

$$\text{Concentration of solution} = \frac{90}{200} \times 100 = 45 \text{ by strength}$$

Solubility

The maximum amount of a solute that can be dissolved in 1 litre of a solution at a specified temperature is known as the solubility of that solute in that solvent. The solubility of a solute in a specific solvent at a given temperature is equivalent to the maximum number of grams of solute needed to saturate 100 grams of the solvent at this temperature. It can be given in g/L or mol/L.

A substance having high solubility is known as a soluble substance. For example, NaCl. A substance having less solubility is known as a sparingly soluble substance.

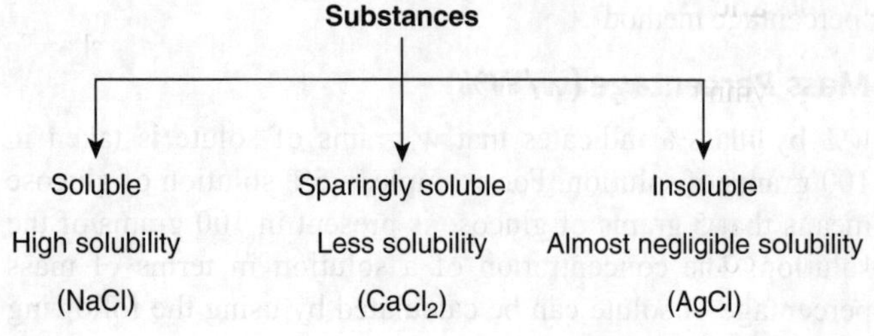

Factors Affecting Solubility

The solubility depends upon the following factors.

(i) Nature of solute

- Polar compounds (HF, NH_3) dissolve more in polar solvents such as water whereas non-polar compounds (F_2, Cl_2) dissolve more in non-polar solvents such as CCl_4, CS_2.
- Gases having more intermolecular forces of attraction between molecules are more soluble. For example, CO_2 is more soluble than oxygen in water.

(ii) Nature of solvent: Solvents having high value of dielectric constant, such as water, can dissolve ionic and polar compounds more than solvents with low dielectric constant.

(iii) Size of solute particles: Solutes with smaller sized particles are more soluble as more surface area is exposed to the solvent. For example, powdered sugar dissolves more than granules in water.

(iv) Stirring: Stirring brings more amount of solvent in contact with the solute and thus increases the solubility.

(v) Pressure: The solubility of solids in liquids remains unaffected by the changes in pressure.

(vi) Temperature: The solubility of most of the solids in water increases with an increase in temperature while that of liquids and gases decreases with increase in temperature (Fig. 9.1).

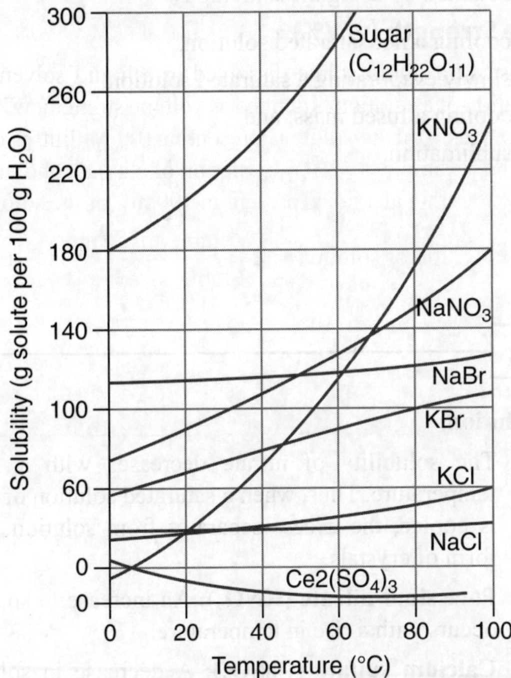

Fig. 9.1 Variation of solubility with temperature

Solubility Curve

For a given substance, the variation of solubility with temperature can be shown by means of solubility curves. Solubility is plotted along the Y-axis and temperature along the X-axis. A solubility curve is a line graph that plots a change of solubility of a substance in a solvent with the change of temperature. Solubility curves are useful in determining the solubility of a substance at a specific temperature and in finding the effect of cooling of hot solutions. For example, the solubility of KNO_3 in grams at 40°C and 50°C are 63 and 84 respectively.

Exothermic Substance

In the case of exothermic substances such as CaO, hydrated calcium sulfate ($CaSO_4 \cdot 2H_2O$) and slaked lime $Ca(OH)_2$, the solubility decreases with an increase in temperature; that is, solubility is inversely proportional to the temperature.

Endothermic Substances

In the case of endothermic substances like $NaNO_3$, KNO_3 and KBr, the solubility increases with an increase in temperature; that is, solubility is directly proportional to the temperature.

In the case of sodium chloride, the solubility increases slightly with an increase in temperature.

Anomalous Solubility

In the case of salts like globular salt ($Na_2SO_4 \cdot 10H_2O$), the solubilities first increase and then decrease with an increase

in temperature. In the case of Glauber's salt, the maximum solubility is at 32.8°C and beyond that, it decreases. This is why it is hydrous below 32.8°C and anhydrous above it (Fig. 9.2).

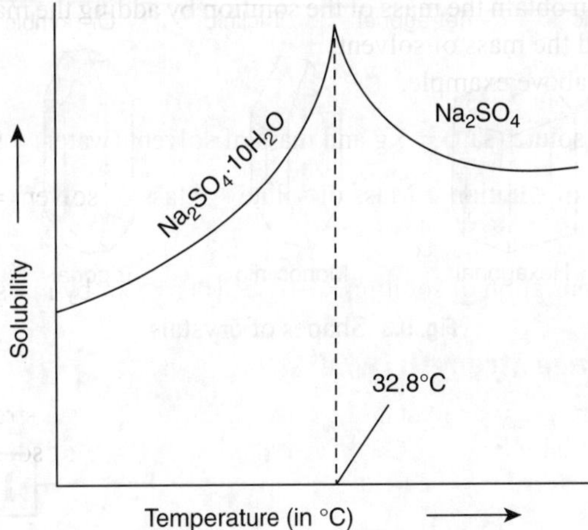

Fig. 9.2 Solubility of Glauber's salt

Effect of Pressure and Temperature on Solubility of Gases in Water (Liquid)

The solubility of a gas in a liquid is decided mainly by pressure and temperature.

Effect of Pressure

The solubility of gases in liquids increases on increasing the pressure and decreases on decreasing the pressure. According to Henry's law, at a given temperature the solubility of a gas is directly proportional to the partial pressure of the gas on the surface of the liquid.

$$\text{Solubility } (w \text{ or } m) \propto \text{Partial pressure of the gas } (P_g)$$

Effect of Temperature

The solubility of solids in liquids generally increases on increasing the temperature and decreases on decreasing the temperature. The solubility of gases in liquids generally decreases on increasing the temperature; and increases on decreasing the temperature as the dissolution of a gas in a solvent is exothermic (heat is evolved).

The solubility of CO_2 in soda water is maximum when it is tightly packed (high pressure) in a container and kept at low temperature. On opening the soda water bottle, the dissolved gas rapidly bubbles out as the pressure on the surface of the water decreases.

Crystals and Crystallisation

A crystal is a homogeneous solid of definite geometrical shape having symmetrically arranged smooth plane surfaces which meet, forming the sharp edges.

Crystallisation is a process by which crystals of a substance can be obtained by cooling a hot saturated solution. All crystals of a pure compound have a similar shape but those of different compounds may differ in shapes. Crystals exist in the following shapes (Fig. 9.3).

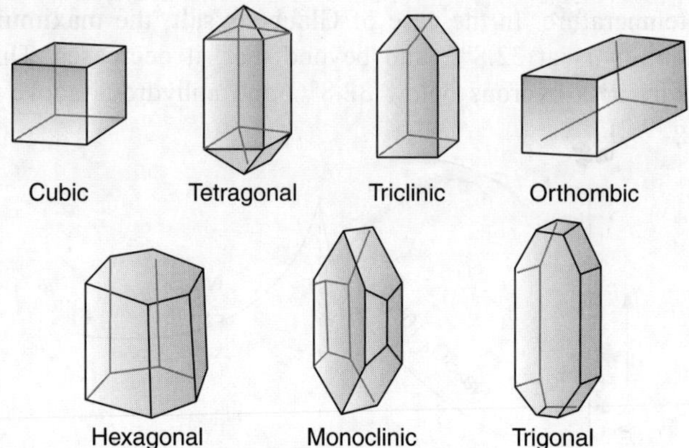

Cubic Tetragonal Triclinic Orthombic

Hexagonal Monoclinic Trigonal

Fig. 9.3 Shapes of crystals

In nature also many crystals are formed, a few of which are recovered from the earth as minerals—potassium nitrate, ruby, sapphire, diamond and common salt. The beautiful colours of some crystals are due to traces of water or impurities present in them. For example, the red colour of ruby is due to a trace of mercury oxide present in its crystals. In a laboratory, crystals can be obtained by using these methods:

- by cooling a hot saturated solution,
- by slowly evaporating a saturated solution,
- by cooling a fused mass, and
- by sublimation.

Illustrations

1. The solubility of copper sulfate in water at 20°C is 20.7 and that of potassium chloride in water at 20°C is 34.

   Solution: Potassium chloride has more solubility in water as it is more ionic in nature and has strong ionic bonds while copper sulphate has low solubility in water as it has less ionic nature. Copper sulphate is soluble in water due to hydrogen bonding.

2. (a) Explain why a hot saturated solution of potassium nitrate forms crystals as it cools.

   (b) What is the effect of temperature on solubility of KNO_3 and $CaSO_4$ in water?

Solution:

(a) The solubility of nitrate decreases with a fall in temperature. Thus, when a saturated solution of nitrate is cooled, the excess separates from solution, in the form of crystals.

(b) **Potassium nitrate (KNO_3):** An increase in solubility occurs with a rise in temperature.

   Calcium sulfate ($CaSO_4$): A decrease in solubility occurs with a rise in temperature.

Anhydrous and Hydrated Substances

Anhydrous

Anhydrous literally means 'no water'. In chemistry, substances without water are labelled anhydrous. The term is most often applied to crystalline substances after the water of crystallisation is removed. Anhydrous can also refer to the gaseous form of some concentrated solutions or pure compounds. For example, NaCl, KNO_3, anhydrous copper sulfate ($CuSO_4$), potassium permanganate ($KMnO_4$) and ammonium chloride (NH_4Cl).

Hydrated

A hydrate is a substance that contains water or its constituent elements. The chemical state of the water varies widely between different classes of hydrates, some of which were so labelled before their chemical structure was understood. This water of crystallisation gives the crystals shape and also colour. For example,

- Blue vitriol (hydrated copper sulfate) is $CuSO_4 \cdot 5H_2O$ and blue in colour.

- Washing soda (sodium carbonate decahydrate) is $Na_2CO_3 \cdot 10H_2O$ and white in colour.
- Epsom salt (magnesium sulfate heptahydrate) is $MgSO_4 \cdot 7H_2O$ and white in colour.
- Glauber's salt (sodium sulfate decahydrate) is $Na_2SO_4 \cdot 10H_2O$ and white in colour.
- Green vitriol [Iron(III) sulfate heptahydrate] is $FeSO_4 \cdot 7H_2O$ and bluish green in colour.
- Plaster of Paris (calcium sulfate semihydrate) is $CaSO_4 \cdot \frac{1}{2} H_2O$ and white in colour.
- Gypsum (hydrated calcium sulfate) is $CaSO_4 \cdot 2H_2O$ and colourless.

Water of Crystallisation

We have seen in hydrated salts that a definite quantity of water is associated with these salts. This fixed amount of water associated with such compounds is known as water of crystallisation or water of hydration. For example, in $CuSO_4 \cdot 5H_2O$, 5 molecules of water are associated. These water molecules are loosely associated with the salt and can be easily removed by heating above 100°C.

Determination of Water of Crystallisation

Take a calculated weight of crystals of a hydrated salt in a china dish and heat it above 100°C. Stop heating when the weight of the residue becomes constant or unchangeable. Now dry the substance and weigh it.

Weight of crystals at room temperature $= x$ g
Weight of residue after heating at 101°C $= y$ g
Weight of water $= (x - y)$ g
Now x g of crystals contain water $= (x - y)$ g

So, percentage of water of crystallisation $= \dfrac{x - y}{x} \times 100$

Conversion of Hydrated Salt into Anhydrous Salt

When the water of crystallisation is removed from a hydrated salt, it changes into an anhydrous salt. It is possible by direct heating, heating in dry and hot air, heating under vacuum and by using dehydrating agents such as concentrated H_2SO_4 and P_2O_5.

$$\underset{\text{Hydrated blue colour salt}}{CuSO_4 \cdot 5H_2O} \underset{\text{cool}}{\overset{\text{Heat}}{\rightleftharpoons}} \underset{\text{Anhydrous white colour salt powder}}{CuSO_4 + 5H_2O}$$

Efflorescence and Deliquescence

Efflorescence

When a compound loses its water of crystallisation on exposure to dry air, the phenomenon is known as *efflorescence* and such compounds are called *efflorescent substances*. The compound loss its crystalline shape and crumbles into the powder form. It is possible only when the vapour pressure of the hydrated crystals is more than the atmospheric vapour pressure, that is, it is minimised in humid conditions.

$$\text{Efflorescence} \propto \text{Temperature}$$

Higher the temperature of the air, higher is the efflorescence as the air absorbs more water with an increase of temperature and decreasing moisture. For example,

- Epsom salt (magnesium sulfate heptahydrate) on exposure to dry air becomes a monohydrate.

$$MgSO_4 \cdot 7H_2O \xrightarrow{\text{dry air}} MgSO_4 \cdot H_2O + 6H_2O$$

- Glauber's salt (sodium sulfate decahydrate $Na_2SO_4 \cdot 10H_2O$), on exposure to dry air changes into powdered anhydrous sodium sulfate.

$$Na_2SO_4 \cdot 10H_2O \xrightarrow{\text{dry air}} Na_2SO_4 + 10H_2O$$

- Washing soda (hydrated sodium carbonate), on exposure to dry air becomes a monohydrate.

$$Na_2CO_3 \cdot 10H_2O \xrightarrow{\text{dry air}} Na_2CO_3 \cdot H_2O + 9H_2O$$

Deliquescence

The process by which a substance absorbs moisture from the atmosphere until it dissolves in the absorbed water and forms a saturated solution is known as deliquescence and such a substance is known as a deliquescent substance. Deliquescence occurs when the vapour pressure of the solution that is formed is less than the partial pressure of water vapour in the air. That is, it is minimum during dry conditions. For example, caustic soda (NaOH), caustic potash (KOH), calcium chloride ($CaCl_2$), magnesium chloride ($MgCl_2$) and zinc chloride ($ZnCl_2$).

- Table salt (having impurities of $MgCl_2$, $CaCl_2$) turns moist and can form a solution on exposure to air, especially in the rainy season. It is important to note that pure NaCl is not a deliquescent substance.
- Such types of impurities can be removed by passing a current of dry HCl gas through a saturated solution of such an affected salt such as table salt, and pure NaCl is produced as a precipitate which can be recovered by filtering and washing with water and alcohol respectively.

Hygroscopy

Hygroscopy is the phenomenon of attracting and holding water molecules by a substance via either *absorption* or *adsorption* from the surrounding environment, which is usually at normal or room temperature. Such a substance is called a *hygroscopic substance* and it has a tendency to take up moisture from the air. It can be in the solid or liquid state. For example, quick lime (CaO), concentrated sulfuric acid (H_2SO_4), phosphorus pentaoxide (P_2O_5) and silica gel (SiO_2).

Dehydration of Solid, Liquid and Gaseous Substances

Solid

A solid substance can be dried by keeping it in a desiccator with a suitable drying agent like anhydrous $CaCl_2$ (Table 9.1).

Liquid

A liquid substance can be dried by keeping it over anhydrous $CaCl_2$ at room temperature overnight and then removing the solid by filtration.

Gas

A gas can be dried by passing it through concentrated sulfuric acid which can easily extract water from it. It is possible by using a drying bulb or a U-tube.

NH_3 being basic cannot be dried by concentrated sulfuric acid however, it can be dried by using quick lime (basic dehydrating agent).

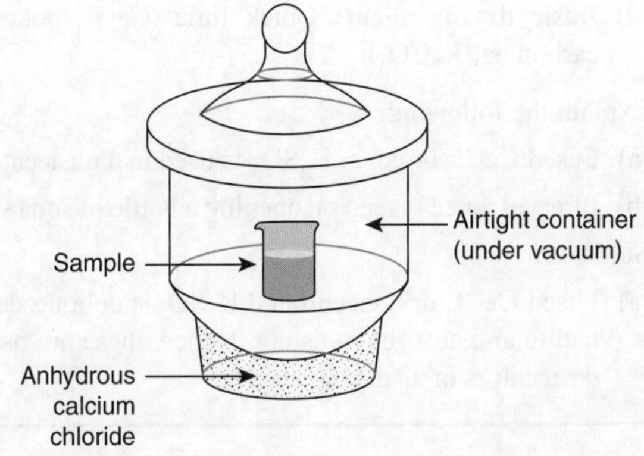

Table 9.1 Drying and dehydrating agents

Drying agent	Dehydrating Agent
A substance that can quickly absorb moisture from other substances without reacting with them is known as a drying agent or desiccating agent or desiccant. For example, all hygroscopic substances can act like drying agents, quick lime (CaO), concentrated sulfuric acid (H_2SO_4), phosphorus pentaoxide (P_2O_5) and silica gel (SiO_2).	A substance that can remove water of crystallisation or chemically combined water from other substances is known as a dehydrating agent. Dehydrating agents can remove hydrogen and oxygen in the ratio 2 : 1, in the form of water. For example, concentrated sulfuric acid (H_2SO_4), phosphorus pentaoxide (P_2O_5) and alumina (Al_2O_3).
It represents a physical change as no change of composition occurs.	It represents a chemical change as change of composition occurs or new substances are formed. $$CuSO_4 \cdot 5H_2O \xrightarrow{\text{Conc. } H_2SO_4}$$ blue $$CuSO_4 + 5H_2O$$ (white)
It is used for drying gases like Cl_2, HCl and SO_2 and is also used in desiccators to keep substances dry.	It can be used to prepare substances like CO and sugar charcoal. $$C_{12}H_{22}O_{11} \xrightarrow{\text{Conc. } H_2SO_4}$$ (Cane sugar) $$12C + 11H_2O$$ $$HCOOH \xrightarrow{\text{Conc. } H_2SO_4}$$ $$CO + H_2O$$

Illustrations

1. Give the names and formulae of two substances in each case.

   (a) Hydrated substance

   (b) Anhydrous crystalline substance

   (c) Drying agent

   (d) Basic drying agent

   Solution: The names and formulae of two substances in each case are as follows:

   (a) **Hydrated substance:** Blue vitriol ($CuSO_4 \cdot 5H_2O$), washing soda ($Na_2CO_3 \cdot 10H_2O$).

   (b) **Anhydrous crystalline substance:** Ammonium chloride (NH_4Cl), potassium permanganate ($KMnO_4$).

   (c) **Drying agent:** Concentrated sulfuric acid (H_2SO_4), phosphorus pentaoxide (P_2O_5).

   (d) **Basic drying agent:** Quick lime (CaO), potassium carbonate (K_2CO_3).

2. Explain the following.

   (a) Fused $CaCl_2$ or conc. H_2SO_4 is used in a desiccator.

   (b) Effervescence is seen on opening a bottle of soda water.

   Solution:

   (a) Fused $CaCl_2$ or concentrated H_2SO_4 is deliquescent in nature and absorbs moisture. Hence, these are used in desiccators or as drying agents.

   (b) Carbon dioxide is dissolved in soda water under pressure. On opening the bottle, the pressure on the surface of the water suddenly decreases, therefore, the solubility of CO_2 in water decreases and the gas rapidly bubbles out.

3. What happens when the given substances are exposed to air?

   (i) Sodium chloride, (ii) iron, (iii) conc. sulfuric acid, (iv) table salt and (v) sodium carbonate crystals.

   In which of these substances will there be

   (a) increase in mass

   (b) decrease in mass

   (c) no change in mass

   Solution:

   (a) Increase in mass: Iron and conc. sulfuric acid (due to formation of hydrated salts).

   (b) Decrease in mass: Sodium carbonate crystals (due to loss of water).

   (c) No change in mass: Sodium chloride (pure NaCl does not lose or gain water).

Soft and Hard Water

Soft Water

Soft water gives lather easily with soap and it is water containing sodium salts. It is suitable for daily use. For example, rain water, distilled water.

Hard Water

Hard water does not give lather with soap due to the presence of bicarbonates, chlorides, sulfates, magnesium and calcium. For example, sea water, water of rivers, lakes.

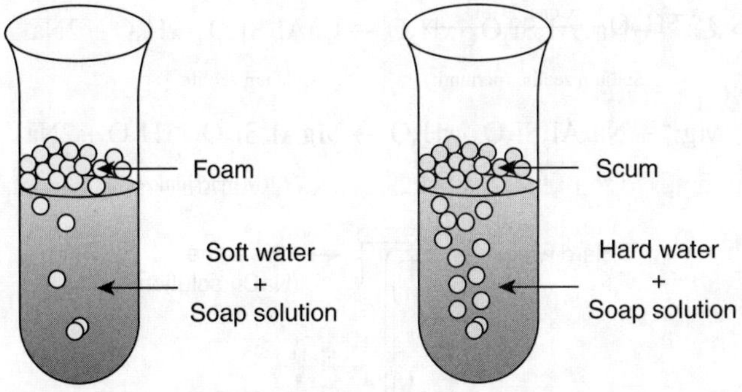

Reason for Hardness of Water

Water of some rivers and springs contains dissolved minerals. Due to the presence of insoluble salts of Ca^{+2}, Mg^{+2}, Fe^{+2}, water does not produce lather with soap easily because soaps are sodium or potassium salts of higher fatty acids like sodium palmitate ($C_{15}H_{31}COONa$), sodium stearate ($C_{17}H_{35}COONa$). When hard water having Ca^{2+}, Mg^{2+} ions is treated with soap solution, a precipitate of calcium or magnesium salts of the higher fatty acids is obtained which is insoluble in water.

$$Ca^{2+} + 2C_{15}H_{31}COONa \rightarrow (C_{15}H_{31}COO)_2Ca \downarrow + 2Na^+$$

Sodium palmitate (soap) soluble in water → Calcium palmitate (insoluble in water)

$$Mg^{2+} + 2C_{17}H_{35}COONa \rightarrow (C_{17}H_{35}COO)_2 Mg \downarrow + 2Na^+$$

Sodium stearate (soap) soluble in water → Magnesium stearate (insoluble in water)

Hence no lather can be produced till all the calcium or magnesium ions get precipitated.

Hardness is expressed in ppm of $CaCO_3$ as follows:

$$\text{Degree of hardness} = \frac{\text{wt. of } CaCO_3 \text{ in g}}{10^6 \text{ g of } H_2O}$$

$$1\ CaCO_3 \equiv 1\ MgCl_2 \equiv 1\ MgSO_4 \equiv 1 CaCl_2 \equiv 1\ CaSO_4$$

Types of Hardness

Water has two types of hardness—simple or temporary hardness and permanent hardness.

Simple or Temporary Hardness

It is due to the bicarbonates of Mg^{+2}, Ca^{+2} and the removal of hardness is called *softening of water*. Softening can be done as follows.

(a) By boiling: When water is boiled for about 15 minutes, the soluble bicarbonates decompose into their insoluble carbonates and carbon dioxides which can be removed by filtration or decantation. As a result, water becomes soft.

$$Ca(HCO_3)_2 \xrightarrow{\Delta} CaCO_3 \downarrow + CO_2 + H_2O$$

Calcium bicarbonate (Soluble) → Calcium carbonate (Insoluble)

$$Mg(HCO_3)_2 \xrightarrow{\Delta} MgCO_3 \downarrow + CO_2 + H_2O$$

Magnesium bicarbonate (Soluble) → Magnesium carbonate (Insoluble)

(b) By Clark's method: A calculated quantity of lime $[Ca(OH)_2]$ is added to remove temporary hardness in water. Insoluble carbonates are formed which get precipitated and can be easily removed by filtration.

$$Ca(HCO_3)_2 + Ca(OH)_2 \leftrightarrow 2CaCO_3 \downarrow + 2H_2O$$
$$Mg(HCO_3) + 2Ca(OH)_2 \leftrightarrow 2CaCO_3 \downarrow + Mg(OH)_2 \downarrow + 2H_2O$$

Insoluble

Permanent Hardness

It is due to the presence of soluble chlorides and sulfates of Ca^{+2} and Mg^{+2}. Permanent hardness is removed by the following methods.

(a) **Soda lime method:** Permanent hardness of calcium ions can be removed by washing soda but not of magnesium salts. When a calculated amount of washing soda is added to hard water, insoluble carbonates are formed which get precipitated and can be removed by filtration.

$$CaSO_4 + Na_2CO_3 \leftrightarrow CaCO_3 \downarrow + Na_2SO_4$$

Insoluble

(b) **By calgon method (masking or hiding of ions):** Calcium and magnesium ions are hidden or masked by calgon reagent [sodium hexa meta phosphate, $Na_2[Na_4(PO_3)_6]$ or sodium poly meta phosphate $(NaPO_3)x$. $(x > 100)$

$$CaCl_2 + Na_2[Na_4(PO_3)_6] \leftrightarrow Na_2[Ca_2(PO_3)_6] + 4NaCl$$

$$MgCl_2 + Na_2[Na_4(PO_3)_6] \leftrightarrow Na_2[Mg_2(PO_3)_6] + 4NaCl$$

These complex salts remain dissolved in water and do not cause hardness as the ions Ca^{2+} and Mg^{2+} are not free to cause any hardness.

(c) **Ion exchange method:** It is a modern method for softening of water. Ca^{2+} and Mg^{2+} ions are exchanged by those ions which do not cause hardness. For this purpose, ions exchangers are used.

By Ion Exchange Resin (Organic Ion Exchangers)

Cation exchange resins ($RCOOH$ or $R\text{–}SO_2OH$) remove Ca^{2+} and Mg^{2+} ions from hard water while anion exchange resins (OH^-, NH_2^-) remove Cl^-, SO_4^{-2} from hard water (Fig. 9.4). Resins which exchange cations for H^+ ions are known as cation exchangers and are shown as resin-H^+ while the resins which exchange anions are known as anion exchangers and are shown as resin-OH^-.

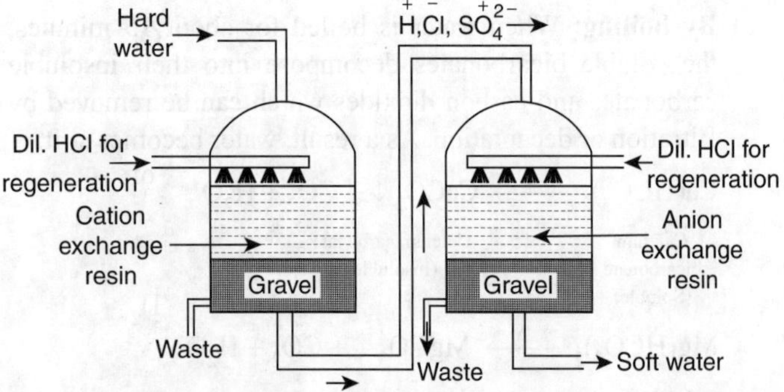

Fig. 9.4 Removal of hardness by organic ion exchangers

$$Ca^{2+} + 2\,\text{resin-}H^+ \rightarrow Ca(\text{resin})_2 + 2H^+$$

$$Mg^{2+} + 2\,\text{resin-}H^+ \rightarrow Mg(\text{resin})_2 + 2H^+$$

Cation exchanger

$$Cl^- + \text{resin-}OH^- \rightarrow \text{resin–}Cl^- + OH^-$$

$$SO_4^{2-} + 2\,\text{resin-}OH^- \rightarrow (\text{resin})_2\text{–}SO_4^{2-} + 2OH^-$$

Anion exchanger

These H^+ and OH^- combine to form water molecules which are called deionised or demineralised water.

$$H + OH^- \rightarrow H_2O$$

Inorganic Cation Exchangers (Permutit Process)

Some complex inorganic salts have the property of exchanging Ca^{2+} and Mg^{2+} ions from hard water for Na^+ ions to remove the hardness. Example, Permutit is artificial zeolite or sodium alumino orthosilicate $[Na_2Al_2Si_2O_8 \cdot xH_2O]$ which exchanges calcium and magnesium ions from hard water by giving Na^+ ion to hard water (Fig. 9.5).

$$Ca^{2+} + Na_2Al_2Si_2O_8 \cdot xH_2O \rightarrow CaAl_2Si_2O_8 \cdot xH_2O + 2Na^+$$

Sodium zeolite (permutit)   Calcium zeolite

$$Mg^{2+} + Na_2Al_2Si_2O_8 \cdot xH_2O \rightarrow MgAl_2Si_2O_8 \cdot xH_2O + 2Na^+$$

Magnesium zeolite

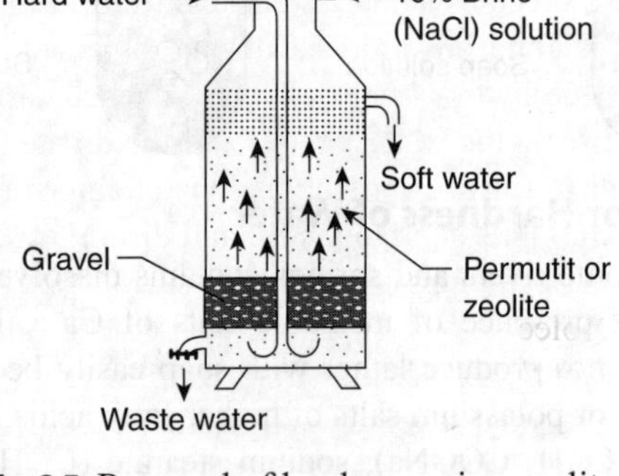

Fig. 9.5 Removal of hardness using zeolites

Advantages and Disadvantages of Hard Water

Hard water has a few advantages along with a few disadvantages.

Advantages

Due to the presence of dissolved salts, it has a few advantages.

- The presence of dissolved salts makes water tasty. That is why hard water is used in the manufacture of beverages and wines.
- Hard water has calcium and magnesium salts which are essential for the growth of our bones and teeth.
- Hard water avoids and prevents lead poisoning of water in lead pipes, as calcium sulfate present in hard water forms an insoluble lead sulfate layer inside these pipes.

Disadvantages

Due to the presence of dissolve salts, it has a few disadvantages.

Not Suitable for Producing Steam

Boilers having narrow copper tubes surrounded by fire are used to produce steam. When cold water enters these tubes, it is quickly changed into steam but the dissolved solids in water cannot be vaporised and they get deposited on the inner walls of the tube (Fig. 9.6). Slowly the bore of the tube gets narrower and narrower as a result formation of steam is retarded and the boiler may even burst.

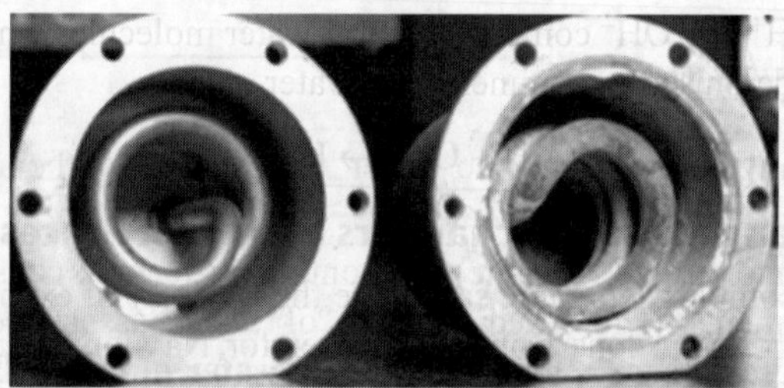

Fig. 9.6 Boiler scale formed due to hard water

Furring of Tea Kettle

Furring of tea kettles is caused by the sediment formed on its walls by regular boiling of hard water due to the formation of insoluble carbonates of calcium and magnesium.

Unfit for Washing

Hard water is not suitable for washing as soap does not give lather properly due to the formation of soap curd or scum. Soap is a sodium salt of higher fatty acids like steric acid with a formula $C_{17}H_{35}COONa$.

$$2C_{17}H_{35}\underset{\text{Soap}}{COONa} + Ca(HCO_3)_2 \rightarrow$$

$$\underset{\text{Soap curd}}{Ca(C_{17}H_{35}COO)_2} \downarrow + 2NaHCO_3$$

Synthetic detergents like Surf, Ariel are better to be used in the place of soap as they are more soluble in water and are almost unaffected by hard water and do not form scum.

KNOWLEDGE BOOSTER

Cleaning action of soap: On dissolving soap or detergent in water, the molecules gather as clusters to form micelles with their tails sticking inwards and the heads outwards. During cleansing, the hydrocarbon tails get attached to the oil and dirt. On stirring the water, the oil and the dirt tend to lift off from the dirty surface and get dissociated into fragments. As a result, other tails stick to the oil and dirt. Now the solution contains small globules of oil and dirt surrounded by soap and detergent molecules. The negatively charged heads present in water prevent these globules from forming aggregates. As a result, the oil and dirt get removed from the clothes when they are rinsed with water.

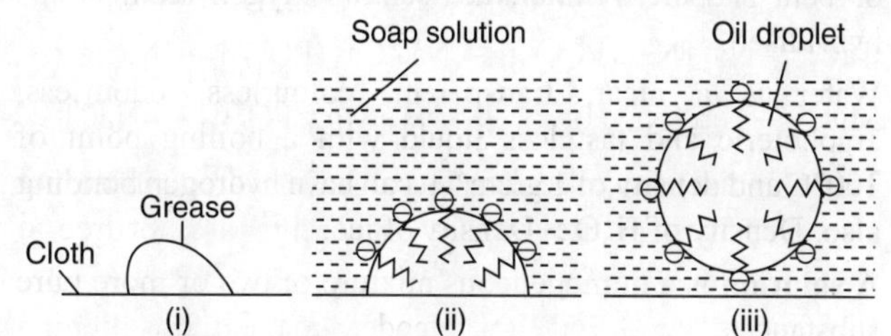

(i) Grease or oil on surface of cloth.
(ii) Stearate ions arranged around the grease or oil droplet.
(iii) Grease or oil droplet surrounded by stearate ions (ionic micelle formed)

Illustrations

1. **How are the resins regenerated after prolonged usage?**

   Solution: After prolonged usage, permutit or resins lose their activity and are hence they need to be activated. The exhausted zeolite is regenerated by treating with sodium chloride solution.

   $$CaZ + 2NaCl \rightarrow Na_2Z + CaCl_2$$

   Here Z is the zeolite.

2. Write the equations to show what happens when temporary hard water is

   (a) boiled, (b) treated with slaked lime.

   Solution: The reactions are as follows:

   (i) On boiling, temporary hardness is removed as follows:

   $$\underset{\substack{\text{Calcium} \\ \text{bicarbonate} \\ \text{(Soluble)}}}{Ca(HCO_3)_2} \xrightarrow{\Delta} \underset{\substack{\text{Calcium carbonate} \\ \text{(Insoluble)}}}{CaCO_3} \downarrow + CO_2 + H_2O$$

   $$\underset{\substack{\text{Magnesium} \\ \text{bicarbonate} \\ \text{(Soluble)}}}{Mg(HCO_3)_2} \xrightarrow{\Delta} \underset{\substack{\text{Magnesium carbonate} \\ \text{(Insoluble)}}}{MgCO_3} \downarrow + CO_2 + H_2O$$

   (ii) On treating with slaked lime, hardness of water is removed and the reactions are as follows:

   $$Ca(HCO_3)_2 + Ca(OH)_2 \leftrightarrow 2CaCO_3 \downarrow + 2H_2O$$

   $$Mg(HCO_3) + 2Ca(OH)_2 \leftrightarrow \underset{\text{Insoluble}}{2CaCO_3 \downarrow + Mg(OH)_2 \downarrow} + 2H_2O$$

CHAPTER AT A GLANCE

- **Water** is the most abundant and most important natural resource in the biosphere.
- The chemical name of water is dihydrogen oxide and its molecular formula is H_2O.
- Almost **70%** of our body weight is due to water.
- **Water** can exist in all the three physical states—solid (ice), liquid (water) and gas (steam).
- Water is a **universal solvent** due to its high dielectric constant, high liquid range and ability to dissolve most of the compounds.
- A water molecule consists of two hydrogen atoms joined to an oxygen atom by covalent bonds. It has an **angular** or bent shape. In water, the central oxygen atom is sp^3 hybridised.
- Water is a clear, transparent, colourless, odourless, amphoteric and tasteless liquid with a boiling point of $100°C$ and density of $1\ g/cm^3$. It can have hydrogen bonding also. Density of H_2O > Density of ice.
- A **solution** is a homogeneous mixture of two or more pure substances.
- **Unsaturated solution:** An unsaturated solution is a solution in which a solvent is capable of dissolving some more solute at a given temperature.
- **Saturated solution:** A saturated solution can be defined as a solution in which a solvent is not capable of dissolving any more solute at a given temperature.
- **Supersaturated solution:** A supersaturated solution contains more than the maximum amount of solute that is capable of being dissolved at a particular temperature. When the solvent is reduced, the extra solute will crystallise quickly.
- A solution in which the particles of the solute can be broken down to such a fine state that they cannot be seen even under a powerful microscope is known as a **true solution**.
- The **concentration** of a solution is the amount of solute present in a given amount. Mass or volume of solution, or the amount of solute dissolved in a given mass or volume of solvent.
- The maximum amount of a solute that can be dissolved in 1 litre of a solution at a specified temperature is known as the **solubility** of that solute in that solvent. Solubility is affected by the nature of solute and solvent, stirring and temperature.

- **Anhydrous** means 'no water'. Substances without water are labelled anhydrous.
- A **hydrate** is a substance that contains water or its constituent elements. The fixed amount of water associated with hydrated compounds is known as **water of crystallisation** or water of hydration.
- When **water of crystallisation** is removed from a hydrated salt it changes into an anhydrous salt.
- When a compound loses its water of crystallisation on exposure to dry air, the phenomenon is known as **efflorescence** and such compounds are called **efflorescent substances**.
- The process by which a substance absorbs moisture from the atmosphere until it dissolves in the absorbed water and forms a saturated solution is known as **deliquescence** and such a substance is known as a **deliquescent substance**.
- **Hygroscopy** is the phenomenon of attracting and holding water molecules by a substance via either absorption or adsorption from the surrounding environment, which is usually at normal or room temperature and such a substance is called a **hygroscopic substance**.
- A substance that can quickly absorb moisture from other substances without reacting with them is known as a **drying agent** or **desiccating agent** or **desiccant.**
- A substance that can remove water of crystallisation or chemically combined water from other substances is known as a **dehydrating agent**.
- **Soft water** gives lather easily with soap. It is water containing sodium salts.
- **Hard water** does not give lather with soap due to the presence of bicarbonates, chlorides, sulfates, magnesium and calcium.
- **Simple** or **temporary hardness** is due to bicarbonates of Mg^{+2}, Ca^{+2} and the removal of hardness is called softening of water. It is possible by boiling water or by Clark's method using $Ca(OH)_2$.
- **Permanent hardness** is due to the presence of soluble chlorides and sulfates of Ca^{+2} and Mg^{+2} and it is removed by the soda lime method, by using zeolites, calgon and resin.

PRACTICE QUESTIONS

Analyse Your Concepts (School Exam Based)

Fill in the Blanks

Instructions: Complete the following statements with an appropriate word/term to be filled in the blank spaces.

1. Water can exist in _______________ physical states.

2. Water occupies nearly _______________ of the earth's area.

3. The specific heat capacity of 1 g of water is _______________ joules.

4. Salts with water of crystallisation are known as _________.

5. The composition of O_2 in dissolved water is _______________ than in air.

6. Anhydrous calcium chloride is a _______________ agent.

7. Solubility of a gas in water is _______________ proportional to the temperature.

8. Potassium nitrate has _______________ crystal shape.

9. The conical pillar which grows downwards from the roof is called _______________.

10. Only moisture is removed in the process known as _______________.

11. For exothermic substances, solubility is _______________ proportional to the temperature.

12. The adding of a crystal of a pure substance into its saturated solution for crystallisation is known as _______________.

True or False

Instructions: Read the following statements and write your answer as true or false.

1. Water is considered to be a universal solvent.

2. Water has low specific heat.

3. Rain water and distilled water do not contain dissolved salts.

4. Temporary hardness of water can be removed by boiling.

5. Water has maximum density at 40°C.

6. Calcium oxide retains its physical state on exposure to air.

7. Distilled water and boiled water have no taste.

8. A solution is heterogenous in nature.

9. Tincture of iodine is a solution of iodine in alcohol.

10. Solubility of a solid or liquid generally does not depend upon pressure.

11. Silica gel is a hygroscopic substance.

12. Potassium nitrate can have water of crystallisation also.

Very Answer Short Type Questions

1. What is the composition ratio of H : O in water by volume?

2. Write the importance of the suitability of CO_2 and O_2 in water.

3. Write any three factors which affect the solubility of a solid solute in a solvent.

4. What causes the violence associated with torrential rain?

5. Define Henry's law.

6. Solubility of NaCl at 40°C is 36.5 g. What is meant by this statement?

7. What are drying or dessicating agents? Give examples.

8. Write the methods by which a hydrated salt can be made anhydrous.

9. What is the use of solubility of oxygen and carbon dioxide in the water?

10. A hot saturated solution of sodium nitrate forms crystals as it cools. Why?

11. Name the substances which cause (a) temporary hardness, (b) permanent hardness in water.

12. What is a soap, what is it used for?

13. What is the advantage of a detergent over a soap?

14. Explain why water is an excellent liquid to be used in cooling systems.

Short Answer Type Questions

1. Why is water considered a compound?

2. How is air dissolved in water different from ordinary air?

3. How is aquatic life benefited by the fact that water has maximum density at 4°C?

4. Define crystallisation and state the different methods of crystallisation.

5. In what way are the properties of water different from the properties of the elements from which it is formed? Discuss.

6. Explain what observations you can understand from the given diagram.

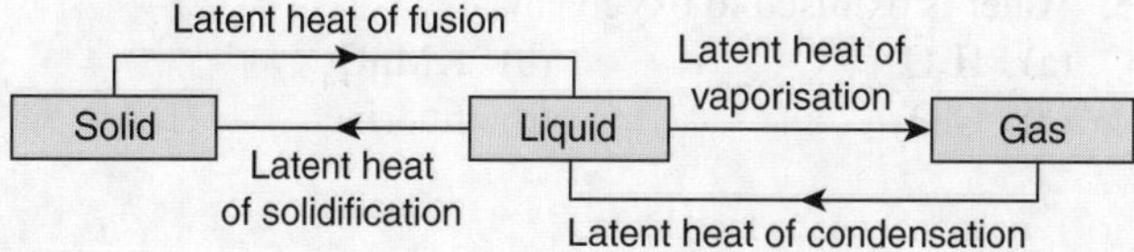

7. Why does table salt become sticky on exposure during the rainy season?

8. Explain the effect of pressure on the solubility of gases in water with an example.

9. Write the importance of dissolved impurities in water.

10. Write any two ways by which a saturated solution can be changed to an unsaturated solution.

11. What is the permutit method, how can it be used for softening hard water?

12. What are stalagmites and stalactites? How are they formed?

13. Which test will you carry out to find out if a given solution is saturated, unsaturated or supersaturated?

Long Answer Type Questions

1. Explain why:
   (a) Burns caused by steam are more severe than burns caused by boiling water.
   (b) Rivers and lakes do not freeze easily.
   (c) When distilled water is kept in a sealed bottle for a long time, it leaves etchings on the surface of the glass.

2. (a) Which property of water enables it to modify the climate?
   (b) Density of water varies with temperature. What are its consequences?
   (c) What is the effect of impurities present in the water on the melting point and boiling point of water?

3. What would you observe when crystals of copper(II) sulfate and iron(II) sulfate are separately heated in two test tubes?

4. (a) Define the following terms.
     (i) Soft water,
    (ii) hard water,
   (iii) temporary hard water,
    (iv) permanent hard water.
   (b) What are the advantages of soft water and hard water?

5. (a) If you are given some copper sulfate crystals, how would you proceed to prepare its saturated solution at room temperature?
   (b) How can you show that your solution is really saturated?

COMPETITION WINDOW (OBJECTIVE TYPE)

[For NEET, JEE (Main and Advanced), NTSE, KVPY and Olympiads]

Topic-wise MCQs

Water

1. Select the incorrect statement regarding water.
   (a) it is the most important natural resource
   (b) it can exist all three physical states
   (c) it has maximum density at 25°C
   (d) its specific heat value is 80 cal/g

2. The boiling point of water is exceptionally high because
   (a) there is a covalent bond between H and O
   (b) water molecule is not linear
   (c) water molecule is linear
   (d) water molecules associate due to hydrogen bonding

3. The H–O–H angle in the water molecule is about
   (a) 90°   (b) 105°
   (c) 135°   (d) 180°

4. When two ice cubes are pressed over each other, they unite to form one cube. Which of the following forces is responsible to hold them together?
   (a) van der Waals forces (b) covalent attraction
   (c) ionic interaction   (d) hydrogen bond formation

5. Water is oxidised to oxygen by
   (a) H_2O_2 (b) $KMnO_4$
   (c) ClO_2 (d) fluorine

6. Reaction of potassium with water is
   (a) hydrolysis   (b) absorption
   (c) exothermic   (d) endothermic

7. Which of the following is not an advantage of air dissolved in water?
   (a) it is good for marine life
   (b) it helps in photosynthesis
   (c) it provides taste to water
   (d) in making shells of snails

Solution and solubility

8. Which of the following is correct about a solution?
   (a) it is a homogenous mixture
   (b) solute is in less amount than solvent in a binary solution
   (c) solutions having water as solvent are aqueous solutions
   (d) all are correct

9. Brass is which type of solution?
   (a) solid in liquid   (b) solid in gas
   (c) solid in solid   (d) liquid in solid

10. Which of the following is incorrectly matched?

Type of solution		Examples
(a) Liquid in liquid	-	Water and alcohol
(b) Solid in liquid	-	sugar in water
(c) Solid in solid	-	Alloys
(d) Liquid in liquid	-	Soda water

11. The maximum amount of solute is present in
    (a) unsaturated solution
    (b) saturated solution
    (c) supersaturated solution
    (d) same in all these

12. Which of the following factors does not affect the solubility of solids in liquids generally?
    (a) size of solute particle   (b) temperature
    (c) stirring   (d) pressure

13. Solubility of a gas is maximum when
    (a) pressure is high   (b) temperature is low
    (c) temperature is high   (d) both (a) and (b)

14. For which of the following substances does the solubility increase with an increase in temperature?
    (a) CaO
    (b) $Ca(OH)_2$
    (c) KNO_3
    (d) gypsum

15. 25 g of sugar is present in 1.225 kg of water. The concentration of this solution is
    (a) 1%   (b) 2%
    (c) 3%   (d) 4%

Hydrated and anhydrous substances, efflorescence and deliquescence, drying and dehydrating agents

16. The shape of ferrous sulfate is
    (a) cubic   (b) octahedral
    (c) prismatic   (d) rhombohedral

17. The number of water molecules present in blue vitriol and green vitriol are respectively
    (a) 2, 5   (b) 5, 5
    (c) 5, 7   (d) 7, 5

18. Which of the following is not an anhydrous substance?
    (a) sucrose
    (b) potassium permanganate
    (c) gypsum
    (d) ammonium chloride

19. Impure table salt turns moist in air due to impurities of
    (a) KNO_3
    (b) $MgCl_2$
    (c) $CaCl_2$
    (d) both (b) and (c)

20. Ammonia can be dried by using
    (a) CaO
    (b) H_2SO_4
    (c) SiO_2
    (d) P_2O_5

21. Which of the following is a dehydrating agent?
    (a) Conc. H_2SO_4
    (b) CaO
    (c) Al_2O_3
    (d) both (a) and (c)

Hard water

22. A type of water which contains soluble salts of Ca and Mg is known as
    (a) soft water   (b) heavy water
    (c) conductivity water   (d) hard water

23. Which of the following ions will cause hardness in a water sample?
    (a) Ca^{2+}
    (b) Na^+
    (c) Cl^-
    (d) K^+

24. The process used for the removal of hardness of water is
    (a) Baeyer   (b) Hoope
    (c) Calgon   (d) Serpeck

25. Sodium hexa-metaphosphate is known as
    (a) permutit   (b) calgon
    (c) nitrolim   (d) natalite

26. Water softening by Clarke's process uses
    (a) potash alum
    (b) calcium bicarbonate
    (c) calcium hydroxide
    (d) sodium bicarbonate

27. Which one of the following is used for reviving the exhausted permutit?
    (a) 10% NaCl solution
    (b) 10% $MgCl_2$ solution
    (c) 10% $CaCl_2$ solution
    (d) HCl solution

28. The chemical formula of zeolite is __________
    (a) $Na_2Al_2Si_2O_8 \cdot xH_2O$
    (b) $Na_2[Na_4(PO_3)_6]$
    (c) $Ca_2Al_2Si_2O_8$
    (d) $K_2Al_2Si_2O_8 \cdot xH_2O$

29. What is the formula of calgon?
    (a) $MgSO_4$
    (b) Na_3PO_4
    (c) $Mg_3(PO_4)_2$
    (d) $(NaPO_3)_6$

30. By which of the following processes can the permanent hardness of water be removed?
    (a) washing soda
    (b) soda lime
    (c) sodium chloride
    (d) sodium bicarbonate

Miscellaneous

1. Reacting with water, an active metal produces
   (a) oxygen   (b) nitric acid
   (c) a base   (d) none of these

2. Anhydrous cobalt chloride can be used to remove moisture from the surroundings. Which of the following renders the sample reusable?
   (a) exposure to dry air
   (b) keeping in contact with another efflorescent substance
   (c) exposure to high humid air
   (d) keeping in air-tight container

3. When NaCl is added to water, the columbic force of attractions between
   (a) Na^+ and partially charged oxygen of water becomes very weak
   (b) Cl^- and partially charged hydrogen of water increase

(c) partially charged hydrogen and oxygen of water increase

(d) radicals present in NaCl increase

4. Which one of the following processes will produce hard water?
(a) saturation of water with $CaCO_3$
(b) addition of Na_2SO_4 to water
(c) saturation of water with $MgCO_3$
(d) saturation of water with $CaSO_4$

5. Which of the following does not show much variation of solubility with the change of temperature?
(a) potassium chloride
(b) potassium sulfate
(c) sodium chloride
(d) sodium sulfate

6. Which of the following changes take place due to the addition of aluminium sulfate to muddy water?
(a) adsorption of mud on aluminium sulfate crystals results in isolation of mud from water
(b) a part of the mud becomes soluble in water
(c) muddy water becomes free from germs and bacteria
(d) coagulation of mud particles takes place

7. When a small crystal of a soluble salt is added to a supersaturated solution,
(a) a saturated solution is formed
(b) an unsaturated solution is formed
(c) some amount of solute is precipitated
(d) both (a) and (c)

8. The formula of exhausted permutit is
(a) $K_2Al_2Si_2O_8.xH_2O$
(b) $CaAl_2Si_2O_8.xH_2O$
(c) $Na_2Al_2Si_2O_8.xH_2O$
(d) $CaB_2Si_2O_8.xH_2O$

9. Which among the following is used as a desiccating agent?
(a) calcium oxide
(b) calcium nitrate
(c) calcium hydroxide
(d) magnesium chloride

10. Which of the following statements is true?
(a) solubility of the gases in the deep sea is more than that at its surface
(b) solubility of the gases in water decreases with increase in temperature
(c) solubility of the gases in water decreases with decreases in pressure
(d) all of the above

11. Which among the following has water of crystallisation?
(a) common salt
(b) calcium hydroxide
(c) washing soda
(d) baking soda

12. 1.25 litres of ethyl alcohol is present in 5 litres of aqueous solution of alcohol. Find the volume per cent.
(a) 25%
(b) 50%
(c) 75%
(d) 85%

13. Which of the following are inversely proportional to the solubility of a gas in a liquid?
(a) temperature
(b) pressure
(c) Henry's constant K_H
(d) both (a) and (c)

14. Cane sugar on heating with concentrated H_2SO_4 gives sugar, charcoal and water. It is a case of
(a) drying
(b) dehydration
(c) deliquescence
(d) efflorescence

Advanced and Olympiads

Single Choice

1. Which of the following set represents only correct statements?
(i) Boiling point of water increases in the presence of impurities
(ii) Latent heat of vaporisation of water is 540 cal/g
(iii) Water has a low dielectric constant
(iv) Rain water and distilled water do not have dissolved solids so no concentric rings are formed

(a) (i), (ii)
(b) (ii), (iii)
(c) (i), (ii), (iii)
(d) (i), (ii), (iv)

2. Polyphosphates are used as water softening agents because they
(a) form soluble complexes with cationic species
(b) precipitate cationic species
(c) precipitate anionic species
(d) form soluble complexes with anionic species

3. Which of the following sets represents only correct statements?
(i) In a true solution, the size of solute particles is about 1 Å
(ii) True solutions can scatter light
(iii) In solids and liquids, solubility is not much dependent on pressure
(iv) Glauber's salt shows anomalous solubility

(a) (i), (ii)
(b) (ii), (iii)
(c) (i), (ii), (iii)
(d) (i), (iii), (iv)

4. The reagent commonly used to titrimetrically determine hardness of water is
(a) disodium salt of EDTA
(b) sodium thiosulfate
(c) sodium citrate
(d) oxalic acid

5. Which of the following sets represents only correct statements?
   (i) Gypsum on heating can give plaster of Paris
   (ii) Gaseous HCl is anhydrous
   (iii) Pure table salt is highly hydrated
   (iv) In dehydration new substances are also formed
   (a) (i), (ii), (iv)   (b) (ii), (iii)
   (c) (i), (ii), (iii)   (d) (i), (iii), (iv)

6. The conical pillar which grows downwards from the roof and upward from the floor of the cave are respectively known as,
   (a) stalagmite, stalagmite
   (b) stalactite, stalagmite
   (c) stalagmite, stalactite
   (d) stalactite, stalactite

Multiple Choice

7. Some of the properties of water are described below. Which of them is/are not correct?
   (a) water is known to be a universal solvent
   (b) hydrogen bonding is present to a large extent in liquid water
   (c) there is no hydrogen bonding in the frozen state of water
   (d) frozen water is heavier than liquid water

8. Which of the following is/are correct?
   (a) temporary hardness of water is due to the presence of bicarbonates of calcium and magnesium
   (b) permutit is artificial zeolite
   (c) H_2O_2 acts as an oxidising agent in the following reaction:
   $$Cl_2 + H_2O_2 \rightarrow O_2 + 2HCl$$
   (d) H_2O_2 is used as a bleaching agent for delicate textiles

9. The reaction of H_2O with X liberates a gaseous product. Which of the following is/are X.
   (a) PbO_2 (b) $KMnO_4/H^+$
   (c) PbS   (d) Cl_2

10. Hardness of water may be temporary or permanent. Permanent hardness is due to the presence of
   (a) chlorides of Ca and Mg in water
   (b) sulfates of Ca and Mg in water
   (c) hydrogen carbonates of Ca and Mg in water
   (d) carbonates of alkali metals in water

11. Which one of the following statements about zeolites is/are true?
   (a) they have open structure which enables them to take up small molecules.
   (b) they are used as cation exchangers.
   (c) zeolites are aluminosilicates having a three-dimensional network.
   (d) They cannot remove permanent hardness.

12. The solubility of which of the following pairs will decrease with an increase in temperature?
   (a) $Ca(OH)_2$, KOH
   (b) KNO_3, KBr
   (c) Sugar, KNO_3
   (d) $CaSO_4$, $Ca(OH)_2$

Matrix Matching

13. Match the following

Column I	Column II
(a) potable water	(p) drinking water
(b) temporary hardness	(q) $CaSO_4$
(c) permanent hardness	(r) $Mg(HCO_3)_2$
(d) heavy water	(s) D_2O

14. Match the following

List I	List II
(a) zeolite	(p) $Na_2[Na_4(PO_3)_6]$
(b) calgon	(q) RCOOH
(c) cation exchange resin	(r) $Na_2Al_2Si_2O_8 \cdot xH_2O$
(d) anion exchange resin	(s) $(RNH_3)OH$

Integer Type

15. In blue vitriol, $CuSO_4 \cdot XH_2O$, what is the value of X?

16. How many of the following compounds have water of crystallisation of more than two?
Plaster of Paris, gypsum, epsom salt, green vitriol, washing soda, Glauber's salt, white vitriol.

17. How many of these compounds are hygroscopic substances?
Conc. H_2SO_4, P_2O_5, CaO, SiO_2, NH_4Cl, $Na_2CO_3 \cdot 10H_2O$, epsom salt.

18. How many of the following substance are deliquescent in nature?
P_2O_5, CaO, NaOH, KOH, $MgCl_2$, $ZnCl_2$, $CaCl_2$, $FeCl_3$.

19. How many of the following are anhydrous substances?
Potassium nitrate, gypsum, common salt, sugar, potassium permanganate, ammonium chloride, potash alum, plaster of Paris

20. 75 g of glucose is present in 2.5 kg of water. The concentration of this solution is X%. Find the value of X.

Answer Keys

Fill in the Blanks

1. three
2. 70%
3. 4.2
4. hydrated salts
5. more
6. dehydrating
7. inversely
8. prismatic
9. stalactite
10. drying
11. inversely
12. seeding

True or False

1. T
2. F
3. T
4. T
5. F
6. T
7. T
8. F
9. T
10. T
11. T
12. F

Very Answer Short Type Questions

1. In water, the composition volume ratio of hydrogen and oxygen is H : O that is 2 : 1.

2. CO_2 and O_2 add taste to water for drinking purposes.

3. The three factors on which the solubility of a solid depends are temperature, nature of the solid and nature of solvent.

4. The sudden release of the latent heat of condensation causes the violence associated with torrential rain.

5. According to Henry's law, 'at any given temperature, the mass of a gas dissolved in a fixed volume of a liquid or solution is directly proportional to the pressure on the surface of a liquid'.

6. The solubility of NaCl at 40°C is 36.5 g. That is, 36.5 g of NaCl dissolves in 100 g of water at a temperature of 40°C.

7. These are the substances which can readily absorb moisture from other substances without chemically reacting with them. For example, phosphorus pentoxide (P_2O_5), quick lime (CaO).

8. A hydrated salt can be made anhydrous by heating or by exposing it to dry air.

9. Oxygen dissolved in water is used by aquatic animals for respiration and survival. CO_2 dissolved in water is used by aquatic plants to prepare their food by photosynthesis.

10. A hot saturated solution of sodium nitrate forms crystals as it cools, as its solubility decreases with a decrease in temperature.

11. (a) Hydrogen carbonates of calcium and magnesium make water temporarily hard.
    (b) Sulfates and chlorides of magnesium and calcium make water permanently hard.

12. Soap is chemically a sodium salt of stearic acid $(C_{17}H_{35}COOH)$ and has the formula $C_{17}H_{35}COONa$. Soap is mainly used for domestic washing purposes.

13. A detergent readily forms lather even with hard water and it is more soluble in water than a soap.

14. Water is an excellent liquid to be used in cooling systems due to its ability to absorb large quantities of heat, that is, its specific heat = 4.2 J/g°C, so it can be used in cooling systems as a cooling agent.

Short Answer Type Questions

1. Water is considered as a compound because it is made up of two elements, hydrogen and oxygen, combined in a ratio of 1 : 8 by mass.
   Mass ratio of elements H_2O
   H : O, $2 \times 1 : 16 \times 1 = 1 : 8$
   (Atomic mass of H = 1, O = 16)
   Also, the components of water cannot be separated by physical methods but can be separated by electrolysis of water.

2. Oxygen is more soluble in water than nitrogen. Air dissolved in water contains a higher percentage of oxygen (30%–35%) and in ordinary air it is only 21%. In this way, air dissolved in water is different from ordinary air.

3. The property of anomalous expansion of water enables aquatic life to exist because the water freezes on the surface of the water body, but it is still liquid below the ice layer.

4. **Crystallisation:** It is the process by which crystals of a substance separate out on cooling its hot saturated solution. In the laboratory, crystals may be obtained by the following methods:
   - by cooling a hot saturated solution gently
   - by cooling a fused mass
   - by sublimation, and
   - by evaporating a saturated solution slowly.

5. The properties of water are different from the properties of elements from which it is formed

Property	Water	Elements (oxygen and hydrogen)
Nature	It is a clear, colourless, odourless, tasteless and transparent liquid.	These are colourless, odourless, tasteless and non-poisonous gases.
Solubility	It can dissolve many things and is known as a universal solvent.	Oxygen and hydrogen are slightly soluble in water.

Property	Water	Elements (oxygen and hydrogen)
Density	Pure water has maximum density at 4°C.	Oxygen is heavier than air while hydrogen is the lightest gas.

6. From this diagram, the following observations can be made.

   (i) When solid changes to the liquid, it absorbs heat equal to the latent heat of fusion and when a liquid changes into the solid, it loses heat equal to the latent heat of solidification.

   (ii) When a liquid changes into the gas, it absorbs heat equal to the latent heat of vaporisation and when a gas condenses into the liquid, it loses heat equal to the latent heat of condensation.

7. Table salt becomes sticky on exposure during the rainy season as table salt generally contains a small percentage of magnesium chloride, as an impurity. As these impurities absorb moisture from the air due to their deliquescent nature, it gets wet in the rainy season and becomes sticky.

8. With an increase in pressure, the solubility of a gas in water increases. For example, the solubility of carbon dioxide in water under normal atmospheric pressure is low, but when the water surface is subjected to higher pressure, a lot more of CO_2 gas gets dissolved. Similarly, in the case of soda water, on opening the bottle, the dissolved gas rapidly bubbles out, since the pressure on the surface of the water suddenly decreases.

9. The importance of dissolved impurities in water are:
   - dissolved minerals and salts are essential for the growth of plants,
   - dissolved salts add taste to water, and
   - dissolved salts and minerals in water provide essential minerals required for our body.

10. The two methods are:
   (i) A saturated solution can be changed to an unsaturated solution by heating.
   (ii) A saturated solution can be changed to an unsaturated solution by adding more solvent.

11. See text part.

12. See text part.

13. (i) A solution in which more of solute can be dissolved at a given temperature is an unsaturated solution.
   (ii) A solution in which no more solute can be dissolved at a given temperature is a saturated solution at that temperature.
   (iii) A solution in which some solute separates on cooling slightly is a super saturated solution.

Long Answer Type Questions

1. (a) Burns caused by steam are more severe than burns caused by boiling water as 1 g of steam contains 2268 J more energy than 1 g of boiled water.

   (b) Rain water does not leave concentric rings when boiled because rain water does not contain dissolved solids.

   (c) When distilled water is kept in a sealed bottle for a long time, it leaves etching on the surface of glass as the substances which are insoluble in water, actually dissolve in minute traces in water. Even when we drink water from a glass, an extremely small amount of glass dissolves in water, so we can see the etching on the surface of glass when a bottle of distilled water (sealed for a long time) is poured into the glass.

2. (a) Specific heat.

   (b) Water has an unusual physical property. When cooled, it first contracts in volume, as do other liquids, but at 4°C (maximum density), it starts expanding, and continues to do so till the temperature reaches 0°C, the point at which it freezes into ice.

   The property of anomalous expansion of water enables marine life to exist in the colder regions of the world, because even when the water freezes on the top, it is still liquid below the ice layer.

   (c) Due to the presence of impurities in water, the melting point of water decreases while the boiling point of water increases.

3. When copper(II) sulfate ($CuSO_4.5H_2O$) crystals are heated in a hard glass test tube, the following observations are made.

   (i) The crystals are converted into a powdery substance.

   (ii) The crystals lose their blue colouration on further heating.

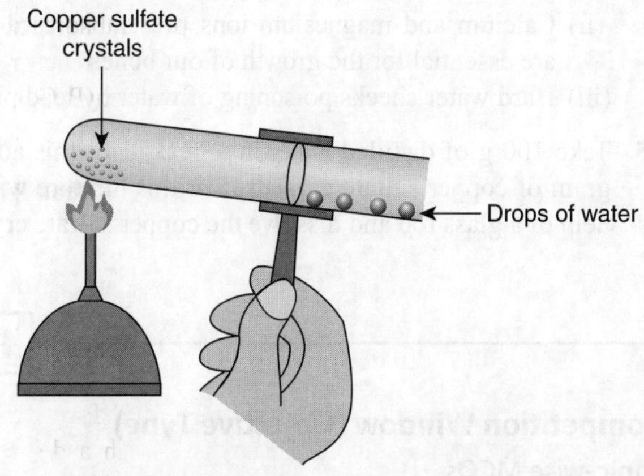

   (iii) Steaming vapours are produced inside the tube which condense near the mouth of the tube to form a colourless liquid.

   (iv) On further heating, steam escapes from the mouth of the tube and water gets collected in a beaker placed under the mouth of the tube.

(v) On further heating, the residue changes to a white powder and steam stops coming out.

$$CuSO_4 \cdot 5H_2O \xrightarrow{\Delta} CuSO_4 + 5H_2O$$
$$\text{Blue} \qquad\qquad \text{White}$$

Action of heat on iron(II) sulfate (FeSO$_4$.7H$_2$O)

When iron(II) sulfate is heated in a test tube, the following observations are made.

(i) The crystals crumble to white powder and a large amount of steam and gas are given out.

(ii) On strong heating, a brown residue of ferric oxide (Fe$_2$O$_3$) is produced and a mixture of SO$_2$ and SO$_3$ is given off.

$$FeSO_4 \cdot 7H_2O \xrightarrow{\Delta} FeSO_4 + 7H_2O$$

$$2FeSO_4 \xrightarrow[\text{(Brown residue)}]{\text{Ferric oxide}} Fe_2O_3 + SO_3 + SO_3$$

4. (a) Water is said to be soft water if it readily gives foam with soap.

 (b) Water is said to be hard water if it does not readily give foam with soap.

 (c) Water that contains only hydrogen carbonates of calcium and magnesium is called temporary hard water.

 (d) Water that contains sulfates and chlorides of magnesium and calcium is known as permanent hard water.

Advantages of soft water

(i) With soft water, less soaps and cleansing agents are consumed, which will save money.

(ii) Soft water does not leave deposits of minerals on pipes which will make plumbing work easy.

(iii) Clothes washed using soft water last long and remain bright.

Advantages of hard water

(i) Presence of salts and ions in hard water makes it tasty.

(ii) Calcium and magnesium ions present in hard water are essential for the growth of our bones.

(iii) Hard water checks poisoning of water by lead pipes.

5. Take 100 g of distilled water in a beaker. To this add one gram of copper sulfate crystals. Stir this mixture with the help of a glass rod and dissolve the copper sulfate crystals.

Similarly, go on dissolving more of copper sulfate, (1 g at a time) with constant and vigorous stirring. A stage is reached when no more copper sulfate dissolves. It is called a saturated solution at this temperature.

Take some of this saturated solution of copper sulfate in a test tube and add some copper sulfate crystals. The crystals do not dissolve but settle down. This indicates that the solution is really saturated.

Competition Window (Objective Type)

Topic-wise MCQs

1. (c)	2. (d)	3. (b)	4. (d)	5. (d)
6. (c)	7. (c)	8. (d)	9. (c)	10. (d)
11. (c)	12. (d)	13. (d)	14. (c)	15. (b)
16. (b)	17. (c)	18. (c)	19. (d)	20. (a)
21. (d)	22. (d)	23. (a)	24. (c)	25. (b)
26. (c)	27. (a)	28. (a)	29. (d)	30. (b)

Miscellaneous

1. (c)	2. (a)	3. (b)	4. (d)	5. (c)
6. (d)	7. (d)	8. (b)	9. (a)	10. (d)
11. (c)	12. (a)	13. (d)	14. (b)	

Advanced and Olympiads

Single Choice

1. (d)	2. (a)	3. (d)	4. (a)	5. (a)
6. (b)				

Multiple Choice

7. (cd)	8. (abd)	9. (abd)	10. (ab)	11. (abc)
12. (ad)				

Matrix Matching

13. (a)—(p), (b)—(r), (c)—(q), (d)—(s)

14. (a)—(r), (b)—(p), (c)—(q), (d)—(s)

Integer Type

15. 5	16. 5	17. 4	18. 6	19. 5
20. 3				

<hr>

Hints and Solutions

Competition Window (Objective Type)

Topic-wise MCQs

1. Water has maximum density at 4°C and not at 25°C.

2. Association of water molecules due to hydrogen bonding makes its boiling point exceptionally high.

4. Hydrogen bonds are formed between water molecules in ice.

5. Fluorine oxidises water to O$_2$.

$$2F_2 + 2H_2O \rightarrow O_2 + 4HF$$

6. Potassium reacts with H$_2$O liberating H$_2$ and heat.

7. The taste of water is due to impurities of dissolved salts in water.

9. Brass is a solid solution having 70% copper (solvent) and 30% zinc (solute).

10. Soda water is a solution of gas in liquid type and not liquid in liquid type.

14. In the case of CaO, Ca(OH)$_2$ and gypsum, the solubility will decrease as they are exothermic substances and increases in the case of only KNO$_3$ (endothermic) with the increase of temperature.

15. Mass of solute (sugar) = 25 g

    Mass of solvent (water) 1.225 kg = 1225 g

    $$\text{Mass percentage} = \frac{\text{Mass solute} \times 100}{\text{Mass of solution}}$$

    $$= \frac{25}{1225} \times 100 = 2\% \text{ solution}$$

16. The shape of ferrous sulfate crystals is octahedral.

17. Blue vitriol is CuSO$_4$.5H$_2$O and green vitriol is FeSO$_4$.7H$_2$O. Hence, the number of water molecules are 5, 7 respectively.

18. Gypsum (CaSO$_4$.2H$_2$O) is a hydrated substance.

19. Table salt turns moist in air due to impurities of MgCl$_2$ and CaCl$_2$.

20. Ammonia can be dried by using basic drying agents like CaO.

21. Both conc. H$_2$SO$_4$ and Al$_2$O$_3$ are dehydrating agents while CaO is a drying agent.

26. In Clarke's process, quick lime is added to hard water which produces Ca(OH)$_2$ which further converts bicarbonates into insoluble carbonates.

Miscellaneous

4. MgCO$_3$ and CaCO$_3$ are water insoluble. CaSO$_4$ dissolves in water adding Ca^{2+} ions which are responsible for producing hardness in water.

10. Solubility of gases in water decreases with a decrease in pressure and an increase in temperature.

11. Washing soda has water of crystallisation.

12. $$\text{Volume percent} = \frac{\text{Volume of solute}}{\text{Volume of solution}} \times 100$$

    $$= \frac{1.25}{5} \times 100 = 25\%$$

Advanced and Olympiads
Single Choice

1. Only statement (iii) is wrong as water has a high dielectric constant.

2. Polyphosphates are used as water softeners because these form soluble complexes with cationic species (Ca^{+2} and Mg^{+2}) present in hard water. The complex calcium and magnesium ions do not form any precipitate with soap and hence water readily produces lather with soap solution.

    $$2Ca^{+2} + Na_2[Na_4(PO_3)_6] \rightarrow Na_2[Ca_2(PO_3)_6] + 4Na^+$$

    $$2Mg^{+2} + Na_2[Na_4(PO_3)_6] \rightarrow Na_2[Mg_2(PO_3)_6] + 4Na^+$$

    Soluble complex

3. A true solution does not scatter light due to the small size of particles.

5. Pure table salt (NaCl) is not hydrated.

Multiple Choice

7. In frozen water or ice there is hydrogen bonding and it is less heavy than water.

8. H$_2$O$_2$ acts like a reducing agent and reduces Cl$_2$ into HCl.

9. X is an oxidant and it can be PbO$_2$, KMnO$_4$/H$^+$ and Cl$_2$.

11. Zeolites can remove permanent hardness of water.

12. For exothermic substances [Ca(OH)$_2$, KOH, CaSO$_4$], the solubility decreases with an increase in temperature while it increases for endothermic substances (KNO$_3$, KBr, sugar).

Integer Type

15. Blue vitriol is CuSO$_4$.5H$_2$O.

16. Epsom salt, green vitriol, washing soda, glauber salt, white vitriol have more than two water molecules as water of crystallisation (see text part).

17. Conc. H$_2$SO$_4$, P$_2$O$_5$, CaO, SiO$_2$ are hygroscopic substances while the rest are efflorescent substances.

18. NaOH, KOH, MgCl$_2$, ZnCl$_2$, CaCl$_2$, FeCl$_3$ are deliquescent in nature.

19. Potassium nitrate, common salt, sugar, potassium permanganate, ammonium chloride are anhydrous while the rest are hydrated.

20. Mass of solute (sugar) = 75 g

    Mass of solvent (water) 2.5 kg = 2500 g

    $$\text{Mass percentage} = \frac{\text{Mass solute} \times 100}{\text{Mass of solution}}$$

    $$= \frac{75}{2500} \times 100 = 3\% \text{ solution}$$

LEARNING OBJECTIVES

After studying this unit, you will be able to understand:
- The occurrence and position of hydrogen
- Its similarities with alkali metals and halogens
- The various methods of preparation
- Its physico-chemical properties
- Hydrides and their types

Introduction

It is the first, lightest, smallest, non-metallic element in the periodic table. It does not have any neutrons. It has maximum specific heat. It exists in the H^+ (hydronium) and H^- (hydride) states.

H —
- H^+: Exists in H_2O as H_3^+O
- H^-: (Metal hydride)

Discovery

Hydrogen was discovered by Henry Cavendish and he prepared it by the reaction of iron with dil. acid. He also introduced its elementary nature and said that it gives water on burning in air. He called it inflammable air due to its combustible nature. Lavoisier named it hydrogen. ('Hydro' means water and 'gen' means creating.)

KNOWLEDGE BOOSTER

Hydrogen
Symbol = H
Relative atomic mass = 1
Molecular formula = H_2
Relative molecular mass = 2
Valency = 1
Atomic number = 1
Electronic configuration: 1 (*K* shell)
Position in periodic table: 1st Period; Group 1

Occurrence

It is the most abundant element in the universe (70%) and the third most abundant element on the surface of the earth. It is found in traces in the earth's crust (0.98%) in the free state and in the atmosphere (0.01%). Volcanic gases also contain 0.025% of hydrogen. In the combined state, it is present in acids, alkalis, hydrocarbons, petroleum, and so on. For example, in water it is 11.1% by weight.

Position in the Periodic Table

Hydrogen is the first element in the periodic table with an atomic number of 1. As it has only 1 valence electron, it is placed in the first group and the first period of the periodic table. Since an atom of hydrogen contains only one electron in its *K*-orbit, it has an equal tendency to lose or to gain one electron to get a stable configuration. Therefore, hydrogen belongs neither to the first group nor to the seventeenth group and its position is not certain even now.

Resemblance of Hydrogen with Alkali Metals (Group 1)

(i) **Electronic configuration:** It has same electronic configuration as the alkali metals as it has only one valence electron (ns^1).

H ($Z = 1$): 1 or $1s^1$

Li ($Z = 3$): 2, 1 or $1s^2, 2s^1$

Na ($Z = 11$): 2, 8, 1 or $1s^2, 2s^2, 2p^6, 3s^1$

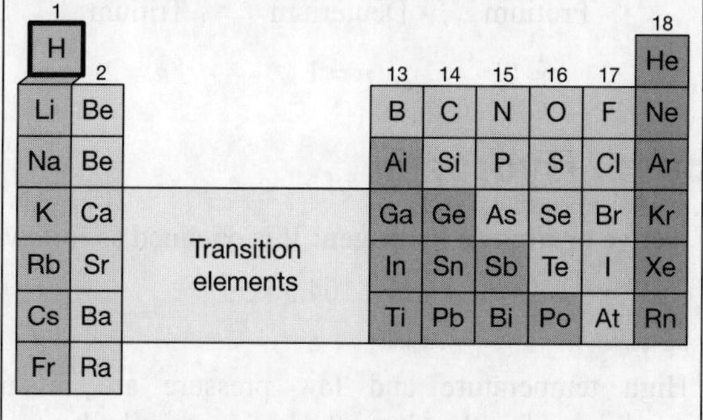

The position of hydrogen in the periodic table

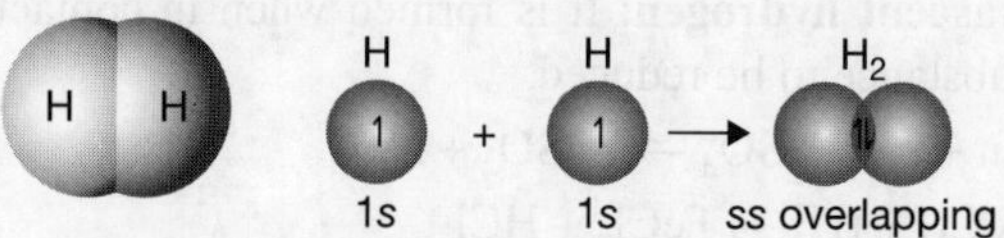

(ii) **Formation of monopositive ion:** Hydrogen forms a cation (H^+) like the M^+ ions formed by the alkali metals by losing one electron.

$$H \rightarrow H^+ + e^- \qquad Li \rightarrow Li^+ + e^-$$

(iii) **Electropositive character:** It is electropositive in nature like these metals.

(iv) **Valency and oxidation state:** Hydrogen is monovalent like the alkali metals and in most of its compounds it exists in the +1 oxidation state like the alkali metals.

For example, $\overset{+1}{H}Cl$, $\overset{+1}{Li}Cl$

(v) **Affinity for non-metals:** Like the alkali metals, hydrogen has a great affinity for non-metals and it combines with them to form a number of compounds such as oxides and halides.

$$H_2 + Cl_2 \rightarrow 2HCl$$

$$2Na + Cl_2 \rightarrow 2NaCl$$

(vi) **Reducing property:** It is a good reducing agent like these metals and can reduce many compounds. As it can be easily oxidised by losing the valence electron, it can act as a good reducing agent.

For example, $CuO + H_2 \xrightarrow{\Delta} Cu + H_2O$

$$B_2O_3 + 6K \xrightarrow{\Delta} 2B + 3K_2O$$

(vii) **Burning:** Hydrogen burns in air or oxygen with a pop sound and produces water. Alkali metals also burn vigorously in air to give their oxides and peroxides.

$$2H_2 + O_2 \rightarrow 2H_2O$$

$$H_2 + O_2 \rightarrow \underset{\text{Hydrogen peroxide}}{H_2O_2}$$

$$2Na + \frac{1}{2}O_2 \rightarrow Na_2O$$

$$2Na + O_2 \rightarrow \underset{\text{Sodium peroxide}}{Na_2O_2}$$

Resemblance with Halogens

(i) **Electronic configuration:** Hydrogen has a similar electronic configuration to that of halogens as it needs just one electron to get the next inert gas configuration.

H $(Z = 1)$: 1 or $1s^1$

F $(Z = 9)$: 2, 7 or $1s^2, 2s^2, 2p^5$

(ii) **Non-metallic nature:** It shows non-metallic nature like the halogens and is a bad conductor of heat and electricity.

(iii) **Atomicity and physical state:** It can exist in the diatomic state (H_2) like the halogens (X_2) as a gaseous molecule.

(iv) **Electronegative nature:** Like the halogens, hydrogen is also an electronegative element and it also forms mononegative ions by gaining an electron.

$$H + e^- \rightarrow H^-$$

$$F + e^- \rightarrow F^-$$

(v) **Ionisation energy:** It has a value of ionisation energy close to that of the halogens.

For example,

Element:	H	F	Cl	Br
IE:	1310	1681	1255	1121 kJ mol^{-1}

(vi) **Oxidation state:** Like the halogens it can show the −1 oxidation state, that is, it can form the H^- ion like the X^- formed by the halogens.

For example, $\underset{\text{Sodium hydride}}{\overset{+1 \quad -1}{Na\ H}}$ $\underset{\text{Sodium chloride}}{\overset{+1 \quad -1}{Na\ Cl}}$

(vii) **Reaction with metals:** Hydrogen reacts with alkali and alkaline earth metals to form hydrides similar to the halogens, which react with these metals and form halides. For example, NaH and NaCl, CaH_2 and $CaCl_2$.

(viii) **Reaction with non-metals:** Hydrogen reacts with non-metals such as C, Si, Ge, and so on, to form covalent compounds just like the halogens.

Hydrides: CH_4 SiH_4 GeH_4

Halides: CCl_4 $SiCl_4$ $GeCl_4$

KNOWLEDGE BOOSTER

- Hydrogen atom has only one orbit while both alkali metals and halogens have two or more orbits.
- Oxide of hydrogen (H_2O) is neutral while the oxides of alkali metals (Na_2O) are strongly basic and those of halogens (Cl_2O_7, BrO_3) are strongly acidic.

(ix) **Liberation at anode during electrolysis:** Hydrogen is liberated at the anode when compounds like LiH, NaH and CaH_2 are subjected to electrolysis in the molten state. This is similar to the halogens which are also released at the anode during the electrolysis of their compounds like NaX, KX and PbX_2.

Isotopes of Hydrogen

Hydrogen has three isotopes. This has been confirmed by the mass spectrographic studies of hydrogen. They are:

$_1H^1$	$_1H^2$ or D	$_1H^3$ or T
Protium	Deuterium	Tritium
$n = 0$	$n = 1$	$n = 2$

Types of Hydrogen

(i) **Active or atomic hydrogen:** It is obtained as follows:

$$\underset{\text{Molecular}}{H_2} \xrightarrow{\text{Electric arc}} \underset{\text{Atomic}}{2H} - 104.5 \text{ kcal}$$

High temperature and low pressure are favourable conditions for the formation of atomic hydrogen. The half-life of active hydrogen is 0.3 s.

(ii) **Nascent hydrogen:** It is formed when in contact with a substance to be reduced.

$$Zn + \text{dil. } H_2SO_4 \rightarrow ZnSO_4 + 2[H]$$

$$FeCl_3 + [H] \rightarrow FeCl_2 + HCl$$

It is more reactive and a stronger reducing agent than ordinary hydrogen. Nascent hydrogen cannot be isolated like atomic hydrogen. It has less reducing power than atomic hydrogen.

(iii) Occluded hydrogen: Some metals (Pd, Pt, Au, Ni) can adsorb hydrogen on their surface and this is called occlusion.

(iv) Ortho and para hydrogen (nuclear spin isomers): Dihydrogen has two nuclear spin isomers known as ortho and para dihydrogen. In ortho hydrogen the nuclear spin is in the same direction but in para hydrogen they are in opposite directions.

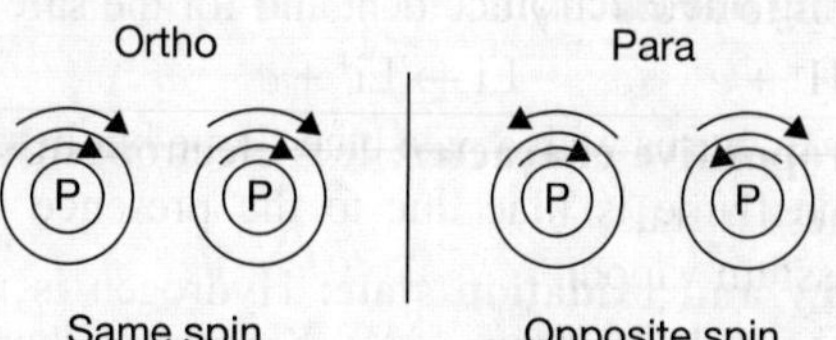

$$\text{Nuclear spin} = \frac{1}{2} + \frac{1}{2} = 1 \qquad \text{Nuclear spin} = \frac{1}{2} + \left(-\frac{1}{2}\right) = 0$$

Illustrations

1. (a) Where does hydrogen occur in the free state?

(b) Write an equation to show the electronegative nature of hydrogen.

Solution:

(a) Hydrogen occurs in the free state in traces in the earth's crust and atmosphere.

(b) Hydrogen reacts with calcium to form the ionic hydride, CaH_2.

$$Ca + H_2 \rightarrow CaH_2$$

2. Write the (a) atomicity of the hydrogen molecule and (b) the oxidation state shown by hydrogen.

Solution:

(a) Atomicity of hydrogen is 2.

(b) Hydrogen can show +1, 0 and –1 oxidation states.

$$\text{For example,} \quad \underset{+1}{H\,X}, \quad \underset{0}{H_2}, \quad \underset{-1}{Ca\,H_2}$$

Methods of Preparation of Dihydrogen

1. With metals: Many metals react with H_2O at different temperatures to release hydrogen.

KNOWLEDGE BOOSTER

Reactivity series of metals

On the basis of reaction with water, metals can be arranged in the decreasing order of reactivity. This arrangement of metals in the decreasing order of their reactivity in the form of a series is known as the activity or reactivity series of metals.

K, Na, Ca, Mg, Al, Zn, Fe, Pb [H], Cu, Hg, Ag, Au

Hence, potassium being the most reactive metal is placed at the top of the list and the least reactive metal being gold is placed at the bottom of the list.

Hydrogen although a non-metal, is included in this series because it can form a positive ion. It would occupy a position based on the formation of its positive ion.

The ability of metals to reduce water to hydrogen decreases on going down the series.

$$K > Na > Ca > Mg > Al > Zn \dots\dots\dots$$

Lead and the metals that are further below in the activity series: No reaction with water, even when the metal is hot and steam is used.

2. With very active metals: Metals such as Na, K and Ca react with cold water to form hydrogen and their hydroxides. These reactions are exothermic in nature.

3. With sodium: Sodium has a density of 0.97 g/mL and melting point of 97°C; on putting into water, it floats and melts later, giving a silvery globule which darts about on the surface of the water. Then the reaction occurs vigorously to give hydrogen and NaOH.

$$2Na + 2H_2O \rightarrow 2NaOH + H_2$$

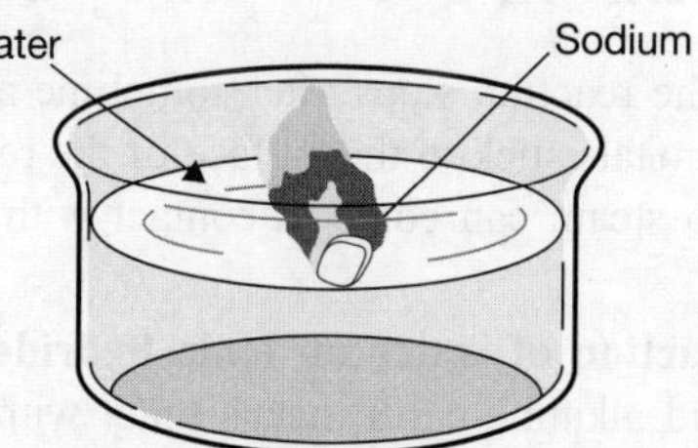

4. With potassium: Potassium has a density of 0.86 g/mL and melting point of 62°C; on putting into water, it floats and melts later on giving a silvery globule which darts about on the surface of the water. Then the reaction occurs vigorously to give hydrogen and KOH.

$$2K + 2H_2O \rightarrow 2KOH + H_2$$

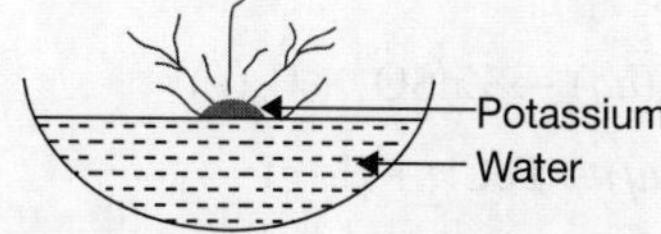

- Reaction with potassium is more vigorous and exothermic than with sodium. It is better to use the amalgam of these metals to avoid any accident and for the safe preparation of H_2.
- Pure H_2 burns with a pale blue flame but here, the colour of the flame is lilac due to the presence of traces of potassium vapour.
- Due to high reactivity of Na, K with air, they are kept in kerosene.

5. **With calcium:** On putting in water, calcium reacts less vigorously than Na to give hydrogen and calcium hydroxide.

$$Ca + 2H_2O \rightarrow Ca(OH)_2 + H_2$$

- The solutions of these metal hydroxides are strongly alkaline and turn red litmus into blue.

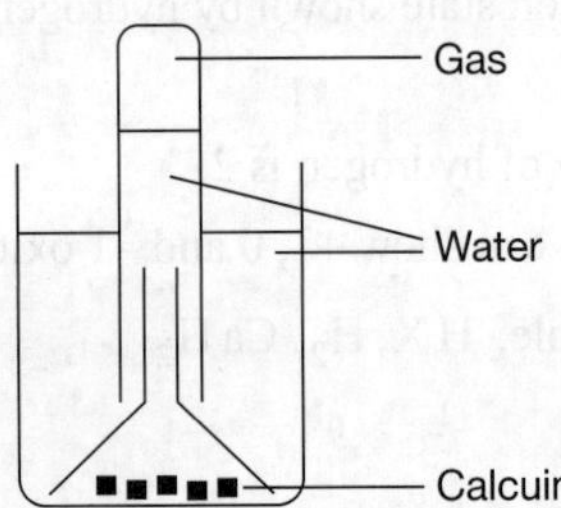

6. **With less active metals:** Metals like Mg, Al and Zn, react with H_2O on heating to form hydrogen.

$$Mg + H_2O \xrightarrow{\Delta} MgO + H_2$$

Magnesium burns in steam with intense white light to give H_2 and white ash or MgO. MgO crumbles on heating.

7. When steam is passed over metals like Al, Fe, Co, Ni and Sn, hydrogen is formed.

$$3Fe + 4H_2O \underset{\text{Steam}}{\xrightarrow{\text{High temp.}}} Fe_3O_4 + 4H_2$$

$$2Al + 3H_2O \xrightarrow{800°C} Al_2O_3 + 3H_2$$

- Here the reaction stops after sometime as the oxides of these metals stick to the surface of the respective metals and no steam can come in contact with the rest of the metal.

8. **By the action of water on ionic hydrides:** Hydrides of alkali and alkaline earth metals react with water to form hydrogen.

$$NaH + H_2O \rightarrow NaOH + H_2$$

$$CaH_2 + 2H_2O \rightarrow Ca(OH)_2 + 2H_2$$

9. **From acids:** When dilute acids are treated with those metals which are above hydrogen in the electrochemical series, hydrogen is formed.

For example,

$$Zn + H_2SO_4(aq) \rightarrow ZnSO_4 + H_2(g)$$

$$Fe + 2HCl(aq) \rightarrow FeCl_2 + H_2(g)$$

10. **From alkalis:** When metals like Zn, Al, Sn are treated with alkalis like NaOH and KOH, hydrogen is formed.

For example,

$$2Al + 2NaOH + 2H_2O \rightarrow 2NaAlO_2 + 3H_2$$

Sodium meta aluminate

$$Zn + 2NaOH \rightarrow Na_2ZnO_2 + H_2$$

Sodium zincate

Oxides and hydroxides of Zn, Pb, Al, being amphoteric in nature, can react with both acids and bases to give hydrogen along with their salts.

Laboratory Preparation of Hydrogen

Reactants: The reactants used are granulated zinc [having impurity of copper (catalyst)], dilute hydrochloric acid or dilute sulfuric acid.

Procedure: Take some pieces of granulated zinc in a flask (Fig. 10.1) and pour dilute HCl or H_2SO_4 into it. After the reaction, hydrogen is released.

$$Zn + 2HCl(dil) \rightarrow ZnCl_2 + H_2 \uparrow$$

$$Zn + 2H_2SO_4(dil) \rightarrow ZnSO_4 + H_2 \uparrow$$

Observation: The reaction will gradually start in the form of effervescence and evolution of hydrogen gas. When all the air from the apparatus has been expelled, hydrogen gas is collected over water.

Collection of hydrogen: Hydrogen is collected by the downward displacement of water.

- It is virtually insoluble in water. About 20 mL of hydrogen dissolves in 1 litre of water under normal conditions.
- It forms an explosive mixture with air and so it cannot be collected by downward displacement of air even though it is lighter than air.

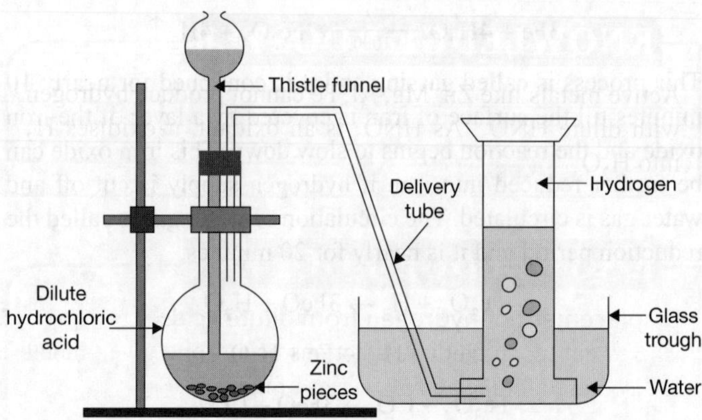

Fig. 10.1 Preparation of hydrogen gas

Impurities Present with Hydrogen

It has impurities of hydrogen sulfide (H_2S), sulfur dioxide (SO_2), oxides of nitrogen, phosphine (PH_3), arsine (AsH_3), carbon dioxide and water vapour. These impurities can be removed from hydrogen by passing it through the following.

(i) Silver nitrate solution (for removing arsine and phosphine).

$$AsH_3 + 6AgNO_3 \rightarrow Ag_3As + 3AgNO_3 + 3HNO_3$$

$$PH_3 + 6AgNO_3 \rightarrow Ag_3P + 3AgNO_3 + 3HNO_3$$

(ii) Lead nitrate solution (for removing hydrogen sulfide).

$$Pb(NO_3)_2 + H_2S \rightarrow PbS + 2HNO_3$$

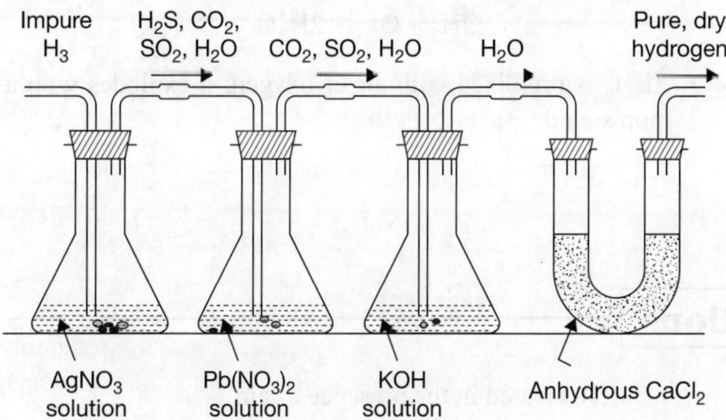

(iii) Caustic potash solution (for removing sulfur dioxide, carbon dioxide and oxides of nitrogen).

$$SO_2 + 2KOH \rightarrow K_2SO_3 + H_2O$$

$$CO_2 + 2KOH \rightarrow K_2CO_3 + H_2O$$

$$2NO_2 + 2KOH \rightarrow KNO_2 + KNO_3 + H_2O$$

(iv) Drying agents like fused $CaCl_2$ or P_2O_5 (for removing water vapour). Pure and dried hydrogen gas is collected over mercury as there is no reaction between them.

- Here concentrated sulfuric acid is not used as a drying agent as it reacts with hydrogen.

$$H_2SO_4 + H_2 \rightarrow 2H_2O + SO_2$$

Precautions: A mixture of hydrogen and air explodes violently on bringing it near a flame. So, the following precautions are necessary.

- The apparatus must be airtight in order to avoid any leakage.
- There must not be any flame burning near the apparatus.
- Hydrogen gas must be collected only after the escape of all air from the apparatus.
- The end of the thistle funnel must be dipped under an acid in order to prevent the gas from escaping from the funnel.

Electrolysis of Acidulated Water

Highly pure hydrogen can be prepared by the electrolysis of water having a small amount of an acid or a base.

$$2H_2O \xrightarrow{\text{Electrolysis}} 2H_2 + O_2$$

$$H_2O \rightleftharpoons H^+ + OH^-$$

At the cathode: $H^+ + e^- \rightarrow H$

 $H + H \rightarrow H_2(g)$

At the anode: $4OH^- \rightarrow 4OH + 4e^-$

 $4OH \rightarrow 2H_2O + O_2\ (g)$

Pure hydrogen is formed at the cathode and pure oxygen is formed at the anode respectively (Fig. 10.2).

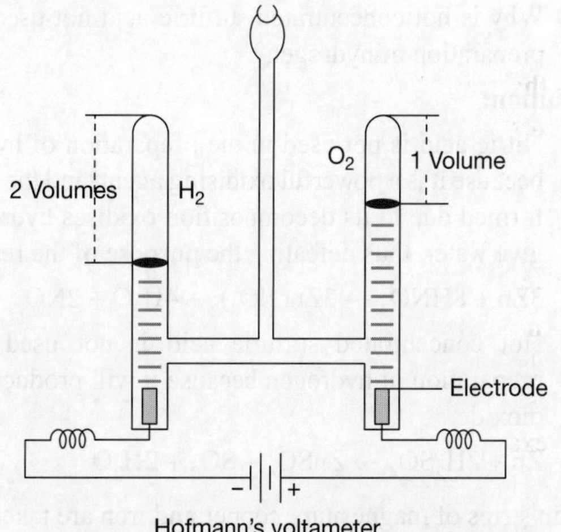

Fig. 10.2 Preparation of hydrogen from acidulated water

Industrial Methods

Bosch and Lane methods are used to prepare hydrogen.

Bosch Method

This method includes two steps.

(i) In the first step, steam is passed over hot coke around 1000°C in a converter (furnace) to get water gas.

$$C + H_2O \xrightarrow{1000°C} \underset{\text{Water gas}}{CO + H_2}$$

(ii) In the second step, water gas mixed with steam is passed over the catalyst heated up to 450°C to get hydrogen.

$$\underset{\text{Water gas}}{CO + H_2} + H_2O \xrightarrow[450°C]{Fe_3O_4 + Cr_2O_3} 2H_2 + CO_2 \uparrow$$

The first step is endothermic and the second step is exothermic.

Removal of CO_2: CO_2 can be separated from hydrogen by passing the mixture through cold water under 30 atm pressure where CO_2 dissolves leaving behind H_2.

Or

When the mixture is passed through caustic potash solution, CO_2 is removed as K_2CO_3.

$$2KOH + CO_2 \rightarrow K_2CO_3 + H_2O$$

Removal of CO: To remove CO, the mixture is passed through ammoniacal cuprous chloride solution where CO is removed leaving behind H_2.

$$CuCl + CO + 2H_2O \rightarrow CuCl \cdot CO \cdot 2H_2O$$

Lane Method

Super-heated steam is passed over iron at a high temperature to get hydrogen. Iron can be regenerated by passing water gas over the iron oxide.

$$3Fe + 4H_2O \xrightarrow[1000°]{\Delta} Fe_3O_4 + 4H_2 \uparrow$$

This process is called gassing and it is continued for nearly 10 minutes till the surface of iron is covered by a layer of the iron oxide and the reaction begins to slow down. This iron oxide can be further reduced into iron if hydrogen supply is cut off and water gas is circulated. The circulation of water gas is called the reduction period and it is nearly for 20 minutes.

$$Fe_3O_4 + H_2 \rightarrow 3FeO + H_2O$$
$$FeO + H_2 \rightarrow Fe + H_2O$$
$$Fe_3O_4 + CO \rightarrow 3FeO + CO_2$$
$$FeO + CO \rightarrow Fe + CO_2$$

Vyeno Method

$$2Al + 2KOH + 2H_2O \rightarrow 2KAlO_2 + 3H_2$$

Water Gas Shift Reaction

$$CO + H_2O \xrightarrow[\text{Fe and Co oxides}]{673 \text{ K}} CO_2 + H_2$$

Test of Hydrogen

(i) It is combustible. However, it does not support combustion.

(ii) It burns silently in air or O_2 with a pale blue flame to give water.

$$2H_2 + O_2 \rightarrow 2H_2O$$

- If H_2 is premixed with air or oxygen, it explodes with a pop sound.

Illustrations

1. (a) Why is nitric acid not used in the preparation of hydrogen?

(b) Why is hot concentrated sulfuric acid not used in the preparation of hydrogen?

Solution:

(a) Nitric acid is not used in the preparation of hydrogen because it is a powerful oxidising agent, and the oxygen formed due to its decomposition oxidises hydrogen to give water, thus defeating the purpose of the reaction.

$$3Zn + 8HNO_3 \rightarrow 3Zn(NO_3)_2 + 4H_2O + 2NO$$

(b) Hot concentrated sulfuric acid is not used in the preparation of hydrogen because it will produce sulfur dioxide.

$$Zn + 2H_2SO_4 \rightarrow ZnSO_4 + SO_2 + 2H_2O$$

2. Thin strips of magnesium, copper and iron are taken.

(a) Write down what happens when these metals are treated as follows:

(i) Heated in the presence of air.

(ii) Heated with dil. HCl and added to an aqueous solution of zinc sulfate.

(iii) Added to an aqueous solution of zinc sulfate.

(b) Arrange these metals in the increasing order of reactivity.

Solution:

(a) (i) When we heat magnesium, copper and iron in the presence of air they form their respective oxides.

(ii) Magnesium and iron react with HCl liberating hydrogen and forming their respective salts.

(iii) Magnesium will displace zinc from zinc sulfate solution as magnesium is more reactive than zinc in the activity series of metals.

(b) Cu < Fe < Mg

Physical Properties

(i) It is a colourless, tasteless and odourless gas.

(ii) It is the lightest among all the known substances and has a density of 0.0899 g/cm^3.

(iii) It is less soluble in water.

(iv) It is highly inflammable and should be handled with care.

(v) Its melting and boiling points are 13.957 K and 20.39 K respectively.

Property	H_2
Molecular mass	2.016
Boiling point (K)	20.39
Enthalpy of dissociation (kJ mol^{-1}) (298.2 K)	435.88
Enthalpy of fusion (kJ mol^{-1})	0.117
Enthalpy of vaporisation (kJ mol^{-1})	0.904
Internuclear distance (pm)	74.14
Melting point (K)	13.957

Chemical Properties

It is not quite reactive at ordinary temperatures due to its high bond dissociation energy (436 kJ mol^{-1}) but at higher temperatures it becomes more reactive.

(i) **Combustion:** Hydrogen undergoes combustion in air or O_2 (with an almost invisible pale blue flame) giving water through an exothermic reaction.

$$2H_2 + O_2 \rightarrow 2H_2O + 485 \text{ kJ}$$

(ii) **Reaction with halogens:** It reacts with halogens and forms HX.

$$H_2(g) + Cl_2(g) \xrightarrow{\text{light}} 2HCl(g)$$

$$H_2(g) + Br_2(g) \xrightarrow[\text{catalyst}]{620 \text{ K}} 2HBr(g)$$

(iii) **Reaction with nitrogen:** Hydrogen combines with nitrogen in a 3 : 1 ratio at 490°C and 200 atm pressure in the presence of iron catalyst to produce ammonia.

$$N_2(g) + 3H_2(g) \rightleftharpoons 2NH_3(g)$$

(iv) **Reaction with non-metals of oxygen family:** It reacts with oxygen, sulfur, selenium and tellurium to form their corresponding compounds under favourable conditions.

$$2H_2 + O_2 \xrightarrow{654°C} 2H_2O$$

$$H_2 + S \xrightarrow[\text{Pressure}]{427°C} H_2S$$

(v) **Reaction with carbon:** It combines with carbon at 1147°C to form methane.

$$C(s) + 2H_2(g) \xrightarrow{1147°C} CH_4(g)$$

However, in the presence of an electric arc, it gives acetylene.

$$2C(s) + H_2(g) \xrightarrow[3027°C]{\text{Electric arc}} C_2H_2(g)$$

(vi) **Reaction with carbon monoxide:** It reacts with carbon monoxide at 427°C and at 200 atm. pressure in the presence of the catalyst ZnO/Cr$_2$O$_3$ to give methanol.

$$CO + 2H_2 \xrightarrow[\text{700 K, 200 atm}]{ZnO/Cr_2O_3} CH_3OH$$

(vii) **Reaction with metals:** It reacts with many metals at high temperature to form metal hydrides. For example,

$$2Li + H_2 \rightarrow 2LiH$$

$$Ca + H_2 \rightarrow CaH_2$$

(viii) **Reducing properties:** It can reduce metal oxides into metals. For example,

$$Fe_3O_4 + 4H_2 \rightarrow 3Fe + 4H_2O$$

$$Ag_2O + H_2 \rightarrow 2Ag + H_2O$$

(ix) **Hydrogenation reactions:** Hydrogen is used for these reactions in the presence of catalysts like Ni, Pd and Pt. For example,

$$\text{Vegetable oil} + H_2 \xrightarrow{\Delta,\ Ni} \text{Vegetable ghee}$$

$$-C\equiv C- \xrightarrow{H_2/Ni} \quad C=C \quad \xrightarrow{H_2} \quad C-C$$

Alkyne · Alkane

Uses of Hydrogen

(i) **In airships, balloons:** In airships and balloons, a mixture of 15% hydrogen and 85% helium is used as hydrogen is highly inflammable.

(ii) **In the preparation of compounds:** It is used to prepare many useful compounds such as methyl alcohol and ammonia.

(iii) **As a cryogenic fluid:** It is used to produce low temperatures—as a cryogenic fluid.

(iv) **In the manufacture of synthetic petrol:** Hydrogen is used in the manufacture of synthetic petrol by heating with coal and heavy oils under very high pressure in the presence of a catalyst.

(v) **Oxygen–hydrogen torch:** A mixture of H_2 and O_2 when burnt in a specially designed apparatus is known as an oxygen–hydrogen torch. It produces a very high temperature of nearly 2500°C. This flame is used for cutting, welding metals, melting platinum and quartz, and also for fusing alumina to get rubies and sapphires (gems).

(vi) **Atomic hydrogen torch:** It creates a very high temperature up to 2800°C which is used for welding, for alloys having metals like W, Mn, Cr, and so on.

(vii) **As a fuel:** Hydrogen has a high heat of combustion; so, it can be used as a fuel in the form of coal gas, water gas. Liquid hydrogen when mixed with liquid oxygen can be used as a fuel for rockets.

KNOWLEDGE BOOSTER

Hydrides

Hydrogen can form binary hydrides of MH_x or M_mH_n types with nearly all elements except noble gases and metals of Groups 7, 8 and 9 (hydride gap).

Types of hydrides

Gibbs classified hydrides into the following types.

- **Saline or ionic hydrides (MH_n):** These are ionic or salt like hydrides which are formed by alkali metals and alkaline earth metals except Be and Mg. (Be and Mg do not form these due to their small size.)
- **Molecular or covalent hydrides:** These hydrides are formed by all *p*-block non-metallic elements except the zero group elements. Such hydrides have a general formula MH_{8-n} (n = number of valence electrons). For example, CH_4, NH_3, HX and so on.
- **Interstitial or metallic hydrides:** These are formed by the transition and inner transition elements. At elevated temperatures they absorb hydrogen which easily fits into the interstitial sites of their lattices. For example, ScH_2, LaH_2, VH.

Illustrations

1. What do you observe when hydrogen gas is passed through a soap solution?

    Solution: When hydrogen gas is passed through a soap solution, soap bubbles filled with hydrogen fly high and burst. This behaviour proves that hydrogen is lighter than air.

2. Under what conditions can hydrogen be made to combine with

    (a) nitrogen and (b) chlorine?

    Name the products in each case and write the equation for each.

    Solution:

    (a) Three volumes of hydrogen and one volume of nitrogen react at 450–500°C and 200–900 atm in the presence of finely divided iron catalyst with molybdenum as a promoter to give ammonia.

    $$N_2 + 3H_2 \rightleftharpoons 2NH_3$$

(b) Equal volumes of hydrogen and chlorine react slowly in diffused sunlight to form hydrogen chloride.

$$H_2 + Cl_2 \rightarrow 2HCl$$

3. (a) Write the reaction between carbon monoxide and hydrogen.

    (b) Write the reaction between carbon and hydrogen under an electric arc.

    (c) Convert C_2H_2 into C_2H_6 by hydrogenation.

Solution:

(a) $CO + 2H_2 \xrightarrow[\text{700 K, 200 atm}]{\text{ZnO/Cr}_2\text{O}_3} CH_3OH$

(b) $2C(s) + H_2(g) \xrightarrow[\text{3027°C}]{\text{Electric arc}} C_2H_2(g)$

(c) $HC \equiv CH + H_2 \xrightarrow[\text{Ni}]{\Delta} H_2C = CH_2 \xrightarrow[\text{Ni}]{\Delta} CH_3 - CH_3$

CHAPTER AT A GLANCE

- **Hydrogen** is the first, lightest, smallest, non-metallic element in the periodic table.
- It is the most abundant element in the universe (70%) and the third most abundant element on the surface of the globe.
- As hydrogen has only 1 valence electron, it is placed in the first group and the first period of the **periodic table**.
- Hydrogen resembles the alkali metals in electronic configuration, formation of monopositive ions, electropositive nature, valency and oxidation state (+1), reducing property and so on.
- Hydrogen resembles the halogens in electronic configuration, non-metallic nature, atomicity (2), physical state as gas, electronegative nature, high ionisation energy, –1 oxidation state and so on.

- Some metals (Pd, Pt, Au, Ni) can adsorb hydrogen on their surface and it is called **occlusion**.
- Dihydrogen gas has two nuclear spin isomers known as **ortho** and **para dihydrogen**. In ortho hydrogen, the nuclear spin is in the same direction but in para hydrogen it is in the opposite direction.
- **Highly pure hydrogen** can be prepared by the electrolysis of water having small amount of an acid or a base.
- **Electrolysis** is the passing of a direct electric current through an electrolyte producing chemical reactions at the electrodes and decomposition of the materials.
- An **electrolyte** is a substance that produces an electrically conducting solution when dissolved in a polar solvent, such as water.

- An **electrode** is a solid electric conductor that carries electric current into non-metallic solids, liquids, gases, plasmas or vacuums.
- An **electrolytic cell** uses electrical energy to drive a non-spontaneous redox reaction.
- In the laboratory, **hydrogen** is prepared by treating zinc with dil. HCl or H_2SO_4. Concentrated sulfuric acid is not used as a drying agent as it reacts with hydrogen. Pure zinc is not used as it is non-porous, so impure granulated zinc is used which becomes porous due to the presence of some impurities.
- Metals like Na, K, Ca react with cold water to form hydrogen and their hydroxides. These reactions are exothermic in nature.
- Pure H_2 burns with a pale blue flame.
- Due to the high reactivity of Na, K with air they are kept in kerosene.
- Active metals like Zn, Mg, Al, Fe cannot produce hydrogen with dilute HNO_3 since HNO_3 oxidises H_2 into H_2O.
- Bosch method and Lane method are used to prepare hydrogen in industries.
- Hydrogen burns silently in air or O_2 with a pale blue flame to give water.

- In airships and balloons, a mixture of 15% hydrogen and 85% helium is used as hydrogen is highly inflammable.
- Hydrogen is used in the manufacture of synthetic petrol by heating with coal and heavy oils under very high pressure in the presence of a catalyst.
- A mixture of H_2 and O_2 when burnt in a specially designed apparatus is known as an oxygen–hydrogen torch. It produces a very high temperature of nearly 2500°C. This flame is used for cutting and welding metals.
- Hydrogen can form binary hydrides of MH_x or M_mH_n types with nearly all elements excepts noble gases and metals of Groups 7, 8 and 9 (**hydride gap**).
- **Ionic** or **salt like hydrides** are formed by alkali metals and alkaline earth metals except Be and Mg.
- **Molecular** or **covalent hydrides** are formed by all *p*-block non-metallic elements except zero group elements. Such hydrides have a general formula MH_{8-n} (*n* = number of valence electrons).
- **Interstitial** or **metallic hydrides** are formed by transition and inner transition elements by absorption of hydrogen at elevated temperatures which easily fits in the interstitial sites of their lattices.

PRACTICE QUESTIONS

Analyse Your Concepts (School Exam Based)

Fill in the Blanks

Instructions: Complete the following statements with an appropriate word/term to be filled in the blank spaces.

1. Hydrogen was discovered by _______________.

2. Hydrogen is a _______________ atomic gaseous molecule.

3. Oxidation states of hydrogen can be + ... and – _______________.

4. Hydrogen has a _______________ value of ionisation energy.

5. Sodium hydroxide + zinc $\rightarrow$ hydrogen + _______________.

6. _______________ metal is preferred for collecting hydrogen from hot water.

7. _______________ metal is preferred for collecting hydrogen from steam.

8. Calcium + water $\rightarrow$ calcium hydroxide + _______________.

9. _______________ metal is preferred for collecting hydrogen from cold water.

10. When CuO reacts with hydrogen, _______________ is reduced and _______________ is oxidised to _______________.

11. Hydrogen is _______________ soluble in water.

12. A metal _______________, hydrogen in the activity series gives hydrogen with _______________ acid.

13. Hydrogen has a _______________ specific heat value.

14. Hydrogen reacts with nitrogen at 490°C and 200 atm pressure to give _______________.

True or False

Instructions: Read the following statements and write your answer as true or false.

1. Hydrogen is the most abundant element in the universe.

2. Water has 2% by weight of hydrogen.

3. Hydrogen can form both H^+ and H^-.

4. Hydrogen is a good oxidising agent.

5. Sodium amalgam reacts smoothly with cold water.

6. Hydrogen adsorbed on the surface of Pt is known as occluded hydrogen.

7. Ag and Au can easily form hydrogen on reaction with dilute acids.

8. Potassium reacts less violently than sodium with water.

9. Hydrogen in the presence of potassium vapour burns with a lilac coloured flame.

10. Zn and Mg can produce hydrogen easily with HNO_3.

11. Hydrogen is a non-inflammable gas.

12. Hydrogen is used to convert vegetable oil into ghee.

13. CaH_2 (hydrolith) is an ionic hydride.

14. *d*-block elements form interstitial hydrides.

Very Short Answer Type Questions

1. How did the name hydrogen originate?

2. Write some sources of hydrogen.

3. Why is hydrogen called 'inflammable air'?

4. How will you justify the position of hydrogen in the periodic table?

5. How does hydrogen differ from (a) alkali metals, (b) halogens? Give at least one case of each.

6. How can you say hydrogen shows dual nature?

7. Which test should be done before collecting hydrogen in a gas jar?

8. Give equations to express the reaction between
   (a) calcium and water
   (b) steam and red-hot iron

9. Hydrogen can be prepared from zinc by using water. Write the equation.

10. Hydrogen can be prepared from zinc by using alkali. Write the equation.

11. Why is purified and dry hydrogen collected over mercury?

12. Why should the end of the thistle funnel be dipped under acid?

13. Based on the reactions of water on metals, arrange these metals in the increasing order of reactivity. Iron, sodium magnesium, zinc, calcium.

14. If the following are kept in closed vessels at over 400°C, what will happen to them?
    (a) Iron filings and steam
    (b) Hydrogen and magnetic oxide of iron

15. Why is helium used in airships and balloons?

Short Answer Type Questions

1. Give the general group study of hydrogen with reference to
   (a) valence electrons   (b) burning
   (c) reducing power

2. Compare hydrogen with alkali metals on the basis of:
   (a) reaction with oxygen   (b) oxide formation
   (c) ion formation   (d) reducing power

3. Compare hydrogen and halogens on the basis of:
   (a) physical state   (b) ion formation
   (c) valency   (d) reaction with oxygen

4. Which metal is preferred for the preparation of hydrogen
   (a) from water?   (b) from acid?

5. Answer the following:
   (a) Write the reaction of steam with red hot iron.
   (b) Why is this reaction considered a reversible reaction?
   (c) How can the reaction proceed continuously?

6. Hydrogen can be manufactured by the Bosch process.
   (a) Give the equations involved with conditions.
   (b) How can you obtain hydrogen from a mixture of hydrogen and carbon monoxide?

7. A small piece of calcium metal is put into a small trough containing water. There is effervescence and white turbidity is formed.
   (a) Write the name of gas formed in the reaction. How will you test this gas?
   (b) Write a balanced equation for the reaction.
   (c) What will you observe when a few drops of red litmus solution are added to the turbid solution?

8. By applying the activity series, name a metal which shows the following properties.
   (a) That reacts readily with cold water.
   (b) That displaces hydrogen from hot water.
   (c) That forms a base which is insoluble in water.
   (d) That displaces hydrogen from dilute HCl.

9. Under what conditions can hydrogen be made to combine with
   (a) sulfur   (b) oxygen?
   Name the products in each case and write the equation for each.

10. What is an oxygen–hydrogen torch and for what is used?

Long Answer Type Questions

1. (a) Why are zinc and aluminium considered to have a unique nature? Give balanced equations to support your answer.
   (b) Write balanced equations for the following.
     (i) Iron reacts with dil. HCl
    (ii) Aluminium reacts with fused sodium hydroxide
   (iii) Lead reacts with potassium hydroxide
    (iv) Zinc reacts with caustic soda solution

2. For the laboratory preparation of hydrogen, give the following:
   (a) materials used   (b) chemical equation
   (c) method of collection   (d) fully labelled diagram

3. Give reasons for the following.
   (a) Hydrogen is collected by the downward displacement of water and not of air, even though it is lighter than air.
   (b) A candle brought near the mouth of a jar containing hydrogen gas starts burning but is extinguished when pushed inside the jar.
   (c) Apparatus for the laboratory preparation of hydrogen should be airtight and away from a naked flame.

(d) Potassium and sodium are not used for reaction with dilute hydrochloric acid or dilute sulfuric acid in the laboratory preparation of hydrogen.

4. Write the names of impurities present in hydrogen prepared in the laboratory and how can these impurities be removed from hydrogen?

5. When hydrogen is passed over a black solid compound P, the products are 'a colourless liquid' and 'a reddish-brown metal Q.'
Substance Q is divided into two parts, each placed in separate test tubes.

Dilute HCl is added to one part of substance Q and dilute HNO_3 to the other.
(a) Name the substances P and Q.
(b) Give two tests for the colourless liquid formed in the experiment.
(c) What happens to substance P when it reacts with hydrogen?
(d) Write an equation for the reaction between hydrogen and substance P.
(e) Is there any reaction between substance Q and dilute hydrochloric acid?
Give proper reasons for your answer.

COMPETITION WINDOW (OBJECTIVE TYPE)

[For NEET, JEE (Main and Advanced), NTSE, KVPY and Olympiads]

Topic-wise MCQs

Preparation

1. Which of the following is incorrect regarding hydrogen?
   (a) it does not have any neutron
   (b) it can exist as H^+ and H^-
   (c) it can show +1 and −1 oxidation states
   (d) it can act both as an oxidant and a reductant

2. Which of the following is not a similarity between hydrogen and alkali metals?
   (a) atomicity
   (b) physical state
   (c) reaction with non-metals
   (d) both (a) and (b)

3. Which of the following compounds is possible when hydrogen burns in air?
   (a) H_2O (b) H_2O_2
   (c) both (a) and (b)   (d) no compound is formed

4. Hydrogen from HCl can be prepared using
   (a) P   (b) Mg
   (c) Hg   (d) Cu

5. Metal hydride on treatment with water gives
   (a) hydrogen   (b) acid
   (c) H_2O (d) H_2O_2

6. The component present in greater proportion in water gas is
   (a) CO   (b) H_2
   (c) CO_2 (d) CH_4

7. Which of the following metals is not used in the lab preparation of hydrogen?
   (a) calcium   (b) iron
   (c) aluminium   (d) sodium

8. The metal that cannot displace hydrogen from dil. HCl is
   (a) Fe   (b) Cu
   (c) Zn   (d) Mg

9. Aluminium reacts with conc. HCl and conc. NaOH to liberate the gases _______ respectively.
   (a) H_2 and H_2 (b) O_2 and O_2
   (c) O_2 and H_2 (d) H_2 and O_2

10. What gas is liberated when alkaline formaldehyde solution is treated with H_2O_2?
    (a) CH_4 (b) H_2
    (c) CO_2 (d) O_2

Properties of hydrogen

11. Which of the following statements about hydrogen is incorrect?
    (a) It is an inflammable gas
    (b) It is the lightest gas
    (c) It is not easily liquefied
    (d) It is a strong oxidising agent

12. Hydrogen burns in air or oxygen with a/an
    (a) invisible pale blue flame   (b) red flame
    (c) golden flame   (d) blue flame

13. Ortho and para hydrogen differ
    (a) in their molecular weights
    (b) in the number of protons
    (c) in the nature of spin of electrons
    (d) in the nature of spin of protons

14. Which metal adsorbs hydrogen?
    (a) Al   (b) Fe
    (c) Pd   (d) K

15. The composition of the nucleus of deuterium is
    (a) $1e^-$ and $1p$ (b) $1e^-$ and $1n$
    (c) $1n$ and $1p$ (d) $2p$ and $1e^-$

16. Helium is preferred to hydrogen for filling balloons because it is
    (a) lighter than air
    (b) almost as light as hydrogen
    (c) non-combustible
    (d) inflammable

17. A metal oxide that is reduced by hydrogen is
    (a) Al_2O_3 (b) CuO
    (c) CaO (d) Na_2O

18. For the reaction $PbO + H_2 \rightarrow Pb + H_2O$, which of the following statements is wrong?
    (a) H_2 is the reducing agent
    (b) PbO is the oxidising agent
    (c) PbO is oxidised to Pb
    (d) H_2 is oxidised to H_2O

19. Hydrogen will not reduce
    (a) heated cupric oxide   (b) heated ferric oxide
    (c) heated stannic oxide (d) heated aluminium oxide

20. Which of the following could act as a propellant for rockets?
    (a) liquid hydrogen + liquid nitrogen
    (b) liquid nitrogen + liquid oxygen
    (c) liquid hydrogen + liquid oxygen
    (d) liquid oxygen + liquid argon

21. Which of the following hydrides is an electron precise hydride?
    (a) B_2H_6 (b) NH_3
    (c) H_2O (d) CH_4

22. Only one element of _______ forms the hydride.
    (a) group 6 (b) group 7
    (c) group 8 (d) group 9

Miscellaneous

1. Hydrogen resembles the halogens in many respects for which several factors are responsible. Of the following factors, which one is most important in this respect?
   (a) Its tendency to lose an electron to form a cation.
   (b) Its tendency to gain a single electron in its valence shell to attain a stable electronic configuration.
   (c) Its low negative electron gain enthalpy value.
   (d) Its small size.

2. Which metal gives hydrogen with all of the following—water, acids, alkalis?
   (a) Fe (b) Zn
   (c) Mg (d) Pb

3. Why does the H^+ ion always get associated with other atoms or molecules?
   (a) Ionisation enthalpy of hydrogen resembles that of alkali metals.
   (b) Its reactivity is similar to that of halogens.
   (c) It resembles both alkali metals and halogens.
   (d) Loss of an electron from the hydrogen atom results in a nucleus that is very small in size when compared to other atoms or ions. Due to its small size, it cannot exist free.

4. Which of the properties of interstitial hydrides is correct?
   (a) they give rise to metals fit for fabrication
   (b) they generally form non-stoichiometric species
   (c) they can be used as hydrogenation catalysts
   (d) the hydrogen dissolved in titanium improves its mechanical properties

5. In the following compound, H is covalently bonded in the case of
   (a) CaH_2 (b) NaH
   (c) SiH_4 (d) BaH_2

6. Tritium is a radioactive isotope of hydrogen. It emits
   (a) α-particles (b) β-particles
   (c) γ-rays (d) neutrons

7. Hydrogen is evolved by the action of cold dil. HNO_3 on
   (a) Fe (b) Cu
   (c) Mg (d) Zn

8. Elements which show a unique property in the preparation of hydrogen are
   (a) Na, K, Li (b) Mg, Ca, Ba
   (c) Al, Zn, Pb (d) Fe, Cu, Ag

9. Which of the following reactions increases the production of dihydrogen from synthesis gas?
   (a) $CH_4(g) + H_2O(g) \xrightarrow[\text{Ni}]{1270\ K} CO(g) + 3H_2(g)$
   (b) $C(s) + 3H_2O(g) \xrightarrow{1270\ K} CO(g) + H_2(g)$
   (c) $CO(g) + H_2O(g) \xrightarrow[\text{catalyst}]{673\ K} CO_2(g) + H_2(g)$
   (d) $C_2H_6 + 2H_2O \xrightarrow[\text{Ni}]{1270\ K} 2CO + 5H_2$

10. Elements of which of the following group(s) of the periodic table do not form hydrides?
    (a) Group 7, 8, 9 (b) Group 13
    (c) Groups 15, 16, 17 (d) Group 14

11. Which of the following is a use of hydrogen?
    (a) in the preparation of artificial petrol from coal
    (b) in the extraction of metals like copper
    (c) in oxygen–hydrogen torch
    (d) all of these

12. The adsorption of hydrogen is possible on the surface of which metal?
    (a) Pt (b) Pd
    (c) Ni (d) all of these

13. Which of the following metal oxides can be reduced by hydrogen?
    (a) CuO (b) PbO
    (c) Fe_2O_3 (d) all of these

14. Which of the following react most violently with dil. acids to give hydrogen?
    (a) Na (b) K
    (c) Ca (d) Mg

15. In which of the following pairs are both metals below hydrogen in the activity series?
    (a) Ca, Cu (b) Mg, Ag
    (c) Cu, Ag (d) Al, Ag

Advanced and Olympiads

Single Choice

1. Which of the following reactions is an example of the use of water gas in the synthesis of other compounds?

(a) $CH_4(g) + H_2O(g) \xrightarrow[\text{Ni}]{1270\ K} CO(g) + H_2(g)$

(b) $CO(g) + H_2O(g) \xrightarrow[\text{catalyst}]{673\ K} CO_2(g) + H_2(g)$

(c) $C_nH_{2n+2} + nH_2O(g) \xrightarrow[\text{Ni}]{1270\ K} nCO + (2n+1)H_2$

(d) $CO(g) + 2H_2(g) \xrightarrow[\text{catalyst}]{\text{Cobalt}} CH_3OH(l)$

2. Metal hydrides are ionic, covalent or molecular in nature. Among LiH, NaH, KH, RbH, CsH, the correct order of increasing ionic character is

(a) LiH > NaH > CsH > KH > RbH

(b) LiH < NaH < KH < RbH < CsH

(c) RbH > CsH > NaH > KH > LiH

(d) NaH > CsH > RbH > LiH > KH

3. 2 g of aluminium is treated separately with excess of dilute H_2SO_4 and excess of NaOH. The ratio of the volumes of hydrogen evolved is

(a) $1:1$ (b) $1:2$

(c) $2:1$ (d) $2:3$

4. Consider the following statements:

I. Atomic hydrogen is obtained by passing hydrogen through an electric arc.

II. Hydrogen gas will not reduce heated aluminium oxide.

III. Finely divided palladium absorbs a large volume of hydrogen gas.

IV. Pure nascent hydrogen is best obtained by reacting Na with C_2H_5OH.

Which of the above statements is/are correct?

(a) I alone   (b) II alone

(c) I, II and III   (d) II, III and IV

5. Match the following

List I	List II
1. Hydrogenation of oils	(i) Fe filings
2. Haber's process	(ii) Ni
3. Nuclear fusion	(iii) hydrogen
4. Most abundant in the universe	(iv) solar energy

The correct matching is

	1	2	3	4
(a)	(iv)	(iii)	(ii)	(i)
(b)	(iv)	(iii)	(i)	(ii)
(c)	(ii)	(i)	(iv)	(iii)
(d)	(iii)	(i)	(iv)	(ii)

Multiple Choice

6. Select the correct statements regarding hydrogen.

(a) It is the most abundant element in the universe.

(b) Hydrogen forms the maximum number of compounds in total.

(c) Hydrogen has less electronegativity like alkali metals.

(d) Like halogens, hydrogen has a high value of ionisation energy.

7. Metals which liberate $H_2(g)$ from acids are

(a) Fe   (b) Zn

(c) Mn   (d) Cu

8. Select the incorrect statements.

(a) Hydrogen is separated from CO by passing the mixture through caustic potash solution.

(b) All metals react with acids to give hydrogen.

(c) Hydrogen is dried by passing it through conc. H_2SO_4.

(d) Very dilute nitric acid reacts with iron to produce hydrogen.

9. Which of the following statements about hydrogen are not true?

(a) It exists as a diatomic molecule.

(b) It has one electron in the outermost shell.

(c) It can lose an electron to form a cation which can exist freely.

(d) It forms a large number of ionic compounds by losing an electron.

10. Dihydrogen can be prepared on a commercial scale by different methods. In its preparation by the action of steam on hydrocarbons, a mixture of CO and H_2 gas is formed. This is known as ________.

(a) water gas

(b) syngas

(c) producer gas

(d) industrial gas

Matrix Matching

11. Match the following

Column I	Column II
(a) H	(p) Used in the name of perhydrol
(b) H_2	(q) Can be reduced to dihydrogen by NaH
(c) H_2O	(r) Can be used in hydroformylation of olefins
(d) H_2O_2	(s) Can be used for cutting and welding

12. Match the following

Column I	Column II
(a) $2Al + 2KOH + 2H_2O \rightarrow$ $2KAlO_2 + 3H_2$	(p) Vivification
(b) $Fe_3O_4 + 4CO \rightarrow 3Fe + 4CO_2$	(q) Involved in Lane's process
(c) $Fe_3O_4 + 4H_2 \rightarrow 3Fe + 4H_2O$	(r) Uyeno's method
(d) $C + H_2O(g) \rightarrow CO(g) + H_2$	(s) Bosch process

Integer Type

13. If the most common oxidation states of hydrogen in its compounds are $+X$ and $-Y$, the sum of $X + Y$ is?

14. How many of these metals can replace hydrogen from dilute acids?
K, Na, Mg, Zn, Fe, Cu, Ag, Hg, Au

15. How many of these metals can replace hydrogen from dilute acids with violent reactions?
K, Na, Mg, Zn, Fe, Ca, Ag, Hg, Pb

16. Which of the following metals cannot be used in the laboratory method of hydrogen preparation?
Zn, Fe, Pb, Al, Ca, Mg, Na, K

17. Which of the following metal oxides cannot be reduced to their corresponding metals?
Na_2O, K_2O, MgO, CaO, Al_2O_3, CuO, Fe_2O_3

Answer Keys

Fill in the Blanks

1. Cavendish
2. di
3. 1
4. high
5. sodium zincate
6. magnesium
7. iron
8. hydrogen
9. sodium
10. CuO, H_2, H_2O
11. sparingly
12. above, dilute hydrochloric
13. high
14. ammonia

True or False

1. T
2. F
3. T
4. F
5. T
6. T
7. F
8. F
9. T
10. F
11. F
12. T
13. T
14. T

Very Short Answer Type Questions

1. The name hydrogen (hydro = water, genate = forming) originated from its ability to form water.

2. Hydrogen is found in the earth's crust and in the atmosphere in the free state. It is found in organic compounds in the combined state.

3. Hydrogen is a highly combustible gas; hence, it is called inflammable air.

4. Hydrogen has an atomic number of 1, and it exists in the first group and the first period of the periodic table. Hence it is the first element of the periodic table.

5. (a) Oxides of alkali metals are basic, whereas oxides of hydrogen are neutral.
   (b) The hydrogen atom has only one shell, but halogens have two or more shells.

6. Hydrogen resembles the alkali metals of Group 1 and the halogens of Group 17, hence it shows dual nature.

7. Hydrogen burns with a pop sound, hence it should be tested with a burning flame. If the burning produces a pop sound, it confirms the production of hydrogen gas.

8. (a) $Ca + 2H_2O \rightleftharpoons Ca(OH)_2 + H_2$
   (b) $3Fe + 4H_2O \rightleftharpoons Fe_3O_4 + 4H_2$

9. $Zn + H_2O \rightarrow ZnO + H_2$

10. $Zn + 2NaOH \rightarrow Na_2ZnO_2 + H_2$

11. Purified and dry hydrogen if collected over mercury will not have any impact.

12. The end of the thistle funnel should be dipped under acid to prevent the escape of gas from the thistle funnel.

13. Iron < zinc < magnesium < calcium < sodium.

14. (a) Iron oxide is formed with the evolution of hydrogen gas.
    (b) Heated magnetic oxide of iron gets reduced by hydrogen.

15. In airships and balloons, a mixture of 15% hydrogen and 85% helium is used as hydrogen is highly inflammable.

Short Answer Type Questions

1. (a) Hydrogen has 1 valence electron.
   (b) Hydrogen burns with oxygen to form oxides of hydrogen with a pop sound.
   $$2H_2 + O_2 \rightarrow 2H_2O$$
   (c) Hydrogen acts as a reducing agent.
   $$CuO + H_2 \rightarrow Cu + H_2O$$

2. (a) Both alkali metals and hydrogen burn in air to form oxides.
   (b) Hydrogen and alkali metals can form oxides on reaction with oxygen.

(c) Hydrogen and alkali metals can form a cation by the loss of an electron.

(d) Both hydrogen atoms and alkali metals are reducing agents.

3. (a) Both, the halogens and hydrogen, are gases.

(b) Both of them form anions as they are short of 1 electron to attain a stable configuration.

(c) Both hydrogen and the halogens have a valency of 1.

(d) Hydrogen reacts with oxygen to form neutral oxides; halogens react with oxygen to form acidic oxides.

4. (a) For the preparation of hydrogen from water, reactive metals like potassium, sodium and calcium are preferred.

(b) For the preparation of hydrogen from acid, magnesium, aluminium, zinc and iron are preferred.

5. (a) $3Fe + 4H_2O \rightleftharpoons Fe_3O_4 + 4H_2$

(b) This reaction is reversible. If the hydrogen formed is not removed the iron oxide formed is reduced back to iron. Thus, the reaction can proceed continuously.

6. (a) $C + H_2O \xrightarrow{1000°C} (CO + H_2) - \Delta$

$(CO + H_2) + H_2O \xrightarrow[450°C]{Fe_2O_3} CO_2 + 2H_2 + \Delta$

(b) We can obtain hydrogen from a mixture of hydrogen and carbon monoxide by passing the mixture through an ammoniacal cuprous chloride solution in order to dissolve any uncombined carbon monoxide.

$CuCl + CO + 2H_2O \rightarrow CuCl \cdot CO \cdot 2H_2O$

7. (a) It is hydrogen gas and it can be tested using litmus paper. It turns blue from red.

(b) $Ca + 2H_2O \rightarrow Ca(OH)_2 + H_2$

(c) The solution turns blue.

8. (a) Sodium (Na)

(b) Magnesium (Mg)

(c) Calcium (Ca)

(d) Zinc (Zn)

9. (a) Hydrogen gas on passing through molten sulfur reacts to give hydrogen sulfide.

$H_2 + S \rightarrow H_2S$

(b) Hydrogen burns in the presence of an electric spark with a 'pop' sound in oxygen and with a blue flame, forming water.

$2H_2 + O_2 \rightarrow 2H_2O$

10. A mixture of H_2 and O_2 when burnt in a specially designed apparatus is known as an oxygen–hydrogen torch. It produces a very high temperature of nearly 2500°C. This flame is used for cutting and welding metals, melting platinum and quartz, and also for fusing alumina to get rubies and sapphires (gems).

Long Answer Type Questions

1. (a) Zinc and aluminium are considered to have a unique nature for the following reasons.

(i) They can react with acids and alkali to form hydrogen and soluble salts.

$Zn + 2NaOH \rightarrow Na_2ZnO_2 + H_2$

$2Al + 6NaOH \rightarrow 2Na_2Al\,O_3 + 3H_2$

(ii) They react with both bases and acids to give their salts and water.

$ZnO + 2HCl \rightarrow ZnCl_2 + H_2O$

$ZnO + 2NaOH \rightarrow Na_2ZnO_2 + H_2O$

(b) (i) $Fe + 2HCl \rightarrow FeCl_2 + H_2$

(ii) $2Al + 6NaOH \rightarrow 2Na_2AlO_3 + 3H_2$

(iii) $Pb + 2KOH \rightarrow K_2PbO_2 + H_2$

(iv) $Zn + 2NaOH \rightarrow Na_2ZnO_2 + H_2$

2. (a) Granulated zinc, dilute HCl or dil. H_2SO_4

(b) $Zn + HCl \rightarrow ZnCl_2 + H_2$

(c) It is collected by the downward displacement of water.

(d) For the diagram, see the text part.

3. (a) Hydrogen is collected by the downward displacement of water because it is insoluble in water and it forms an explosive mixture with air.

(b) Hydrogen is a combustible gas but it does not support the combustion process. When a candle is brought near the mouth of a jar containing hydrogen, it burns with the help of the oxygen in the atmosphere. When the candle is pushed inside the jar, oxygen supply gets cut off and hence the fire gets extinguished.

(c) Apparatus for the laboratory preparation of hydrogen should be airtight and away from a naked flame because hydrogen forms an explosive mixture with air.

(d) Potassium and sodium are not used for reaction with dilute hydrochloric acid or dilute sulfuric acid in the laboratory preparation of hydrogen as both potassium and sodium react violently with acids.

4. See the text part.

5. (a) Here P is CuO and Q is Cu.

(b) Blue and red litmus paper, when dipped in the colourless liquid, do not change colour. This confirms that the liquid formed is neutral and is water. It changes white anhydrous copper sulfate to a blue salt.

(c) Black copper oxide P on heating with hydrogen reduces copper oxide to reddish-brown copper, and itself gets oxidised to water. Hydrogen is a strong reducing agent and removes oxygen from less active metals. That is, it removes oxygen from heated metal oxides when passed over them and itself gets oxidised to water.

(d) $CuO + H_2 \rightarrow \Delta\ Cu + H_2O$

(e) $Cu + HCl \rightarrow$ No reaction

As copper is less reactive than hydrogen, it cannot displace it from HCl.

Competition Window (Objective Type)

Topic-wise MCQs

1. (d)	**2.** (d)	**3.** (c)	**4.** (b)	**5.** (a)
6. (b)	**7.** (a)	**8.** (b)	**9.** (a)	**10.** (b)
11. (d)	**12.** (a)	**13.** (d)	**14.** (c)	**15.** (c)
16. (c)	**17.** (b)	**18.** (c)	**19.** (d)	**20.** (c)
21. (d)	**22.** (a)			

Miscellaneous

1. (b)	**2.** (b)	**3.** (d)	**4.** (b)	**5.** (c)
6. (b)	**7.** (c)	**8.** (c)	**9.** (c)	**10.** (a)
11. (d)	**12.** (d)	**13.** (d)	**14.** (b)	**15.** (c)

Advanced and Olympiads

Single Choice

1. (d)	**2.** (b)	**3.** (a)	**4.** (c)	**5.** (c)

Multiple Choice

6. (abd)	**7.** (abc)	**8.** (bcd)	**9.** (cd)	**10.** (ab)

Matrix Matching

11. (a)—(s); (b)—(r); (c)—(q); (d)—(p)

12. (a)—(r); (b)—(p, q); (c)—(p, q); (d)—(s)

Integer Type

13. 0	**14.** 5	**15.** 3	**16.** 7	**17.** 5

Hints and Solutions

Competition Window (Objective Type)

Topic-wise MCQs

1. It can be only a reductant and not an oxidant.

2. Hydrogen is diatomic and a gas while alkali metals are monoatomic solid metals.

5. $MH_2 + 2H_2O \rightarrow M(OH)_2 + 2H_2$

6. Water gas contains about 50 volumes of H_2, 40 volumes of CO and 5 volumes of CO_2 and N_2 and others.

7. The answer is (a) calcium because it is expensive.

8. Cu is below hydrogen in the electrochemical series and unable to displace hydrogen from HCl.

12. Hydrogen on burning in air gives a pale blue flame.

13. Ortho and para hydrogen have different nuclear spins.

19. Hydrogen will not reduce heated aluminium oxide because the Al–O bond is very strong.

20. A mixture of liquid hydrogen and liquid oxygen is used as a propellant for rockets.

Miscellaneous

15. Both Cu and Ag are below hydrogen in the electrochemical series.

Advanced and Olympiads

Single Choice

3. $2Al + 3H_2SO_4 \rightarrow Al_2(SO_4)_3 + 3H_2$

$2Al + 2NaOH + 2H_2O \rightarrow 2NaAlO_2 + 3H_2$

So, the ratio of volumes of hydrogen evolved is 1 : 1.

Multiple Choice

6. Hydrogen has more electronegativity (2.1) while alkali metals have low electronegativity values.

7. Cu being less electropositive than H, cannot liberate hydrogen from acids.

8. The metals which are above hydrogen in the activity series give hydrogen when they react with an acid.

Hydrogen is dried by passing through calcium chloride, caustic potash and phosphorous pentoxide.

Very dilute nitric acid reacts with magnesium and manganese to produce hydrogen and not with iron.

Integer Type

13. Hydrogen has +1 and –1 as the common oxidation states

$X = +1, Y = -1$

Sum of $X + Y = (+1) + (-1) = 0$

14. K, Na, Mg, Zn, Fe can replace hydrogen from dilute acids.

15. K, Na and Ca can replace hydrogen from dilute acids with violent reaction.

16. Fe, Pb, Al, Ca, Mg, Na, K cannot be used in the laboratory method of preparation of hydrogen.

17. Na_2O, K_2O, MgO, CaO, Al_2O_3 cannot be reduced into their corresponding metals by hydrogen as these metals have very strong affinity towards oxygen.

Atmospheric Pollution

Introduction

The term 'environment' simply means the surroundings and the conditions in which we live. Our environment constitutes the air, water, soil, atmosphere, and the plants and animals around us. Environmental chemistry deals with the changes in chemical dynamics taking place around us and the damage it causes to the environment. The environment consists of four segments:

(i) atmosphere,
(ii) hydrosphere,
(iii) lithosphere, and
(iv) biosphere.

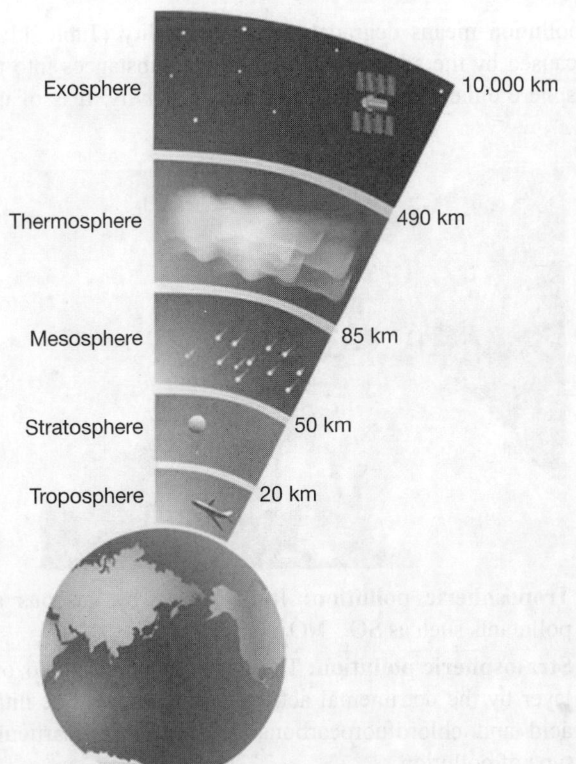

Fig. 11.1 Layers of the atmosphere

Atmosphere

It is the layer of air enveloping the earth at a height of around 1600 km and consists of many gases (Fig. 11.1, Table 11.1). The major components of the atmosphere are nitrogen, oxygen and water vapour. The minor components are argon and carbon dioxide, the trace components include gases like methane. The atmosphere maintains the heat balance on the earth by absorbing electromagnetic radiations coming from the sun and transmitting ultraviolet, visible and infrared radiations.

Hydrosphere

It contains all types of water resources such as oceans, seas, rivers, reservoirs, lakes, polar ice caps and ground water.

Lithosphere

The landmass available on the earth's surface excluding the water bodies constitute this segment of the environment. It is the solid component of the earth consisting of soil, rocks and mountains. The solid, thick, uppermost part of the earth is called the crust. The inner layers of the lithosphere contain minerals and the deep inner layers contain natural gas and oil.

Biosphere

It includes all living organisms and their interactions with the environment—atmosphere, hydrosphere and lithosphere. For example, the levels of oxygen and carbon dioxide depend on plants.

Environmental Pollution and Pollutants

The word pollution is derived from a Latin word 'pollutes' which means 'made dirty'. Environmental pollution is the effect of undesirable changes in our surroundings that have a harmful effect on human beings, animals and plants. Environmental pollution is caused by the addition of any undesirable substance to air, water or soil, naturally or by human activity to such an extent that its adverse effects are observed on human beings, animals and plants. A pollutant is a substance whose presence in undesirably higher concentrations causes pollution, in turn

Table 11.1 Regions of atmosphere

	Troposphere[a]	Stratosphere[b]	Mesosphere	Thermosphere
Altitude (km)	0–10	10–50	50–85	85–500
Temperature range	15–56°C	56–2°C	2–92°C	92–1200°C
Gases/species	N_2, O_2, CO_2, water vapour	N_2, O_2, O_2	N_2, O_2, O_2^+, NO^+	O_2^+, O^+, NO^+

adversely affecting the environment. Hg, Pb, CO and SO_2 are some examples of pollutants. A contaminant is a substance which is not present in nature but is introduced into the environment by human activity and has adverse effects on the environment. For example, methyl isocyanate (MIC). A *receptor* is the medium affected by the pollutant. Human eyes are the receptors for the smoke released by automobiles which cause irritation to eyes.

Sink is the medium which interacts with the long-lived pollutant and removes pollution. Sea water acts as a sink for carbon dioxide. The main causes of pollution are the increase in population and depletion of natural resources, industrialisation, urbanisation and deforestation.

Sources of Pollution

On the basis of their origin, pollutants have two sources—natural and man-made.

Natural Sources

The following two are the natural sources of air pollutants.

(a) Volcanoes: Volcanoes release large amounts of air pollutants such as carbon monoxide, sulfur dioxide, hydrogen sulfide, chlorine, hydrocarbons and particulates.

(b) Decaying vegetation: Microbial action on organic matter in the soil releases pollutants such as nitrous oxide (N_2O). Forest fires release poisonous gases such as carbon monoxide. Winds and dust storms carry particulate matter such as sand and dust.

Man-made Sources of Air Pollutants

- Automobiles use petrol and diesel as fuel. The incomplete combustion of these fuels releases carbon monoxide, sulfur dioxide, nitrogen oxides and hydrocarbons and particulates like lead (now, petrol is lead free).
- Factories release carbon dioxide, sulfur dioxide, nitrogen monoxide and particulates.
- Industrial processes release different types of air pollutants, depending upon the type of process involved. For example, coal power plants release carbon monoxide, sulfur dioxide, ash and smoke. Fertiliser industries mainly release nitrogen oxides and ammonia.
- Burning of plastics releases carbon monoxide and other harmful gases.
- Decay of crop residue in rural areas is the main source of carbon monoxide and methane.

Types of Pollutants

(i) Primary pollutants: These pollutants, after their formation, enter the environment and remain there unchanged. For example, NO, NO_2, CO, SO_2.

(ii) Secondary pollutants: They are formed as a result of the chemical reactions between primary pollutants present in the atmosphere and those in the hydrosphere. For example, peroxy acylnitrate (PAN).

(iii) Biodegradable pollutants: These pollutants can be easily decomposed with the help of microorganisms and are not harmful. For example, domestic sewage, cow dung.

(iv) Non-biodegradable pollutants: These pollutants cannot be decomposed and their presence is very harmful for animals and human beings. For example, DDT, mercury.

Types of Pollution

Some of the major types of pollutions are—air, water, soil and noise.

Air pollution

Air pollution means degradation of air quality (Table 11.2). It is caused by the addition of undesirable substances into the atmosphere either naturally or by human activity. It is of two types.

(i) Tropospheric pollution: It is caused by gaseous air pollutants such as SO_2, NO_2, CO_2 or H_2S.

(ii) Stratospheric pollution: The damage done to the ozone layer by the detrimental action of compounds like nitric acid and chlorofuorocarbons constitute this particular type of pollution.

Table 11.2 Components of non-polluted dry air

Pure air components	By volume (% proportion)	Concentration (ppm)
Nitrogen	78.09	780900
Oxygen	20.94	209400
Inert gases		
Argon	0.93	9300
Neon		18
Helium		5
Krypton		1
Xenon		1
Carbon dioxide	0.03	315
Methane		1
Hydrogen		0.5
Natural pollutants		
Oxides of nitrogen		0.52
Ozone		0.52

Some common air pollutants are:

- oxides of carbon such as CO and CO_2,
- oxides of nitrogen such as N_2O, NO,
- compounds of sulfur such as SO_2, SO_3, H_2S,
- chlorofluorocarbons (freons),
- hydrocarbons such as methane and butane,
- metals such as lead and mercury,
- organic pollutants such as benzopyrene, biocides.

Carbon Monoxide (CO) as Air Pollutant

It is one of the most poisonous air pollutants. Most of the air pollution is because of automobiles in which the carbon fuels undergo incomplete combustion and liberate carbon monoxide.

$$2C + O_2 \rightarrow 2CO$$

Due to degradation of organic substances, methane gas will be liberated which oxidises into CO.

$$2CH_4 + 3O_2 \rightarrow 2CO + 4H_2O$$

Effect: Carbon monoxide combines with hemoglobin to form carboxyhemoglobin. It binds the hemoglobin nearly 200 times more strongly than O_2. Therefore, hemoglobin loses its oxygen-carrying capacity. As a result, the heart and brain suffer from a lack of oxygen and it may lead to paralysing normal brain action and even death.

$$\text{Hemoglobin} + CO \rightarrow \text{Carboxyhemoglobin}$$

Control of CO Pollution

CO pollution can be controlled by applying the following methods.

- Using electrically powered automobiles.
- Using pollution control devices in automobiles.
- Using gasoline as fuel as burning of gasoline does not give CO.

$$2C_8H_{18} + 25O_2 \rightarrow 16CO_2 + 18H_2O$$
$$\text{Octane}$$

- Using fuels like compressed natural gas (CNG) and liquified natural gas (LNG) and alcohols.
- Using catalytic converters like finely divided platinum which oxidise CO into CO_2.

$$CO \xrightarrow[{[O]}]{Pt} CO_2 + H_2O$$

Oxides of Nitrogen as Air Pollutants

Oxides of nitrogen such as N_2O, NO and NO_2 are liberated into the air during the combustion of fossil fuels, burning of fuel in internal combustion engines and also when N_2 and O_2 in the atmosphere react due to electric discharge due to lightning during storms.

$$N_2 + O_2 \rightarrow 2NO$$
$$2NO + O_2 \rightarrow 2NO_2$$
$$NO + O_3 \rightarrow NO_2 + O_2$$

Effect:

- Higher concentration of NO_2 damages the leaves of plants, retards the rate of photosynthesis, causes acute respiratory disease in children, and so on.
- It causes irritation in the mucous membrane.
- In sunlight, NO_2 oxidises hydrocarbons into photochemical smog which causes eye irritation, asthma attacks and nasal and throat infections.
- The irritant red haze in the high traffic places is due to oxides of nitrogen.
- Nitrogen oxide coming out of the supersonic jets directly enters the stratosphere and decomposes the ozone present there.

$$NO + O_3 \rightarrow NO_2 + O_2$$
$$NO_2 + O_3 \rightarrow NO + 2O_2$$

Control: It is possible by using electrical vehicles and using catalytic converters like Pt or Pd.

$$2NO \xrightarrow{Pt} N_2 + O_2$$
$$2NO_2 \xrightarrow{Pt} N_2 + 2NO_2$$

Sulfur Dioxide (SO₂) as Air Pollutant

SO_2 gas is released into the atmosphere directly by burning sulfur or by roasting sulfide ores or by burning fuels containing the sulfur.

$$S + O_2 \rightarrow SO_2$$

Effect:

- SO_2 causes headache, irritation of eyes, vomiting, respiratory tract diseases and may even cause death due to respiratory failure.
- SO_2 bleaches the chlorophyll and thus prevents photosynthesis.
- Higher concentration of SO_2 leads to stiffness of flower buds.
- It destroys vegetation and weakens building materials.
- It mixes with smoke and fog to form harmful smog.
- It also causes acid rain by forming H_2SO_4 as follows:

$$2SO_2(g) + O_2(g) \rightarrow 2SO_3(g)$$

$$SO_3(g) + H_2O(l) \rightarrow H_2SO_4(aq)$$

Control: It can be controlled by minimising combustion of sulfur-containing fuels such as coal and oil.

Hydrogen Sulfide (H_2S) as Air pollutant

H_2S is produced by the decay of organic matter such as vegetables, sewage and industrial effluents.

Effect: It causes nausea, irritation of the eyes and throat and also destroys vegetable matter.

Smoke and Smog

A pollutant that is a combination of the oxides of nitrogen and sulfur and of partially oxidised hydrocarbons and their derivatives produced by industries and automobiles forms a dark, thick, dust and soot laden fog and is called *smog*. Smog is noxious and irritating. It reduces visibility, induces respiratory troubles and can cause death by suffocation also. Smog particles can cause asthma and other lung diseases. Cotton dust can cause lung fibrosis.

KNOWLEDGE BOOSTER

Polychlorinated biphenyls (PCBS) are chemical compounds used as fluids in transformers and capacitors. Being resistant to oxidation, these are released into the atmosphere as vapours. They mix with rain water and thus contaminate water. These have been found to be carcinogenic and are thus a source of pollution.

- Carcinogenic benzopyrene is released into the air by diesel engines. They cause cancer and harm plants. Methane gas is released into air by the degradation of biomass.
- Peroxyacyl nitrate (PAN) and peroxy benzoyl nitrate (PBN) are the air pollutants forming smog (smoke + fog) known as photochemical smog. The London smog which caused the death of several people, was formed by the above pollutants.

Illustration

1. Between CO and CO_2, which is a more dangerous air pollutant and why?

   Solution: CO is more dangerous than CO_2 as it combines with hemoglobin in the blood, forming a stable complex— carboxyhemoglobin. It destroys the oxygen carrying capacity of blood to tissues.

Effects of Air Pollutants

Acid Rain

The term *acid rain* is used to describe all precipitation, that is, rain, snow, fog and dew, which are more acidic than normal water. According to Robert August, the pH of rain water is normally 5.6 due to the dissolution of CO_2 in it.

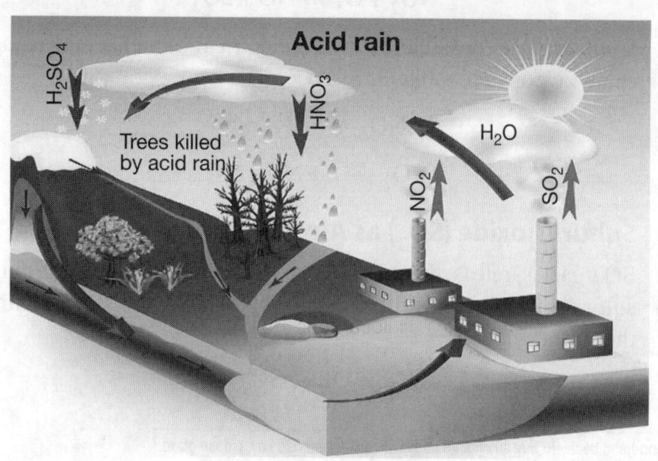

$$CO_2 + H_2O \rightleftharpoons H_2CO_3 \rightleftharpoons H^+ + HCO_3^-$$

Composition and Cause of Acid Rain

Acid rain is a result of the presence of acids such as HNO_3 and H_2SO_4 in polluted air. Acid rain is due to the large-scale emission of acidic gaseous oxides (SO_2, NO_2) into the atmosphere by thermal power plants, automobiles and industries. The oxides of nitrogen and sulfur dissolve in rain water forming nitric acid and sulfuric acid, resulting in acid rain. The reactions involved are:

HNO_3 formation

$$N_2 + O_2 \xrightarrow{\Delta} 2NO$$

$$2NO + O_2 \rightarrow 2NO_2$$

$$4NO_2 + 2H_2O + O_2 \rightarrow 4HNO_3$$

H_2SO_4 formation

$$S + O_2 \xrightarrow{\Delta} SO_2$$

$$SO_2 + \frac{1}{2}O_2 \rightarrow SO_3$$

$$SO_3 + H_2O \rightarrow H_2SO_4$$

The pH of the acid rain is a about 4–5.

Effects of Acid Rain

Acid rain changes the pH of the soil which removes basic minerals in soil (calcium and potassium), it affects the fertility of the soil. Loss of fertility of the soil damages forests and causes loss of nutrients from plants as a result of which their leaves also get damaged.

- It is toxic to vegetation and plant life.
- The life of buildings is considerably reduced by acid rain. For example, the Taj Mahal is affected due to the reaction between the rain and marble ($CaCO_3$).

$$CaCO_3 + H_2SO_4 \rightarrow CaSO_4 + CO_2 + H_2O$$

- It corrodes water pipes due to which heavy metals (Fe, Cu, Pb) mix in the water and cause toxic effects.
- It has serious ecological impacts as it affects water bodies also by making the water of rivers and lakes acidic, thus, disturbing aquatic life.
- It affects human life also as in high concentrations, it causes breathing problems, other respiratory disorders and injuries to the lung tissues.

Reducing the Impacts of Acid Rain

The impacts of acid rain can be reduced by controlling the root cause of acid rain—by reducing the emission of SO_2 and NO_2 into the atmosphere. It is possible by using fuels having low sulfur content. Now, a scrubber type device can also be used for absorbing these oxides (SO_2) (Fig. 11.2).

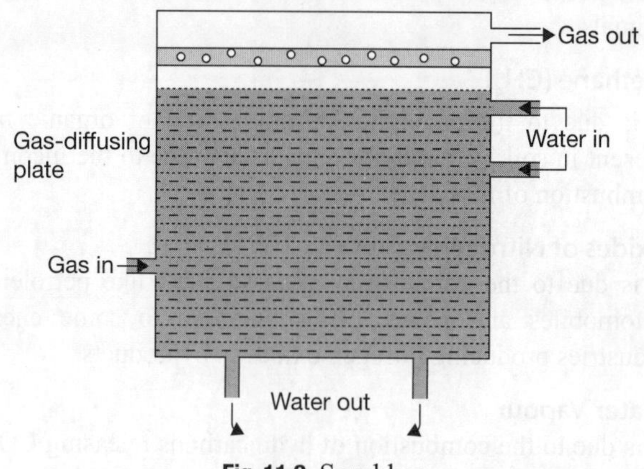

Fig. 11.2 Scrubber

Illustrations

1. When do we call rainwater as acid rain?

   Solution: When the presence of acids such as H_2CO_3, H_2SO_4, HNO_3 decreases the pH value of rain below 5.6, we call it acid rain.

2. Why is the pH of acid rain sometimes found to be as low as 2?

   Solution: The pH of rainwater is neutral but when the rain combines with the carbon dioxide in the atmosphere to form carbonic acid, it reduces the pH of rainwater to 3.5–5.6 which sometimes may reach as low as 2.

Greenhouse Effect or Global Warming

Greenhouse Effect

The phenomenon of abnormal heating up of the earth's surface due to the presence of excess of greenhouse gases which trap solar radiation is called *greenhouse effect* (Fig. 11.3). The name *greenhouse* originates from the glass structures known as greenhouses that are used to grow green plants in the colder regions of the earth. The glass structure allows sunlight to enter it but does not allow the radiated heat to escape, hence heating up the greenhouse.

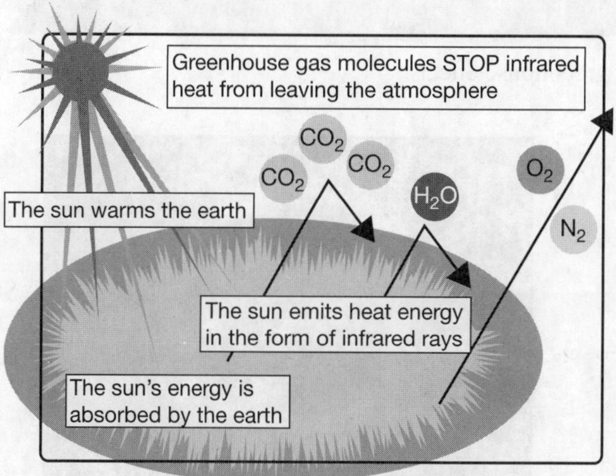

Fig. 11.3 The greenhouse effect

Greenhouse Gases

Greenhouse gases are those that absorb and emit infrared radiation in the wavelength range emitted by the earth. Carbon dioxide is the main greenhouse gas. Some other greenhouse gases are methane, CFCs (chlorofluorocarbons), O_3, NO and water vapour (Fig. 11.4).

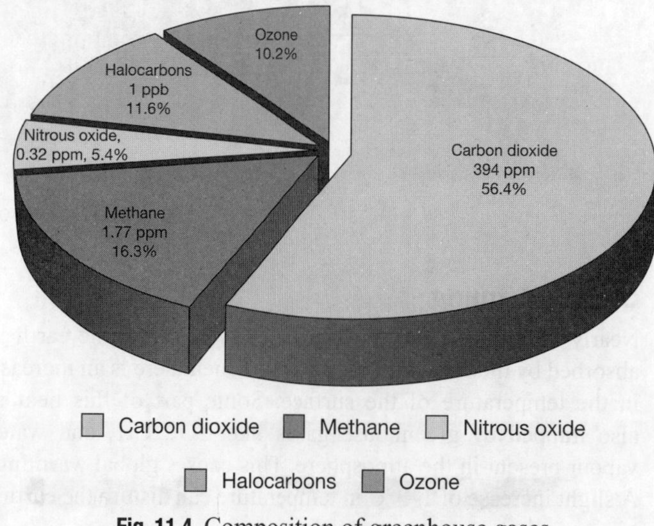

Fig. 11.4 Composition of greenhouse gases

- The composition of these gases differs slightly in different countries and in different time intervals.

Sources of Greenhouse Gases

The major sources of greenhouse gases are the following.

Carbon Dioxide (CO_2)

It is due to the burning of fossil fuels like coal, petroleum, and industrial processes involving the manufacture of lime, biological decay of plants and respiration by human beings and animals.

Methane (CH_4)

It is due to the anaerobic decomposition of organic matter present in soil, water and sediments and due to the incomplete combustion of fossil fuels.

Oxides of Nitrogen

It is due to the combustion of fossil fuels like petroleum in automobiles and power plants and due to some chemical industries producing nitrogen oxides as byproducts.

Water Vapour

It is due to the combustion of hydrocarbons releasing CO_2 and H_2O.

Chlorofluorocarbons (CFCs)

The most common source of CFCs are refrigerants, but fire suppression systems for aircrafts and aerosols also emit CFCs into the atmosphere.

Mechanism of Greenhouse Effect

Greenhouse effect occurs in these six steps (Fig. 11.5):

Step 1: Solar radiation reaches the earth's atmosphere and some part of this is reflected back into space.

Step 2: The rest of the sun's energy is absorbed by the land and the oceans, heating the earth.

Step 3: Heat radiates from the earth towards space.

Step 4: Some part of this heat is trapped by greenhouse gases in the atmosphere, keeping the earth warm enough to sustain life.

Step 5: Human activities such as burning of fossil fuels, agriculture and land clearing are increasing the amount of greenhouse gases released into the atmosphere.

Step 6: This involves trapping of extra heat and causing the earth's temperature to rise.

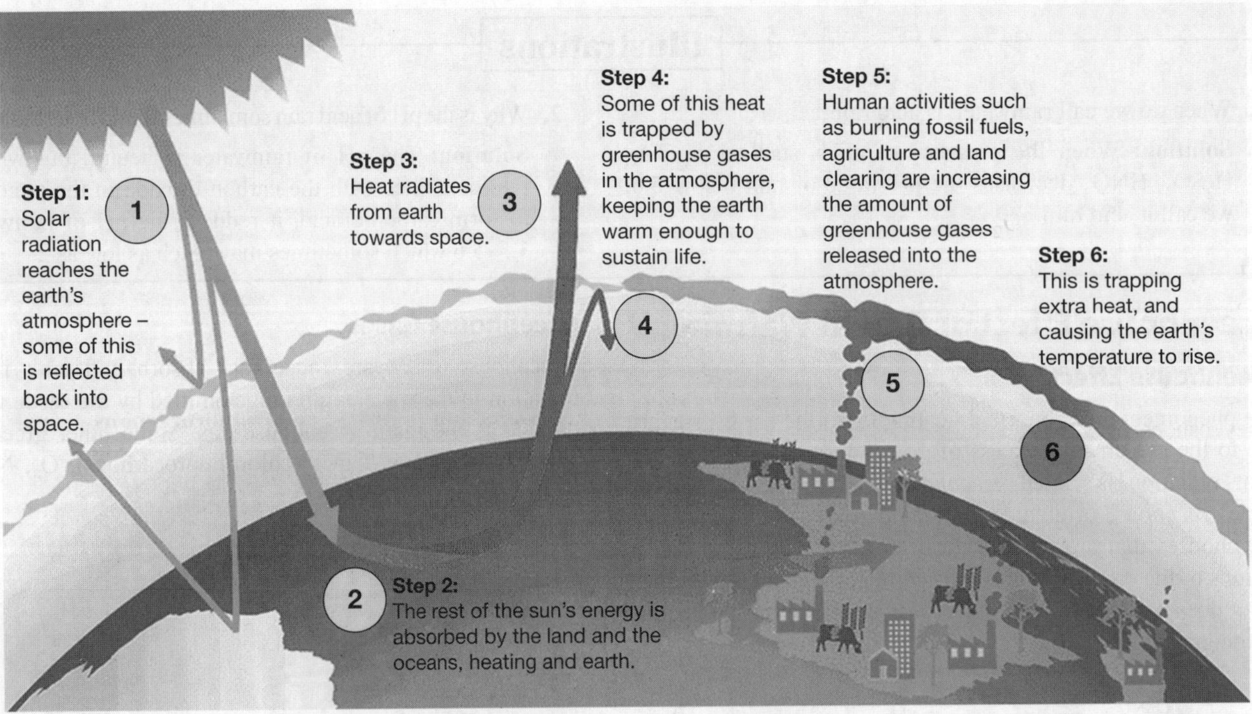

Fig. 11.5 Steps involved in greenhouse effect

Global Warming

Nearly three-fourths of the sun's energy reaching the earth is absorbed by the earth's surface due to which there is an increase in the temperature of the surface. Some part of this heat is also trapped by greenhouse gases like CO_2, CH_4 and water vapour present in the atmosphere. This causes global warming. A slight increase of 1–3°C in temperature can disturb the earth's environment.

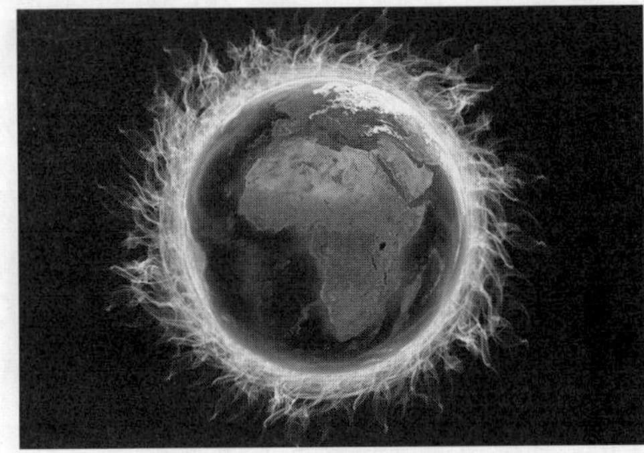

Effect of Global Warming

Some of the effects of global warming are as follows:

- **Rise in sea level:** Due to melting of glaciers and polar ice caps, there is an increase in the sea level. It causes flooding in coastal areas like the Maldives and India.
- **Decrease in water on earth:** Due to global warming, more water evaporates from water bodies; so there is a decrease in the amount of water and an increase in water vapour leading to greenhouse effect.
- **Unseasonal rains:** Due to global warming, a change in the rain pattern is observed which affects trees and plants in the forests which in turn disturbs wildlife.

- **Effects on agriculture:** Due to rapid depletion of surface water by global warming, the agriculture sector is also affected.

Methods of Reducing Global Warming

A few important steps can be taken to reduce global warming.

- Minimise the use of automobiles based on petroleum and replace them with bicycles, electric vehicles and promote the use of the public transport system.
- More planting of trees must be done.
- Check or avoid burning of wood and dry leaves.
- Avoid release of carbon dioxide during industrial processes.

Illustration

1. Write some major drawbacks of global warming.

Solution: Due to global warming, the average global temperature will increase to a level which may lead to melting of polar ice caps and flooding of low-lying areas. It may also increase cases of dengue, malaria and yellow fever.

Depletion of Ozone Layer

Ozone

Ozone is a light bluish gas (O_3) that is present in the stratosphere (the upper layer of atmosphere). It is formed by the action of UV-rays of the sun on oxygen.

$$3O_2 \rightarrow 2O_3$$

Function of Ozone Layer

It acts like a blanket in the atmosphere about 16 km from the earth's surface and it prevents the UV-rays from reaching the earth by absorbing and converting them into infrared rays. It protects the life on earth from the harmful effect of UV-rays as they can cause skin cancer and destroy many organisms necessary for life.

High energy UV-radiation (far-UV) helps in breaking the O_2 molecule into O-atoms as follows:

$$O_2 + UV \rightarrow O + O$$

Oxygen atom reacts with O_2 molecules to give ozone.

$$O + O_2 \rightarrow O_3$$

The overall reaction is $3O_2 + UV \rightarrow 2O_3$

This ozone produced here absorbs UV radiations of longer wavelength giving rise to an O_2 molecule and an O-atom.

$$O_3 \rightarrow O_2 + O$$

Depletion of Ozone Layer

Depletion refers to the decrease in the quantity of ozone present in the upper layer of the atmosphere. It leads to the formation of an ozone hole due to which UV-rays of the sun reach the earth. In 1980, the biggest ozone layer hole was seen above Antarctica (Fig. 11.6). Under normal conditions, an equilibrium is setup between the generation and destruction of ozone which keeps the concentration of ozone constant. In the presence of pollutants like the chlorine free radical ($\dot{C}l$) or NO_2, the ozone layer gets depleted as ozone reacts with these pollutants.

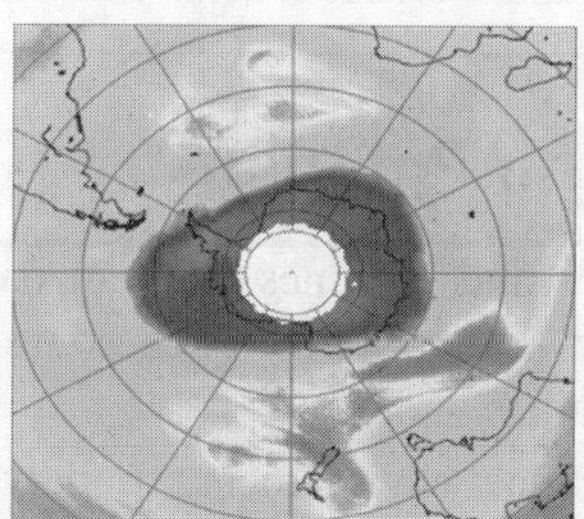

Fig. 11.6 Ozone hole above Antarctica

(i) Depletion by chlorofluorocarbons (CF_2Cl_2): Chlorofluorocarbons (freons), absorb the ultraviolet radiation and get photolysed to liberate chlorine atoms. The chlorine atoms catalyse the decomposition of ozone resulting in the depletion of the ozone layer (Fig. 11.7).

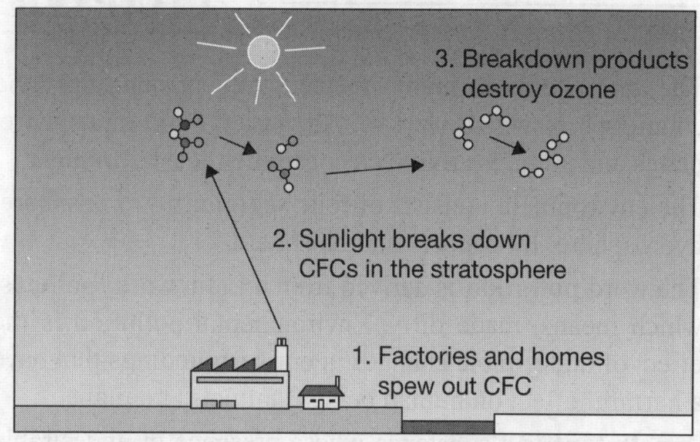

Fig. 11.7 Process of destruction of ozone

$$CF_2Cl_2 \xrightarrow{hv} CF_2Cl + \overset{\circ}{C}l$$

$$CFCl_3 \xrightarrow[uv]{hv} CFCl_2 + \overset{\circ}{C}l$$

$$Cl + O_3 \rightarrow ClO^o + O_2$$

$$ClO^o + [O] \rightarrow Cl^o + O_2$$

(ii) Aviation fuel: Burning of aviation fuel emits a large quantity of NO and other gases into the atmosphere. The presence of oxides of nitrogen in the atmosphere increases the decomposition of ozone as follows:

$$NO + O_3 \rightarrow NO_2 + O_2$$

$$O_2 \xrightarrow{h\nu} O + O$$

$$NO_2 + O \rightarrow NO + O_2$$

It has been noticed that one molecule of CFC can destroy more than one thousand O_3 molecules in the stratosphere. Due to this reaction, a huge ozone hole was created in the ozone layer over Antarctica. However, in other parts of the stratosphere, the ozone hole is not observed due to the fact that both ClO and Cl^o can be consumed as follows:

$$ClO^o + NO_2 \rightarrow ClONO_2$$

$$Cl^o + CH_4 \rightarrow \overset{o}{C}H_3 + HCl$$

This locking of ClO^o by NO_2 and Cl^o by CH_4 is known as *scavenging*.

Some chemicals used in industries, such as CH_3Cl (methyl chloride), CH_3Br (methyl bromide), CCl_4 (carbon tetrachloride) and gases causing global warming also deplete the ozone layer.

Effect of Depletion of Ozone Layer

Due to the damage done to the ozone layer, the ultraviolet light from the sun directly falls on the earth, causing skin cancer, irritation to the eyes and is harmful to vegetation.

Illustration

1. Why is an ozone hole formed over Antarctica?

Solution: It was observed that a unique set of conditions was responsible for the formation of the ozone hole. In summer, NO_2 and CH_4 react with ClO^o and Cl^o forming chlorine that sinks preventing much O_3 depletion. In winter, polar stratospheric clouds are formed over Antarctica which provide a surface for reactions to occur. These are responsible for O_3 depletion.

Controlling Air Pollution

- Microorganisms and enzymes that can naturally degrade pollutants must be developed. Such control measures are called bio-remedies.
- Clean and green, environment friendly (eco-friendly) technology must be developed.
- By dissolving or absorbing harmful gases and chemicals.
- Large-scale plantation of trees must be carried out.
- Industries must be situated far away from urban areas.
- Use of vehicles should be limited.
- Petrol without lead and diesel with less sulfur must be used.
- Automobiles must be fitted with tune ups (for high air–fuel ratio) and catalytic convertors (to change CO into CO_2 and NO_2 into N_2).
- Bagasse and rice husk should not be used as fuel.
- Fly ash (coal-thermal plants) should be removed by the wet method and should be used in building materials.
- Industrial wastes can be checked by using tall chimneys, wet scrubbers, cyclone collectors, bag filters, electrostatic precipitators, and so on.

CHAPTER AT A GLANCE

- The term **'environment'** means, the surroundings and conditions in which we live. Air, water, soil, atmosphere, plants and animals around us constitute the environment.
- The environment consists of **four segments**—atmosphere, hydrosphere, lithosphere and biosphere.
- The word **pollution** is derived from a Latin word 'pollutes' which means, made dirty. Environmental pollution is the effect of undesirable changes in our surroundings that have a harmful effect on human beings, animals and plants.
- A **pollutant** is a substance whose presence in undesirably higher concentrations causes pollution in turn, adversely affecting the environment.
- A **contaminant** is a substance which is not present in nature but is introduced into the environment by human activity and has adverse effects on the environment.
- **Pollutants** are of the following types—primary, secondary, biodegradable and non-biodegradable.
- **Air pollution** denotes the degradation of air quality. It is caused by the addition of undesirable substances into the atmosphere either naturally or by human activity.
- **Carbon monoxide (CO)** is one of the most poisonous air pollutants. Most of the air pollution is because of automobiles in which the carbon fuels undergo incomplete combustion and liberate carbon monoxide. Carbon monoxide combines with hemoglobin to form carboxyhemoglobin, thus reducing the oxygen carrying capacity of hemoglobin.
- **Oxides of nitrogen** (N_2O, NO and NO_2) are liberated into air during the combustion of fossil fuels, burning of fuel in internal combustion engines and also when N_2 and O_2 in the atmosphere react due to electric discharge due to lightning.

Higher concentration of NO_2 damages leaves of plants, retards rate of photosynthesis and causes acute respiratory disease in children.

- **Sulfur dioxide (SO_2)** is released into the atmosphere directly by burning sulfur, by roasting sulfide ores or by burning fuels containing sulfur. SO_2 causes headache, irritation of eyes, vomiting, respiratory tract diseases and may even cause death due to respiratory failure.

- **Hydrogen sulfide (H_2S)** is produced by the decay of organic matter such as vegetables, sewage and industrial effluents. It causes nausea, irritation of eyes and throat and also destroys vegetation.

- **Smoke and smog:** A pollutant that is a combination of the oxides of nitrogen and sulfur and of partially oxidised hydrocarbons and their derivatives produced by industries and automobiles forms a dark, thick, dust and soot-laden fog and is called smog. Smog is noxious and irritating. It reduces visibility, induces respiratory troubles and can cause death by suffocation.

- The term **acid rain** is used to describe all precipitation – rain, snow, fog and dew – which is more acidic than normal water.

- **Acid rain** results due to the presence of acids like HNO_3, H_2SO_4 in polluted air. Acid rain is due to the large-scale emission of acidic gaseous oxides (SO_2, NO_2) into the atmosphere by thermal power plants, automobiles and industries.

- The phenomenon of abnormal heating up of the earth's surface due to the presence of excess of greenhouse gases which trap solar radiation is called **greenhouse effect**.

- **Greenhouse gases** are those that absorb and emit infrared radiation in the wavelength range emitted by the earth. Carbon dioxide is the main greenhouse gas. Some other greenhouse gases are methane, CFCs (chlorofluorocarbons), O_3, NO and water vapour.

- **Global warming:** Nearly three-fourths of the sun's energy reaching the earth is absorbed by the earth's surface due to which there is an increase in the temperature. Some part of this heat is also trapped by greenhouse gases like CO_2, CH_4 and water vapour present in the atmosphere. This causes global warming. A slight increase of 1–3°C in temperature can disturb the earth's environment.

- **Depletion of ozone layer:** Depletion of ozone present in the upper layer of the atmosphere leads to the formation of an ozone hole due to which the UV rays of the sun reach the earth and cause skin cancer.

- **Controlling air pollution**

  (i) Microorganisms and enzymes (bio-remedies) which can naturally degrade pollutants must be developed.

  (ii) Clean and green, environment friendly (eco-friendly) technology must be developed.

  (iii) By dissolving or absorbing harmful gases and chemicals.

PRACTICE QUESTIONS

Analyse Your Concepts (School Exam Based)

Fill in the Blanks

Instructions: Complete the following statements with an appropriate word/term to be filled in the blank spaces.

1. In Latin, the word 'pollutes' means __________ .

2. The pollutants such as NO_2, SO_2 and SO_3 dissolved in the moisture of air are the cause of __________.

3. Locking of chlorine monoxide and free radical of chlorine is known as __________ .

4. Excessive release of carbon dioxide into the atmosphere is the cause of the __________ effect which causes global warming.

5. Smoke particles can cause asthma and other __________ diseases.

6. The ozone layer prevents the harmful __________ radiation of the sun from reaching the earth.

7. Carbon monoxide reacts with hemoglobin to form __________ .

8. Decrease of the concentration of ozone in the stratosphere is the cause of the formation of __________ holes.

9. The pH of rain water is in between 3.5 and __________ .

10. Ozone depletion is mainly caused by the active __________ atoms generated from CFCs in the presence of UV radiation.

True or False

Instructions: Read the following statements and write your answer as true or false.

1. Pollutants are toxic and harmful substances.

2. Volcanoes release large amounts of chlorofluorocarbons.

3. Fluorides can cause destruction of vegetation.

4. Cotton dust produces lung fibrosis.

5. Impact of acid rain can be reduced by using scrubbers.

6. Oxygen is a greenhouse gas.

7. Ozone is a light yellowish gas.

8. Acetic acid is a component of acid rain.

9. Acid rain can cause severe fever.

10. Existence of coastal countries like the Maldives are in danger due to global warming.

Very Short Answer Type Questions

1. Name any four gaseous pollutants.

2. Name the natural sources of air pollution.

3. Name the man-made sources of air pollution.

4. Why does rainwater have a pH of less than 7?

5. Write the significance of growing plants in greenhouses.

6. Name some particulate pollutants.

7. Why is cigarette smoking injurious to health?

8. Define the term ppm.

9. Name any two gases which are responsible for the formation of acid rain.

10. Write any two effects of ozone layer depletion.

11. Write an advantage of CNG (Compressed Natural Gas) as a fuel.

Short Answer Type Questions

1. Define the following terms.
   (a) Pollution
   (b) Pollutant
   (c) Air pollution

2. Write any three compounds of sulfur that cause air pollution and also write the harmful effects of these compounds.

3. Write the main sources and effects of carbon monoxide.

4. Explain the formation of acid rain due to:
   (i) oxides of sulfur
   (ii) oxides of nitrogen

5. Write few ways of reducing the presence of greenhouse gases.

6. Write only one effect each of the following pollutants on living beings.
   (a) Fluorides
   (b) Nitrogen oxides
   (c) Mercury compounds
   (d) Lead
   (e) Smog
   (f) Smoke particles

7. Define air pollution. Why does it occur?

8. Write the origin and health impacts of smog.

9. What are the major air pollutants? Explain how carbon monoxide pollutes our environment.

10. Explain how CFCs deplete the ozone layer?

Long Answer Type Questions

1. How do the oxides of nitrogen enter the atmosphere? Also write their harmful effects.

2. (a) Give the mechanism of action of carbon monoxide.
   (b) How can you control carbon monoxide poisoning?

3. (a) What are the causes of acid rain?
   (b) Give the impact of acid rain:
      (i) on plants, (ii) on soil and (iii) on water bodies.
   (c) How does a scrubber help in reducing the formation of acid rain?

4. (a) What do you understand by the greenhouse effect?
   (b) What are greenhouse gases? How are they responsible for global warming?
   (c) State the sources and effects of the following gases:
      (i) carbon dioxide, (ii) methane and (iii) water vapour.

5. (a) Explain the formation of ozone in the atmosphere.
   (b) What is the function of ozone in the atmosphere?
   (c) State the chemicals responsible for ozone layer destruction.

COMPETITION WINDOW (OBJECTIVE TYPE)

[For NEET, JEE (Main and Advanced), NTSE, KVPY and Olympiads]

Topic-wise MCQs

Environment

1. The gaseous envelope around the earth is known as the atmosphere. The lowest layer of this extends up to 10 km above the sea level. This layer is the __________.
   (a) stratosphere
   (b) troposphere
   (c) mesosphere
   (d) hydrosphere

2. Biosphere includes
   (a) plants and animals
   (b) rocks and minerals
   (c) atmosphere and lithosphere
   (d) water sources

3. The major components of the atmosphere are
   (a) CO_2 and N_2
   (b) O_3 and SO_2
   (c) CO and CO_2
   (d) N_2 and O_2

4. The uppermost region of the atmosphere is
   (a) troposphere
   (b) exosphere
   (c) stratosphere
   (d) ionosphere

5. The point of temperature inversion between the troposphere and the ionosphere is called

    (a) mesopause   (b) stratopause
    (c) ionopause   (d) tropopause

6. Hydrosphere includes various forms of water as
    (a) polar ice caps and ground water
    (b) oceans and lakes
    (c) sea and rivers
    (d) all of these

7. Which of the following is a contaminant?
    (a) CO   (b) CO_2
    (c) SO_2 (d) MIC

8. Ozone layer is present in the
    (a) stratosphere   (b) troposphere
    (c) mesosphere   (d) exosphere

Air pollution

9. The medium which interacts with the long-lived pollutant is
    (a) source   (b) receptor
    (c) sink   (d) all

10. Major sources of NO_2 pollutants are
    (a) natural gas
    (b) gasoline
    (c) combustion of coal and oil
    (d) all of these

11. Which of the following is a primary pollutant?
    (a) PAN   (b) aldehydes
    (c) CO   (d) H_2SO_4

12. Which of the following is not regarded as a pollutant?
    (a) CO_2 (b) O_3
    (c) NO_2 (d) hydrocarbons

13. Which of the following is not an air pollutant?
    (a) N_2O (b) N_2
    (c) CO   (d) NO

14. The industrial usage of fluorocarbons is very high because
    (a) they are unstable
    (b) they are gases
    (c) they can be manufactured cheaply
    (d) their reactivity is high

15. SO_2 is a dangerous air pollutant and harms plant life. What changes in the plant indicate its toxic effects?
    (a) falling of leaves   (b) bleaching of leaves
    (c) darkening of leaves   (d) withering of leaves

16. Photochemical smog occurs in warm, dry and sunny climate. One of the following is not amongst the components of photochemical smog, identify it.
    (a) NO_2 (b) O_3
    (c) SO_2 (d) unsaturated hydrocarbon

17. Dinitrogen and dioxygen are the main constituents of air but these do not react with each other to form oxides of nitrogen because __________.
    (a) the reaction is endothermic and requires very high temperature

    (b) the reaction can be initiated only in the presence of a catalyst
    (c) oxides of nitrogen are unstable
    (d) N_2 and O_2 are unreactive

18. The pollutants which come directly into the air from sources are called primary pollutants. Primary pollutants are sometimes converted into secondary pollutants. Which of the following are secondary air pollutants?
    (a) CO
    (b) hydrocarbons
    (c) peroxyacetyl nitrate
    (d) NO

Depletion of ozone layer, acid rain, greenhouse effect

19. Ozone hole refers to
    (a) hole in the ozone layer
    (b) reduction in the thickness of the ozone layer in the stratosphere
    (c) reduction in the thickness of the ozone layer in the troposphere
    (d) increased concentration of ozone

20. Ozone hole is maximum over
    (a) Africa   (b) Europe
    (c) Antarctica   (d) India

21. Excessive release of carbon dioxide in the atmosphere is the cause of
    (a) depletion of ozone
    (b) formation of polar vertex
    (c) global warming
    (d) formation of smog

22. Which of the following statements is wrong?
    (a) Ozone is not responsible for the greenhouse effect.
    (b) Ozone can oxidise the sulfur dioxide present in the atmosphere to sulfur trioxide.
    (c) Ozone hole is the thinning of the ozone layer present in the stratosphere.
    (d) Ozone is produced in the upper stratosphere by the action of UV rays on oxygen.

23. Ozone is an important constituent of the stratosphere because it
    (a) removes poisonous gases of the atmosphere by reacting with them
    (b) destroys bacteria which are harmful to human life
    (c) prevents the formation of smog over large cities
    (d) absorbs ultraviolet radiation which is harmful to human life

24. Acid rain
    (a) retards the growth of trees
    (b) affects big marble constructions
    (c) results in the loss of flora and fauna
    (d) all of these

25. Inhalation of air polluted with carbon monoxide is dangerous because
    (a) CO combines with O_2 dissolved in the blood
    (b) CO combines with hemoglobin of blood
    (c) CO removes water from the body and causes dehydration
    (d) CO causes coagulation of proteins in the body

26. A decrease in the amount of ozone in the stratosphere is called depletion of ozone and it is caused by
    (a) UV radiations of the sun
    (b) use of CFCs
    (c) excessive use of detergents
    (d) use of polychlorinated biphenyls

27. Which of the following gases is not a greenhouse gas?
    (a) CO   (b) O_3
    (c) CH_4 (d) H_2O vapour

28. Which of the following is not a greenhouse gas?
    (a) water vapour   (b) CO_2
    (c) O_2 (d) CH_4

29. The greenhouse effect is
    (a) rise in temperature of the earth
    (b) rise in pressure on the earth
    (c) decrease in oxygen content of the earth's atmosphere
    (d) decrease in CO_2 content of the earth's atmosphere

30. Which of the following is responsible for depletion of the ozone layer in the upper strata of the atmosphere?
    (a) freons   (b) ferrocene
    (c) fullerenes   (d) polyhalogens

Miscellaneous

1. Ozone in the stratosphere is depleted by
   (a) C_7F_{16} (b) C_6F_6
   (c) CF_2Cl_2 (d) $C_6H_6Cl_6$

2. In Antarctica, ozone depletion is due to the formation of the compound,
   (a) SO_2 and SO_3 (b) chlorine, nitrate
   (c) acrolein   (d) formaldehyde

3. Which of the following statements is not true about classical smog?
   (a) Its main components are produced by the action of sunlight on emissions of automobiles and factories.
   (b) Produced in cold and humid climate.
   (c) It contains compounds of reducing nature.
   (d) It contains smoke, fog and sulfur dioxide.

4. Which of the following statements about photochemical smog is wrong?
   (a) It has a high concentration of oxidising agents.
   (b) It has a low concentration of oxidising agents.
   (c) It can be controlled by controlling the release of NO_2, hydrocarbons, ozone, and so on.
   (d) Plantation of some plants such as pines helps in controlling photochemical smog.

5. Which of the following statements is correct?
   (a) Ozone hole is a hole formed in the stratosphere from which ozone oozes out.
   (b) Ozone hole is a hole formed in the troposphere from which ozone oozes out.
   (c) Ozone hole is the thinning of the ozone layer of the stratosphere at some places.
   (d) Ozone hole means vanishing of ozone layer around the earth completely.

6. A contaminant is
   (a) a pollutant released from industries
   (b) a pollutant
   (c) a component originally not present in environment but released into the environment by human activity
   (d) a pollutant released into the environment during natural calamites

7. Ozone layer of the stratosphere requires protection from the indiscriminate use of
   (a) aerosols and high-flying jets
   (b) balloons
   (c) pesticides
   (d) atomic explosions

8. There is a possibility of melting of polar ice caps and increase in the level of sea water due to
   (a) greenhouse effect
   (b) acid rain
   (c) depletion of ozone layer
   (d) any one of these

9. Incomplete combustion of petrol or diesel in automobile engines can be best detected by testing the fuel gases for the presence of
   (a) sulfur dioxide
   (b) nitrogen dioxide
   (c) carbon monoxide
   (d) carbon monoxide and water vapour

10. Among the following, all except which causes pollution?
    (a) nuclear power plant   (b) thermal power plant
    (c) hydroelectric plant   (d) automobiles

11. $CFCl_3$ is responsible for the decomposition of ozone into oxygen. Which of the following reacts with ozone to form oxygen?
    (a) Cl_2 (b) Cl^-
    (c) F^- (d) $Cl°$

Answer Keys

Fill in the Blanks

1. made dirty
2. acid rain
3. scavenging
4. greenhouse effect
5. lung
6. ultraviolet
7. carboxyhemoglobin
8. ozone
9. 5.6
10. chlorine

True or False

1. T
2. F
3. T
4. T
5. T
6. F
7. F
8. F
9. F
10. T

Very Short Answer Type Questions

1. Carbon monoxide, nitrogen oxide, sulfur dioxide and hydrogen sulfide.

2. Natural sources of air pollution are volcanoes, decaying vegetation, forest fires and dust storms.

3. Man-made sources of air pollution are automobiles, factories, industrial processes and decay of crop residue in rural areas.

4. Carbon dioxide in the atmosphere combines with rainwater to give carbonic acid (H_2CO_3) which is acidic in nature. So, the pH of the rainwater is less than 7.

5. Greenhouse gases increase the temperature of the earth to the optimum level. Temperature moderation provides a controlled environment for plants to grow and thrive.

6. Dust, smoke, mist, spray and fumes are examples of some particulate pollutants.

7. Cigarette smoking is injurious to health because it increases the environmental pollution and smoking causes asthma, respiratory diseases and cancer in humans.

8. ppm is parts per million and it describes the concentration of a substance.

9. SO_2 and NO_2 are gases responsible for acid rain.

10. Effects of ozone layer depletion are:
   - it causes respiratory problems, and
   - it damages plants and trees.

11. CNG is a complete fuel, and it burns without soot and emission of greenhouse gases.

Short Answer Type Questions

1. (a) Presence or introduction of a harmful or poisonous substance into the environment is known as pollution.
   (b) Toxic or harmful substances that have an adverse effect on the environment and living beings are known as pollutants.
   (c) Degradation of air quality due to an increase in the concentration of harmful contaminants is known as air pollution.

2. Hydrogen sulfide, sulfur dioxide and sulfur trioxide are the compounds of sulfur that cause air pollution.
   Harmful effects of sulfur compounds are:
   - Hydrogen sulfide causes nausea and irritates the eyes and throat.
   - Sulfur dioxide affects crop yield, causes damage to the lungs and causes irritation to human eyes.
   - Sulfur trioxide combines with water to form H_2SO_4 which causes acid rain.

3. The main sources of carbon monoxide are incomplete combustion of fuels in homes, factories and automobiles.
   The harmful effects caused by carbon monoxide are:
   - it decreases the oxygen-carrying capacity of the lungs,
   - a high amount of carbon monoxide paralyses normal brain function.

4. (a) Sulfur dioxide and sulfur trioxide react with water to form H_2SO_4 which causes acid rain.
   $$S + O_2 \rightarrow SO_2$$
   $$2SO_2 + O_2 \rightarrow 2SO_3$$
   $$SO_3 + H_2O \rightarrow H_2SO_4$$
   (b) Nitrogen dioxide combines with water to form a mixture of nitrous acid and nitric acid which causes acid rain.
   $$2NO_2 + H_2O \rightarrow HNO_2 + HNO_3$$

5. These are a few steps to reduce the presence of greenhouse gases.
   - Use of automobiles should be reduced by using public transport, bicycles and electric vehicles.
   - Trees should be planted and grown.
   - Burning of dry leaves and wood should be avoided.
   - Smoking should be avoided.
   - Educate people about the harmful effects caused by the greenhouse effect.

6. (i) Fluorides adversely affect teeth and bones.
   (ii) Nitrogen oxides cause cancer in humans.
   (iii) Mercury compounds cause Minamata disease.
   (iv) Lead impairs our metabolic process, thereby affecting our health.
   (v) Smog can cause respiratory diseases and also leads to suffocation.
   (vi) Smoke particles cause asthma and other respiratory diseases.

7. See text part.

8. Pollutants which are a combination of oxides of nitrogen and sulfur, and partially of hydrocarbons produced by industries and automobiles. This forms a dark, thick soot laden fog known as smog.
   Impacts of smog are as follows:
   - it is noxious and irritating,
   - it reduces visibility,

- it causes respiratory problems,
- it can cause suffocation and death.

9. Major air pollutants are:
   sulfur dioxide, hydrogen sulfide, fluorides, nitrogen oxides, carbon monoxide, lead, cotton dust particles, smog, nitric oxide and sulfur trioxide.

   Carbon monoxide combines with atmospheric oxygen to increase carbon dioxide concentration in the atmosphere. This will adversely affect the environment by leading to global warming and ozone depletion.

10. CFCs are decomposed by UV rays to highly reactive chlorine, which is produced in its atomic form.

$$CF_2Cl_2(g) \xrightarrow{\text{UV rays}} CF_2Cl(g) + Cl(g)_{\text{(free radical)}}$$

These free radicals of chlorine react with ozone to form chlorine monoxide.

$$Cl(g) + O_3(g) \rightarrow ClO(g) + O_2(g)$$

This causes ozone depletion.

Long Answer Type Questions

1. Oxides of nitrogen enter the atmosphere in the following ways:
   (i) On the burning of fuels in furnaces, the temperature increases. At high temperature, nitrogen and oxygen present in the air combine to form oxides of nitrogen.
   (ii) Oxides of nitrogen are produced during the burning of fuel in an internal combustion engine. They enter the atmosphere as exhaust gases.
   (iii) During thunderstorms, nitric oxide is formed by the reaction between atmospheric nitrogen and oxygen in the presence of electric discharge.
   (iv) Nitric oxide further reacts with atmospheric oxygen and ozone to form nitrogen dioxide.

   The harmful effects caused by the oxides of nitrogen are:
   (i) Nitrogen oxides cause irritation of the mucous membrane.
   (ii) Large concentrations of nitrogen oxide cause lung problems in humans.
   (iii) Nitrogen oxide oxidises hydrocarbons in the presence of sunlight, which causes eye irritation, asthma attacks, nasal and throat infections.
   (iv) It causes injuries to vegetation by damaging leaves.

2. (a) When we inhale carbon monoxide, it combines with hemoglobin (which carries oxygen to tissues) forming carboxyhemoglobin. Hemoglobin binds carbon monoxide 200 times more strongly than it does oxygen. This reduces the oxygen carrying capacity of the blood (see text part).
   (b) We can control carbon monoxide poisoning by taking the following measures.
   - By switching over from internal combustion engines to electrically powered cars or vehicles.
   - Many pollution control devices are now installed in cars. Most of these devices help in reducing pollution by burning gasoline completely. Complete combustion of gasoline produces only carbon dioxide and water vapour.

   $$2C_8H_{18} + 5O_2 \rightarrow 16CO_2 + 18H_2O$$

   - By using substitute fuels for gasoline—natural gas in both compressed (CNG) and liquefied (LNG) forms is now increasingly being used as fuel. Alcohols are other feasible substitutes.
   - By using catalytic converters.
     (i) Nitrogen oxide is reduced to nitrogen and oxygen in the presence of finely divided platinum or palladium as a catalyst.

   $$2NO \xrightarrow{Pt} N_2 + O_2$$
   $$2NO_2 \xrightarrow{Pt} N_2 + 2O_2$$

     (ii) Carbon monoxide changes to carbon dioxide in the presence of finely divided platinum as a catalyst.

   $$CO \xrightarrow[\text{[O]}]{Pt} CO_2 + H_2O$$

3. (a) Formation of mineral acids like carbonic acid (H_2CO_3), nitric acid (HNO_3), nitrous acid (HNO_2) and sulfuric acid (H_2SO_4) and their mixture with rain is the main cause of acid rain.
   (b) Impact on acid rain are as follows:
     (i) Acid rain causes nutrient loss and causes damage to leaves.
     (ii) Acid rain results in calcium and potassium loss from the soil which affects the soil fertility.
     (iii) Acid rains make the water acidic which will affect the aquatic life adversely.
   (c) The impact of acid rain can be reduced by checking the root cause of acid rain, that is, by reducing the emission of the oxides of sulfur and nitrogen. This can be done by using coal or oil with low sulfur content. We can also reduce such emissions by using a scrubber—a device that absorbs gaseous pollutants. A scrubber is used for removing sulfur dioxide from a smokestack. It usually consists of a fine spray of water and gas rising from the stack is passed through the scrubber and the water absorbs the sulfur dioxide. Thus, the formation of this constituent of acid rain is reduced. (See figure in text part.)

4. See text part.

5. See text part.

Competition Window (Objective Type)

Topic-wise MCQs

1. (b)	**2.** (a)	**3.** (d)	**4.** (b)	**5.** (d)
6. (d)	**7.** (d)	**8.** (a)	**9.** (c)	**10.** (d)
11. (c)	**12.** (a)	**13.** (b)	**14.** (c)	**15.** (b)
16. (c)	**17.** (a)	**18.** (c)	**19.** (b)	**20.** (c)
21. (c)	**22.** (a)	**23.** (d)	**24.** (d)	**25.** (b)
26. (b)	**27.** (a)	**28.** (c)	**29.** (a)	**30.** (a)

Miscellaneous

1. (c)	**2.** (b)	**3.** (a)	**4.** (b)	**5.** (c)
6. (c)	**7.** (a)	**8.** (a)	**9.** (c)	**10.** (c)
11. (d)				

Hints and Solutions

Competition Window (Objective Type)

Topic-wise MCQs

4. The uppermost region of the atmosphere is the exosphere.

5. Tropopause is the point of temperature inversion between troposphere and ionosphere.

8. Ozone layer is present in the stratosphere (at an altitude of 25–0 km).

11. CO is a primary pollutant.

12. CO_2 is generally not regarded as a pollutant.

13. N_2 is not an air pollutant.

20. The ozone hole is maximum over Antarctica.

27. Greenhouse gases absorb solar energy near the earth's surface and radiate it back to the earth.

28. O_2 is not a greenhouse gas.

Miscellaneous

7. Aerosols use CFCs and high-flying jets release NO which are responsible for the depletion of the ozone layer.

9. Incomplete combustion of petrol or diesel in automobile engines produce CO.

Appendices

Some Useful Conversion Factors

Common Units of Mass and Weight

1 pound = 453.59 grams

1 pound = 453.59 grams = 0.45359 kilogram
1 kilogram = 1000 grams = 2.205 pounds
1 gram = 10 decigrams = 100 centigrams
     = 1000 milligrams
1 gram = 6.022×10^{23} atomic mass units or u
1 atomic mass unit = 1.6606×10^{-24} gram
1 metric tonne = 1000 kilograms = 2205 pounds

Common Units of Volume

1 quart = 0.9463 litre
1 litre = 1.056 quarts

1 litre = 1 cubic decimetre = 1000 cubic
centimetres = 0.001 cubic metre
1 millilitre = 1 cubic centimetre = 0.001 litre
     = 1.056×10^{3} quart
1 cubic foot = 28.316 litres = 29.902 quarts
     = 7.475 gallons

Common Units of Energy

1 joule = 1×10^{7} ergs
1 thermochemical calorie
     = 4.184 joules
     = 4.184×10^{7} ergs
     = 4.129×10^{-2} litre-atmospheres
     = 2.612×10^{19} electron volts
1 ergs = 1×10^{-7} joule = 2.3901×10^{-8} calorie
1 electron volt = 1.6022×10^{-19} joule
     = 1.6022×10^{-12} erg

Common Units of Length

1 inch = 2.54 centimetres (exactly)

1 mile = 5280 feet = 1.609 kilometres
1 yard = 36 inches = 0.9144 metre
1 metre =100 centimetres = 39.37 inches
     = 3.281 feet
     = 1.094 yards
1 kilometre = 1000 metres = 1094 yards
     = 0.6215 mile
1 Angstrom = 1.0×10^{-8} centimetre
     = 0.10 nanometre
     = 1.0×10^{-10} metre
     = 3.937×10^{9} inch

Common Units of Force and Pressure

1 atmosphere = 760 millimetres of mercury
     = 1.013×10^{5} pascals
     = 14.70 pounds per square inch
1 bar = 10^{5} pascals
1 torr = 1 millimetre of mercury
1 pascal = $1 \ kg/ms^{2}$ = $1 \ N/m^{2}$

Temperature

SI Base Units: kelvin (K)

K = −273.15 C
K = C + 273.15
F = 1.8(C) +32

$$°C = \frac{°F - 32}{1.8}$$

Physical Constants

Quantity	Symbol	Traditional Units	SI Units
Acceleration of gravity	g	980.6 cm/s	9.806 m/s
Atomic mass unit (1/12 the mass of ^{12}C atom)	amu or u	1.6606×10^{-24} g	1.6606×10^{-27} kg
Avogadro constant	N_A	6.022×10^{23} particles/mol	6.022×10^{23} particles/mol
Bohr radius	a_o	0.52918 Å 5.2918×10^{-9} cm	5.2918×10^{-11} m
Boltzmann constant	k	1.3807×10^{-16} erg/K	1.3807×10^{-23} J/K
Charge-to-mass ratio of electron	e/m	1.7588×10^{8} coulomb/g	1.7588×10^{11} C/kg
Electronic charge	e	1.60219×10^{-19} coulomb 4.8033×10^{-19} esu	1.60219×10^{-19} C
Electron rest mass	m_e	9.10952×10^{-28} g 0.00054859 u	9.10952×10^{-31} kg
Faraday constant	F	96,487 coulombs/eq 23.06 kcal/volt. eq	96,487 C/mol e$^-$ 96,487 J/V.mol e$^-$
Gas constant	R	$0.8206 \dfrac{\text{L atm}}{\text{mol K}}$ $1.987 \dfrac{\text{cal}}{\text{mol K}}$	$8.3145 \dfrac{\text{kPa dm}^3}{\text{mol K}}$ 8.3145 J/mol.K
Molar volume (STP)	V_m	22.710981 L/mol (mostly taken as 22.4 L/mol)	22.710981×10^{-3} m^3/mol 22.710981 dm^3/mol
Neutron rest mass	m_n	1.67495×10^{-24} g 1.008665 u	1.67495×10^{-27} kg
Planck constant	h	6.6262×10^{-27} ergs	6.6262×10^{-34} J s
Proton rest mass	m_p	1.6726×10^{-24} g 1.007277 u	1.6726×10^{-27} kg
Rydberg constant	R_∞	3.289×10^{15} cycles/s 2.1799×10^{-11} erg	1.0974×10^{7} m^{-1} 2.1799×10^{-18} J
Speed of light (in a vacuum)	c	2.9979×10^{10} cm/s (186.281 miles/second)	2.9979×10^{8} m/s

$\pi = 3.1416$ $2.303\, R = 4.576$ cal/mol K $= 19.15$ J/mol K
$e = 2.71828$ $2.303\, RT$ (at 25 C) $= 1364$ cal/mol $= 5709$ J/mol
$\ln X = 2.303 \log X$

Elements, their Atomic Numbers and Molar Masses

Element	Symbol	Atomic Number	Molar mass/ (g mol^{-1})	Element	Symbol	Atomic Number	Molar mass/ (g mol^{-1})
Actinium	Ac	89	227.03	Francium	Fr	87	(223)
Aluminium	Al	13	26.98	Gadolinium	Gd	64	157.25
Americium	Am	95	(243)	Gallium	Ga	31	69.72
Antimony	Sb	51	121.75	Germanium	Ge	32	72.61
Argon	Ar	18	39.95	Gold	Au	79	196.97
Arsenic	As	33	74.92	Hafnium	Hf	72	178.49
Astatine	Al	85	210	Hassium	Hs	108	(269)
Barium	Ba	56	137.34	Helium	He	2	4.00
Berkelium	Bk	97	(247)	Holmium	Ho	67	164.93
Beryllium	Be	4	9.01	Hydrogen	H	1	1.0079
Bismuth	Bi	83	208.98	Indium	In	49	114.82
Bohrium	Bh	107	(264)	Iodine	I	53	126.90
Boron	B	5	10.81	Iridium	Ir	77	192.2
Bromine	Br	35	79.91	Iron	Fe	26	55.85
Cadmium	Cd	48	112.40	Krypton	Kr	36	83.80
Caesium	Cs	55	132.91	Lanthanum	La	57	138.91
Calcium	Ca	20	40.08	Lawrencium	Lr	103	(262.1)
Californium	Cf	98	251.08	Lead	Pb	82	207.19
Carbon	C	6	12.01	Lithium	Li	3	6.94
Cerium	Ce	58	140.12	Lutetium	Lu	71	174.96
Chlorine	Cl	17	35.45	Magnesium	Mg	12	24.31
Chromium	Cr	24	52.00	Manganese	Mn	25	54.94
Cobalt	Co	27	58.93	Meitneium	Mt	109	(268)
Copper	Cu	29	63.54	Mendelevium	Md	101	258.10
Curium	Cm	96	247.07	Mercury	Hg	80	200.59
Dubnium	Db	105	(263)	Molybdenum	Mo	42	95.94
Dysprosium	Dy	66	162.50	Neodymium	Nd	60	144.24
Einsteinium	Es	99	(252)	Neon	Ne	10	20.18
Erbium	Er	68	167.26	Neptunium	Np	93	(237.05)
Europium	Eu	63	151.96	Nickel	Ni	28	58.71
Fermium	Fm	100	(257.10)	Niobium	Nb	41	92.91
Fluorine	F	9	19.00	Nitrogen	N	7	14.0067

(Continued)

Element	Symbol	Atomic Number	Molar mass/ (g mol^{-1})	Element	Symbol	Atomic Number	Molar mass/ (g mol^{-1})
Nobelium	No	102	(259)	Silver	Ag	47	107.87
Osmium	Os	76	190.2	Sodium	Na	11	22.99
Oxygen	O	8	16.00	Strontium	Sr	38	87.62
Palladium	Pd	46	106.4	Sulphur	S	16	32.06
Phosphorus	P	15	30.97	Tantalum	Ta	73	180.95
Platinum	Pt	78	195.09	Technetium	Tc	43	(98.91)
Plutonium	Pu	94	(244)	Tellurium	Te	52	127.60
Polonium	Po	84	210	Terbium	Tb	65	158.92
Potassium	K	19	39.10	Thallium	T1	81	204.37
Praseodymium	Pr	59	140.91	Thorium	Th	90	232.04
Promethium	Pm	61	(145)	Thulium	Tm	69	168.93
Protactinium	Pa	91	231.04	Tin	Sn	50	118.69
Radium	Ra	88	(226)	Titanium	Ti	22	47.88
Radon	Rn	86	(222)	Tungsten	W	74	183.85
Rhenium	Re	75	186.2	Ununbium	Uub	112	(277)
Rhodium	Rh	45	102.91	Ununnilium	Uun	110	(269)
Rubidium	Rb	37	85.47	Unununium	Uuu	111	(272)
Ruthenium	Ru	44	101.07	Uranium	U	92	238.03
Rutherfordium	Rf	104	(261)	Vanadium	V	23	50.94
Samarium	Sm	62	150.35	Xenon	Xe	54	131.30
Scandium	Sc	21	44.96	Ytterbium	Yb	70	173.04
Seaborgium	Sg	106	(266)	Yttrium	Y	39	88.91
Selenium	Sc	34	78.96	Zinc	Zn	30	65.37
Silicon	Si	14	28.08	Zirconium	Zr	40	91.22

Some Important Common Monoatomic Ions

+1 Charge Name of ion	Formula	+2 Charge Name of ion	Formula	+3 Charge Name of ion	Formula
Lithium ion	Li^+	*Lead (II) ion	Pb^{2+}	Arsenic ion	As^{3+}
		Cadmium ion	Cd^{2+}	Bismuth ion	Bi^{3+}
		Magnesium ion	Mg^{2+}	Antimony ion	Sb^{3+}
Sodium ion	Na^+	*Copper (II) ion	Cu^{2+}	Scandium ion	Sc^{3+}
Potassium ion	K^+	Strontium ion	Sr^{2+}	Chromium (III) ion	Cr^{3+}
Silver ion	Ag^+	Iron (II) ion (Ferrous ion)	Fe^{2+}	Iron (III) ion (Ferric ion)	Fe^{3+}
Copper ion (Cuprous ion)	Cu^+	Barium ion	Ba^{2+}	Aluminium ion	Al^{3+}
		Cobalt ion	Co^{2+}	Auric ion	Au^{3+}
Aurous	Au^+	Manganese (II) ion	Mn^{2+}		
		*Mercury (I) ion	Hg_2^{2+}		
		Zinc ion	Zn^{2+}		

−1 Charge Name of ion	Formula	−2 Charge Name of ion	Formula	−3 Charge Name of ion	Formula
Chloride ion	Cl^-	Sulfide ion	S^{2-}	Phosphide ion	P^{3-}
Fluoride ion	F^-			Boride ion	B^{3-}
Bromide ion	Br^-	Oxide ion	O^{2-}	Nitride ion	N^{3-}
Iodide ion	I				

Some Common Important Polyatomic Ions

−1 Charge Name of ion	Formula	−2 Charge Name of ion	Formula	−3 Charge Name of ion	Formula
Nitrate ion	NO_3^-	Sulfite ion	SO_3^{2-}		
Hydrogen carbonate or bicarbonate ion	HCO_3^-	Carbonate ion	CO_3^{2-}	Phosphate ion	PO_4^{3-}
		Manganate ion	MnO_4^{2-}	Arsenate ion	AsO_4^{3-}
Hydrogen sulfate or (bisulfate ion)	HSO_4^-	Thiosulfate ion	$S_2O_3^{2-}$	Arsenite ion	AsO_3^{3-}
		Silicate ion	SiO_3^{2-}		
Hydroxide ion	OH^-	Sulfate ion	SO_4^{2-}	Phosphite ion	PO_3^{3-}
Acetate ion	CH_3COO^-	Oxalate ion	$C_2O_4^{2-}$		
Chlorate ion	ClO_3^-	Chromate ion	CrO_4^{2-}	Borate ion	BO_3^{3-}
Nitrite ion	NO_2^-	Dichromate ion	$Cr_2O_7^{2-}$	Ferricyanide ion	$[Fe(CN)_6]^{3-}$

(Continued)

−1 Charge Name of ion	Formula	−2 Charge Name of ion	Formula	−3 Charge Name of ion	Formula
Permanganate ion	MnO_4^-	Hydrogen phosphate ion	HPO_4^{2-}		
Ammonium ion	NH_4^+				
Cyanide ion	CN^-				
Hypophosphite ion	$H_2PO_2^-$				− 4 Charge	
Metaaluminate ion	AlO_2^-			Carbide ion	C^{4-}
				Ferrocyanide ion	$[Fe(CN)_6]^{4-}$

Important Compounds

1. Ammonal is a mixture of ammonium nitrate and Al powder (NH_4NO_3 + Al). It is used as an explosive.
2. Alum is $(NH_4)_2SO_4 \cdot Al_2(SO_4)_3 \cdot 24H_2O$. It is used as mordant by dyers of clothes. Potash alum is $K_2SO_4 \cdot Al_2(SO_4)_3 \cdot 24H_2O$.
3. Aqua fortis is nitric acid, HNO_3.
4. Antichlor is sodium thiosulfate, $Na_2S_2O_3 \cdot 5H_2O$. It is also called hypo.
5. Aqua-regia is a mixture of conc. HNO_3 and conc. HCl in the ratio of 1 : 3. It is also known as kingly water.
6. Baking soda is sodium bicarbonate, $NaHCO_3$.
7. Barytes is barium sulfate, $BaSO_4$.
8. Brine is sodium chloride (NaCl) solution
9. Blue vitriol is copper sulfate, $CuSO_4 \cdot 5H_2O$.
10. Bone ash is mainly calcium phosphate, $Ca_3(PO_4)_2$.
11. Borax is the name of sodium tetraborate hydrate $Na_2B_4O_7 \cdot 10H_2O$. Borax ($Na_2B_4O_7$) is also called tincal.
12. In the nitrate test, the brown ring is $FeSO_4 \cdot NO$ or $[Fe(H_2O)_5NO]SO_4$.
13. Cuprite is Cu_2O.
14. Calomel is Hg_2Cl_2.
15. Caustic potash is KOH.
16. Caustic soda is sodium hydroxide, NaOH.
17. Chile saltpeter is sodium nitrate, $NaNO_3$.
18. Cinnabar is HgS.
19. Carbonic acid is hydrogen carbonate, H_2CO_3.
20. Carbolic acid is phenol, C_6H_5OH.
21. Carborundum is silicon carbide, SiC.
22. Copper glance is Cu_2S.
23. Carbogen is a mixture of 1% CO_2 and O_2. It is used as antidote for for CO poisoning.
24. Corrosive sublimate is mercuric chloride, $HgCl_2$.
25. Corundum is aluminium oxide, Al_2O_3.
26. Chromyl chloride is CrO_2Cl_2.
27. Cyanogen is C_2N_2.
28. Dead burnt plaster is anhydrous $CaSO_4$.
29. Dry ice is solid carbon dioxide (CO_2).
30. Epsom salt is the name of magnesium sulfate, $MgSO_4 \cdot 7H_2O$.
31. Eka aluminium is gallium.
32. Fluorspar is CaF_2.
33. Freon is CCl_2F_2.
34. Formalin is 40% formaldehyde (HCHO).
35. Foul air is nitrogen, N_2. It is also called azote.
36. Fluorine is called super halogen.
37. Fulminating gold is $Au(NH_2) = NH$.
38. Grain alcohol is ethyl alcohol, C_2H_5OH.
39. Grape sugar is dextrose, $C_6H_{12}O_6$.
40. Glauber's salt is the name of sodium sulfate, $Na_2SO_4 \cdot 10H_2O$.
41. Gypsum is calcium sulfate, $CaSO_4 \cdot 2H_2O$.
42. Green vitriol is ferrous sulfate, $FeSO_4 \cdot 7H_2O$.
43. Gammexane is benzene hexachloride (BHC), $C_6H_6Cl_6$.
44. Gun powder is a mixture of sulfur, charcoal and nitre.
45. Graham's salt is $(NaPO_3)_3$.
46. Hydrolith is calcium hydride, CaH_2.
47. Halite is common rock salt (NaCl).
48. Horn silver is AgCl.
49. King of chemicals is H_2SO_4.
50. Limestone is calcium carbonate, $CaCO_3$.
51. Lunar caustic is silver nitrate, $AgNO_3$.
52. Milk of magnesia is magnesium hydroxide, $Mg(OH)_2$.
53. Marshall's acid is persulfuric acid, $H_2S_2O_8$.
54. Milk of lime is calcium hydroxides, $Ca(OH)_2$. It is also called slaked lime.
55. Magnesite is $MgCO_3$.

56. Marsh gas or fire damp is CH_4.
57. Mohr's salt is $FeSO_4 \cdot (NH_4)_2SO_4 \cdot 6H_2O$.
58. Muriatic acid is hydrochloric acid, HCl
59. Nessler's reagent is K_2HgI_4. It contains $HgCl_2$, KI and NaOH. The ion present in it is HgI_4^{2-}.
60. Norweigian saltpetre is basic calcium nitrate, $Ca(NO_3)_2$.
61. Oil of vitriol is sulfuric acid, H_2SO_4.
62. Oleum is fuming sulfuric acid, concentrated $H_2SO_4 + SO_3$.
63. Plaster of Paris is calcium sulfate hemihydrate, $CaSO_4 \cdot \frac{1}{2}H_2O$.
64. Philosopher's wool is zinc oxide, ZnO.
65. Perhydrol is 30% H_2O_2.
66. Prussic acid is HCN.
67. Quick lime is calcium oxide, CaO.
68. Quartz is silicon dioxide, SiO_2.
69. Quick silver is mercury, Hg.
70. Rectified spirit is 95% ethyl alcohol, C_2H_5OH.
71. Red lead is lead tetroxide, Pb_3O_4. It is also called minium.
72. Rock salt is NaCl.
73. Salt cake is sodium sulfate, Na_2SO_4.
74. Sand is silicon dioxide, SiO_2.
75. Smelting salt is $(NH_4)_2CO_3$.
76. Soda lime is a mixture of NaOH and CaO.
77. Soda ash is sodium carbonate, Na_2CO_3.
78. Sodamide is $NaNH_2$.
79. Spirit of wine is C_2H_5OH.
80. Spirit of salt is HCl.
81. Stranger gas is xenon (Xe).
82. Sugar of lead is lead acetate, $(CH_3COO)_2Pb$.
83. Super phosphate of lime contains $Ca(H_2PO_4) \cdot H_2O$ and $2CaSO_4 \cdot 2H_2O$.
84. Syvine is KCl.
85. Tear gas is chloropicrin, CCl_3NO_2.
86. TEL is tetra ethyl lead, $Pb(C_2H_5)_4$.
87. Thomas slag is calcium phosphate, $Ca_3(PO_4)_2$.
88. Tincture of iodine is I_2 and KI solution in alcohol.
89. Tinstone or cassiterite is SnO_2.
90. Vinegar is dilute acetic acid, CH_3COOH.
91. Washing soda is Na_2CO_3.
92. Water glass is sodium silicate, Na_2SiO_3.
93. White lead is $Pb(OH)_2 \cdot 2PbCO_3$.
94. White vitriol is zinc sulfate, $ZnSO_4 \cdot 7H_2O$.
95. Wood spirit is CH_3OH.
96. Zincite is ZnO.